ADVANCES IN ROBOT KINEMATICS AND COMPUTATIONAL GEOMETRY

ADVANCES IN ROBOT KINEMATICS AND COMPUTATIONAL GEOMETRY

Edited by

Jadran Lenarčič
"Jozef Stefan" Institute,
University of Ljubljana,
Slovenia

and

Bahram Ravani
University of California,
Davis, California, U.S.A.

SPRINGER-SCIENCE+BUSINESS MEDIA, B.V.

A C.I.P. Catalogue record for this book is available from the Library of Congress.

DOI 10.1007/978-94-015-8348-0

Printed on acid-free paper

Preface

Recently, research in robot kinematics has attracted researchers with different theoretical profiles and backgrounds, such as mechanical and electrical engineering, computer science, and mathematics. It includes topics and problems that are typical for this area and cannot easily be met elsewhere. As a result, a specialised scientific community has developed concentrating its interest in a broad class of problems in this area and representing a conglomeration of disciplines including mechanics, theory of systems, algebra, and others.

Usually, kinematics is referred to as the branch of mechanics which treats motion of a body without regard to the forces and moments that cause it. In robotics, kinematics studies the motion of robots for programming, control and design purposes. It deals with the spatial positions, orientations, velocities and accelerations of the robotic mechanisms and objects to be manipulated in a robot workspace. The objective is to find the most effective mathematical forms for mapping between various types of coordinate systems, methods to minimise the numerical complexity of algorithms for real-time control schemes, and to discover and visualise analytical tools for understanding and evaluation of motion properties of various mechanisms used in a robotic system.

The purpose of the book is to present the recent advances in robot kinematics and computational geometry in kinematics. The articles were submitted by authors from sixteen countries. Among these, the reader will find the most outstanding names in the field. All of the fifty three articles were peer reviewed by at least two independent reviewers. They each represent an original contribution that hasn't been published elsewhere. The book is divided into twelve sections that were identified as the prevalent areas of the contemporary research in kinematics and computational geometry applied to robots and mechanisms. The introductory article is written by Professor Bernard Roth from Stanford University, USA.

The articles of this book were reported and discussed at the fourth workshop on Advances in Robot Kinematics in conjunction with the first workshop on Computational Geometry in Kinematics organised by the Institute "Jožef Stefan" in Ljubljana (Slovenia), July 4-6, 1994, under patronage of the International Federation for the Theory of Machines and Mechanisms. It should be emphasised that, since the first submission of the articles, the whole procedure of reviewing, preparing the final manuscripts, editing and publishing took approximately six months. We trust, therefore, that this book presents the actual state of the art in the fields of robot kinematics and computational geometry in kinematics. In various articles on the current research, the reader will find interesting personal views, scientific conclusions, as well as speculations and attempts to foresee new scientific problems and research directions. We are grateful to the authors of the articles for their collaboration in this project and their valuable contributions. We are also indebted to the reviewers for their timely review of the articles, and to the personnel at Kluwer Academic Publishers for their excellent technical and editorial support.

J. Lenarčič and B. Ravani, Editors

Table of Contents

Introduction

B. Roth
Computational Advances in Robot Kinematics

Computational Advances in Robot Kinematics

Bernard Roth

Department of Mechanical Engineering
Stanford University
Stanford, CA 94305-4021, USA

Abstract – This paper presents an overview of the current state–of–the–art in solving the sets of nonlinear equations which arise in robot kinematics. It reviews some recent results and the history and fundamental concepts of the most widely used computational methods. The paper deals mainly with equations with numerical coefficients, since for such equations methods have been developed which, in principle, will allow for the determination of all solutions.

I. Introduction

Computations arise in robot kinematics in connection with the analysis, design and control of robot devices. The computations that are usually considered to be interesting, rather than mundane, are the ones which involve sets of nonlinear equations. This paper deals with the solution of sets of nonlinear equations which arise in robot kinematics. It reviews recent advances in numerical methods applied to the solution of problems in robot kinematics, and assess the methods' past achievements, current utility and future potential.

The basic kinematic equations come from either loop closure equations or joint constraint equations. In either case the results are inherently nonlinear, since kinematic joints and constraints lead to mathematical descriptions in terms of trigonometric functions and their products. Interestingly, it is the variables themselves, and not their derivatives, that cause the nonlinearities. In kinematics, the derivatives of the motion variables are inherently linear. This fact accounts for the reason that, computationally, it is much easier to study a robotic device's velocities, accelerations and force/torque transmissions at a given position than it is to determine the position itself.

From the beginning, it was clear to researchers that the problems associated with determining the position kinematics were the most challenging from a computational point of view. The position kinematics problems divide into easy and hard ones. In general the easy problems are the ones which have one or two solutions, while the hard ones involve four or more solutions. For the easy ones the only interesting questions have to do with minimizing computer time; we will not discuss such questions here. The hard problems are the inverse kinematics of series chains, the direct kinematics of the in–parallel chains, and both the direct and inverse kinematics of some hybrid chains. There are also design problems in robot kinematics that are hard in the sense that they involve finding multiple solutions, since they are described by sets of nonlinear equations.

This paper touches on several of the basic issues involved in solving the hard problems. Primarily, the paper reviews the use of continuation and elimination methods. It then provides a summery of the state–of–the–art in solving the (hard) robot position kinematics problems.

A. J. Lenarčič and B. B. Ravani (eds.), Advances in Robot Kinematics and Computationed Geometry, 7–16.

II. Basic Tools for Numerical Solutions

There are two very productive basic tools for solving nonlinear kinematics problems. These are the continuation (or homotopy) methods and the elimination methods. These methods will be discussed in this paper, and some of the more important considerations that govern their use will be mentioned. There are other tools which are highly thought of by some researchers, but are not as yet proven to be generally effective in robotic kinematics. These include Gröbner bases, simulated annealing and genetic algorithms. These will be mentioned briefly. On the other hand, there is a set of tools that are so ubiquitous and so fundamental, that they could easily be omitted from this discussion on the grounds of triviality. Instead we begin our discussion with these, the "trivial" trigonometric tools.

A. Basic Trigonometric Facts

Next to matrix multiplication and scalar arithmetic, the most common computational tools rely upon simple algebra to manipulate equations based on graphical constructions and the sine and cosine laws of triangles. The following are basic facts about the trigonometry of equations with unknowns in terms of sines and cosines of angles. Although they are common knowledge among many researchers, they are repeated here since, in my experience, they are not as universally known as they should be.

1.) The angle variable may be extracted from the sine or cosine function in an unambiguous manner by substitution of the tangent half–angle formulas

$$\cos\alpha = \frac{1-x^2}{1+x^2}, \quad \sin\alpha = \frac{2x}{1+x^2}, \quad \text{here } x = \tan\frac{\alpha}{2}.$$

This substitution converts equations original in sine and cosine into algebraic relationships in the variable x. For each real solution of x, an unambiguous value for α in the range $0 \le \alpha < 2\pi$ follows from $\alpha = 2\arctan(x)$, if we evaluate the arc tangent in the interval $0, \pi$.

2.) If, instead of the substitution in 1., the equation is left in terms of sines and cosines, it can be converted into algebraic functions by the transformations $\cos\alpha = x$, $\sin\alpha = y$. In this case we need to add the auxiliary equation $x^2+y^2=1$. Here, however, knowledge of either x or y alone does not yield an unambiguous value for α. In fact when $-1 \le x \ge 1$ and $-1 \le y \ge 1$, the arccos(x) and the arcsin(y) each yield two values of α in the range $0 \le \alpha < 2\pi$. If we know both x and y, then a unique value of α in the range $0 \le \alpha < 2\pi$ is established. It is useful to use the arc tan2 function: $\alpha = \text{arctan2}(y,x)$. This, two-argument, inverse–tangent function yields a single value of α in the range $0 \le \alpha < 2\pi$.

Alternatively, a single value of α, in the range $-\pi < \alpha < \pi$, can be obtained by using:

$$\alpha = 2\arctan\left(\frac{y}{1+x}\right)$$

3.) From the foregoing, if A, B and C denote known coefficients:

i) a single equation of the form $A\cos\alpha + B\sin\alpha = C$, if it has real solutions, yields two values of α in the interval $0 \le \alpha < 2\pi$; this is the case even when $C = 0$ and/or $A = 0$ or $B = 0$.

ii) two commensurate simultaneous equations, both of the form given in i), yield only a single value of α in the interval $0 \le \alpha < 2\pi$.

Although these concepts are very elementary, they are at times not properly recognized. Mostly though, these are correctly applied and form the backbone of most other methods.

B. Iterative Methods

The most commonly used iterative methods are variants of either the Newton-Raphson, steepest descent, or conjugate gradient methods. All of these methods require an initial guess at a solution, and, if the initial guess is not close to a solution, they tend to diverge, or converge very slowly, or converge to an unacceptable solution. Furthermore, nonlinear problems have more than one solution, and these iterative methods produce only one of these — the solution "closest" to the initial guess. Once one solution is obtained, it is not any easier to obtain additional ones, and repetitive applications of iterative methods may not yield all solutions.

In order to overcome these difficulties, a technique known as the bootstrap method was developed explicitly for mechanical systems in the early 1960s [1]. This iterative procedure has been improved and is now known as the homotopy or continuation method. It is widely used in other fields as well as in robot kinematics (for example, [2,3,4]). The advantage of the continuation method is that it does not require a priori knowledge of an approximate solution, and it yields all possible solutions. The disadvantages are that it is an iterative numerical procedure which (1) gives little or no symbolic information about how the design parameters influence the solutions and (2) can require dealing with large numbers of unwanted solutions at infinity.

The continuation method is usually applied to systems of equations by using a set of known solutions to a starting problem. The starting problem is a set of simpler equations having the same degree structure, i.e., the same number of solutions, as the set we actually want solved. If $F(a,x) = 0$ represents a set of equations with x as a vector of unknown variables, and a a vector of known parameters, and if this set of equations has m solutions, then there are m vectors x that satisfy this set of equations. The continuation method starts with a set of equations $G(b,x^*) = 0$ which has the same number, m, of solutions as the original equations and for which all m sets of x^* are known. Then the coefficients of the reference problem are slowly perturbed so that they gradually become the coefficients of the actual set of equations to be solved.

So we create new sets of equations where for the ith set, ${}^iG({}^ib,{}^ix^*) = 0$; the coefficients ${}^ib$ are obtained as the ith step–change of b in the direction of a. The steps can be taken in a variety of ways, the simplest uses uniform increments obtained by stepping a parameter t_i from 0 to 1 in small increments (say, 10, 100, 1000 or more steps, depending on the ease of convergence from the initial guess) according to ${}^ib = (1-t_i)b + t_ia$. The known solution sets x^* are used as initial guesses for the first intermediate problem ${}^1G({}^1b,{}^1x^*) = 0$. The m solution vectors ${}^1x^*$ obtained for this problem, are very good initial guesses for the next intermediate problem ${}^2G({}^2b,{}^2x^*) = 0$. All m ${}^2x^*$ will be found provided t_i is chosen small enough so that the coefficients ${}^2b$ are close enough to ${}^1b$. We continue in this way, using the solution to the previous problem as the initial guess for the next problem, until we reach $t_i=1$, at which point we have the m solutions of our original set of equations.

By breaking the problem into an ordered series of sub problems, where the solution to the previous problem is a good initial guess for the current problem, the limitation of the initial guess requirement is bypassed. In theory, the method always generates good initial guesses for each subsequent step..

The two most important concepts that are characteristic of the advances in continuation methods in robot kinematics are 1) the use of complex numbers to represent real parameters and variables, and 2) the use of n–homogenous variables, rather than a single homogenous variable.

The use of complex quantities is important, since it turns out that many solutions start off either real or complex and then change their character, i.e., the real become complex and the complex become real as the iterations move from the staring set of equations toward the actual set. If we restrict the iteration to the reals, we cannot obtain convergence once the solution moves to the complex, and we cannot even begin the full procedure unless we have an example with all real solutions, which is rare.

The use of n–homogeneous variables is a very important development. Recently it was shown how this dramatically reduces the number of extraneous solutions [5] in a robot kinematics problem. What is involved here is the associating of variables in groups, and then transforming each group into a separate homogeneous set of variables. If one has n homogeneous groups the problem is said to be in an n–homogeneous formulation.

At first glance this would seem to be a strange procedure, since making each set homogenous introduces a new variable for that set; a normal one-homogeneous system introduces only one new variable, whereas an n–homogeneous formulation introduces n new variables. However, since the number of solutions can be predicted by computing a number known as the multihomogeneous Bezout number [6] of the system, the efficacy of this concept can easily be check. The Bezout number is determined in the following way:

For a set of m equations, we arrange the variables into n groups named α_j ($j = 1,..,n$), and we note that the jth group is of degree d_{ij} in the ith equation. We form a sum over j, separately for each equation, of the products $d_{ij}\alpha_j$. This yields m linear polynomials in the α_j with their coefficients as known integers. Finally, we form the product of all these m linear polynomials, and obtain a nonlinear polynomial in the α_j. The multihomogeneous Bezout number is the coefficient of the term of this polynomial in which the degree of each α_j equals the number of variables in the jth group.

From this it can be proved that introducing a n–homogeneous formulation can reduce the Bezout number, the reason for this is that a judicious choice of groupings minimizes the number of solutions that the system has at infinity [6].

If one wants all the solutions, it is important that each of the m solution sets remains distinct throughout the iteration. In this regard a very important concept has to do with the choice of the starting solutions $\boldsymbol{x}^*$. There exist some theorems which state that if all m sets of the $\boldsymbol{x}^*$ are the solutions of equations with randomly generated coefficients, $\boldsymbol{b}$, the m sets of solutions can be tracked independently. There are standard techniques to easily obtain starting solutions of systems defined with (pseudo) random coefficients [6].

Although it seems conceptually simple, there are several important subtleties to the continuation method. One of the most important consideration is the form that the equation–set is put in before starting the use of the continuation method. There is an important trade–off between eliminating variables and equations, at the price of increased complexity of the remaining equations, and leaving equations in their original form, even though they represent a larger set of equations and unknowns. The final word has not been spoken on this issue, but clearly elimination is to be avoided when the process of elimination of variables introduces extraneous roots.

The continuation methods have proven to be very effective. Two of the most sought after pieces of information in robot kinematics have come to us from this method. The proof that the inverse kinematics of a general 6R manipulator has 16 solutions [7] and the proof that the so–called generalized Stewart Platform (i.e., the general 6/6, in–parallel platform linkage) has

40 solutions for the direct kinematics problem [5] were first obtained by the continuation method. Most importantly, these were done while the true numbers were still in doubt, and after some "proofs" of incorrect or misleading numbers had been obtained by other methods. This speaks very highly of this method. It has also been applied to various robot design problems [8].

Actually, the continuation method can be used in conjunction with any iterative technique. The Newton–Raphson method is used because of its efficiency, but it is not necessary. Steepest descent, conjugate gradient, Monte Carlo, simulated annealing, genetic algorithms and several other iterative methods have all been used in robotics. They have tended to be applied more to optimization type problems rather than to determining the solutions of sets of nonlinear equations with finite numbers of solution sets. However there is no reason they could not be applied to iterative solutions of these types of equations. To date though there does not seem to be any evidence that these other methods will prove to be, in general, more effective than the continuation methods based on Newton–Raphson type iterations.

C. Methods that Determine a Simple Set with the Same Ideal

Gröbner bases are a simple set of equations which have the same ideal as the original set. Gröbner bases seem to have first been published in 1964, under the name of standard bases [9]. They were then renamed and applied to various problems including the solution of multivariate polynomial sets [10] and geometrical theorem proving [11,12]. To date this method has proved to be too inefficient for the types of problems we are interested in. In difficult problems it suffers from exploding intermediate results and excessive computational time. In response to these difficulties, some ingenious variants and computer codes have been devised. As a result of these the method has been used effectively in problems where the results could be deduced by solving an integer example. The results so far have mainly enabled the count of the number of solutions theoretically possible, although even in this domain I am unaware of any case where a genuinely new result was first obtained by these methods. They have been useful in providing further verification of some results previously derived by other methods [13]. In any case, some of the ideas behind the method are generally useful, and since the method has its strong proponents, time will tell if does lead to new knowledge in the area of robot kinematics.

D. Elimination Methods

The elimination methods seem to be the most widely used and productive analytical techniques. The original ideas come from Cayley and were published in 1848 [14]. All elimination methods require forming equations known as eliminants or resultants. An early summary of various types of eliminants and resultants can be found in the book by Salmon [15]. More modern treatments appear in [16,17,18]. The elimination methods will, in theory, lead to a solution of any system of multivariate polynomial equations. However, since the method is based on a set of equations which are sufficient and not necessary, in practice the method can only be applied to relatively simple equations — beyond these it explodes in complexity and introduces large numbers of extraneous solutions. In the 1980s there was an increased interest in improving these methods in order to make them practical computational tools especially in regard to robot planning [19].

In kinematics, the most widely used elimination method is dialytic elimination. The advantages of the dialytic elimination method, and the problems with it have been discussed at some length in a recent paper [20]. The main advantage is that, in principle, dialytic elimination, like other elimination methods, can yield:

1. All the solutions to a set of nonlinear equations.
2. An analytical means to determine the influence of the system parameters on the number and character of the solutions.

Unfortunately, the practical reality often leaves one short of these ideal attainments. In the following we first briefly describe the dialytic method, then point out its major limitation.

There are six basic steps in using the dialytic elimination method to solve a nonlinear set of equations [20]:

1. Rewrite equations with one variable suppressed.
2. Define the remaining power products as new linear unknowns.
3. Use the original equations to manufacture new linear equations so as to have as many linearly independent homogeneous equations as linear unknowns.
4. Set the determinant of the coefficient matrix to zero, and obtain a polynomial in the suppressed variable. (If interested in only numerical solutions, this step can be omitted if we calculate eigenvalues in Step 5.)
5. Determine the roots of the characteristic polynomial of the matrix, or the eigenvalues of a derived matrix. (This yields all possible values for the suppressed variable.)
6. Substitute (one of the roots or eigenvalues) for the suppressed variable and solve the linear system for the remaining unknowns. Repeat this for each value of the suppressed variable.

Step 3 is the most crucial step. In the traditional dialytic method proposed by J. J. Sylvester [15], the new equations are manufactured by multiplying the original set by powers of one or more of the variables. This is workable and effective for equations of low degree, or for small numbers of equations. However, in many cases of interest in robot kinematics this makes the problem unmanageably large. Therefore, it is very desirable to determine ways to manufacture new linearly independent equations from the original equations in a manner which does not introduce new power products, or at worst introduces a small number of new power products. In cases where this can been done the method is very useful, and in these restricted areas it is a basic tool which has enjoyed much success.

III. Position Kinematics

Historically, the original work on the solution of robot kinematics equations involved the problems that we now call the direct and inverse (position) kinematics and the direct and inverse velocity problems. The first work in this area was in the dissertation of Pieper [21,22]. Although this work is widely quoted for its introduction of Denavit and Hartenberg coordinates into robotics, and also its solution of the inverse kinematics problem when there are three intersecting adjacent axes (as in a manipulator with a three–axis wrist), it includes several other results of more direct interest to our present discussion. Most notably it contains the use of iterations for determining the inverse kinematics, by use of both Newton–Raphson and what we would now call an inverse Jacobian method, and it also contains several examples of the first use of eliminants in robotics. Almost forgotten is the fact that it contains the first published example of 16 real solutions of the inverse kinematics problem — a result obtained with the aid of elimination theory.

Robot position kinematics problems are analogous to classical linkage analysis problems. There has been much effort associated with the solution of loop–closure equations. One of the first works to look at this problem in a generic form was due to Chace [23]. He used a vector method for systems with four or less links. Since then there have been many attempts at dealing with larger numbers of links. The most extensive body of work is due to Duffy and his coworkers [24,25]. They developed an innovative method base on spherical images of linkages. Their methods are applicable to both closed–loop and open kinematic chains. They rely extensively on the use of dialytic elimination.

Recently a lot of progress has been made on the hard problems for series and in–parallel manipulators. The inverse kinematics of all series chains are now known and the direct kinematics of the common in–parallel chains are also known. The current surge of results and publications is probably traceable to two pivotal events: The first was a luncheon talk by Professor F. Freudenstein at the 1972 ASME Mechanisms Conference, and a subsequent paper based on this talk, in which he put forward the idea that the position analysis of the closed series–loop seven revolute (7R) chain was the most challenging unsolved problem. He called it "the Mount Everest of Kinematics." The second event was the publication by Lee (actually Li) and Liang in 1988 [26] of the first complete, non iterative, solution of the inverse kinematics of the 7R linkage. This initial work and related studies were also published in a short monograph [27].

After the Lee and Liang publication, Raghavan and Roth [28] developed the same solutions by a new systematic method based on elimination theory and its basic ideas. They then went on to elaborate their method to include all six–degree–of–freedom open–loop series chains, and by implication all closed–loop, series, one–degree–of–freedom mechanisms [29,30]. What we now know is that a series, six–degree–of–freedom chain has at most 16 ways to reach a pose if the chain is composed of all revolute joints (6R) or has one prismatic joint and five revolute joints (5R1P). If the chain has two prismatic joints (4R2P) the maximum drops to 8. While if it has three prismatic joints (3R3P) the number drops to 2 [29,30]. Furthermore, any time three neighboring axes intersect or are parallel, the number is at most 8. By using special structural restrictions, such as parallel and intersecting joint axes, it is also possible to get 2, 4, 8, and 12 maximum assembly variants at each pose [31]. Most recently, the methods of Raghavan and Roth have also led to studies of special manipulator geometry which lead to overconstrained mechanisms [32].

There have since been other variants discussed in the literature. For example, there is a study by Weiss [33] which concludes that resultant methods are better suited for the inverse kinematics problem than Gröbner basis techniques. There has also been a lot of progress in making the inverse kinematics algorithm into a fast computational tool. This work has centered on the use of eigenvalue formulations [34–37].

For the in–parallel mechanisms it has been proven that the general in–parallel platform mechanism (i.e., the mechanism with six extensible legs in–parallel between the ground and platform) link has 40, at most, different poses associated with a set of leg lengths [5]. It is known that for special geometries this same mechanism gives 4, 8, 12, 16, 24 and 40 poses [38–44].

There is currently a lot of interest in different in-parallel and hybrid configurations [45,46], and various new designs are being developed with different numbers of links. While these are all of potential interest, the major outstanding problem in all of manipulator direct and inverse kinematics is finding a method to determine the actual 40th degree polynomial first proved to exist by Raghavan [5]. It is almost certain this will be done by use of elimination theory. After that the concern will turn to developing general theories for solving the types of nonlinear equations which arise in the analysis and design of other in-parallel and hybrid manipulator systems.

IV. Conclusions

The most effective methods for the solution of the sets of nonlinear equations in robot kinematics have been discussed. The most important methods to date are the continuation method, which is a numerical iteration, and the elimination methods, which allow for symbolic as well as numerical development of a polynomial which characterizes the solution set. Both of these methods are, in theory, universally applicable to robot kinematics, and both have yielded very strong results in recent years. Currently the continuation method is the more limited of the two.

There is now a lot of interest in these methods, and various researchers are developing analytical, graphical [47] and numerical improvements. Therefore, it is to be expected that their current limitations will be circumvented. In addition other not yet proven methods may gain efficiency and become of wider utility. On the whole, the future of this subject seems as bright as its recent past.

V. Acknowledgment

The financial support of the National Science Foundation is appreciated.

VI. References

[1] B. Roth and F. Freudenstein, "Synthesis of Path-Generating Mechanisms by Numerical Methods," *Journal of Engineering for Industry, Trans. ASME*, v.85, pp.298-307 (1963).

[2] C.W. Wampler, A.P. Morgan and A.J. Sommese, "Numerical Continuation Methods for Solving Polynomial Systems Arising in Kinematics," *Journal of Mechanical Design, Trans. ASME*, v.112, pp.59-68 (1990).

[3] G.K. Starns and D.R. Flugrad Jr., "Five-Bar Path Generation Synthesis by Continuation Methods," *Mechanism Synthesis and Analysis*, eds. M. McCarthy, S. Derby and A. Pisano, DE. Vol. 25, ASME, pp.353-360 (1990).

[4] A.P. Morgan, *Solving Polynomial Systems Using Continuation for Scientific and Engineering Problems*, Prentice-Hall, Englewood Cliffs, NJ (1986).

[5] M. Raghavan, "The Stewart Platform of General Geometry Has 40 Configurations," *Journal of Mechanical Design, Trans. ASME*, v.115, 2, pp.277-282 (1993).

[6] C. Wampler, A. Morgan and A. Sommese, "Numerical Continuation Methods for Solving Polynomial Systems Arising in Kinematics," *Journal of Mechanical Design, Trans. ASME*, v.112, pp.59–68 (1990).

[7] L.W. Tsai and A. Morgan, "Solving the Kinematics of the Most General Six– and Five–Degree–of–Freedom Maniipulators by Continuation Methods," *Journal of Mechanisms, Transmissions, and Automation in Design, Trans. ASME*, v. 107, pp. 189–200 (1985).

[8] M. Raghavan and B. Roth, "On the Design of Manipulators for Applying Wrenches," *Proceedings of the 1989 IEEE International Conference on Robotics and Automation*, vol. 1, pp.438-443 (1989).

[9] H. Hironaka, "Resolution of Singularities of an Algebraic Variety Over a Field of Characteristic Zero: 1,2," *Annals of Mathematics*, v.79, pp.109-326 (1964).

[10] B. Buchberger, "Gröbner Bases: An Algorithmic Method in Polynomial Ideal Theory," Chapter 6 in *Multidimensional System Theory*, ed. N. K. Bose. D. Reidel Publishing Co. (1985).

[11] D. Kapur, "Geometry Theorem Proving Using Gröbner Bases, *Journal of Symbolic Computation*, v.2 (1986).

[12] B. Kutzler and S. Stifter, "On the Application of Buchberger's Algorithm to Automated Geometry Theorem Proving," *Journal of Symbolic Computation*, v.2 (1986).

[13] D. Lazard, "On the Representation of Rigid–Body Motions and its Application to Generalized Platform Manipulators," *Computational Kinematics* (J. Angeles, G. Hommel and P. Kovács, Eds.) Kluwer Academic Publishers, Dordrecht, pp.175–183 (1993).

[14] A. Cayley, "On the Theory of Elimination," *Cambridge and Dublin Mathematics Journal*, v.3 (1848).

[15] G. Salmon, *Higher Algebra*, 5th Edition (1885), Chelsea Publishing Co., NY (1964).

[16] B. Van der Waerden, *Modern Algebra*, vol.2, Frederick Unger Publishing Co. (1964).

[17] J. Jouanolou, *Singularities rationelles du Resultant*, Lecture Notes in Mathematics, No. 732, Springer-Verlag (1978).

[18] W. Fulton, *Intersection Theory*, Springer-Verlag (1984).

[19] J.F. Canny, *The Complexity of Robot Motion Planning*, MIT Press, Cambridge, MA (1987).

[20] B. Roth, "Computations in Kinematics," *Computational Kinematics* (J. Angeles, G. Hommel and P. Kovács, Eds.) Kluwer Academic Publishers, Dordrecht, pp.3–14 (1993).

[21] D.L. Pieper, *The Kinematics of Manipulators Under Computer Control*, Ph.D. dissertation, Dept. of Mechanical Engineering, Stanford University (1968).

[22] D.L. Pieper and B. Roth, "The Kinematics of Manipulators Under Computer Control," *Proc. Second Int. Congress on the Theory of Machines and Mechanisms*, Zakopane, Poland, v. 2, pp.159–169 (1969).

[23] M. Chace, "Vector Analysis of Linkages," Journal of Engineering for Industry, Trans. ASME, v.85, pp.289-297 (1963).

[24] J. Duffy, *Analysis of Mechanisms and Robot Manipulators*, Halstead Press (John Wiley & Sons) New York (1980).

[25] J. Duffy and C. Crane, "A Displacement Analysis of the General Spatial Seven-Link, 7R Mechanism," Mechanism and Machine Theory, v.15, pp.153-169 (1980).

[26] H.Y. Lee and C.G. Liang, Displacement Analysis of the General Spatial 7–Link 7R Mechanism, " *Mechaism and Machine Theory*, v.23, pp.219–226 (1988).

[27] C.G. Liang, H.Y. Lee and Q.Z. Liao, *Analysis of Spatial Linkages and Robot Mechaisms*, Beijing University of Post & Telecommunications Publishing House, Beijing (1988)

[28] M. Raghavan and B. Roth, "Kinematic Analysis of the 6R Manipulator of General Geometry," *Robotics Research, The Fifth International Symposium*, eds. H. Miura and S. Arimoto, MIT Press, pp.263-270 (1990).

[29] M. Raghavan and B. Roth, "A General Solution of the Inverse Kinematics of All Series Chains," *Proceedings Eight CISM-IFToMM Symposium on Theory and Practice of Robots and Manipulators*, Hermes, Paris (1991).

[30] M. Raghavan and B. Roth, "Inverse Kinematics of the General 6R Manipulator and Related Linkages," *Journal of Mechanical Design, Trans. ASME* (1992).

[31] C. Mavroidis and B. Roth, "Structural Parameters which Reduce the Number of Manipulator Configurations," Trans. *ASME, J. of Mech. Des.*, v.115, pp. 3–10 (1994).

[32] C. Mavroidis and B. Roth, "Analysis and Synthesis of Overconstrained Mechanisms," Accepted for publication in the proceedings of the ASME Mechanisms Conference (1994).

[33] J. Weiss, "Resultant Methods for the Inverse Kinematics Problem," *Computational Kinematics* (J. Angeles, G. Hommel and P. Kovács, Eds.) Kluwer Academic Publishers, Dordrecht, pp.41–52 (1993).

[34] D. Manocha and J.F. Canny, "Real Time Inverse Kinematics for General 6R Manipulators," *Proceedings of the 1992 IEEE International Conference on Robotics and Automation* , v.1, pp.383–389 (1992).

[35] D. Manocha and Y. Zhu, "A Fast Algorithm and System for the Inverse Kinematics of General Serial Manipulators," *Proceedings of the 1994 IEEE International Conference on Robotics and Automation* (1994).

[36] M. Ghazvini, "Reducing the Inverse Kinematics of Manipulators to the Solution of a Generalized Eigenproblem," *Computational Kinematics* (J. Angeles, G. Hommel and P. Kovács, Eds.) Kluwer Academic Publishers, Dordrecht, pp.15–26 (1993).

[37] D. Kohli and M. Osvatic, *The Inverse Kinematics of General Serial Manipulators*, monograph published by the authors, Milwaukee, (1992).

[38] K.H. Hunt and E.J.F. Primrose "Assembly Configurations of Some In–parallel –actuated Manipulators," *Mech. and Mach. Th.*, v.28, 1, pp.31–42 (1993).

[39] C. Innocenti and V. Parenti–Castelli "Direct Position Analysis of the Stewart Platform Mechanism," *Mech. and Mach. Theory*, v.25, 6, pp.611–621 (1990).

[40] J.–P. Merlet, *Les robots parallèles*, Hermès, Paris (1990).

[41] D. Lazard and J–.P Merlet, "The (true) Stewart Platform has 12 Configurations" *Proceedings of the 1994 IEEE International Conference on Robotics and Automation* , (1994).

[42] W. Lin, J. Duffy and M. Griffis, "Forward Displacement Analyses of the 4-4 Stewart Platforms," *Mechanism Synthesis and Analysis*, eds. M. McCarthy, S. Derby and A. Pisano, DE. Vol. 25, ASME, pp.263-270 (1990).

[43] C. Innocenti and V. Parenti–Castelli "Closed–Form Direct Position Analysis of the 5–5 Parallel Mechanism," *Journal of Mechanical Design, Trans. ASME*, v.115, pp.515–521 (1993).

[44] H.Y. Lee and B. Roth, "A Closed–Form Solution of the Forward Displacement Analysis of a Class of In–Parallel Mechanisms," *Proceedings of 1993 IEEE International Conference on Robotics and Automation*, Atlanta, Georgia, v. 1, pp. 720–724 (1993).

[45] K.J. Waldron, B. Roth and M. Raghavan, "Kinematics of a Hybrid Serial-Parallel Manipulator System," *Journal of Dynamic Systems, Tans. ASME*, v.111, pp.211-221 (1989).

[46] M. Lee, M. and D.K. Shah, "Kinematic Analysis of a Three Degrees of Freedom In-Parallel Actuated Manipulator, *Proceedings of the 1987 IEEE International Conference on Robotics and Automation*, Vol. 1, pp.345-350 (1987).

[47] K.E. Zanganeh and J. Angeles, "The Semigraphical Solution of the Direct Kinematics of General Platform Manipulators," *Computational Kinematics* (J. Angeles, G. Hommel and P. Kovács, Eds.) Kluwer Academic Publishers, Dordrecht, pp.165–174 (1993).

1. Workspace and Trajectory Analysis

Characterizing the Workspace of the Spherical Image of Cooperating Robots

A. P. Murray J. M. McCarthy

Department of Mechanical Engineering
University of California Irvine
Irvine, California 92717

Abstract - In this paper we consider the set of orientations available to the workpiece held by a pair of cooperating robot arms. The rotational terms in the kinematics equations of the system define constraints on the orientation of the workpiece that can be viewed as a spherical linkage, called the spherical image of the robot system. We determine the differential volume element for this orientation workspace and examine two cases, a near planar spherical $6R$, and a $6R$ with orthogonal joint axes which is a model for the case of cooperating PUMA Robots.

I. Introduction

Two robots holding the same workpiece form a closed kinematic chain with cylindric joints that are constrained either to pure rotation (revolute) or pure sliding (prismatic). Duffy [4] shows that the first step in the analysis of chains of this type is the derivation of the constraint equations arising from the angular dimensions of the chain, that is the twist angles between each joint axis and the rotations about these axes. These equations describe the kinematics of a revolute jointed linkage having the same angular dimensions but constructed such that the axes of the joints intersect in the same point. This linkage characterizes the orientation constraints that arise from the angular properties of the cooperating robot system and is termed its *spherical image*. The orientation workspace of the system is also limited by the linear dimensions of the closed chain, therefore this is a first step in the complete characterization of the workspace of these systems.

The reachable workspace of cooperating robots which form the fingers of a mechanical hand was studied by Kerr and Roth [7]. Ge [5] used the Clifford algebra of projective space, or dual quaternions, to define manifolds characterizing both positions and orientations available to the workpiece held by cooperating robots. Bodduluri [1] evaluated the volume of possible positions and orientations of the workpiece held by two planar $3R$ robots as a function of their base separation, Fig. 1.

An experimental system of cooperating PUMA robots has been developed by Xi et al. [10]. Cooperating planar $3R$ robots have been demonstrated by Hsu [6], and Paljug and Yun [9]. Dooley and McCarthy [3] formulate the equations of motion of a cooperating planar $3R$ robots using Ge's Clifford algebra formulation to plan paths for the system.

In this paper we present a simplified version of Ge's derivation and generalize it to obtain the differential volume element for cooperating spherical robots. We then compute the volume of orientations available to the workpiece held by a pair of cooperating

A. J. Lenarčič and B. B. Ravani (eds.), Advances in Robot Kinematics and Computationed Geometry, 19–28.

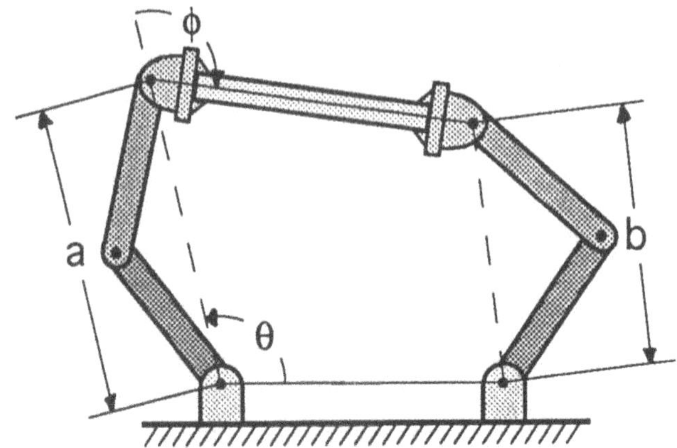

Figure 1: Cooperating planar robots.

spherical $3R$ robots that are "near planar" in angular dimensions. The results are seen to parallel closely the planar results. We then consider the case of $3R$ robots that have twist angles of 90°.

II. Cooperating Planar Robots

A. *RPR planar robot*

In this section we derive the constraint manifold for a single planar robot and its differential volume element so we can determine the volume of available positions of the workpiece. The robot has revolute joints at the base and wrist. We consider the distance between these two joints to be the third degree of freedom represented by a prismatic (P) joint, so the robot is an RPR open chain. This degree of freedom need not, physically, be a prismatic joint, it can be controlled by the elbow joint in $3R$ robot, for example.

Let the position of the workpiece frame M relative to the base frame F be represented by a 2×2 rotation matrix $[A(\psi)]$ and a 2×1 translation vector $\mathbf{d} = (d_x,\, d_y)^{\mathrm{T}}$. The Clifford algebra element associated with this tranformation is the four dimensional vector $\mathbf{D}$ with components given by (McCarthy [8]):

$$\begin{aligned}
D_1 &= \left(\frac{d_y}{2}\right)\sin\left(\frac{\psi}{2}\right) + \left(\frac{d_x}{2}\right)\cos\left(\frac{\psi}{2}\right), \\
D_2 &= -\left(\frac{d_x}{2}\right)\sin\left(\frac{\psi}{2}\right) + \left(\frac{d_y}{2}\right)\cos\left(\frac{\psi}{2}\right), \\
D_3 &= \sin\left(\frac{\phi}{2}\right),
\end{aligned} \tag{1}$$

$$D_4 = \cos\left(\frac{\phi}{2}\right).$$

The kinematics equations of the RPR chain define the position of the frame M relative to F in terms of the base and wrist rotations θ and ϕ, and the distance a between them. When computed using Clifford algebra elements we obtain the parameterized "constraint" manifold:

$$\mathbf{D}(\theta, a, \phi) = \left\{ \begin{array}{c} \left(\frac{a}{2}\right)\cos\left(\frac{\theta-\phi}{2}\right) \\ \left(\frac{a}{2}\right)\sin\left(\frac{\theta-\phi}{2}\right) \\ \sin\left(\frac{\theta+\phi}{2}\right) \\ \cos\left(\frac{\theta+\phi}{2}\right) \end{array} \right\}. \tag{2}$$

This manifold defines the set of positions reachable by the workpiece held by the robot. The Jacobian of this mapping is the 4×3 matrix

$$\mathbf{J} = \left[\begin{array}{ccc} \frac{\partial \mathbf{D}}{\partial \theta} & \frac{\partial \mathbf{D}}{\partial a} & \frac{\partial \mathbf{D}}{\partial \phi} \end{array} \right] \tag{3}$$

and the differential volume element of the manifold is given by (Do Carmo [2]):

$$dV = \sqrt{\det[\mathbf{J}^T\mathbf{J}]}\; d\theta da d\phi = \left(\frac{a}{8}\right) d\theta da d\phi \tag{4}$$

The integration of this volume element over the range of angles θ and ϕ and the distance a defines the size of the manifold of reachable positions for this robot. The dimension of this volume is m^2, where m is the unit of length along the links.

B. *Cooperating planar RPR robots*

The workspace of a cooperating system is the intersection of the constraint manifolds of the two individual robots. The boundaries in the constraint manifold of the first robot caused by the presence of the second robot is obtained using the loop equation of the closed chain formed by the two robot system. In the case of cooperating RPR robots, this is the constraint equation of a four bar linkage with variable crank lengths a and b:

$$A\cos\phi + B\sin\phi + C = 0 \tag{5}$$

where

$$\begin{aligned} A &= 2ah - 2gh\cos\theta, \\ B &= 2gh\sin\theta, \\ C &= g^2 + h^2 + a^2 - 2ag\cos\theta - b^2. \end{aligned} \tag{6}$$

Here g is the distance between the base joints of the two robots and h is the distance between the wrist joints.

We now determine the differential volume element of the cooperating robot system. It is convenient to use the lengths a and b of the two robots and the base rotation θ of the first robot as the independent variables. Thus, the parameter ϕ in Eq. (4) now depends on b as defined by Eq. (5), so we compute:

$$\frac{d}{db}(A\cos\phi + B\sin\phi + C) = -A\sin\phi\frac{d\phi}{db} + B\cos\phi\frac{d\phi}{db} - 2b = 0 \tag{7}$$

and obtain

$$d\phi = \frac{-2b\ db}{A\sin\phi - B\cos\phi}. \tag{8}$$

Before substituting this relation into Eq. (4), we note that

$$A\sin\phi - B\cos\phi = \sqrt{A^2 + B^2 - C^2} \tag{9}$$

which eliminates the variable ϕ. The resulting differential volume element is

$$dV = \frac{ab}{8\sqrt{A^2 + B^2 - C^2}}\ d\theta da db \tag{10}$$

The term $A^2 + B^2 - C^2$ in the denominator expands to become

$$A^2 + B^2 - C^2 = -(d^2 - (h+b)^2)(d^2 - (h-b)^2), \tag{11}$$

where

$$d^2 = a^2 + g^2 - 2ag\cos\theta. \tag{12}$$

d is the diagonal of the quadrilateral formed by the four revolute joints of the cooperating robot system. The limits on the range of motion of the first robot imposed by the second robot are seen to occur when this diagonal is equal to the size of the workpiece h plus or minus the length b of the second robot.

Figure 2 shows the result of a numerical integration of this volume for the case of two RPR robots holding a workpiece of size 1 with lengths that vary over $0 \le a, b \le 2$, and for a range of base separations $0 \le g \le 5$. This reproduces Bodduluri's [1] results.

III. Spherical Image of Cooperating Robots

Consider the pair of cooperating $3C$ robots shown in Fig 3. The spherical image of this system is the $6R$ spherical chain shown in Fig 4. Closed chains which have a $6R$ spherical image are termed Group 3 mechanisms by Duffy [4].

A. *The spherical RPR robot*

We now derive results for the spherical image of cooperating robots that parallel the planar results above. Consider a spherical open chain with revolutes for the base and wrist joints, and let the angular distance between these joints be variable. This is the spherical equivalent to the planar RPR robot.

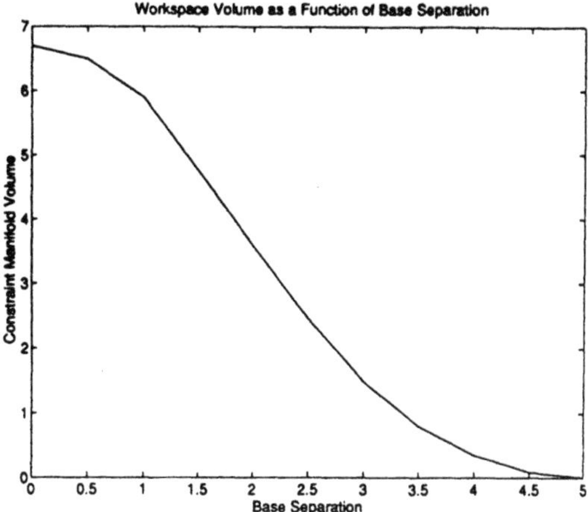

Figure 2: Volume of positions for cooperating planar RPR robots.

Let the orientation of a frame M in a workpiece held by an RPR spherical chain relative to a fixed frame F be defined by the 3×3 rotation matrix $[A]$. With each rotation matrix we can identify an axis of rotation $\mathbf{S}$ and a rotation angle ψ and define the associated element of the Clifford algebra (quaternion) as the four dimensional vector $\mathbf{D}$ with components:

$$\begin{aligned} D_1 &= S_x \sin\left(\frac{\psi}{2}\right), \\ D_2 &= S_y \sin\left(\frac{\psi}{2}\right), \\ D_3 &= S_z \sin\left(\frac{\psi}{2}\right), \\ D_4 &= \cos\left(\frac{\psi}{2}\right). \end{aligned} \tag{13}$$

The kinematics equations of the spherical RPR chain define the position of the workpiece frame M relative to F in terms of the base and wrist rotation angles θ and ϕ, and the variable angular length α. In the Clifford algebra, the set of orientations reachable by the workpiece is the parameterized "constraint" manifold:

$$\mathbf{D}(\theta, \alpha, \phi) = \left\{ \begin{array}{c} \sin\left(\frac{\alpha}{2}\right) \cos\left(\frac{\theta-\phi}{2}\right) \\ \sin\left(\frac{\alpha}{2}\right) \sin\left(\frac{\theta-\phi}{2}\right) \\ \cos\left(\frac{\alpha}{2}\right) \sin\left(\frac{\theta+\phi}{2}\right) \\ \cos\left(\frac{\alpha}{2}\right) \cos\left(\frac{\theta+\phi}{2}\right) \end{array} \right\} \tag{14}$$

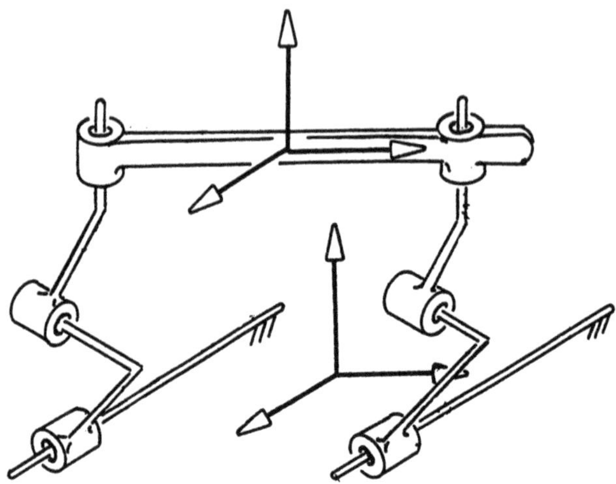

Figure 3: A pair of cooperating $3C$ robots.

As in the planar case, we compute the 4×3 Jacobian $\mathbf{J} = [\partial\mathbf{D}/\partial\theta, \partial\mathbf{D}/\partial\alpha, \partial\mathbf{D}/\partial\phi]$, and obtain the differential volume element for this manifold as:

$$dV = \sqrt{\det[\mathbf{J}^T\mathbf{J}]}\; d\theta d\alpha d\phi = \left(\frac{\sin\alpha}{8}\right) d\theta d\alpha d\phi. \tag{15}$$

Integration of this differential volume over the ranges of values for θ, ϕ, and α yields the volume of orientations available to the workpiece of the spherical RPR robot. For the case, $0 \le \theta \le 2\pi$, $0 \le \alpha \le \pi$, and $0 \le \phi \le 2\pi$, we obtain the volume of the set of all orientations as $\pi^2 = 9.88$.

The variation in distance between the revolute joints in the spherical RPR can be implemented using a revolute joint that is orthogonal to the plane containing the base and wrist joints. This is the configuration of the typical orthogonal $3R$ robot wrist.

B. Cooperating spherical RPR robots

The set of positions available to the workpiece held by two spherical RPR robots is the intersection of the constraint manifolds of the two individual robots. To study this intersection, we consider the limits on the constraint manifold of the first robot imposed by the second. This is done by considering the variable lengths α and β of the two robots and the base rotation angle θ of the first robot as the independent variables of the system. Thus, the wrist rotation angle ϕ of the first robot is constrained by the presence of the second robot. This constraint is the loop equation of the spherical quadrilateral formed by the base and wrist joints of the system:

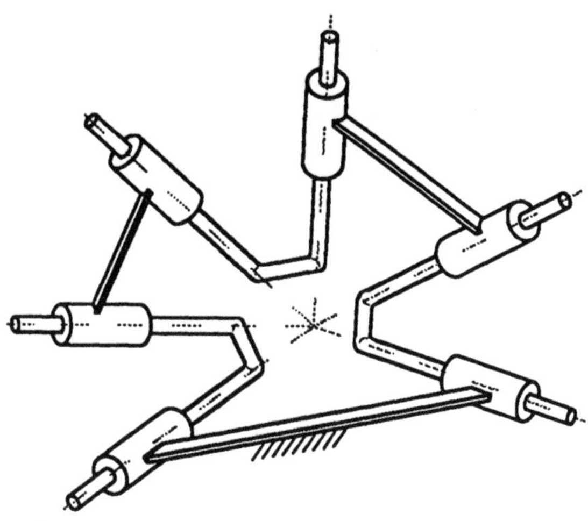

Figure 4: The $6R$ spherical image.

$$A\cos\phi + B\sin\phi + C = 0 \tag{16}$$

where

$$\begin{aligned} A &= \sin\gamma\sin\eta\cos\alpha\cos\theta - \sin\alpha\sin\eta\cos\gamma, \\ B &= -\sin\gamma\sin\eta\sin\theta, \\ C &= \cos\gamma\cos\alpha\cos\eta + \sin\gamma\cos\eta\sin\alpha\cos\theta - \cos\beta. \end{aligned} \tag{17}$$

In this equation γ is the angular distance between the base joints and η is the angular distance between the wrist joints.

The differential volume element for this system is obtained by computing $d\phi/d\beta$ using the equation,

$$\frac{d}{d\beta}(A\cos\phi + B\sin\phi + C) = -A\sin\phi\frac{d\phi}{d\beta} + B\cos\phi\frac{d\phi}{d\beta} + \sin\beta = 0 \tag{18}$$

which yields

$$d\phi = \frac{\sin\beta d\beta}{A\sin\phi - B\cos\phi}. \tag{19}$$

This relation can be simplified by noting that

$$A\sin\phi - B\cos\phi = \sqrt{A^2 + B^2 - C^2}. \tag{20}$$

Substituting this into the Eq. (15), we obtain

$$dV = \frac{\sin\alpha\sin\beta}{8\sqrt{A^2 + B^2 - C^2}}\, d\theta d\alpha d\beta \tag{21}$$

The term $A^2 + B^2 - C^2$ expands in essentially the same way as in the planar case to become

$$A^2 + B^2 - C^2 = -(\cos\delta - \cos(\eta + \beta))(\cos\delta - \cos(\eta - \beta)) \tag{22}$$

where δ is the diagonal of the spherical quadrilateral given by

$$\cos\delta = \cos\theta \sin\alpha \sin\gamma + \cos\alpha \cos\gamma. \tag{23}$$

The limit to the movement of the first robot occurs when this diagonal equals the angular dimension of the workpiece η plus or minus the angular dimension β of the second robot.

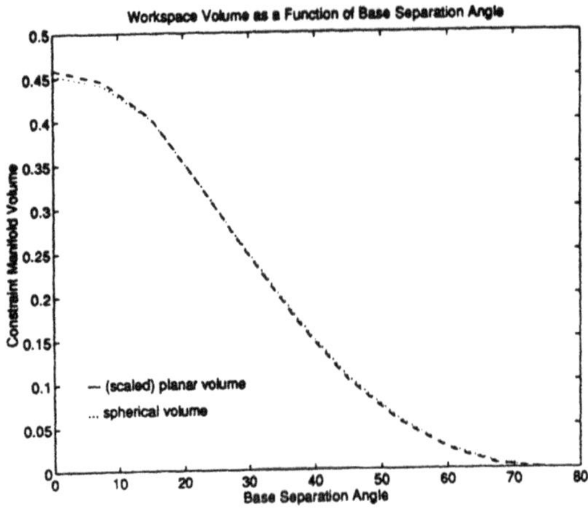

Figure 5: Volume of positions for cooperating spherical RPR robots.

Figure 5 is the result of the numerical integration of this differential volume for the case of a 15° workpiece with the angular lengths of the two robots varying over the range $0° \leq \alpha, \beta \leq 30°$. This volume is evaluated for base separations varying over the range $0° \leq \gamma \leq 75°$. We scaled the planar results obtained above to compare them to these spherical results by considering the planar links as projections of links on a unit sphere. We obtain the relation $m = \tan(\pi/12) \approx \pi/12$, so the volume scales as $\pi^2/144 = 1/14.6$. This scaled version of the planar curve is also shown in Figure 5. The similarity of the two curves arises from the "near planar" dimension of the spherical image.

IV. Orthogonal Spherical Images

Most robots have twist angles that are either 0 or 90°. If the twist angle is zero then spherical images of the two axes collapse into a single axis. Therefore the spherical image of the typical robot is a chain with 90° link lengths which we call an "orthogonal" chain. The RPR robot with a revolute joint orthogonal to the plane of the base and wrist joints is an orthogonal $3R$ chain. We now show that cooperating robots that have

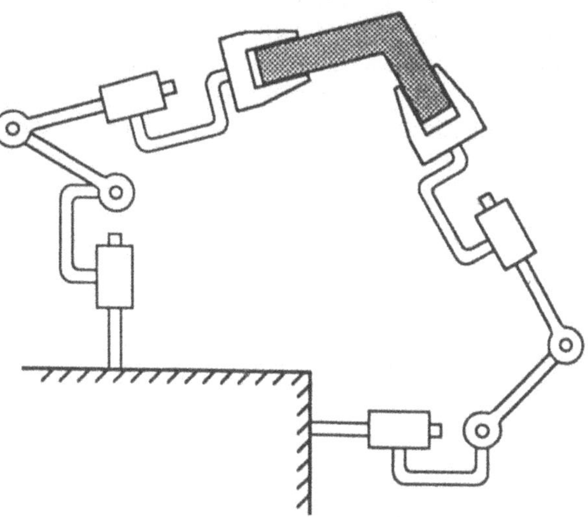

Figure 6: Cooperating PUMA-like robots.

$3R$ orthogonal spherical images can reach all orientations, independent of the angular dimension of the workpiece and the angular orientation of the base joints. Figure 6 shows two cooperating PUMA-like robots that will have this spherical image. Note that the wrists allow rotation around only one axis.

Let $L(\theta, \alpha, \phi)$ and $R(\theta, \alpha, \phi)$ be the kinematics equations of the two orthogonal $3R$ spherical images. Let H_1 be the orientation of the moving frame M in the workpiece relative to the first robot and H_2 be its orientation relative to the second robot. Simlarly G_1 and G_2 are the orientations of the bases of these two robots relative to the fixed frame F. Thus we have the kinematics equations:

$$\begin{aligned} A &= G_1 L(\theta, \alpha, \phi) H_1, \\ &= G_2 R(\theta, \alpha, \phi) H_2, \end{aligned} \tag{24}$$

where A is the orientation of M relative to F.

Since L and R are orthogonal $3R$ robots they can reach any orientation. In particular they can reach $G_1^{-1} A H_1^{-1}$ and $G_2^{-1} A H_2^{-1}$, which means they can reach all orientations independent of the angular dimensions of the workpiece and the base separation.

V. Conclusion

This paper examines the set of orientations available to cooperating robot systems by representing the positions as points in the Clifford algebra of projective space. A simplified derivation for the differential volume element for cooperating planar robots is presented. A similar derivation yields the differential volume element for cooperating spherical robots which arise from the spherical image of general cooperating robot systems.

Of particular interest is the similarity between the workspace volume obtained for cooperating planar robots and that of a near-planar system of spherical cooperating robots. We also show that cooperating orthogonal 3R robots can achieve all orientations of the workpiece, independent of its dimensions or the angular separation of the base joints.

Characterization of the workspace of the spherical image of cooperating robots is a step toward characterization of the workspace of general cooperating robot systems.

References

[1] Bodduluri, R. M. C., *Design and Planned Movement of Multi-degree of Freedom Spatial Mechanisms.* Dissertation, University of California, Irvine, 165pp., 1990.

[2] Do Carmo, M. P., *Riemannian Geometry.* Birkhauser, Boston, 300pp., 1992.

[3] Dooley, J. R., and McCarthy, J. M., On the Geometric Analysis of Optimum Trajectories for Cooperating Robots using Dual Quaternion Coordinates. *Proc. IEEE Int. Conf. on Robotics and Automation.* pp. 1031-1036., 1993.

[4] Duffy, J. *Analysis of Mechanisms and Robot Manipulators.* Edward Arnold, London, 419pp., 1980.

[5] Ge, Q. J., *Kinematic Constraints as Algebraic Manifolds in the Clifford Algebra of Projective Three Space.* Dissertation, University of California, Irvine, 160pp., 1990.

[6] Hsu, P., Adaptive Coordination of a Multiple Manipulator System. *Proc. IEEE Int. Conf. on Robotics and Automation.* pp. 259-264., 1993.

[7] Kerr, J., and Roth, B., Analysis of Multifingered Hands. *Int. J. of Robotics Research.* 4(4): 3-17, 1986.

[8] McCarthy, J. M., *An Introduction to Theoretical Kinematics.* MIT Press, Cambridge, MA, 1990.

[9] Paljug, E., and Yun, X., Experimental Results of Two Robot Arms Manipulating Large Objects. *Proc. IEEE Int. Conf. on Robotics and Automation.* pp. 517-522., 1993.

[10] Xi, N., Tarn, T. J., and Bejczy, A., Event-Based Planning and Control for Multiple Manipulator Systems. *Proc. IEEE Int. Conf. on Robotics and Automation.* pp. 251-258, 1993.

On the Kinematics of Nonsingular and Singular Posture Changing Manipulators

P. WENGER and J. EL OMRI

Laboratoire d'Automatique de Nantes (U.A. CNRS 823)
Ecole Centrale de Nantes, University of Nantes
1, rue la Noë 44072 NANTES - FRANCE

Abstract - Some non redundant robot manipulators have the unusual ability to change posture without meeting a singularity. In this paper, several conditions are stated for classifying manipulators geometry. As a result, a list of 3-DOF singular posture changing manipulators geometries is provided. Both revolute and prismatic joints are considered in the analysis. In the end of the paper, the study is enlarged to the case of 6-DOF manipulators.

I. Introduction

Most manipulators have multiple inverse kinematic solutions; that is, a prescribed location can be reached with several joint configurations, commonly called postures. A typical example is the possibility for a 2-R planar mechanism to reach a point with both postures "elbow up" and "elbow down". For a long time, it has been thought that all non-redundant robots had to pass through a singularity when changing posture (like going through the outstretched singular configuration of the 2-R planar arm). Accordingly, it has been established that the singularities should always divide the jointspace into connected sets where there is one unique inverse kinematic solution (such sets were called aspects in [1], regions in [2] and c-sheets in [3]).

Quite recently, it was pointed out, however, that some manipulators are capable of going from one posture to another while not meeting a singularity [2], [3]. In [2], this property was numerically verified for two 6R manipulators, and the existence of "sub-regions" of uniqueness inverse solutions in the jointspace was intuitively established. In [3], the non-singular posture changing property has been more formally stated, using topological arguments. In [4], a general stringent framework was established for defining a new partition of the jointspace into uniqueness domains. These domains are separated by hypersurfaces formed by the singularities together with "pseudo-singularities" (see [4] for more details). These pseudo-singularities do not exist for usual manipulators. The ability to follow continuous paths in the workspace is significantly modified when the manipulator has pseudo-singularities. In particular, it turns out that a manipulator with pseudo-singularities can change posture while not necessarily leaving the prescribed path. We know, by experience, that all the manipulators commonly used in industry must pass through a singularity when changing posture. Also, we know that all these manipulators have simplifying kinematic conditions such as intersecting, parallel and orthogonal joint axes. On the other hand, it turns out that : 1. there exist nonsingular posture changing manipulators which do have simplifying geometries as well; 2. there exist singular posture changing manipulators with rather complex geometry. In [5], a list of RRR geometries which must pass through a singularity when changing posture was suggested. In this paper, several conditions are stated for classifying manipulator geometries. As a result, a list of singular

J. Lenarčič and B. B. Ravani (eds.), *Advances in Robot Kinematics and Computationed Geometry*, 29–38.

posture changing 3-DOF manipulators geometries is provided. Both revolute and prismatic joints are considered in the analysis.

Generalization to spatial 6R geometries is considered in the end of this paper.

II. Definitions [3]

A. Type-1 and Type-2 Geometries

A manipulator is said to have a *type 1 geometry* if it must pass through a singularity when changing posture. A manipulator with *type 2 geometry* can change posture while not meeting a singularity.

B. Characteristic Surfaces

Let Q denote the reachable jointspace. Q is bounded by the joint limits of the manipulator. Let A_i be an aspect in Q, that is, a connected component of $Q \dot{-} S$ ($\dot{-}$ means the difference between sets) where S is the set of singularities. The *characteristic surfaces* (CS) of A_i, denoted Sc(Ai), are defined as the preimage in A_i of the boundary A_i^*:

$$Sc(A_i) \stackrel{\Delta}{=} f^{-1}(f(A_i^*)) \cap A_i \qquad (2\text{-}1)$$

The characteristic surfaces are "pseudo-singularities" in the sense that they form a set of configurations which are non-singular, but which place the manipulator's tip at a location also reachable with a singular configuration.

The characteristic surfaces are independent of q_1 only when it is unlimited (the singularities are always independent of q_1). Representation of singularities and characteristic surfaces of 3-DOF geometries can be made, thus, in (q_2, q_3) when q_1 is unlimited.

C. Basic Components and Basic Regions

The characteristic surfaces induce a partition on each aspect into open connected sets called basic components.

Let A_i be an aspect. The basic components of A_i, denoted $\{Ab_{ik}, k \in K\}$, are the connected components of the set $A_i \dot{-} Sc(A_i)$:

$$A_i = (\cup_{k\in K} Ab_{ik}) \cup Sc(A_i) \qquad (2\text{-}2)$$

Let $WA_i=f(A_i)$ and $WAb_{ik}=f(Ab_{ik})$. WAb_{ik} is a connected domain of W, called basic region

$$WA_i = (\cup_{k\in K} Wb_{ik}) \cup f(Sc(A_i)) \qquad (2\text{-}3)$$

The basic components of 3-DOF geometries with first joint unlimited are represented in (q_2, q_3), and their basic regions can be represented in a cross-section of the workspace.

Fig. 1b depicts the basic components and the basic regions for the 3R manipulator depicted in Fig. 1a. This manipulator has orthogonal joint axes.

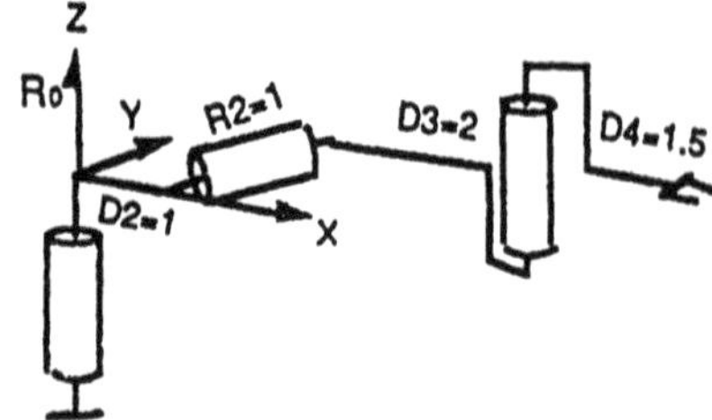

Fig. 1a : 3-R nonsingular posture changing geometry

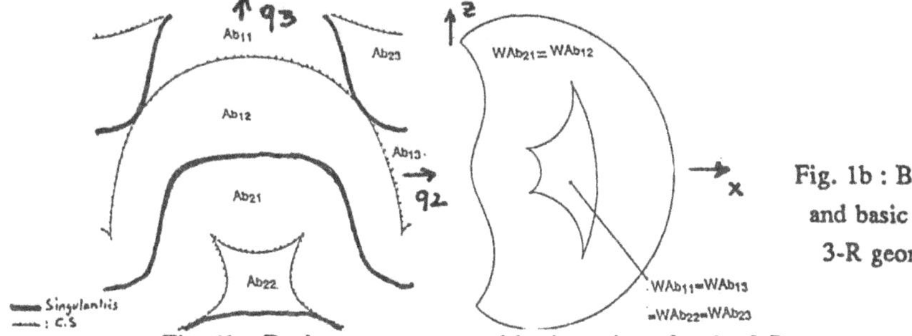

Fig. 1b : Basic components and basic regions for the 3-R geometry above

Fig. 1b : Basic components and basic regions for the 3-R geometry above

This manipulator has two pairs of inverse kinematic solutions. These two pairs are in two different aspects, respectively (see Fig.1c).

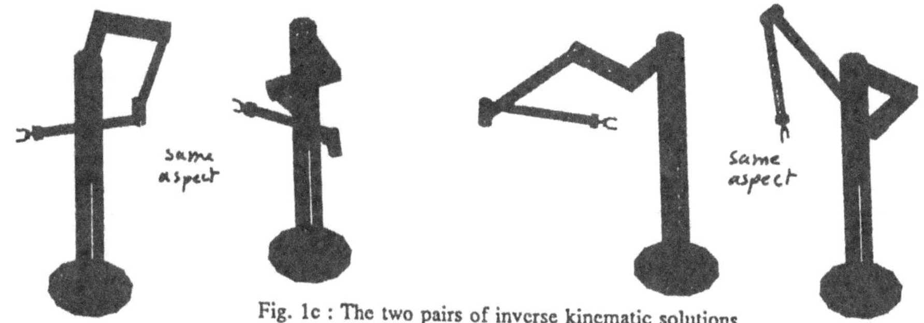

Fig. 1c : The two pairs of inverse kinematic solutions

III. Classification

A. Previous work

Burdick has proposed a classification of RRR-regional geometries into generic and non-generic geometries [5]. The genericity is based on the concept of generic and non-generic singularities introduced in [6]. A regional geometry is said non-generic if there exists a singular configuration q_s satisfying :

$$\frac{\delta DetJ}{\delta q_i}(q_s)=0\ ,\ i=2,3 \qquad (3\text{-}1)$$

DetJ is the determinant of the jacobian. The regional geometries are classified using (3-1).

Burdick made the conjecture that non-generic RRR manipulators with two extra-singularities are type 1. It turns out, unfortunately, that this is not true in all cases as shown in Fig. 2 : there are two inverse solutions in one aspect for this geometry which, however, has two additional singularities. The manipulator geometry considered here is close to that of Fig. 2. In Fig. 2a, we have d_4=2.5m instead of 1.5m in Fig. 2. Since $d_4>d_3$, there are two additional singularities defined by $q_3=\pm\arccos(-d_3/d_4)$, yielding 5 aspects. The point X defined by x=2.9m, y=0, z=0 in frame R_0 has the four following inverse kinematic solutions (in degrees):

$q^1=[-75\ \ 180\ \ 134]^t$, $q^2=[-67\ \ 0\ \ 138]^t$, $q^3=[30\ \ 0\ \ -101]^t$, $q^4=[165\ \ 180\ \ -44]^t$
with q^2 and q^4 in a same aspect.

In the following, we shall propose, first, two separate conditions for classifying all 3---DOF geometries (including revolute and prismatic joints). Then, generalization to spatial 6-DOF geometries will be considered.

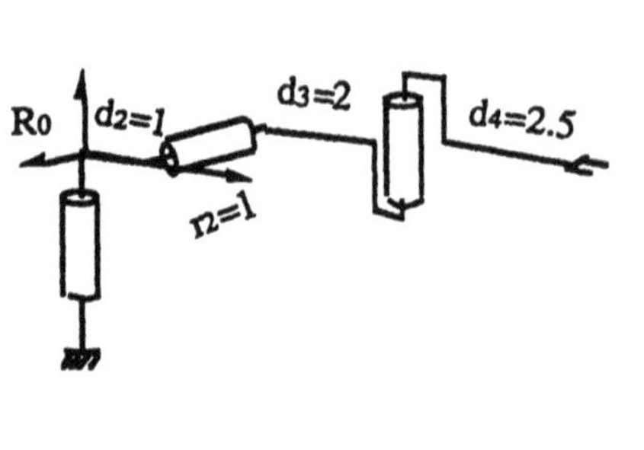

Fig. 2a : A type-2 manipulator geometry with five aspects

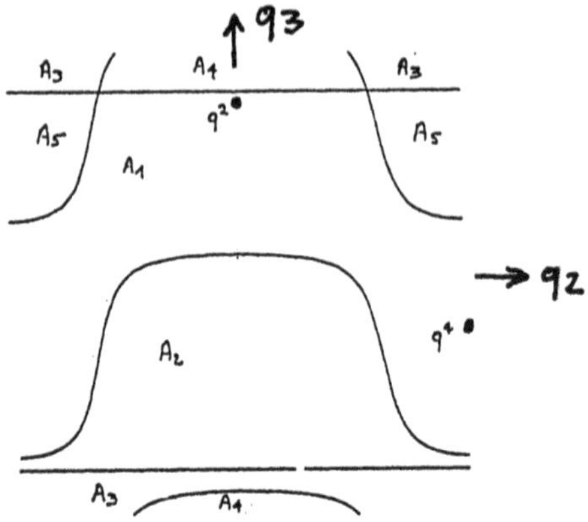

Fig. 2b : Two additional singularities and two solutions in one aspect

B. Two Conditions

Sufficient condition for a regional geometry to be type-1 : All regional geometries having no more than 2 inverse kinematic solutions are type 1.

Proof : Since there are no more than two solutions to the polynomial of the inverse kinematic problem, the change from one solution to another must make the end-effector passing through a point X_0 in the workspace where the polynomial has two equal roots $q'=q''=q_0$. Such a point X_0 must belong to a jacobian surface [7], that is the image of a singularity in the workspace. Since X_0 is the image of an unique joint configuration q_0, there is no way to avoid a singularity when changing solution.

Necessary condition : For a regional geometry with 4 inverse kinematic solutions to be type-1, there must exist extra singularities.

Proof : Assume that we are given a regional geometry with 4 inverse kinematic solutions and no extra singularities. Let X having 4 inverse solutions. Since there are only two singularities dividing the jointspace into two aspects, there must exist one aspect containing at least 2 inverse solutions. Our regional geometry, thus, is not type-1.

Unfortunately, the necessary condition above is not sufficient, as noticed in Fig. 2 above.

C. A List of 3-DOF Type 1 geometries

The kinematic parameters used in this paper are the modified Denavit & Hartenberg parameters introduced in [8]

* Geometries with one inverse solution.

All manipulators with (non-degenerate) PPP geometry have only one inverse kinematic. Thus, they are type-1.

* Geometries with at most two solutions.

- It is well-known that all regional (non-degenerate) geometries with two prismatic joints (i.e. PPR, PRP et RPP) have only 2 inverse kinematic solutions (see for example [9]).
- Some regional geometries with one prismatic joint have only two inverse solutions.

- All the geometries RRR, PRR, RPR and RRP whose polynomial in one of the joint variable is 4th order and has always 2 complex roots, are type 1 since there are, thus, at most 2 inverse solutions. Examples of such geometries are all RRR geometries with the first two joint axes orthogonal and such that $r_2=r_3=0$ and $d_2>d_3>d_4\sin^2(\alpha_3)$ (see Fig. 3).

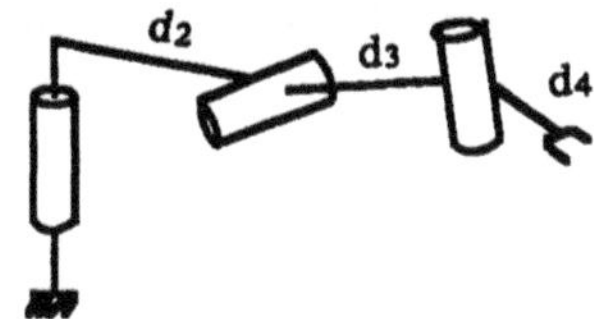

Fig. 3 : RRR geometry with at most two inverse kinematic solutions ($\alpha_2=90°$, $d_2>d_3>d_4\sin^2(\alpha_3)$ and $r_2=r_3=0$)

* Geometries with four solutions

We need to analyse the determinant of the jacobian, Det(J). Det(J) can be easily obtained from J. SYMORO [10] provides automatic symbolic calculation of J. In the most general case, there is no factorized expression for Det(J), and Det(J)=0 yields only two singularities. The special cases of possible factorization of Det(J) have to be found out. This is done by hand by looking for possible geometric simplifications. If the factorization yields additional singularities, the necessary condition stated above tells us that the geometry at hand is candidate for type 1. It is necessary, then, to check whether there is one or two solutions in each aspect. This can be done as follows. The jacobian surfaces (singularities in the workspace) divide the workspace into connected regions. Each of them is, actually, one or several coincident basic regions. Since the number of inverse solutions is constant for all points in each basic region, it is sufficient to identify the number of inverse solutions in each aspect for only one point X in each basic region.

<u>- RRR geometries</u>

The RRR geometries have been extensively studied in [3]. Burdick has proposed an original method for getting the equations of the singularities which does not require an explicit calculation of Det(J). We have used his method for our analysis. It turns out, after verification, that most of the non-generic RRR geometries listed in [5] are actually type-1 :

- 2 parallel joint axes (e.g. $\sin(\alpha_2)=0$ or $\sin(\alpha_3)=0$
- $\cos(\alpha_2) = 0$, $r_2=0$ and $r_3=0$
- 2 intersecting joint axes (e.g. $d_2=0$ or $d_3=0$)
- $\cos(\alpha_2) = 0$, $r_2=0$, $\cos(\alpha_3)=0$

The case $\cos(\alpha_2) = 0$, $r_3=0$ and $d_4>d_3$ mentioned in [5] is not type-1 (see Fig. 2)

For all other regional geometries, we have calculated Det(J), with the help of SYMORO. For more simplicity, J has been calculated at joint 2 and projected in frame 1. The following notations are used :

$sai = \sin(\alpha_i)$, $cai = \cos(\alpha_i)$, $ci = \cos(q_i)$ and $si = \sin(q_i)$.

<u>- PRR geometries</u>

$$J = \begin{pmatrix} s_2sa_2 & -s_3ca_3d_4+sa_3r_3 & -s_3d_4 \\ c_2sa_2 & d_3+c_3d_4 & c_3ca_3d_4 \\ ca_2 & 0 & c_3sa_3d_4 \end{pmatrix}$$

and
$$Det(J) = ca_2d_4(sa_3^2d_4c_3s_3 + sa_3r_3ca_3c_3 + d_3s_3) + c_3sa_2sa_3d_4(c_2(s_3ca_3d_4 - sa_3r_3) + s_2(d_3 + c_3d_4))$$

In the most general case, equation Det(J)=0 yields two singularities defined by $\tan(q_2)=K1(q_3)\pm K2(q_3)$ where K1 and K2 are functions of c_3 and s_3.

The following special cases of factorization have been found :

<u>If $ca_2=0$</u> (the first two joint axes are orthogonal) :

then $\quad Det(J) = c_3sa_2sa_3d_4(c_2(s_3ca_3d_4 - sa_3r_3) + s_2(d_3 + c_3d_4))$

There are two extra singularities defined by : $q3= \pm 90°$

<u>If $sa_3=0$</u> (the last two axes are parallel) :

then $\quad Det(J) = ca_2d_4(d_3s_3)$

In this case, there are only two singularities. On the other hand, we verify that there are only two inverse solutions.

<u>If $d_3=0$</u> (the last two axes are intersecting)

then $\quad Det(J) = d_4c_3\left(ca_2(d_4sa_3^2s_3 + sa_3r_3ca_3) + sa_2sa_3(c_2(s_3ca_3d_4 - sa_3r_3) + s_2c_3d_4)\right)$

There are two additional singularities defined by : $q3= \pm 90°$.

<u>If $sa_2=0$ and $r_3=0$</u>

then $\quad Det(J) = s_3ca_2d_4(sa_3^2d_4c_3 + d_3)$

There are always at least two singularities defined by : $q3=0°$ and $q3=180°$. When $d_3<d_4sa_3^2$, there are two additional branch singularites defined by $c_3=d_3/sa_3^2d_4$). Otherwise, there are no extra singularities, but there are no more than two inverse kinematic solutions.

<u>If $sa_2=0$ and $ca_3=0$</u>

$$Det(J) = s_3ca_2d_4(sa_3^2d_4c_3 + d_3)$$

We have the same results as above.

<u>If $ca_3=0$ and $r_3=0$</u>

$$Det(J) = (d_3+c_3d_4)(c_3sa_2sa_3d_4s_2+s_3ca_2d_4)$$

If, in addition, $d_4>d_3$, there are two additional singularities defined by $q3=\pm Arcos(-d_3/d_4)$. However, there are still two inverse kinematic solutions per aspect.

In summary, the following PRR geometries have been verified to be type-1 :

- First 2 joint axes are orthogonal (e.g. $\cos(\alpha 2)=0$)
- Last 2 joint axes are parallel (e.g. $\sin(\alpha 3)=0$)
- Last 2 joint axes are intersecting (e.g. $d_3=0$)
- $\sin(\alpha 2)=0$ and $r_3=0$
- $\sin(\alpha 2)=0$ and $\cos(\alpha 3)=0$

<u>- RPR geometries</u>

$$J = \begin{pmatrix} V1s_3+W1r_2+T1 & 0 & -s_3d_4 \\ U2c3+V2s_3+W2r_2+T2 & 0 & c_3ca_3d_4 \\ U3c_3+V3s_3+T3 & 1 & c_3sa_3d_4 \end{pmatrix}$$

with $V1=-ca_2ca_3d_4+c_2sa_2sa_3d_4$, $W1=c_2sa_2$,

$T1=ca_2sa_3r_3+c_2sa_2ca_3r_3+s_2ca_2d_2$, $V2=-s_2sa_2sa_3d_4$, $U2=ca_2d_4$

$W2=-s_2sa_2$, $T2=ca_2d_3-s_2sa_2ca_3r_3+c_2ca_2d_2$, $U3=-c_2sa_2d_4$

$V3=s_2sa_2ca_3d_4$, $T3=-sa_2(c_2d_3+d_2+s_2sa_3r_3)$

r_2 is, here, the joint variable, while s_2 and c_2 are constant.

$$Det(J) = r_2(c_3ca_3d_4W1+s_3d_4W2) + c_3ca_3d_4(V1s_3+T1)+s_3d_4(U2c_3+V2s_3+T2)$$

In the most general case, there are only one branch singularity defined by $r_2=K(q3)$ where K is function of $q3$. Since joint 2 is necessarily limited, there are, in general, two aspects (there may be more when joint 2 is strictly restricted).

<u>If $ca_2=0$</u> (the first two joint axes are orthogonal) :

Then, $Det(J) = (sa_3d_4s_3+ca_3r_3+r_2)(c_2c_3ca_3-s_3s_2)$

There are two extra singularities defined by $c_2c_3ca_3-s_3s_2=0$, which yields $q3=K1$ et $q3=K2$, where K1 and K2 are constant.

<u>If $ca_3=0$</u> (the last two axes are orthogonal) :

Then, $Det(J) = s_3d_4(r_2W2+U2c_3+V2s_3+T2)$

There are two additional singularities defined by : $q_3=0°$ and $q_3=180°$.

<u>If $s_2=0$ and $d_3+c_2d_2=0$</u>

$Det(J) = c_3(ca_3d_4(V1s_3+W1r_2+T1)+d_4^2ca_2s_3)$

There are two additional singularities defined by : $q_3=\pm 90°$.

Finally, the following RPR geometries have been verified to be type-1 :

- First 2 joint axes are orthogonal (e.g. $\cos(\alpha_2)=0$)
- Last 2 joint axes are orthogonal (e.g. $\cos(\alpha_3)=0$)
- $\sin(\theta_2)=0$ and $d_3+\cos(\theta_2)d_2=0$

<u>- RRP geometries</u>

$$J = \begin{pmatrix} V1s_3+W1r_2+T1 & -s_3ca_3d_4+sa_3r_3 & -0 \\ U2c_3+V2s_3+W2r_2+T2 & d_3+c_3d_4 & -sa_3 \\ U3c_3+V3s_3+T3 & 0 & ca_3 \end{pmatrix}$$

where U1, W1, T1, U2, V2, W2, T2, U3, V3, T3 are defined as above.

Here r_3 is the joint variable, while s_3 and c_3 are constant.

$$Det(J) = (s_3ca_3d_4-sa_3r_3)(sa_3(U3c_3+V3s_3+T+ ca_3(U2c_3+V2s_3+W2r_2+T2)) + ca_3(V1s_3+W1r_2+T1)(d_3+c_3d_4)$$

In the most general case, there are only one branch singularity defined by $r_3=K(q_2)$ where K is function of q_2. Since joint 3 is necessarily limited, there are, in general, two aspects (there may be more when joint 3 is strictly restricted).

<u>If $d_3+c_3d_4=0$</u>

Then $Det(J) = (s_3ca_3d_4-sa_3r_3)(sa_3(U3c_3+V3s_3+T3)+ca_3(U2c_3+V2s_3+W2r_2+T2))$

There is an extra branch-singularity defined by $r_3=d_3s_3/\mathrm{Tan}(\alpha_3)$ which yields 4 aspects. However, there are still two solutions per aspect.

<u>If $sa_2=0$</u> (the first two axes are parallel) :

Then, $Det(J) = (d_2ca_2ca_3)((d_3+c_3d_4)s_2-c_2(-s_3ca_3d_4+sa_3r_3))$

In this case, assuming $d_3+c_3d_4\neq 0$, there is only one branch-singularity. On the other hand, it can be easily shown that there are only two inverse kinematic solutions.

<u>If $ca_3=0$</u> (the last two axes are orthogonal) :

Then $Det(J) = (U3c_3+V3s_3+T3)(-sa_3r_3)$

There is one extra branch-singularity defined by $r_3=0$ yielding at least (depending on the limits on r_3) four aspects.

<u>If $ca_2=0$ and $r_2=0$ and $s_3=0$</u>

Then, $Det(J)=r_3(-sa_3(sa_3(U3c_3+V3s_3+T3)+ca_3(U2c_3+V2s_3+W2r_2+T2)) +c_2sa_2ca_3^2(d_3+c_3d_4))$

There is one extra branch-singularity defined by $r_3=0$ yielding at least (depending on the limits on r_3) four aspects.

<u>If $d_2=0$</u> (the first two axes are intersecting)

Then, $Det(J) = sa_2(d_3+c_3d_4)(r_3+ca_3r_2)(c_2-s_2(s_3ca_3d_4-sa_3r_3))$

There is one extra branch-singularity defined by $r_3=-ca_3r_2$ yielding four aspects.

In summary, the RRP geometries listed below have been verified to be type 1 :

- First 2 joint axes are parallel (e.g. $\sin(\alpha_2)=0$)
- First 2 joint axes are intersecting (e.g. $d_2=0$)
- Last 2 joint axes are parallel (e.g. $\cos(\alpha_3)=0$)
- $\cos(\alpha_2)=0$ and $r_2=0$ and $\sin(\alpha_3)=0$

The type-1 3-DOF geometries are summarized in following table 1.

PPP	RPP	PRP	PPR	RRR	PRR	RPR	RRP
all	all	all	all	sa_2=0 sa_3=0 d_2=0 d_3=0 ca_2=0, r_2=0 and r_3=0	ca_2=0 sa_3=0 d_3=0 sa_2=0 and r_3=0 sa_2=0 and ca_3=0	ca_2=0 ca_3=0 s_2=0 and $d_3+d_2c_2$=0	sa_2=0 ca_3=0 d_2=0 ca_2=0, sa_3=0 and r_2=0

Tb. 1 : Type-1 regional geometries

D. Generalization to 6-DOF Spatial Geometries

D.1- Spatial Manipulators With Spherical Wrist

Manipulators having their last three revolute joint axes intersecting (a spherical wrist) can be quite simply analysed. It is well-known that the singularities of manipulators with wrist can be decoupled. A wrist has only one singularity which is obtained when joint axes 4 and 6 are coplanar [2]. This singularity only depends on θ_5 and divides the joint space of the wrist into two aspects. Since there is always one single solution in each aspect of the wrist [2], the aspects are the uniqueness domains. Thus, all spatial manipulators geometries with wrist whose regional structure is listed in section above, are also type-1.

D.2- Other Spatial Manipulators

Another class of type-1 spatial geometries can be derived from the 3-DOF type-1 geometries: those with any three consecutive revolute joint axes (z_k,z_{k+1},z_{k+2}) intersecting in one common point. Like for the spatial geometries with wrist, their singularities are decoupled [11]. The additional singularity depends only on q_{k+1}, yielding two additional uniqueness domains in Q. Thus, designing a spatial manipulator by inserting anywhere a set of three consecutive intersecting revolute joint axes into one type-1 3-R geometry listed in section below, will result in a type-1 spatial geometry.

Manipulators with full arbitrary 6-DOF geometry are difficult to analyse in practice since this requires representing at least 4-dimensional spaces. In [2], a 6R spatial manipulator geometry with two solutions in one aspect is depicted having quite simple kinematic parameters (Fig. 4). This manipulator does not have explicit inverse kinematic solutions. With the help of SYMORO and CIMSTATION Robotic CAD software, eight inverse solutions have been found out (Fig. 5). The solutions are in fact a set of four pairs of "equivalent" solutions (i.e. solutions placing the links in apparent identical configurations).

It has been found, using MATHEMATICA software package, that four of these solutions are in a same aspect, the other four solutions are in another aspect (by searching for nonsingular posture changing paths in the jointspace, see Fig. 6).

Fig. 4 : Spatial manipulator presented in [2]

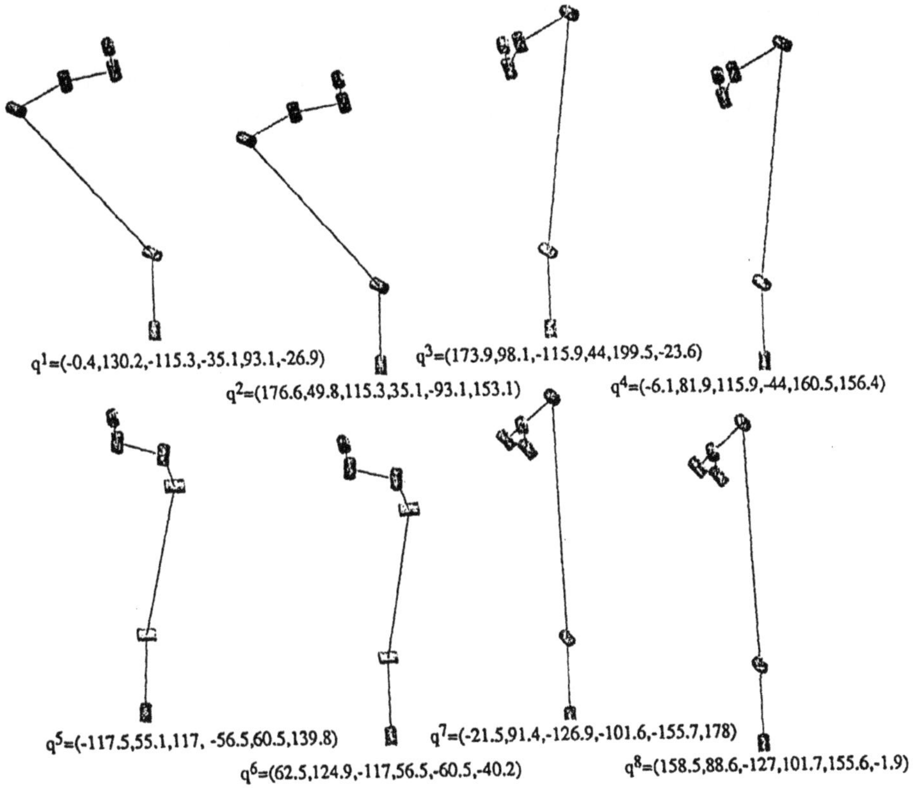

q^1=(-0.4,130.2,-115.3,-35.1,93.1,-26.9)

q^2=(176.6,49.8,115.3,35.1,-93.1,153.1)

q^3=(173.9,98.1,-115.9,44,199.5,-23.6)

q^4=(-6.1,81.9,115.9,-44,160.5,156.4)

q^5=(-117.5,55.1,117, -56.5,60.5,139.8)

q^6=(62.5,124.9,-117,56.5,-60.5,-40.2)

q^7=(-21.5,91.4,-126.9,-101.6,-155.7,178)

q^8=(158.5,88.6,-127,101.7,155.6,-1.9)

Fig. 5 : 8 inverse kinematic solutions

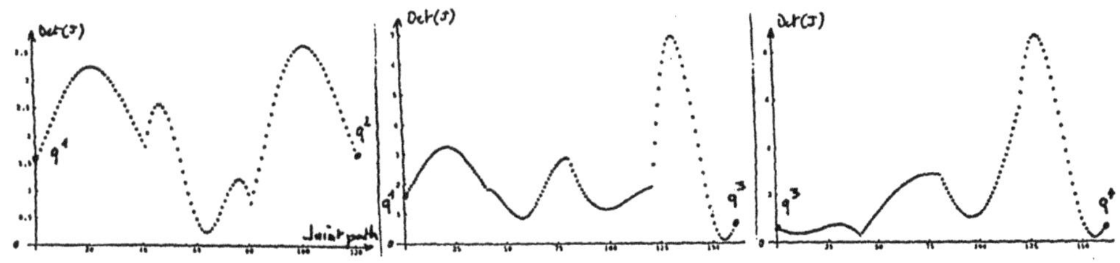

Fig. 6 : 4 solutions q^1,q^2,q^3,and q^4 in one aspect

Figure 6 depicts the value of the determinant in function of intermediate points. The determinant does not vanish, showing that the posture changing path is nonsingular.

This means that this manipulator can take four different postures while never meeting a singularity when going from one to another. This fact confirms the result that this manipulator must have only two aspects [2].

Systematic characterization of 6-DOF type-1 geometries is under study. Clearly, there must exist much stronger simplifying kinematic conditions for a spatial manipulator geometry to be type-1, than for 3-DOF geometries.

IV. Conclusions and Perspectives

This papers addressed a study of manipulators geometries in relation to their capability of changing posture without meeting a singularity. A classification of manipulator geometries was established in this regard. Manipulators with type-1 geometry must pass through a singularity when changing posture. Those having the non-singular posture changing property are called type-2. A list of type-1 geometries was provided, showing that there exist much more type-1 manipulators than those commonly used in industry. All type-1 manipulators have special geometries in the sense that the value of their kinematic parameters must satisfy some particular conditions. On the other hand, these conditions are not limited to simplifying conditions like intersecting, parallel or orthogonal joint axes. Both revolute and prismatic joints were considered in the classification. Not all type-1 geometries, however, were listed in this paper. The list of type-1 geometries has been extended to those 6-DOF geometries having their last three (resp. first three) revolute joint axes intersecting in one common point. In addition, the 6-R type-2 geometry first presented in [2] was found to have height inverse kinematic solutions, and four solutions in one aspect, confirming the result that this manipulator must have only two aspects. Partial results obtained for a 6-DOF geometry with full arbitrary geometry seem to show that a type-2 manipulator may have as many as six solutions in one aspect. More details will be presented in a next paper after verifications. Classification of all spatial 6-DOF geometries requires more general tools than those proposed in this paper. The seek for such suitable tools is under study.

V. References

[1] Borrel P. and Liegeois A. , "A Study of Manipulator Inverse Kinematic Solutions with Application to Trajectory Planning and Workspace Determination, Proc. IEEE Int. Conf. Rob. and Aut., pp1180-1185, 1986.

[2] Parenti C.V. and Innocenti C., "Position Analysis of Robot Manipulator : Regions and Subregions", Advances in Robot Kinematics, pp 150-158, Ljubljana, Yug., 1988.

[3] Burdick J. W., "Kinematic Analysis and Design of Redundant Manipulators", PhD thesis, Stanford University, 1988.

[4] Wenger P., "A New General Formalism for the Kinematic Analysis of all Nonredundant manipulators", IEEE Conf. on Rob. and Aut.,1992, pp 442-447., Nice, France.

[5] Burdick J. W., "A Classification of 3R Regional Manipulator Singularities and Geometries", Proc. IEEE Int. Conf. on Rob. and Aut., pp 2670-2675, Sacramento, California, 1991.

[6] Paï D.K., "Generic Singularities of Robot Manipulators", Proc. IEEE Conf. on Rob. and Aut., pp738-744, 1989.

[7] Kholi D. and Hsu M.S., "The Jacobian Analysis of Workspaces of Mechanical Manipulators", Journal of Mechanisms and Machine Theory, Vol 22, No 3, pp 265-275, 1987.

[8] Khalil W., Kleinfinger J.F., "A New Geometric Notation for Open and Closed Loop Robots", Proc. IEEE Conf. on Robotics and Automation, p : 1174-1179, 1986.

[9] Bennis F., "Contribution to the Geometric and Dynamic Modelling of Robots with Simple and Complex Structure" (in french), PhD Thesis, University of Nantes, France, January 1991.

[10] Khalil W., Bennis F., Chevallereau C., Kleinfinger J.F., "SYMORO : a Software Package for the Symbolic Modelling of Robots", 20th Int. Symp. on ind. Robots, Tokyo (Japan),1989.

[11] Tourassis V. D. and Ang M. H., "Identification and Analysis of Robot Manipulator Singularities", The Int. Journ. of Robtics Research, Vol. 11, No. 3, June 1992, pp248-259.

Determination of the Workspace Boundary of a General n-Revolute Manipulator

Marco Ceccarelli

Dipartimento di Ingegneria Industriale
Università di Cassino
Via Zamosch 43, 03043 Cassino (Fr), Italy

Abstract - The geometrical envelope concept is used in ring and hyper-ring generation to deduce an algebraic formulation for the workspace boundary of a general n-revolute manipulator open chain. The workspace boundary can be obtained from the envelope of a torus family which is traced by the parallel circles cut in the boundary of a generating hyper-ring. In this paper, an algebraic and recursive new algorithm is proposed as a function of the link parameters and the last revolute joint angle of the chain. Cross-section contours has been obtained for an illustrative example to show some peculiarities of general workspace shape of manipulators.

I. Introduction

Manipulator workspace is defined as the region of reachable points by a reference point H on the extremity of a manipulator chain. Great interest is still addressed to workspace evaluation for both analysis and synthesis purposes and several procedures has been proposed in the last decade. Nevertheless, workspace determination may require time-consuming algorithms in general procedures for workspace evaluation by using implicit formulation of the dependence from the geometrical parameters of the chain.

Indeed, manipulators workspace can be determined through its boundary so that an interesting issue is:

- is it possible to formulate a recursive algebraic algorithm for workspace boundary determination of n-revolute joint open chain manipulators?

This paper addresses to this question by proposing an algebraic formulation of an explicit function of the link parameters for the workspace boundary determination as a generalization of a procedure first proposed for 3R manipulators, [1], and then extended specifically up to 5R manipulators, [2,3].

A recursive process for the generation of the workspace volume as a solid of revolution has been recognized in [4] and some results from the common topology of torus, ring and hyper-ring has been deduced, [3,4].

The term hyper-ring is here used to stress revolution operations about more than three axes, which is the case usually referred to the ring geometry.

II. Workspace Boundary Generation

A n-revolute open kinematic chain can be sketched by using the axes Z_i, (i=1,...,n), of the n frames fixed on the links of a nR manipulator with n revolute joints R_i, (i=1,...,n). Thus,

A. J. Lenarčič and B. B. Ravani (eds.), Advances in Robot Kinematics and Computationed Geometry, 39–48.

the geometrical parameters in a general nR manipulator are assumed the link lengths $a_i \neq 0$ (i=1,...,n), the link offsets d_i (i = 2,...,n), and the twist angles $\alpha_i \neq k\,\pi/2$ (i=1,...,n-1), (k=0, ...,4). d_1 may not be considered since it shifts the hyper-ring up and down, only. A reference point H can be selected on the extreme link of a manipulator chain for workspace determination.

A ring is generated by revolving a torus about an axis. Therefore, the ring volume $W_{3R}(H)$ can be thought as the union of the points swept by the torus $T_{R_2R_3}(H)$, which is traced by H due to the mobility in R_2 and R_3 joints, during the θ_1 revolution about Z_1 axis

$$W_{3R}(H) = \bigcup_{\vartheta_1=0}^{2\pi} T_{R_2R_3}(H) \tag{1}$$

Alternatively, a ring $W_{3R}(H)$ can be considered as the union of the tori $T_{R_1R_2}(H)$, which are due to the mobility in R_1 and R_2 joints and are traced by all the parallel circles cut on the generating torus $T_{R_2R_3}(H)$, [1],

$$W_{3R}(H) = \bigcup_{\vartheta_3=0}^{2\pi} T_{R_1R_2}(H) \tag{2}$$

Thus, the boundary $\partial W_{3R}(H)$ of a ring can be thought as the envelope of toroidal surfaces generated by revolution of the generating torus or, alternatively, it can be obtained by an envelope of a torus family traced from the parallel circles of the generating torus, yet, as

$$\partial W_{3R}(H) = \underset{\vartheta_3=0}{\overset{2\pi}{\mathrm{env}}}\, T_{R_1R_2}(H) \tag{3}$$

where "env" expresses an envelope operator.
The characteristics of the ring geometry, previously reviewed, have been recognized topologically common also to a hyper-ring in the form of a "solid hollow ring" shape, [4], on the basis of a consideration of the analytical structure of an iterative revolving process for the hyper-ring generation in nR manipulators. Infact, a (n-j+1)R hyper-ring $W_{(n-j+1)R}(H)$ is generated by revolving a (n-j)R hyper-ring $W_{(n-j)R}(H)$ about Z_j axis. The (n-j+1)R hyper-ring, which is traced by a point H on the last link of an open chain with j+1 consecutive revolute pairs, is the volume swept by the (n-j)R hyper-ring during its revolution, Fig.1.
The generation process of a hyper-ring is a consecutive revolving process of a circle, a torus, a ring, a 4R hyper-ring and so on. This can be expressed through a revolution operator "Rev", [4], in the form

$$W_{(n-j+1)R}\left(H_j\right) = \underset{\vartheta_j=0}{\overset{2\pi}{\mathrm{Re\,v}}}\, W_{(n-j)R}\left(H_{j+1}\right) \qquad j=1,\dots,n-1 \tag{4}$$

where H_j represents reachable points with respect to the j link frame and due to (n-j+1) last revolute joints in the chain.

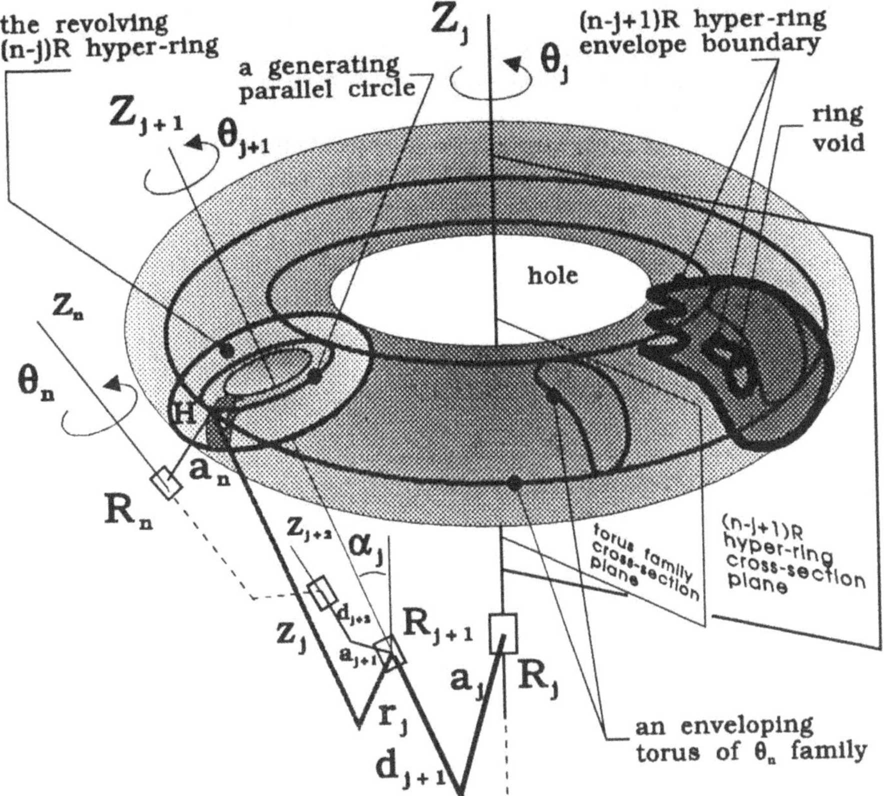

Fig.1. Parameters in the tori envelope procedure for a determination of the boundary of a general (n-j+1)R hyper-ring.

Alternatively, $W_{(n-j+1)R}(H)$ can be considered as the union of a suitable torus family which is traced by the boundary points in the revolving torus, ring ,4R hyper-ring and so on, when they are rotated completely about the first two revolute axes in the corresponding generating sub-chain,

$$W_{(n+1-j)R}\left(H_j\right) = \bigcup_{\vartheta_j,\vartheta_{j+1}=0}^{2\pi} T_{R_jR_{j+1}}\left[\partial W_{(n-j)R}\left(H_{j+1}\right)\right] \tag{5}$$

where $T_{RjRj+1}(H)$, represents the torus generated by revolutions θ_j and θ_{j+1} about the joints axes of R_j and R_{j+1}.

Hence, the boundary $\partial W_{(n-j+1)R}(H)$ of a (n-j+1)R hyper-ring can be described as an envelope of the aforementioned torus family, Fig.1, as

$$\partial W_{(n-j+1)R}\left(H_j\right) = \underset{\vartheta_j,\vartheta_{j+1}=0}{\overset{2\pi}{\text{env}}}\; T_{R_jR_{j+1}}\left[\partial W_{(n-j)R}\left(H_{j+1}\right)\right] \tag{6}$$

Thus, a workspace boundary $\partial W_{nR}(H)$ of a general nR manipulator can be determined by evaluating, recursively by using Eq.(6), the tori envelope from the last ring to the first nR hyper-ring in the chain.

III. An Envelope Formulation for the Hyper-Ring Boundary

The (n-j+1)R hyper-ring boundary can be expressed algebraically when it is thought generated by enveloping the torus family traced by the parallel circles in the boundary of the revolving (n-j)R hyper-ring.

A torus family equation can be expressed with respect to the j link frame, by supposing $C_j \neq 0$ and $\cos\alpha_j \neq 0$, in the form of the radial r_j and axial z_j reaches as

$$(r_j^2 + z_j^2 - A_j)^2 + (C_j z_j + D_j)^2 + B_j = 0 \tag{7}$$

where the torus parameters are a_j, α_j, r_{j+1}, z_{j+1}, and the coefficients are given as

$$\begin{aligned} A_j &= a_j^2 + r_{j+1}^2 + (z_{j+1} + d_{j+1})^2 \\ B_j &= -4a_j^2 r_{j+1}^2 \\ C_j &= 2a_j / s\alpha_j \\ D_j &= -2a_j \left(z_{j+1} + d_{j+1}\right) c\alpha_j / s\alpha_j \end{aligned} \tag{8}$$

in which $s\alpha_j$ and $c\alpha_j$ designate $\sin\alpha_j$ and $\cos\alpha_j$, respectively.

Particular cases with $C_j = 0$ or $\cos\alpha_j = 0$ are not represented by Eq.(7) but specific formulation can be developed when the torus boundary is not generated as an envelope of the revolving circle for: $a_j = 0$, since the torus degenerates in an ellipsoid surface with major axis r_{j+1}; $s\alpha_j = 0$, since the torus degenerates in a circular rim of width r_{j+1}; and $c\alpha_j = 0$, since "common or symmetrical-offset torus", [5], is obtained from the revolving circle of radius r_{j+1}.

Indeed, a torus family is generated by considering r_{j+1} and z_{j+1} as variables, which can be calculated from the boundary points of the revolving (n-j)R hyper-ring. Particularly, r_{j+1} and z_{j+1} can be expressed recursively by using Eq.(7) up to the last revolving torus workspace, or they can be formulated as a function of the revolute joint angles θ_{j+1}, θ_{j+2}, or finally, they can be a function of the last revolute joint angle by using an iterative formulation of the revolving boundary workspace and their torus families up to the extreme ring workspace for the chain. Therefore, r_{j+1}, z_{j+1} can be thought expressed as a function of the last kinematic joint angle θ_n in the nR chain to have a single variable formulation. Thus, θ_n can be considered as a torus family parameter since for each value a torus of the enveloping family can be determined.

The envelope equations of a torus family can be obtained from Eq.(7) and its derivative with respect to the torus family parameter θ_n. After some algebraic manipulations for which $C_j \neq 0$ and $E_j \neq 0$ are required, the ring boundary equations are obtained in the form

$$r_j = \left[A_j - z_j^2 + \frac{(C_j z_j + D_j) G_j + F_j}{E_j} \right]^{1/2}$$

$$z_j = \frac{-F_j G_j \pm Q_j^{1/2}}{(E_j^2 + G_j^2) C_j} - \frac{D_j}{C_j} \tag{9}$$

where the so-called ring coefficients are given as

$$E_j = R_{j+1} + S_{j+1}$$

$$F_j = -2\, a_j^2\, R_{j+1} \tag{10}$$

$$G_j = -2 a_j\, z'_{j+1}\, c\alpha_j / s\alpha_j$$

$$Q_j = -\, E_j^2 \left[F_j^2 + B_j (E_j^2 + G_j^2) \right]$$

with

$$R_{j+1} = \left[(C_{j+1} z_{j+1} + D_{j+1})(E_{j+1} G'_{j+1} - G_{j+1} E'_{j+1}) + F'_{j+1} E_{j+1} - F_{j+1} E'_{j+1} \right] / E_{j+1}^2$$

$$+ G_{j+1} E_{j+1} (C_{j+1} z'_{j+1} + G_{j+1}) / E_{j+1}^2 + E_{j+1} - 2 z_{j+1} z'_{j+1}$$

$$S_{j+1} = 2 (z_{j+1} + d_{j+1}) z'_{j+1} \tag{11}$$

$$z'_{j+1} = \left[\left(\pm 0.5 Q'_{j+1} Q_{j+1}^{-1/2} - F_{j+1} G'_{j+1} - G_{j+1} F'_{j+1} \right) \left(E_{j+1}^2 + G_{j+1}^2 \right) \right.$$

$$\left. + 2 \left(F_{j+1} G_{j+1} \mp Q_{j+1}^{1/2} \right) \left(E_{j+1} E'_{j+1} + G_{j+1} G'_{j+1} \right) \right] / \left(E_{j+1}^2 + G_{j+1}^2 \right)^2 C_{j+1} - G_{j+1} / C_{j+1}$$

The symbol ' is the derivative operator with respect to the torus family parameter θ_n.

Eqs.(11) can be computed through the equation coefficients E_{j+1}, F_{j+1}, G_{j+1}, Q_{j+1} and their derivatives E'_{j+1}, F'_{j+1}, G'_{j+1}, Q'_{j+1}, whose expressions and values can be deduced from r_{j+1} and z_{j+1} formulation. Particularly, the derivatives of the ring coefficients can be calculated, from Eqs.(10), as

$$E'_j = R'_{j+1} + S'_{j+1}$$

$$F'_j = -2a_j^2 R'_{j+1}$$

$$G'_j = -2a_j\, z''_{j+1}\, c\alpha_j / s\alpha_j \tag{12}$$

$$Q'_j = -\,2E_j^2\left[F_j\left(F'_j + E_j^2 + G_j^2\right) + B_j(E_jE'_j + G_jG'_j)\right] + 2Q_jE'_j / E_j$$

The derivatives R'_{j+1}, S'_{j+1} and z''_{j+1} can be computed by means of a further derivative operation from Eqs.(11) and by using, in addition, the second derivatives of the ring equation coefficients. The radial r_{j+1} and axial z_{j+1} reaches, the ring coefficients and derivatives of the (n-j)R workspace boundary may be expressed iteratively, by means of the envelope development, from the (n-j-1)R hyper-ring boundary and so on up to the extreme ring workspace due to the last three revolute joints in the chain. Finally, last ring workspace can be algebraically expressed as a function of the torus family parameter θ_n .

Thus, Eqs. (8)-(12) provide an envelope formulation for the boundary of a general (n-j+1)R hyper-ring as a function of a boundary expression of the generating (n-j)R hyper-ring, according to Eq.(6).

Because of the hypotheses $C_j \neq 0$ and $E_j \neq 0$, the proposed formulation can describe a general nR hyper-ring in which an envelope surface of torus families of the generating boundaries exists. In the case of $\alpha_i = k\,\pi/2$ (i = j,...,n-1; k = 0,...,4) the boundary of the ring and hyper-rings are not generated by any envelope surface, but directly by the smallest and largest parallel circles of the revolving hyper-ring.

IV. An Algorithm for Determination of a Workspace Boundary

The algebraic formulation of Eqs.(8)-(12) can be used to determine numerically the workspace boundary of any subchain up to a n-revolute chain when the geometrical sizes of the links are given. The proposed envelope procedure is recursive according to Eq.(6), from the extremity to the base of the robot. In addition, an evaluation of ring coefficient derivatives for the (n-j+1) hyper-ring requires to have computed higher derivatives of R_{j+2}, S_{j+2} and z'_{j+2}, again according to Eqs.(11)-(12), in previous calculations for the (n-j-2) hyper-ring.

Particularly, looking at Eqs.(11) and (12), R_{j+1}, S_{j+1} and z'_{j+1} can be calculated through E'_{j+1}, F'_{j+1}, G'_{j+1}, Q'_{j+1}, as well as these can be computed through R'_{j+2}, S'_{j+2} and z''_{j+2} and so on recursively, since the implicit expressions (10) of E_{j+1}, F_{j+1}, G_{j+1}, Q_{j+1} , as far as explicit expressions of E_{n-2}, F_{n-2}, G_{n-2}, Q_{n-2} can be computed. This iterative computation can be expressed, from Eqs.(12), in a general form

$$E^k_{j+1} = R^k_{j+2} + S^k_{j+2}$$

$$F^k_{j+1} = -2a^2_{j+1} R^k_{j+2} \qquad k = 0, 1, \ldots, j; \qquad j = 0, 1, \ldots, n\text{-}4 \tag{13}$$

$$G^k_{j+1} = -2a_{j+1}\, z^{k+1}_{j+2}\, c\alpha_{j+1} / s\alpha_{j+1}$$

and, from Eqs.(11), as

$$R^{k}_{j+2} = f^{k}\left(E^{k+1}_{j+2}, F^{k+1}_{j+2}, G^{k+1}_{j+2}\right)$$

$$S^{k}_{j+2} = g^{k}\left(E^{k+1}_{j+2}, F^{k+1}_{j+2}, G^{k+1}_{j+2}\right) \qquad k = 0, 1, \ldots, j;\ j = 0, 1, \ldots, n\text{-}4 \qquad (14)$$

$$z^{k+1}_{j+2} = h^{k}\left(E^{k+1}_{j+2}, F^{k+1}_{j+2}, G^{k+1}_{j+2}\right)$$

where f^{k}, g^{k}, and h^{k}, represent the k-derivatives of the functions f, g and h which are those that express R_{j+1}, S_{j+1} and z'_{j+1} in Eqs.(11), respectively.
Thus, all the abovementioned computations can be numerically evaluated by starting from the computation of derivatives of E_{n-2}, F_{n-2}, G_{n-2}, Q_{n-2} up to (n-3) order. Infact, the ring coefficients for the ring workspace can be algebraically expressed, [1], in the form

$$E_{n-2} = -2a_n (d_{n-1}\, s\alpha_{n-1} c\theta_n + a_{n-1}\, s\theta_n)$$
$$F_{n-2} = 4a^2_{n-2} a_n (a_n\, s^2\alpha_{n-1} s\theta_n c\theta_n + a_{n-1} s\theta_n - d_n s\alpha_{n-1} c\alpha_{n-1} c\theta_n) \qquad (15)$$
$$G_{n-2} = 2a_{n-2} a_n\, c\alpha_{n-2} s\alpha_{n-1} c\theta_n / s\alpha_{n-2}$$

where

$$r_{n-1} = \left[(a_n c\theta_n + a_{n-1})^2 + (a_n s\theta_n c\alpha_{n-1} + d_n s\alpha_{n-1})^2\right]^{1/2}$$
$$z_{n-1} = d_n\, c\alpha_{n-1} - a_n\, s\theta_n\, s\alpha_{n-1} \qquad (16)$$

Then, by differentiating Eqs.(15) the k derivative of the ring coefficients can be expressed as

$$E^{k}_{n-2} = -2a_n\left[a_{n-1} \sin(\theta_n + k\pi/2) + d_{n-1}\, s\alpha_{n-1} \cos(\theta_n + k\pi/2)\right]$$
$$F^{k}_{n-2} = 4a^2_{n-2} a_n\left[0.5\, a_n s^2\alpha_{n-1} \sin(2\theta_n + k\pi/2) + a_{n-1} \sin(\theta_n + k\pi/2)\right.$$
$$\left. - d_n s\alpha_{n-1} c\alpha_{n-1} \cos(\theta_n + k\pi/2)\right] \qquad (17)$$
$$G^{k}_{n-2} = 2a_{n-2} a_n c\alpha_{n-2} s\alpha_{n-1} \cos(\theta_n + k\pi/2) / s\alpha_{n-2}$$

Therefore, a numerical algorithm can be developed to compute the radial r_j and axial z_j reaches of all the envelopes for j from n-2 to 1, that is from the extreme workspace ring up to nR hyper-ring. This can be obtained by scanning the joint angle θ_n from 0 to 2π and calculating at each j the coefficients A_j, B_j, C_j, D_j, E_j, F_j, G_j, R_j, S_j and z'_j, from Eqs.(8), (10), (11) and finally r_j, z_j from Eqs.(9), when the j derivatives of E_j, F_j, G_j, are evaluated from Eqs.(13) by using Eqs.(14) for R_{j+1}, S_{j+1} and z'_{j+1}.

It is to point out that, altough the laborious computations for the expressions of the coefficients, the algorithm is algebraic in nature and, mostly, it is a function of the variable θ_n angle, only.

An illustrative example is reported in Fig.2, where two branches of envelope contours in a cross-section of a workspace boundary are observable: an external one and an internal ones. The external branch of the torus family envelope determines the external boundary of a bulk hyper-ring. The internal branches of the torus family envelope may show several disconnected or intersecting contours in the cross-section. Nevertheless they may indicated one ring void only.

Two types of voids have been individuated. They can be addressed as ring void and apple void, [1]. A ring void is an internal region not belonging to the ring or hyper-ring and, topologically, it is ring shaped. In fact, a ring void is generated by the hole in the revolving torus family, which is delimited by the smallest parallel circle cut in the generating ring or hyper-ring. Therefore, when this is rotated about an axis, this parallel circle rotates as well, so that it generates, once more, a torus which is the boundary surface of a ring void.

Other internal contour loops are determined since the scanning algorithm takes into account also the boundary points of the generating (n-j)R hyper-ring which does not generate any boundary contour in the (n-j+1)R hyper-ring and it also produces further envelope surfaces which do not represent workspace boundary.

Therefore, a certain reasoning is required to individuate the ring void boundary among the internal envelope surfaces and an interactive computer graphics may determinate it as follows. In a ring the internal branch of the envelope consists on three connected contour loops in a cross-section visualization. The middle one has been recognized as the only one that may represent a ring void boundary, since the cusps in the other two are not related with a shape of a toroidal surface tha may delimite a ring, [1]. Further revolutions of a ring and additional computations of the envelope procedure may determinate several internal contour loops in a cross-section visualization of a hyper-ring, as the case in Fig.2. Nevertheless, the basic topology, which has been individuated for the internal envelope loops in a ring workspace, may be recognized yet in the internal contour loops in a hyper-ring cross-section.

Particularly, loops which show more than two cusps cannot be a ring boundary surface and only those with a torus shape can be considered. Moreover, since each of those may delimitate a region of unreachable points the ring void can be determined as the intersection of these regions and the boundary can be identified consequently from the delimiting envelope contour segments.

Some further attention may be required when particular cases for the envelope contour loops occur since some of them may be collapsed in a point or degenerated in a single-loop contour with cusps.

V. Conclusions

An algebraic formulation for the hyper-ring workspace of a general nR manipulator is of great interest for performance evaluation and dimensional design of robots. In fact, an algebraic treatment of manipulators workspace has been recognized useful, [6], and, indeed, some specific applications of the proposed approach in kinematical synthesis, [7], or

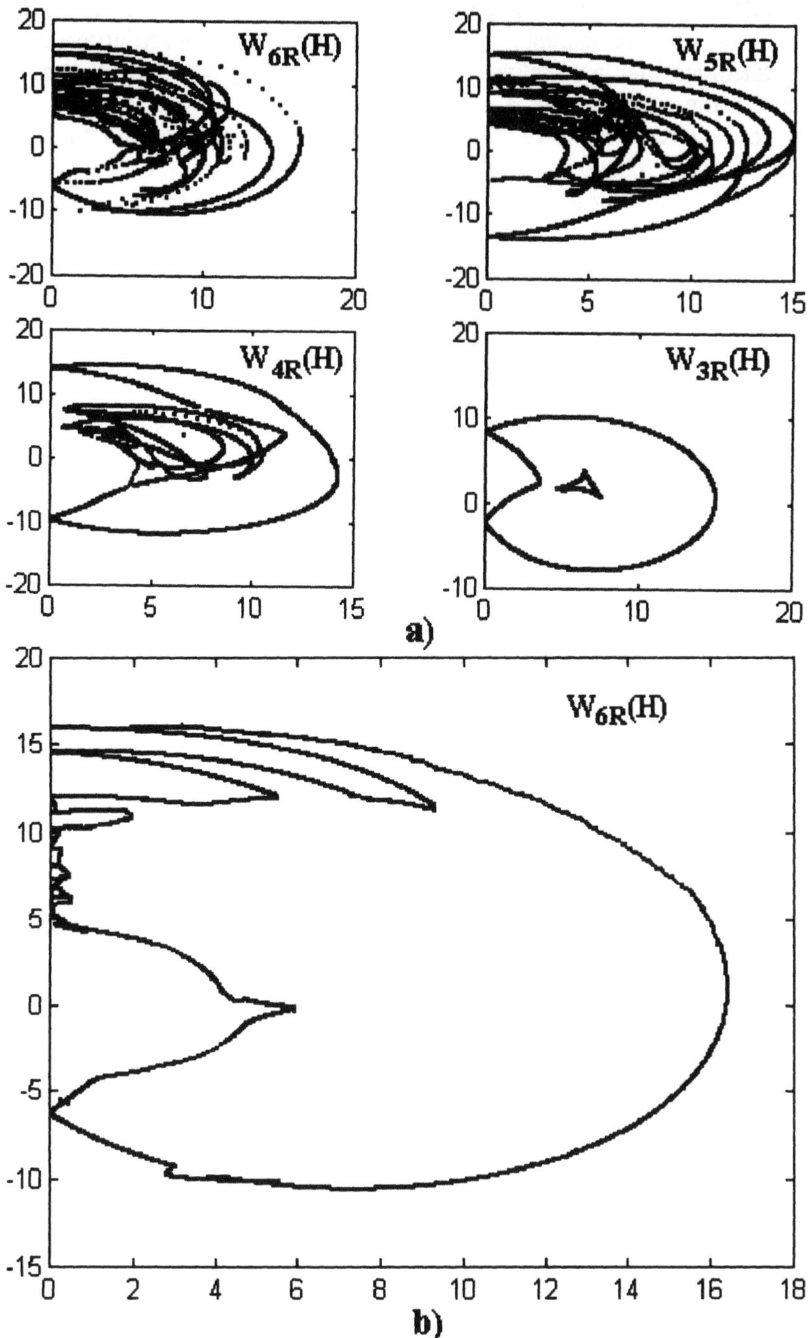

Fig.2. An explanatory example of the proposed envelope procedure for workspace boundary determination for a 6R manipulator with $a_j = j$, $d_j = 1$, $\alpha_j = \pi/4$, (j=1,...,5,6): a) envelope cross-section contours of the generating ring and hyper-rings; b) cross-section of the workspace boundary.

optimization design formulation, [8], have already proved a practical engineering feasibility.

Moreover, since a general manipulator chain may occur also in industrial robots because of machining tolerance and assembling errors in the robot production, a straigthforward formulation for the workspace of a general manipulator chain may be also of practical interest to check the real workspace of industrial robots and, mostly, to investigate the effects of the abovementioned robot construction errors.

In this paper, a hyper-ring is described by means of its boundary which has been thought generated as a geometrical envelope of a suitable torus family. Each torus is thought traced by revolving a parallel circle which is cut on the boundary in the generating (n-j)R hyper-ring, about the first two revolute joint axes in the (n-j+1)R chain, Fig.1. Then, an algebraic formulation for the (n-j+1)R hyper-ring boundary is completely expressed as a function of the geometrical parameters and the last revolute joint angle only.

A suitable algorithm has been developed by using the proposed envelope formulation to obtain a hyper-ring boundary and a numerical example has been reported in the paper to show an application of the procedure and to illustrate a general geometry of hyper-rings.

VI. References

[1] M. Ceccarelli, "On the Workspace of 3R Robot Arms", *Proc. of the Fifth International IFToMM Symposium on Theory and Practice of Mechanisms*, Bucharest, Vol.II-1, 37-46 (1989).

[2] M. Ceccarelli and A. Vinciguerra, "Workspace Evaluation of General 4R Manipulators", *Proc. of 3rd International Workshop on Advances in Robot Kinematics*, Ferrara, 313-317 (1992)

[3] M. Ceccarelli, "A Boundary Formulation for General Hyper-Rings in 5R Manipulators", *Proc. of The Sixth International Conference on Advanced Robotics,* Tokyo, 631-636 (1993).

[4] K.C. Gupta and B. Roth, "Design Considerations for Manipulator Workspace", *ASME Jnl of Mechanical Design*, 104, 704-711 (1982).

[5] E.F. Ficther and K.H. Hunt, "The Fecund Torus, its Bitangent-Circles and Derived Linkages" *Mechanism and Machine Theory*, 10, 167-176 (1975).

[6] F. Freudenstein and E.J.F. Primrose, "On the Analysis and Synthesis of the Workspace of a Three-Link Turning-Pair Connected Robot Arm", *ASME Jnl of Mechanisms, Transmissions and Automation in Design*, 106, 365-370 (1984).

[7] M. Ceccarelli, "A Synthesis Algorithm of Three-Revolute Manipulators by means of the Workspace Contour Algebraic Formulation", *Robotics, Spatial Mechanisms, and Mechanical Systems* (ASME Ed.), DE-Vol.45, 397-403 (1992).

[8] M. Ceccarelli, "Optimal Design and Location of Manipulators", *Proc. of NATO Advanced Study Institute on Computer Aided Analysis of Rigid and Flexible Mechanical Systems,* Troia, Vol.II, 299-310 (1993).

Acknowldgments

The financial support of the Italian Ministero della Università e della Ricerca Scientifica e Tecnologica through funds 40% and 60% is gratefully acknowledged.

Local Pictures for General Two-parameter Planar Motions

C. G. Gibson and W. Hawes [1]
University of Liverpool, PO Box 147, Liverpool L69 3BX, England

C. A. Hobbs
University of Nottingham, University Park, Nottingham NG7 2RD, England

Abstract – Local models are given for the singularities which can appear on the trajectories of general two-dimensional motions of the plane. Versal unfoldings of these model singularities give rise to computer generated pictures describing the family of trajectories arising from small deformations of the tracing point.

I. Introduction

This paper attempts to codify the extremely complex geometry associated to singularities of trajectories for *general* motions of the plane and space, via the ideas of Singularity Theory. First, we seek a list of local models for trajectories. And second, for each local model we seek a qualitative picture of the changes in the local model under arbitrarily small deformations of the tracing point.

The programme has its origin in two publications: a detailed study of the algebraic geometry of the planar 4–bar [1] and the work of Donelan [2] on general properties in Euclidean kinematics. Its first success is the case of one–parameter motions of the plane and space, a principal concern of classical kinematics. A complete solution appears in the third author's thesis [3, 4], resulting in an exhaustive list of ten local models and complete descriptions of their changes under small deformations. An account for an engineering audience appeared in [5], providing a useful introduction to this paper. The mathematics gives rise naturally to the three classical bifurcation curves in the moving lamina .

The situation changes dramatically for two–parameter motions of the plane and space. Complete solutions appeared in [3, 6, 7] using relatively recent results in Singularity Theory. So far as we are aware, these results are totally new to kinematics. Moreover, it takes an exceptional geometric "feel" to guess the local pictures which can arise in the two–parameter case, so there is little question the mathematics has much to offer the working kinematician, especially since the interface is a pictorial one. In the two–parameter planar case there are twenty eight different local models, and six different bifurcation curves. The problem of obtaining local models for n–parameter motions of the plane for $n \geq 3$ has been solved by the first two authors, and will be discussed in [8].

[1] Supported by a grant from the Science and Engineering Research Council.

A. J. Lenarčič and B. B. Ravani (eds.), Advances in Robot Kinematics and Computationed Geometry, 49–58.

Acknowledgements: We wish to acknowledge the considerable help we received from Dr. R. J. Morris and Dr. N. P. Kirk in producing computer renderings for the unfoldings of the corank 2 germs.

II. An Engineering Example

Figure 1 represents the double four–bar, a planar linkage with two degrees of freedom. Associated to the coupler point C is a *coupler region* in the plane together with a curve called the *critical image*, which includes the boundary of the coupler region. (Later we will say more carefully what we mean by the critical image.)

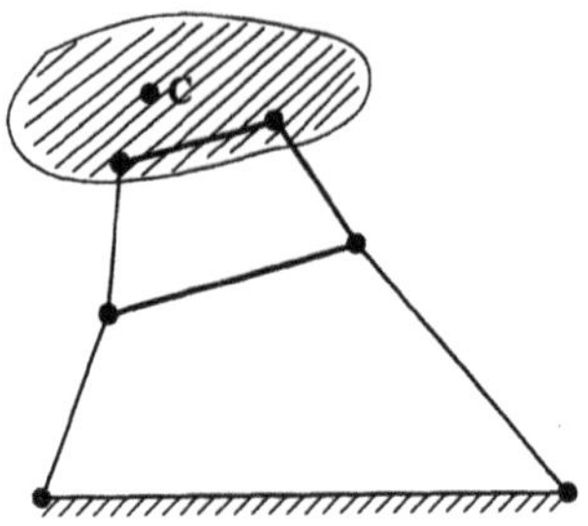

Figure 1: The double four–bar.

We are indebted to Dr. D. Marsh and Ms. Y. Xiang who kindly allowed us to use the computer generated illustration Figure 2 of a typical critical image. P is a general point in the interior of the coupler region: on the critical image itself Q is a general smooth point, R is a node (or transverse self–crossing), and S is a cusp (looks like $x = t^2,\ y = t^3$). Although P, Q, R and S represent different types of local behaviour they share a common feature, that they are preserved by small changes in the link lengths – changing slightly the underlying *motion*, but leaving the position of C fixed in the coupler plane. In robotics this is an important feature of local behaviour, as practical devices can only be constructed to certain (small) tolerances.

However, rather more is true. Suppose we leave the link lengths fixed, but perturb the coupler point C very slightly. Experimentation will verify that the local behaviour at P, Q, R and S still cannot be eliminated by arbitrarily small changes in C. Thus P, Q, R and S are examples of *stable* local behaviour. Because stable local behaviour occurs for most positions of the coupler point C in the coupler plane it is said to be of "codimension 0".

For a given engineering example there will be only finitely many distinct types of critical images which are *generic*, i.e. they exhibit only nodes and ordinary cusps. They represent different types of mechanical behaviour, and arise by taking different positions of C in the coupler plane. Intuitively, we expect transitions from one type to another to take place on *bifurcation curves* in the coupler plane: such transitions are of "codimension 1". And at isolated points on these bifurcation curves we expect more degenerate transitions, are of "codimension 2".

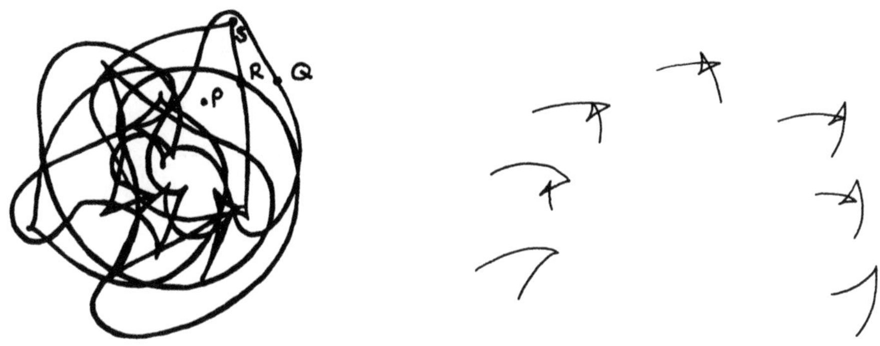

Figure 2: Critical image.

Figure 3: Deformations.

Figure 3 illustrates a codimension 2 transition T for the double four–bar. Suppose we trace a small circle around T in the coupler plane. As the coupler point moves around this circle so the form of the critical image close to T exhibits a complex series of changes. T shares with P, Q, R and S the fundamental property that *it is preserved by small changes in the underlying motion*. That raises the fundamental theoretical question of listing all possible transitions of codimension ≤ 2 that can arise for *general* two–parameter planar motions. Despite the apparent complexity of the pictures there is a clear underlying qualitative pattern, capable of systematic codification. More precisely, for *general* two–parameter planar motions there are only twenty one possible local models of transition types, for each of which a "versal unfolding" gives a complete qualitative description of the changes under small perturbations of the coupler point.

III. Planar Monogerms

A two–parameter motion of the plane is a mapping of a *parameter surface* into the Lie group of proper rigid mappings of the plane. The simplest case is when the parameter surface is least an open subset of a plane. For each choice of tracing point C this yields a *trajectory mapping* from the parameter surface into the fixed lamina. To study the local behaviour of this map we choose a fixed point (the *source*) on the parameter surface mapped to a point (the *target*) in the fixed lamina, and restrict the mapping to an arbitrarily small neighbourhood of the source. We can suppose that the source is the origin in the (x, y)–plane, and the target is the origin in the (X, Y)–plane. The basic feature of the germ is its *corank*, i.e. the corank of the Jacobian matrix (evaluated at the source) of the map $(x, y) \mapsto (X, Y)$ where X, Y are expressed as functions of x, y. Whitney [9] proved that any germ of codimension 0 is equivalent to one of the models in Table 1.

Type	Local Model	Corank
immersion	$X = x, Y = y$	0
fold	$X = x, Y = y^2$	1
cusp	$X = x, Y = xy + y^3$	1

Table 1: The stable planar germs

One can visualize a germ of the form $X = x, Y = f(x, y)$ by considering it as the composite of the mapping $X = x, Y = f(x, y), Z = y$, of the plane to the graph of the function $f(x, y)$ in (X, Y, Z)–space, followed by projection of the graph onto the (X, Y)–plane. Figure 4 illustrates this for the fold and the cusp.

A better understanding of these pictures is obtained via their critical sets. A germ is *critical* when the corank is positive, i.e. the determinant of the Jacobian matrix vanishes. The *critical curve* of the trajectory mapping is the curve of sources on the parameter surface where the germ is critical. And the *critical image* is the image of the critical set under the mapping. For the immersion there is no critical curve or critical image.

Table 2 gives the classification of corank 1 planar germs of codimension ≤ 2 obtained by Gaffney [10] and Kergosien [11]. The meaning of the invariants c, d will be explained later. The critical images of the codimension 1 germs are shown in Figure 5. There are just two germs of corank 2 having codimension 2 given in Table 3.

Type	Local Model	Codim	c	d
lips	$X = x, Y = y^3 + x^2y$	1	2	0
beaks	$X = x, Y = y^3 - x^2y$	1	2	0
swallowtail	$X = x, Y = xy + y^4$	1	2	1
goose	$X = x, Y = y^3 + x^3y$	2	3	0
butterfly	$X = x, Y = xy + y^5 \pm y^7$	2	3	3
gulls	$X = x, Y = xy^2 + y^4 + y^5$	2	3	2

Table 2: Germs of corank 1 and codimension ≤ 2.

Type	Local Model	Codim	c	d
sharksfin	$X = x^2 + y^3, Y = y^2 + x^3$	2	3	2
deltoid	$X = x^2 - y^2 + x^3, Y = xy$	2	3	2

Table 3: Germs of corank 2 and codimension ≤ 2

Figure 4: Stable germs.

Figure 5: Codimension 1 germs.

IV. Planar Multigerms

By a *planar multigerm* of multiplicity m we mean m planar monogerms with distinct sources, and a *common* target. For $m = 1, 2, 3, \ldots$ we refer to monogerms, bigerms, trigerms, and so on. The concept of a multigerm allows us to discuss the local behaviour (at a point in the coupler region) which arises from several distinct points on the parameter surface. For instance the node R in Figure 2 arises from two different configurations of the double four–bar: each configuration produces a fold on the critical images, and these cross transversely. We refer to this as a *node fold*: it is the only *stable* multigerm, i.e. it is preserved by small perturbations of the coupler point, and is said to be of codimension 0.

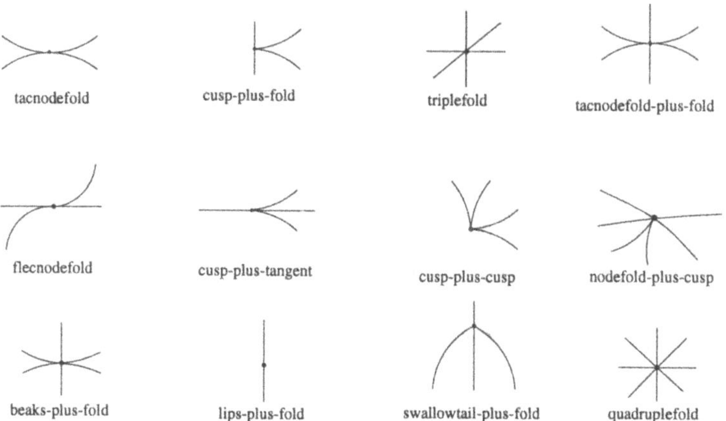

Figure 6: Critical images for multigerm transitions.

We expect multigerm transitions of codimension 1 to take place on curves in the moving lamina, and at isolated points on these curves we expect more degenerate multigerm transitions of codimension 2. It is by no means easy to find local models for planar multigerms. So far as we are aware the first mention (without formal proofs) is in [11], with more detail in [12]: a complete account appears in [3, 6]. In total there are thirteen multigerm types of codimension ≤ 2, for which a formal list of local models appears in [6]. There are nine bigerm types, three trigerm types, and one quadrigerm type: of these only three types (tacnode fold, cusp–plus–fold, triple fold) have codimension 1, whilst the remainder have codimension 2. Figure 6 illustrates the twelve non–stable multigerm types. Each sketch superimposes the critical images of the constituent monogerms.

V. Versal Unfoldings

The basic mental picture introduced above is that for most positions of the coupler point C we expect to see a generic critical image exhibiting only cusps and nodes. And if we move C slightly from such a position we do not expect any change in the qualitative picture. We expect codimension 1 transitions from one type of critical image to another to take place when C lies on a bifurcation curve B. In that case we can view the changes in the generic type by allowing C to move along a (short) path in the moving lamina, crossing B. At isolated points on B we expect the critical image to exhibit codimension 2 transitions. We can view the changes in the generic types

Type	Versal Unfolding	Codim
immersion	$X = x, Y = y$	0
fold	$X = x, Y = y^2$	0
cusp	$X = x, Y = xy + y^3$	0
lips	$X = x, Y = y^3 + x^2y + ay$	1
beaks	$X = x, Y = y^3 - x^2y + ay$	1
swallowtail	$X = x, Y = xy + y^4 + ay^2$	1
goose	$X = x, Y = y^3 + x^3y + ay + bxy$	2
butterfly	$X = x, Y = xy + y^5 \pm y^7 + ay^2 + by^3$	2
gulls	$X = x, Y = xy^2 + y^4 + y^5 + ay + by^3$	2
sharksfin	$X = x^2 + y^3 + ay, Y = y^2 + x^3 + bx$	2
deltoid	$X = x^2 - y^2 + x^3 + ax, Y = xy + bx$	2

Table 4: Versal unfoldings for monogerms of codimension ≤ 2.

by allowing C to move around in a small neighbourhood of such a point. We can exhibit all possible deformations of a singularity of codimension k by embedding it in a k–parameter family, called a *versal unfolding*. In such a family there will be exactly k *unfolding parameters* a, b, ... giving the original singularity when $a = b = \ldots = 0$, and assumed to be arbitrarily small. Versal unfoldings of singularities can be computed by purely algebraic procedures: for the planar monogerms of codimension ≤ 2 they are given in Table 4, whilst for multigerms they can be found in [6].

For the lips, beaks and swallowtail Figure 7 illustrates the qualitative changes, in the critical image as a changes.

LIPS. The critical image for $a = 0$ is a point, which disappears for $a > 0$, and becomes a "lip" shaped curve with two cusps for $a < 0$.

BEAKS. The critical image for $a = 0$ comprises two cusps, with common cusp and cuspidal tangent. For $a > 0$ the critical image separates into two smooth arcs, and for $a < 0$ into two distinct cusps, or "beaks".

SWALLOWTAIL. The critical image for $a = 0$ is a curve which at the origin looks like $x = t^3$, $y = t^4$. For $a > 0$ the critical image is a smooth arc, and for $a < 0$ a "swallowtail" shaped curve exhibiting two cusps and a node.

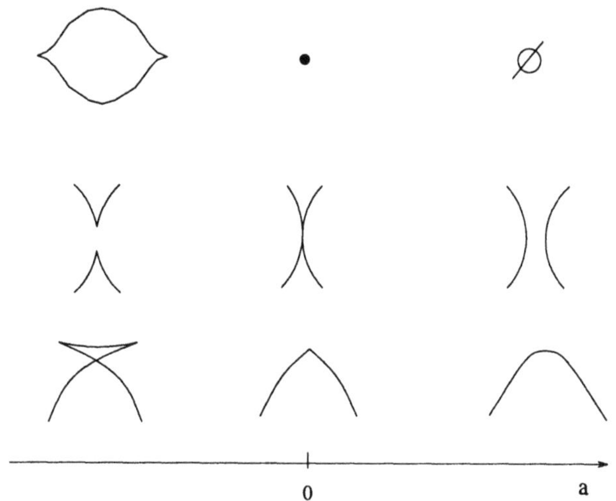

Figure 7: Versal unfoldings for the lips, beaks and swallowtail.

The invariants c and d listed in Table 2 and Table 3 have simple geometric interpretations in terms of versal unfoldings. c is the maximal number of cusps, and d is the maximal number of nodes which can appear in a versal unfolding of the given singularity. The reader is invited to peruse Figure 7 to see that the pictures are consistent with the tabled values of c, d.

The three remaining monogerms of corank 1, namely the goose, butterfly and gulls all have codimension 2, so their versal unfoldings contain two unfolding parameters a, b. Here again one can determine critical curves and images for given values of a, b to obtain qualitative pictures of the changes in the critical image illustrated in Figure 8, Figure 9 and Figure 10. In each case we have drawn a "clock diagram" in the (a, b)–plane giving a series of snapshots of the critical image as (a, b) moves round a small circle centred at the origin. The curves coming into the origin correspond to the bifurcation curves in the moving lamina, that is they represent the loci of (a, b) giving rise to codimension 1 (multi)germs, i.e. lips, beaks or swallowtail types in the monogerm case, tacnode fold or cusp–plus–fold in the bigerm case, and triple fold in the trigerm case. Again, the reader is invited to peruse the figures to see that they are consistent with the tabled values of c, d.

GOOSE. There is one bifurcation curve with a cusp at the origin on which the unfolding exhibits beak transitions.

BUTTERFLY. There are two bifurcation curves on which codimension 1 transitions occur, each having a cusp at the origin, and common cuspidal tangent. On one curve we have swallowtail transitions, and on the second cusp–plus–fold transitions. Figure 3 is an explicit example of a butterfly transition arising from the double four–bar.

GULLS. There are three bifurcation curves on which codimension 1 transitions occur. The first is the b–axis on which beak transitions take place: the second is a smooth cubic tangent to the b–axis at the origin on which swallowtail transitions take place: and the third is a half–parabola in the fourth quadrant on which tacnode folds appear.

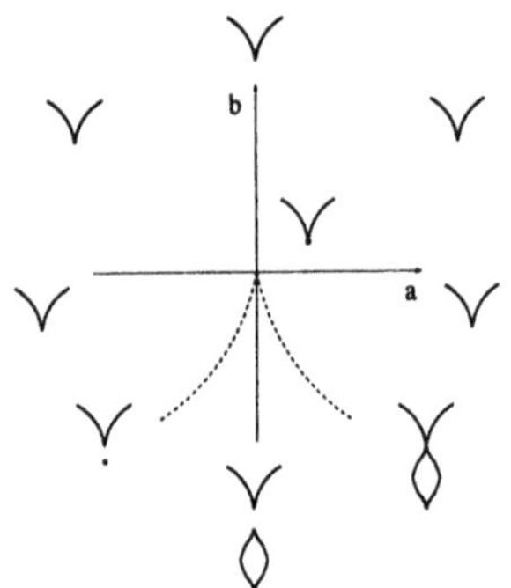

Figure 8: Goose Unfolding.

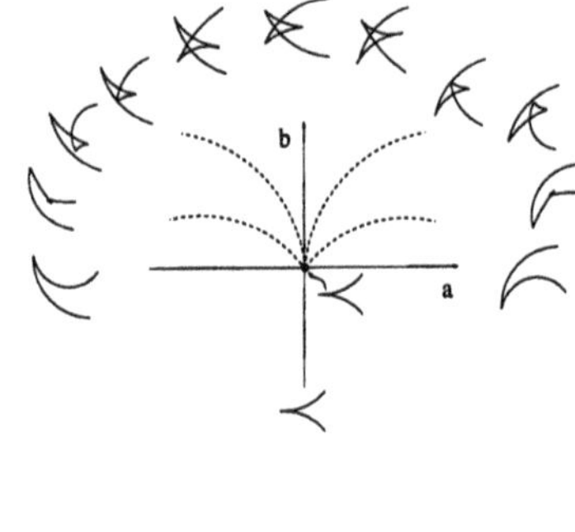

Figure 9: Butterfly Unfolding.

A versal unfolding for the corank 2 germ, the sharksfin is likewise illustrated in Figure 11 by a clock diagram. For the sharksfin there are beaks transitions on the a– and b–axes, and swallowtail transitions on a curve.

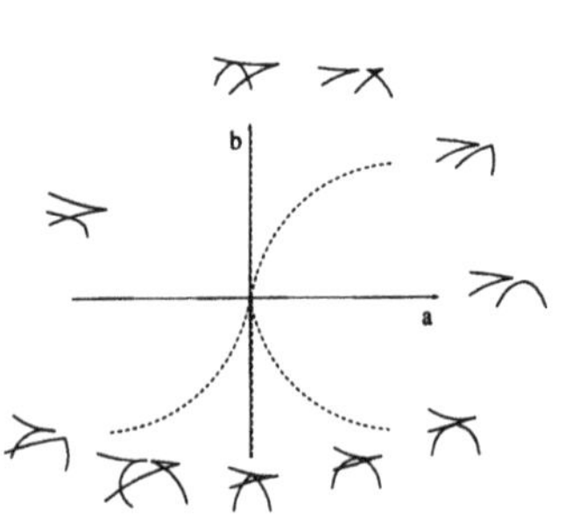

Figure 10: Gulls Unfolding.

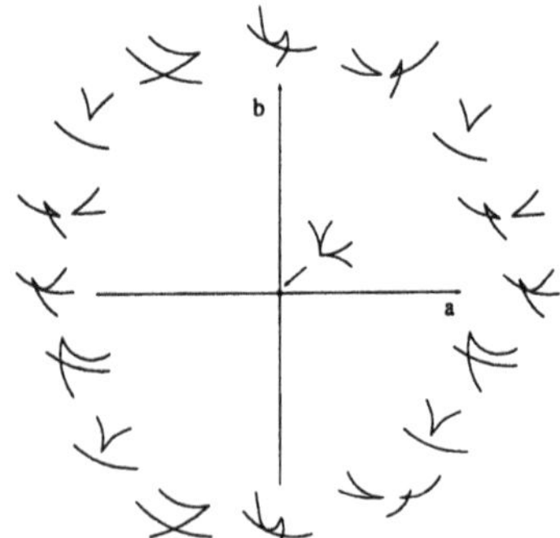

Figure 11: Sharksfin Unfolding.

VI. The Parallel Motion

A very simple geometric example of a planar motion [14, page 438] with two degrees of freedom, which we like to call the *parallel motion*, is obtained as follows. In the fixed lamina we trace a curve Γ, and in the moving lamina we choose a fixed point F. The parallel motion is obtained by insisting that F must always lie on Γ. One degree of freedom arises as F traverses Γ, and the second because at each point on Γ the moving lamina can rotate with one degree of freedom. Suppose we choose a tracing point C in the moving lamina distance d from F. If we fix F at one point of Γ the point C describes a circle of radius d centred at that point. Thus the trajectory region will be the union of all these circles of radius d as F traces out Γ, which is the union of all the curves parallel to Γ and distance $\leq d$ from Γ. Moreover it is easily checked that the critical image is the envelope of this family of circles, so comprises the two parallels to Γ at distance d. The interest of the example is that (provided the curve Γ is sufficiently general) the parallel motion can only exhibit swallowtail transitions on the critical images. Figure 12 illustrates the parallels of the standard parabola $y^2 = 4ax$ in the plane. When $d = a$ the parallel through the focus $(a, 0)$ exhibits a swallowtail transition. As d changes slightly from the value $d = a$ we see the versal unfolding of the swallowtail transition described in Figure 7. For $0 < d < a$ the parallel is a smooth curve which looks rather like the original parabola, but for $a < d$ the parallel exhibits the typical cusp and nodes of the swallowtail transition. In this example the parabola can be replaced by any smooth curve having only simple vertices without changing the result. The swallowtail transition goes a long way towards explaining why it is that garden hoses kink!

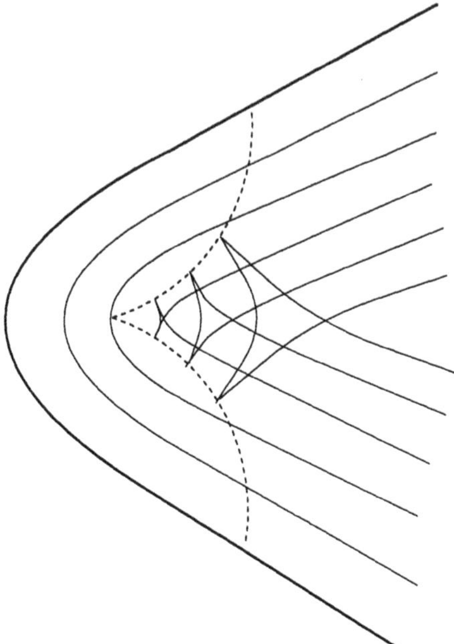

Figure 12: Swallowtail transition exhibited by the parallels of a parabola.

References

[1] C. G. Gibson and P. E. Newstead, "On the Geometry of the Planar 4–Bar Mechanism", *Acta Applicandae Mathematicae, 7*, 113–135 (1986).

[2] P. S. Donelan, "Generic Properties in Euclidean Kinematics", *Acta Applicandae Mathematicae, 12*, 265–286 (1988).

[3] C. A. Hobbs, "Kinematic Singularities of Low Dimension", *Ph.D. Thesis, University of Liverpool* (1993).

[4] C. G. Gibson and C. A. Hobbs, "Local Models for General One–Parameter Motions of the Plane and Space", *Preprint, University of Liverpool* (1992).

[5] C. G. Gibson, "Kinematic Singularities – A New Mathematical Tool", *Third International Workshop on Advances in Robot Kinematics*,209–215, Ferrara, Italy (1992).

[6] C. G. Gibson and C. A. Hobbs, "Local Models for General Two–Parameter Motions of the Plane", *In Preparation* (1994).

[7] C. G. Gibson and C. A. Hobbs, "Local Models for General Two–Parameter Motions of Space", *In Preparation* (1994).

[8] C. G. Gibson and W. Hawes, "Local Models for General Multi–Parameter Motions of the Plane", *In Preparation* (1994).

[9] H. Whitney, "On Singularities of Mappings of Euclidean Space I: Mappings of the Plane to the Plane", *Ann. Math., 62*, 374–410 (1955).

[10] T. Gaffney, "The Structure of $T\mathcal{A}(f)$, classification and an application to differential geometry", *AMS – Proceedings of Symposia in Pure Mathematics, 40 (Singularities) Part 1*, 409–427 (1983).

[11] L. Kergosien, "Topologie Différentielle – La Famille des Projections Orthogonales d'une Surfaces et ses Singularités", *Comptes Rendus des séances de l'Académie des Sciences, 292*, 929–932 (1981).

[12] J. Rieger, "Apparent Contours and their Singularities", *Ph.D. Thesis, University of London* (1988).

[13] J. Rieger, "Families of Maps from the Plane to the Plane", *London Math. Soc., 2*, 351–369 (1987).

[14] O. Bottema and B. Roth, *Theoretical Kinematics*, Dover Publications, New York (1990).

Limited Existence of Three-Dimensional Conformal Mapping in Robots

Veljko Milenkovic

Engineering Consultant
1222 Highway 42, Sturgeon Bay, WI 54235 USA

Paul H. Milenkovic

Associate Professor
Department of Electrical and Computer Engineering
University of Wisconsin-Madison
1415 Johnson Dr., Madison, WI 53706 USA

Abstract – We prove that any conformal mapping between three-dimensional Cartesian spaces is limited to translation, rotation, uniform scaling, reciprocal, or simple combinations of these four operations. This result extends our earlier work on the CRV (Conformal Rotation Vector), which defines a conformal mapping from the non-Euclidean space of rigid body orientations into a Euclidean 3-space. Conformal means that changes in orientation resulting from two consecutive body rotations, about mutually-perpendicular axes through equal infinitesimal angles, are represented by mutually- perpendicular equal-sized vector increments in CRV space. This property makes the CRV particularly useful for motion planning in robots. Because the conformal mapping between CRV space and another Cartesian space is limited, the CRV is essentially unique.

1 Introduction

In order to control a robot, it is advantageous to choose separate specifications for end-effector position and orientation. There is no better choice for position than the 3-vector of Cartesian coordinates. Orientation is not as clear-cut.

We have found reasons to choose different representations of orientation [1,2], depending on the stage of the control process. The most conventional orientation triplet in robotic use is the set of Euler angles. The Euler angles require transcendental functions (sine, cosine) to compute the rotation matrix. Furthermore, they suffer from singularities, becoming indeterminate at special orientations; awkward robot motions may result. Our choices were the 3×3 orthonormal rotation matrix, the unit quaternion [3], and a triplet called the CRV (conformal rotation vector) [4]. Conversion from CRV to either matrix or quaternion is by "add, multiply, divide" only. The quality of CRV-interpolated paths is usually high because of conformality.

We may better understand the different orientation specifications if we model body orientations as points in a space. Any orientation is reachable from another orientation by fixing an axis and rotating by an angle; this is guaranteed by a theorem due to Euler [5]. The axis vector gives the direction of a line connecting two points in orientation space and the angle gives the length of that line.

A. J. Lenarčič and B. B. Ravani (eds.), Advances in Robot Kinematics and Computationed Geometry, 59–68.

Orientation space is a non-Euclidean 3-space curved into a 4th dimension, analogous to the surface of the Earth, a 2-space curved into the 3rd dimension. Distances on the earth's surface can be represented to constant scale only on a globe; the scale of a flat (Euclidean) map of the earth will by necessity vary across the map. A map made by conformal projection, however, preserves all local angles, and the scale in the neighborhood of a point is the same in all directions; the Mercator map is an example.

Relationships between distances and directions in orientation space are correctly modeled by the 3-space surface of a 4-space hypersphere represented by "unit quaternion space." The CRV is a conformal projection of unit quaternion space (the globe) into Euclidean 3-space (the flat map).

Is the CRV unique? For two-dimensional mapping, there exists an infinity of analytic functions of a complex variable; each is conformal [6], and when applied to a conformal geographic map will produce a different but still conformal map. For three-dimensional conformal mapping, we prove the "Fundamental Theorem"

> *The only existing conformal mapping functions from one Cartesian 3-space to another are: rigid-body translation, rigid-body rotation, uniform change of scale, and the negative reciprocal function, singly or in various simple combinations.*

We will also show that the Theorem makes the CRV the only conformally-mapped Euclidean representation of body orientation to within nonessential variations.

2 Definitions and Examples of Conformal Functions

For Cartesian to Cartesian functions, we define the following

Trifunction: Function mapping 3-space $\mathbf{p} = (u, v, w)$ from 3-space $\mathbf{r} = (x, y, z)$ (dependent variables u, v, w are each functions of independent variables x, y, z).

J-matrix: 3×3 matrix mapping $d\mathbf{p} = (du, dv, dw)^T$ from $d\mathbf{r} = (dx, dy, dz)^T$ (or time derivatives $\dot{\mathbf{p}} = (\dot{u}, \dot{v}, \dot{w})^T$ from $\dot{\mathbf{r}} = (\dot{x}, \dot{y}, \dot{z})^T$) according to

$$d\mathbf{p} = \mathbf{J}d\mathbf{r} \qquad (\text{or } \dot{\mathbf{p}} = \mathbf{J}\dot{\mathbf{r}}) \qquad \text{where} \tag{1}$$

$$\mathbf{J} = \begin{bmatrix} u_x & u_y & u_z \\ v_x & v_y & v_z \\ w_x & w_y & w_z \end{bmatrix} \tag{2}$$

and subscripts denote partial derivatives.

Jacobian: $\det \mathbf{J}$ (determinant of a J-matrix) [7].

Scale-orthonormal: A scalar multiple of an orthonormal matrix; dot product of any distinct two of its rows is zero, sum of squares in each row is the same.

Conform: A conformal trifunction; whose J-matrix is scale-orthonormal, and $\det \mathbf{J} > 0$.

Scale: S, local mapping scale of a "conform"; square root of the sum of squares in any J-matrix row, same as cube root of the Jacobian ($\det \mathbf{J}$).

Plainform: Linear "conform", $\mathbf{J}$ = const., S = const.; consists of one, two or all three of: uniform translation, uniform rotation, uniform change of scale.

Uniscale: A "conform" whose scale S = const. (uniform); tentatively, $\mathbf{J}$ is variable.

Variscale: A "conform" whose local scale S varies from point to point.

Reciprocal: The "conform" in Eq. (7), which transforms every 3-vector into one whose magnitude is reciprocal, and direction opposite, of the original one.

The "Fundamental Theorem" may then be stated as follows:

(i) Every uniscale is equivalent to a plainform (i.e. if S = const., then $\mathbf{J}$=const.).

(ii) Every variscale is expressible as the "reciprocal", preceded by some plainform and/or followed by another plainform.

Conformality conditions are imposed on the J-matrix. In the two-dimensional case (from x, y to u, v), the scale-orthonormality and positive determinant conditions are

$$u_x^2 + u_y^2 = v_x^2 + v_y^2; \quad u_x v_x + u_y v_y = 0; \quad \det \mathbf{J} \equiv u_x v_y - u_y v_x > 0, \tag{3}$$

which simplify to the well-known Cauchy-Riemann conditions

$$u_x = v_y,\, u_y = -v_x \quad (i.e\ \partial u/\partial x = \partial v/\partial y,\, \partial u/\partial y = -\partial v/\partial x), \tag{4}$$

which are necessary and sufficient for Eq. (3) to be satisfied. Eq. (4) is an attribute of analytic functions of complex variables [6,8,9] (that possess power series expansions); it also ensures that the mapping is conformal.

The 3-D conformality conditions (which do not simplify as in Eq. (4)) are given by

$$u_x^2 + u_y^2 + u_z^2 = v_x^2 + v_y^2 + v_z^2 = w_x^2 + w_y^2 + w_z^2, \tag{5}$$

$$u_x v_x + u_y v_y + u_z v_z = u_x w_x + u_y v_y + u_z w_z = v_x w_x + v_y w_y + v_z w_z = 0. \tag{6}$$

For example, the "reciprocal" (negative reciprocal trifunction) given by

$$\mathbf{p} = -\mathbf{r}/|\mathbf{r}|^2; \text{ or } u = -x/r^2,\ v = -y/r^2,\ w = -z/r^2; \text{ where } r^2 = x^2 + y^2 + z^2 \tag{7}$$

has "J-matrix"

$$\mathbf{J} = \begin{bmatrix} (x^2 - y^2 - z^2)/r^4 & 2xy/r^4 & 2xz/r^4 \\ 2xy/r^4 & (-x^2 + y^2 - z^2)/r^4 & 2yz/r^4 \\ 2xz/r^4 & 2yz/r^4 & (-x^2 - y^2 + z^2)/r^4 \end{bmatrix}, \tag{8}$$

that satisfies Eqs. (5,6), and the scale is $S = 1/r^2$. This shows that the "reciprocal" given in Eq. (7) is a conform of the "variscale" type.

Two Examples of "conforms" are later considered: (1) the identity uniscale $\mathbf{p} = \mathbf{r}$, and (2) a linear (plainform) modification of the variscale "reciprocal" from Eq. (7), which will serve to remove the singularity in (7) at the origin to facilitate power series expansion. The Examples will be standards of comparison in proving parts (i) and (ii) of the Theorem.

2.1 Conformal Mapping of Orientation - CRV In some robot applications it is advantageous to use conformal, locally "distorsion-free" 3-parameter representation of end-effector orientation. Examples where it helps if orientational scale is locally uniform involve operations along smooth curves drawn on curved surfaces. These are found in robotic arc welding, sealant deposition or spray painting, particularly in automotive applications. The orientation programming strategy should be such that during motion the x- and z-axes of the end effector remain tangent to the path and perpendicular to the surface, respectively. In synthesizing motion by suitable interpolation between discrete points along the desired path, conformality of the mapping space permits more separated (i.e. fewer) taught points, needed to ensure correct tool orientation along the path, compared with the needs if non-conformal parameters such as Euler angles are used. This will save on programming effort and time and on memory storage requirements.

Rigid-body orientation is describable in terms of rotation of a coordinate frame fixed in the body, relative to an "immobile" frame of reference. Ref. [4] defines CRV as a vector aligned with the axis of frame rotation (directed according to the "right-hand" rule), of magnitude $\tan(\theta/4)$ where θ is the rotation angle, and finds it conformal with incremental body rotation. Such CRV corresponding to each orientation is double-valued: Rotation in opposite sense through $\theta' = \theta - 2\pi < 0$ will bring the body to the same orientation. Since $\tan(\theta/4) = -1/\tan(\theta/4)$, the two CRVs representing the same orientation are "negative reciprocals" of each other, hence if one is conformal the other is likewise.

Conformality is evident when the increment of CRV is related to the angular velocity vector $\vec{\omega}$ (velocity in "orientation space"). Expressing the CRV as vector

$$\mathbf{e} = (a, b, c)^T, \quad \text{then } d\mathbf{e} = J\vec{\omega}dt \text{ or } \dot{\mathbf{e}} = \mathbf{J}\vec{\omega}, \tag{9}$$

and the above J-matrix is scale-orthonormal [2]. Eq. (9) resembles (1,2), with $\vec{\omega}$ in the place of the time derivative of a vector (though $\vec{\omega}$ is not a derivative of an existing vector). The ratios among amounts of successive infinitesimal body rotations, and angles between respective axes of rotation, are preserved upon mapping into e-space.

If a function other than CRV existed, which mapped orientation conformally into a 3-parameter space, then this new function would be related to the CRV by some Cartesian to Cartesian conform. The proof offered here guarantees that no conform exists to generate a "new" map of orientation. A displaced, rotated or uniformly scaled CRV hardly qualifies as new, while the "reciprocal" only changes the CRV from one of its branches to the other.

3 Proof of the Fundamental Theorem

Consider an arbitrary "conform": $\mathbf{p}^* = (u^*, v^*, w^*)$ of $\mathbf{r}^* = (x^*, y^*, z^*)$. If it is uniscale ($S$ = const.), we prove (with $\mathbf{r}^* = \mathbf{r}$, unchanged) that there exists a plainform change from $\mathbf{p}^*$ to $\mathbf{p}$ which will transform our "conform" into Example (1), i.e. identity $\mathbf{p} = \mathbf{r}$. If $\mathbf{p}^*$ of $\mathbf{r}^*$ is variscale, however, our proof will show that there exists a plainform from $\mathbf{r}^*$ to $\mathbf{r}$ and a plainform from $\mathbf{p}^*$ to $\mathbf{p}$, to transform the arbitrary "conform" into Example (2), shown in Eq. (10) of Sec. 3.2.

At the origin $\mathbf{r} = (0,0,0)$ the boundary conditions of both "Examples" are

(a) $\mathbf{p} = (0,0,0)$ and

(b) $\mathbf{J} = [1,0,0;\ 0,1,0;\ 0,0,1]$ (identity matrix: scale=1, no rotation).

Another boundary condition, scale gradient at the origin, is different for the two Examples:

(c) $(S_x, S_y, S_z) = (0,0,0)$, for Example (1) or for any other uniscale,
$(S_x, S_y, S_z) = (2,0,0)$, for Example (2) as given by Eq. (10).

When boundary condition (b) holds, then (c) is the same as gradient of u_x, namely (u_{xx}, u_{xy}, u_{xz}).

The proof of the Theorem will consist of two parts, which must prove:

1. Plainforms exist to transform any "conform" (conformal trifunction) into one that satisfies the same above boundary conditions as either Example (1) or (2).
2. A solution of Eqs. (5,6), subject to boundary conditions (a), (b), (c) above, is unique: only one such exists for each version of condition (c).

The fact that a plainform exists to transform any uniscale into the identity proves that the uniscale is the inverse of that plainform, i.e. a plainform itself. Thus, if its $\det \mathbf{J}$ = const., then all elements of $\mathbf{J}$ are constant. Since the result of two or more successive plainforms is obviously uniscale, it is expressible as a single plainform. Furthermore, if one preceeding and/or one following plainform suffices to convert any variscale into the function of Example (2), which in the same way is transformable into the reciprocal (7), then some single plainform before and another after will convert the original variscale directly into the reciprocal, Eq. (7), Q.E.D.

3.1 Matching of Boundary Conditions Part 1. of the proof is quite simple for a uniscale. Since S = const., scale gradient is already $(0,0,0)$, matching that of Example (1). It will remain the same no matter how this uniscale is translated, rotated, or its scale uniformly altered. Simple translation makes $\mathbf{p} = (0,0,0)$ at the origin $\mathbf{r} = (0,0,0)$. Multiplying $\mathbf{p}$ by a constant multiplies the J-matrix by the same, and a suitable scaling constant makes $\det \mathbf{J} = 1$ everywhere. The J-matrix is now orthonormal. Using the transpose (same as inverse) of its value at the origin as constant matrix multiplier (pure rotator) on $\mathbf{p}$ makes $\mathbf{J} = \mathbf{I}$ (identity matrix) at $\mathbf{r} = (0,0,0)$. The uniscale now has correct boundary values (a), (b) and (c) at the origin - see Sec. 3, Example (1).

More has to be done in the case of a variscale. Since S is now variable, its gradient vector will be non-zero at most points. Select one such point R in (x,y,z), where the function is analytic (expandable into a power series). A pre-translation (in x,y,z) exists to make any R coincide with the origin $\mathbf{O}$ of the new x,y,z-coordinates. A post-translation (in u,v,w) exists to make the functional vector $\mathbf{p} = (0,0,0)$ at this origin $\mathbf{O}$.

Consider the surface S = const. which passes through $\mathbf{O}$ in x,y,z. Perform a rigid-body rotation in x,y,z, making the new yz-plane tangent to that surface, so that the scale gradient is in the new x-direction. Then perform a similar fixed rotation in the (u, v, w) space, so that at **O** the x-direction maps into new u, y into new v, z into new w-direction (*i.e* diagonalize **J** at **O**). That is done by normalizing the value of **J** at **O**, transposing it, and using the result as a constant rotation multiplier on **p**. The latest rotation in (u, v, w) has no effect on the direction of the scale gradient vector relative to x, y, z (this vector remains in the x-direction), since S is a scalar function of x, y, z (unaffected by rotation of u, v, w).

Two scale changes (multiplications by constants) are now to be performed, pre-scaling (in x, y, z) and post-scaling (in u, v, w), in order to satisfy two boundary conditions at **O**, namely $S = 1$ and $S_x = 2$. This is always possible, since constant multiplication of either **r** or **p** alters S linearly, while S_x changes as the square of the former but linearly with the latter multiplication. Thus, any variscale, preceeded and followed by suitable plainforms, can be made to satisfy appropriate boundary values (a), (b), (c) of Sec. 3.

3.2 Example (2) and its Expansion Write Eq. (7) with (u', v', w') functions of (x', y', z')

$$x' = x - 1,\ y' = -y,\ z' = -z; \qquad \text{and } u' = u + 1,\ v' = v,\ w' = w.$$

This amounts to the "reciprocal" preceeded by translation along x and 180 degree rotation about x, and followed by translation along the u-axis:

$$\left.\begin{array}{ll} u = (1 - x)D - 1;\ v = yD;\ w = zD; & \text{where} \\ D = [(1 - x)^2 + y^2 + z^2]^{-1} = [1 - (2x - r^2)]^{-1}; & r^2 = x^2 + y^2 + z^2 \end{array}\right\}. \tag{10}$$

Eq. (10) above represents Example (2), with applicable boundary conditions of Sec. 3.

We expand D as $(1 - h)^{-1} = 1 + h + h^2 + h^3 + h^4 + \ldots$ and collect like powers of x, y, z:

$$\left.\begin{array}{l} u = x + U + U' + U'' + \ldots + U^{()} + \ldots \\ v = y + V + V' + V'' + \ldots + V^{()} + \ldots \\ w = z + W + W' + W'' + \ldots + W^{()} + \ldots \end{array}\right\}, \tag{11}$$

where U, V, W are quadratic in x, y, z; U', V', W' are cubic; U'', V'', W'' are quartic. For example $U = x^2 - y^2 - z^2$, $U' = x(x^2 - 3y^2 - 3z^2)$, $U'' = x^4 - 6x^2y^2 - 6x^2z^2 + y^4 + 2y^2z^2 + z^4, \ldots$; $V = 2xy$, $V' = y(3x^2 - y^2 - z^2)$, $V'' = 4xy(x^2 - y^2 - z^2), \ldots$; $W = 2xz$, $W' = z(3x^2 - y^2 - z^2)$, $W'' = 4xz(x^2 - y^2 - z^2), \ldots$. The J-matrix of Eq.(11) is

$$\begin{array}{ccc} 1 + U_x + U'_x + U''_x + \ldots & U_y + U'_y + U''_y + \ldots & U_z + U'_z + U''_z + \ldots \\ V_x + V'_x + V''_x + \ldots & 1 + V_y + V'_y + V''_y + \ldots & V_z + V'_z + V''_z + \ldots \\ W_x + W'_x + W''_x + \ldots & W_y + W'_y + W''_y + \ldots & 1 + W_z + W'_z + W''_z + \ldots \end{array}. \tag{12}$$

The scale (by root square of 1st row of **J**) is $S = 1 + U_x + O(r^2)$ for small r, so that at $(0, 0, 0)$ the gradient has the value

$$(S_x, S_y, S_z) = (U_{xx}, U_{xy}, U_{xz}) = (2, 0, 0). \tag{13}$$

Treating U,V,W; U',..., etc. as unknowns, substituting the elements from Eq. (12) into Eqs. (5,6), performing required multiplications, and collecting terms of like degrees in x,y,z, one obtains five independent equations that are linear in each following set of unknowns, with squares and products of derivatives of previous unknowns as forcing functions (we show up to cubic):

Linear terms (governing quadratic polynomials U, V, W):

$$\left.\begin{array}{lll} U_x - V_y = 0; & U_x - W_z = 0; & \\ U_y + V_x = 0; & U_z + W_x = 0; & V_z + W_y = 0 \end{array}\right\}; \tag{14}$$

Quadratic terms (governing cubic polynomials U', V', W'):

$$\left.\begin{array}{lll} U'_x - V'_y &=& (V_x^2 + V_y^2 + V_z^2)/2 - (U_x^2 + U_y^2 + U_z^2)/2 \\ U'_x - W'_z &=& (W_x^2 + W_y^2 + W_z^2)/2 - (U_x^2 + U_y^2 + U_z^2)/2 \\ U'_y + V'_x &=& -(U_x V_x + U_y V_y + U_z V_z) \\ U'_z + W'_x &=& -(U_x W_x + U_y W_y + U_z W_z) \\ V'_z + W'_y &=& -(V_x W_x + V_y W_y + V_z W_z) \end{array}\right\}; \tag{15}$$

Cubic terms (governing quartic polynomials U'', V'', W''):

$$\left.\begin{array}{lll} U''_x - V''_y &=& (V_x V'_x + V_y V'_y + V_z V'_z) - (U_x U'_x + U_y U'_y + U_z U'_z) \\ U''_x - W''_z &=& (W_x W'_x + W_y W'_y + W_z W'_z) - (U_x U'_x + U_y U'_y + U_z U'_z) \\ U''_y + V''_x &=& -(U_x V'_x + U'_x V_x + U_y V'_y + U'_y V_y + U_z V'_z + U'_z V_z) \\ U''_z + W''_x &=& -(U_x W'_x + U'_x W_x + U_y W'_y + U'_y W_y + U_z W'_z + U'_z W_z) \\ V''_z + W''_y &=& -(V_x W'_x + V'_x W_x + V_y W'_y + V'_y W_y + V'_z W_z + V_z W'_z) \end{array}\right\}. \tag{16}$$

To help in later derivations, we arrange coefficients of homogeneous quadratics $b_0x^2 + b_1xy + b_2xy + b_3y^2 + b_4xz + b_5z^2$; cubics $c_0x^3 + c_1x^2y + c_2x^2z + c_3xy^2 + c_4xyz + c_5xz^2 + c_6y^3 + c_7y^2z + c_8yz^2 + c_9z^3$; etc. in triangles in the form

$$\text{Quadratic} = \begin{array}{ccccc} & & b_0 & & \\ & b_1 & & b_2 & \\ b_3 & & b_4 & & b_5 \end{array} \qquad \text{Cubic} = \begin{array}{ccccccc} & & & c_0 & & & \\ & & c_1 & & c_2 & & \\ & c_3 & & c_4 & & c_5 & \\ c_6 & & c_7 & & c_8 & & c_9 \end{array}$$

Using the above format, terms shown in eqn. (11) are expressible as

$$U = \begin{array}{ccccc} & & 1 & & \\ & 0 & & 0 & \\ -1 & & 0 & & -1 \end{array} \qquad V = \begin{array}{ccccc} & & 0 & & \\ & 2 & & 0 & \\ 0 & & 0 & & 0 \end{array} \qquad W = \begin{array}{ccccc} & & 0 & & \\ & 0 & & 2 & \\ 0 & & 0 & & 0 \end{array}$$

$$U' = \begin{array}{ccccccc} & & & 1 & & & \\ & & 0 & & 0 & & \\ & -3 & & 0 & & -3 & \\ 0 & & 0 & & 0 & & 0 \end{array} \qquad V' = \begin{array}{ccccccc} & & & 0 & & & \\ & & 3 & & 0 & & \\ & 0 & & 0 & & 0 & \\ -1 & & 0 & & -1 & & 0 \end{array} \qquad W'' = \begin{array}{ccccccc} & & & 0 & & & \\ & & 0 & & 3 & & \\ & 0 & & 0 & & 0 & \\ 0 & & -1 & & 0 & & -1 \end{array}$$

and quartics U'', V'', W'' in a similar way, this time with 15 elements in each triangle.

3.3 The Difference Function A putative variscale different from Eq. (11), satifying Eqs. (5,6), and conditions (a), (b), (c) of Sec. 3, is expressed as

$$\left.\begin{array}{lll} u &=& x + (U + A) + (U' + A') + (U'' + A'') + \ldots + (U^{()} + A^{()}) + \ldots \\ v &=& y + (V + B) + (V' + B') + (V'' + B'') + \ldots + (V^{()} + B^{()}) + \ldots \\ w &=& z + (W + C) + (W' + C') + (W'' + C'') + \ldots + (W^{()} + C^{()}) + \ldots \end{array}\right\}. \tag{17}$$

Here U, U', ...; V, V', ...; etc. denote the same polynomials as in Eq. (11), while A, B, C; A', B', ... express the difference between the putative variscale and Example (2), starting with quadratics. Constant and linear terms of this "difference" are zero because of (a) and (b). Since $(U+A)_{xx}$, $(U+A)_{xy}$, and $(U+A)_{xz}$ must satisfy the same conditions as U_{xx}, etc. (Eq. 13), it follows that

$$A_{xx} = A_{xy} = A_{xz} = 0. \tag{18}$$

Quadratics A, B, ... as well as $(U+A)$, ... must satisfy the same Eq. (14) as U,V,W (homogen. equation). It will be shown shortly that Eqs. (14) and (18) make $A \equiv 0$, $B \equiv 0$, $C \equiv 0$.

When the cubics $(U' + A')$, ... are substituted into Eq. (15), vanishing of A, B, C makes the right side the same as when Eq. (15) is used with U', V', W'. This makes the right side (forcing function) zero when applied to A', B', C'. The eqns. thus become homogeneous [10], as are (14). It will be shown that cubics subject to Eqs. (15) vanish identically as well. A similar reasoning applies to quartics A'', B'', C'' in Eqs. (16), and so on. Vanishing of all lower power polynomials of the "difference" makes the forcing functions of eqns. governing the next power vanish, which will be shown to make the next polynomials vanish, and so on. Vanishing of the "difference" makes the solution under stipulated boundary conditions unique. The same argument holds for a uniscale relative to Example (1), where condition (c), Eq. (13) is $(0,0,0)$.

3.4 Rules to Facilitate the Proof With lower terms zero, n-th power polynomials $A^{()}$, $B^{()}$, $C^{()}$ must satisfy Eqs. (14) when substituted in place of U, V, W. Shown below are triangles representing these polynomials. They contain symbols to denote terms to be proven zero by various methods. The three entries marked 0 in A have that value due to Eq. (18). Given these, 1st and 2nd eqn. of (15) make zero the entries marked 0 in B and C, respectively.

RULE 1. Consider one particular term in each of $A^{()}$, $B^{()}$, $C^{()}$, notably

$$A^{()} = ax^{i-1}y^j z^k;\ B^{()} = bx^i y^{j-1} z^k;\ C^{()} = cx^i y^j z^{k-1};\ \text{with } i > 0,\ j > 0,\ k > 0. \tag{19}$$

From the 3rd eqn. in (14), $ja + ib = 0$ (since $jax^{i-1}y^{j-1}z^k + ibx^{i-1}y^{j-1}z^k = 0$); from the 4th eqn. in (14), $ka + ic = 0$; from the 5th eqn. in (14), $kb + jc = 0$. Since i, j, k are all positive, it follows that

$$a = 0; \quad b = 0; \quad c = 0. \tag{20}$$

Entries made zero by RULE 1 (possible combinations of positive i, j, k) are marked "!"

RULE 2. By differentiation and elimination of terms between eqns. of (14) with A, B, C:

$$-A_{xx} = A_{yy} = A_{zz}; \quad B_{xx} = -B_{yy} = B_{zz}; \quad C_{xx} = C_{yy} = -C_{zz}. \tag{21}$$

Starting with any entry proven zero earlier, it follows from (21) that all entries in line located 2 nodes away in any direction (i.e. all 2nd neighbors) become zero. Thus, from entries marked "0" and "!", which are = 0 already, RULE 2 makes 0 all entries marked "&".

$$
\begin{array}{ccc}
A: \begin{array}{c} 0 \\ 0 \quad 0 \\ \& \quad ! \quad \& \end{array} &
B: \begin{array}{c} \& \\ 0 \quad ! \\ 0 \quad 0 \quad \& \end{array} &
C: \begin{array}{c} \& \\ ! \quad 0 \\ \& \quad 0 \quad 0 \end{array}
\end{array}
$$

$$
\begin{array}{ccc}
A': \begin{array}{c} a_1 \\ \& \quad \& \\ a_2 \quad ! \quad a_3 \\ \& \quad ! \quad ! \quad \& \end{array} &
B': \begin{array}{c} \& \\ b_1 \quad ! \\ \& \quad ! \quad ! \\ b_2 \quad \& \quad b_3 \quad \& \end{array} &
C': \begin{array}{c} \& \\ ! \quad c_1 \\ ! \quad ! \quad \& \\ \& \quad c_2 \quad \& \quad c_3 \end{array}
\end{array}
$$

Cubics A', B', C' require a special procedure to solve for entries marked a_n, b_n, c_n; $n = 1, 2, 3$. In conventional notation the entries are

$$
\begin{array}{l}
A' = a_1x^3 + a_2xy^2 + a_3xz^2; \quad B' = b_1x^2y + b_2y^3 + b_3yz^2; \\
C' = c_1x^2z + c_2y^2z + c_3z^3.
\end{array} \tag{22}
$$

From the 1st of (14): $3a_1 = b_1$; $a_2 = 3b_2$; $a_3 = b_3$. From the 2nd of (14): $3a_1 = c_1$; $a_2 = c_2$; $a_3 = 3c_3$. From the 3rd, 4th, 5th: $a_2 = -b_1$; $a_3 = -c_1$; $b_3 = -c_2$. Therefore

$$
\begin{array}{l}
b_1 = 3a_1 = c_1 = -a_3 = -b_3 = c_2 = a_2 = -b_1 = 0 \text{ (since } b_1 = -b_1) \\
b_2 = a_2/3 = 0; \quad c_3 = a_3/3 = 0.
\end{array}
$$

If entries in a triangular subset of 6 or more nodes are zero by virtue of RULE 1. ("!") which is true for quartics (see below), as well as for all polynomials of degree 5 or higher (not shown), then RULE 2. will make all other entries ("&") equal to zero in succession (some being 2nd neighbors of a previously zeroed "&"):

$$
\begin{array}{ccc}
A'': \begin{array}{c} \& \\ \& \quad \& \\ \& \quad ! \quad \& \\ \& \quad ! \quad ! \quad \& \\ \& \quad ! \quad ! \quad ! \quad \& \end{array} &
B'': \begin{array}{c} \& \\ \& \quad ! \\ \& \quad ! \quad ! \\ \& \quad ! \quad ! \quad ! \\ \& \quad \& \quad \& \quad \& \quad \& \end{array} &
C'': \begin{array}{c} \& \\ ! \quad \& \\ ! \quad ! \quad \& \\ ! \quad ! \quad ! \quad \& \\ \& \quad \& \quad \& \quad \& \quad \& \end{array}
\end{array}
$$

The foregoing process, showing successively all sets of polynomials in ascending order to be zero, completes the proof that the "difference" solution vanishes identically.

4 Conclusions

Rotation and reciprocal function commute, and two consecutive rotations can be consolidated into one. Instead of change of scale before, a reciprocal change after the function produces the same effect. In view of these facts, and of the Theorem that has been proven, the most general conform (3-D conformal mapping function) is expressible as a combination of no more than one each of: negative reciprocal, rotation and scaling, and no more than two translations (one before, one after the reciprocal). This most general set of variscale functions is definable as a ten-parameter family. Three numbers give each of two translations, three define rotation, one gives scaling, and none for the reciprocal. Any member of this set will transform a CRV into another variant of the CRV, not to any new conformal map of orientation. The proof offered here is thus of considerable practical significance in robot control, where the CRV has been used [1,2].

For a given body orientation the value of the CRV is not unique, insofar as it depends on the selected attitudes of the body-attached coordinate frame and of the fixed frame of reference. Every general conform will transform a CRV into an entity CRV$''$ that is defined below. Let CRV and CRV$'$ represent the same body orientation but employ different attitudes (orientations) of the two defining coordinate frames. A suitable translation (adding a constant vector to it) and changing its scale (multiplying it by a constant) alters a CRV$'$ into the CRV$''$ in question.

There are 6 degrees of freedom in defining CRV$'$ (specifying two coordinate frames), and 4 more in displacing and scaling CRV$''$ – same total as in a general conform. The foregoing allows one to conclude that no conform exists to change a CRV into something substantially different from a CRV. A repositioned or scaled map is not a map of a different kind. There exists only one kind of conformal map of orientation - namely the CRV.

References

[1] M. Loo, Y. A. Hamidieh, and V. Milenkovic, "Generic Path Control for Robot Applications", Robots 14 Conf. Proc., (SME), 10/49-10/64, Detroit, MI, USA (1990).

[2] V. Milenkovic, "Framework to Facilitate Orientational Motion Planning in Robots," Proc. 3rd Int. Workshop on Advances in Robot Kinematics, 47-53, Ferrara, Italy (1992).

[3] J. Rooney, "A Survey of Representations of Spatial Rotation About a Fixed Point", Environment and Planning, B(4), 185-210 (1977).

[4] V. Milenkovic, "Coordinates Suitable for Angular Motion Synthesis in Robots", Robots VI Conf. Proc., (SME), 407-420, Detroit, MI, USA (1982).

[5] H. Goldstein, Classical Mechanics, Addison-Wesley Publ. Comp., Reading, MA, USA (1959).

[6] E. T. Whitaker and G. N. Watson, Modern Analysis, Cambridge at the University Press (1952).

[7] I. S. Sokolnikoff, Advanced Calculus, McGraw-Hill, New York (1939).

[8] I. S. Sokolnikoff, Mathematical Theory of Elasticity, McGraw-Hill, New York (1956).

[9] V. L. Streeter, Fluid Dynamics, McGraw-Hill, New York (1948).

[10] L. R. Ford, Differential Equations, McGraw-Hill, New York (1933).

2. Computational Geometry in Kinematics

Geometric Analysis of Spatial Rigid-BodyDynamics with Multiple Friction Contacts

J. R. Dooley, Assistant Professor
Department of Mechanical Engineering
Washington University, St.Louis, MO
B. Ravani, Professor
Department of Mechanical, Aerospace and Materials Engineering
University of California, Davis

Abstract - This paper presents a technique for analyzing spatial, dynamic manipulation problems involving multiple frictional contacts. The basis of this technique is to transform the negative contact forces on and the desired motion of a rigid body into instantaneous axes of rotation. These axes are directed lines or spears. The direction is determined by the sense of the rotation about the lines. A force capable of producing the desired motion must lie within the convex hull of the these rotation axes. This condition does not use force magnitudes and is therefore necessary, but not sufficient.

I. Introduction

Robotics and assembly problems often involve frictional contact forces whose magnitudes are difficult to accurately measure. The set of possible directions of these forces may be determined from the geometry of the environment and the configuration of the manipulator and workpiece. The difficulty of obtaining accurate measures of contact forces often requires force selection based on problem geometry. Brost and Mason [1] developed a technique for planar problems. They found possible forces by mapping the lines of the contact forces and the desired motion of the body into points in the dual-force space. This space is a mapping of a line of force or acceleration into an acceleration center. They observed that a force that produces the desired motion must fall within the convex hull of these points. Using line geometry, we extend the results of Brost and Mason to spatial problems. A set of possible forces is determined from the reactions to the contact forces and the desired final motion. A force that has an instantaneous rotation axis in the convex hull of the other rotation axes is necessary to obtain the desired motion.

II. Instantaneous Rotation Axes

Some definitions from line geometry are helpful for understanding the technique presented here. A convenient algebraic representation of a line is its ***Plücker coordinates.***

A. J. Lenarčič and B. B. Ravani (eds.), Advances in Robot Kinematics and Computationed Geometry, 71–80.

Define **s** as the direction of the line and **r** as a vector from the reference origin to a point on the line. Then, the six Plücker coordinates are

$$\hat{\mathbf{L}} = \left\{ \begin{array}{c} \mathbf{l} \\ \mathbf{l}^o \end{array} \right\} = \left\{ \begin{array}{c} \mathbf{s} \\ \mathbf{r} \times \mathbf{s} \end{array} \right\}. \tag{1}$$

The Plücker coordinates can be written as a dual vector, $\hat{\mathbf{L}} = \mathbf{l} + \epsilon \mathbf{l}^o$, where $\epsilon^2 = 0$.

Plücker coordinates are homogenous coordinates. Thus, if $\hat{\mathbf{L}}$ represents a line and $\hat{\lambda}$ is any dual number with a non-zero real part,$\hat{\lambda}\hat{\mathbf{L}}$ represents the same line as $\hat{\mathbf{L}}$. A six-vector, $\mathbf{P} = \{\mathbf{p}, \mathbf{p}^o\}$, can be transformed into normalized Plucker coordinates:

$$\hat{\mathbf{L}} = \mathbf{l} + \epsilon \mathbf{l}^o \tag{2}$$

where

$$\mathbf{l} = \frac{\mathbf{p}}{(\mathbf{p} \bullet \mathbf{p})^{\frac{1}{2}}}, \quad \mathbf{l}^o = \frac{(\mathbf{p} \bullet \mathbf{p})\mathbf{p}^o - (\mathbf{p} \bullet \mathbf{p}^o)\mathbf{p}}{(\mathbf{p} \bullet \mathbf{p})^{\frac{3}{2}}}.$$

The linear combination of two skew lines is a *linear ruled surface*. The linear combination of two intersecting lines is a *planar pencil* of lines. Two lines, $\hat{\mathbf{A}}$ and $\hat{\mathbf{B}}$, intersect, if their Plücker coordinates satisfy the relation $\mathbf{a} \bullet \mathbf{b}^o + \mathbf{a}^o \bullet \mathbf{b} = 0$.

The linear combination of three skew lines forms a *regulus*. A regulus is defined by three linearly independent skew lines not in the regulus that all lines of the regulus intersect. These lines of intersection are the *directorices* of the regulus. The linear combination of three linearly independent, coincident lines is a *bundle*.

The linear combination of four skew lines is a *line congruence*. A line congruence can be defined as all lines that intersect two directorices. The linear combination of five skew lines is a *line complex*. All lines in a line complex are othogonal to the *polar line* to the complex. [2] has further details on linear combinations of lines.

A. *The Instantaneous Acceleration Axis for a Desired Motion*

The instantaneous axis of rotation is found from the change in velocity, $d\mathbf{v}_g$, at the origin of the moving frame and the change in angular velocity, $d\omega$, of the moving frame relative to the fixed frame. A point, A, fixed in the moving frame has differential velocity of $d\mathbf{v}_a = d\mathbf{v}_g + d\omega \times \mathbf{R}$, where $\mathbf{R}$ is the vector from the origin, G, to A. The instantanious axis of rotation satisfies the relation $d\mathbf{v}_a = 0$. A point on the rotation axis relative to the moving frame is given by the vector

$$\mathbf{d}_a = \frac{d\omega \times d\mathbf{v}_g}{d\omega \bullet d\omega}. \tag{3}$$

The direction of the rotation axis is parallel to the direction of the change in angular velocity, $\mathbf{e}_\omega$. The Plücker coordinates representing the rotation axis are then

$$\hat{\mathbf{E}}_a = \left\{ \begin{array}{c} \mathbf{e}_\omega \\ \mathbf{d}_a \times \mathbf{e}_\omega \end{array} \right\}. \tag{4}$$

This axis is a directed line with sense determined from the direction of the change in angular velocity.[3]

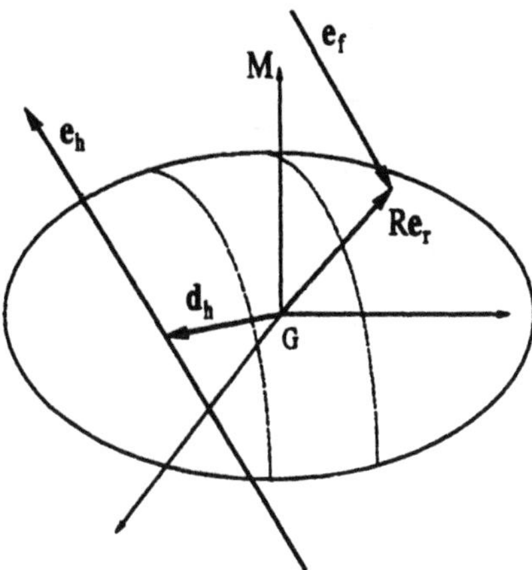

Fig. 1: A rigid body, B, with a force applied at C.

B. The Instantaneous Axis of Rotation of an Impulsive Force

Fig. (1) shows a body, B, with center of mass at G. The body has mass, m, and mass moment of inertia $[J_G]$ at G in the moving frame, M. Apply an impulsive force, $\mathbf{F} = F\hat{\mathbf{e}}_f$, at point C. The position of C is given by $\mathbf{r} = R\hat{\mathbf{e}}_r$.

The change in velocity of the mass center is $d\mathbf{v}_G = \frac{F}{m}\hat{\mathbf{e}}_f$ and in angular velocity is $d\omega = [J_G]^{-1}FR\hat{\mathbf{e}}_r \times \hat{\mathbf{e}}_f$. The change in velocity of a point a distance $\mathbf{h}$ from G is $\mathbf{v}_h = d\mathbf{v}_G + d\omega \times \mathbf{h}$. Set $\mathbf{v}_h$ to zero to get the rotation axis for $\mathbf{F}$. A point on the axis is at

$$\mathbf{d}_h = \frac{([J_G]^{-1}\hat{\mathbf{e}}_r \times \hat{\mathbf{e}}_f) \times \hat{\mathbf{e}}_f}{mR|[J_G]^{-1}\hat{\mathbf{e}}_r \times \hat{\mathbf{e}}_f|^2}, \tag{5}$$

from the origin of M. The direction of the axis is

$$\hat{\mathbf{e}}_h = \frac{[J_G]^{-1}\hat{\mathbf{e}}_r \times \hat{\mathbf{e}}_f}{|[J_G]^{-1}\hat{\mathbf{e}}_r \times \hat{\mathbf{e}}_f|}. \tag{6}$$

The Plücker coordinates defining this line are

$$\hat{\mathbf{L}}_h = \begin{bmatrix} \hat{\mathbf{e}}_h \\ \mathbf{d}_h \times \hat{\mathbf{e}}_h \end{bmatrix}. \tag{7}$$

This axis is a directed line with sense determined from the direction of the rotation caused by $\mathbf{F}$.

III. Implementation to Force Analysis Problems

Transformation of the contact forces and the desired motion into axes of rotation yields a velocity change independent of the angular velocity change. Thus, the impulse-momentum equation can be written as

$$\sum_{i=0}^{n} b_i \hat{\mathbf{L}}_{hi} = \hat{\mathbf{E}}_a. \tag{8}$$

Because $\hat{\mathbf{L}}_{hi}$ and $\hat{\mathbf{E}}_a$ are directed lines, b_i, $i = 0, \cdots, n$, are non-negative scalars. Choose $i = 1, \cdots, n$ as the contact forces. An applied force that is capable of producing the desired motion must satisfy the equation

$$\hat{\mathbf{L}}_{h0} = \frac{1}{b_0}\hat{\mathbf{E}}_a - \sum_{i=1}^{n} a_i \hat{\mathbf{L}}_{hi}, \tag{9}$$

where $a_i = \frac{b_i}{b_0}$, for $i = 1, \cdots, n$. Eq. (9) implies that the applied force must be within the convex hull formed by the negative contact forces and the desired motion profile.

An algorithm for determining a force that may provide the desired motion trajectory requires the following steps.

- Determine the instantaneous rotation axis for the desired motion.
- From the contact forces and limiting friction coefficients determine the ranges of force directions at the points of contact. Map the negative contact forces into instantaneous rotation axes.
- Construct the convex hull, CH, from the instantaneous axes of rotation.
- Select a force that has its axis of rotation, $\hat{\mathbf{L}}_{h0}$, in CH.

Note that this algorithm does not guarantee a force that produces the desired motion, because the convex hull condition is necessary but not sufficient.

To determine an applied force direction $\mathbf{f}$ from an axis of rotation, $\hat{\mathbf{L}}$, note that the force must be perpendicular to $[J_G]\mathbf{l}$ and $\mathbf{l}^o \times \mathbf{l}$. Thus the direction of $\mathbf{f}$ is the cross-product of these two vectors. The sense is determined from the sense of $\hat{\mathbf{L}}$.

A. Contact Force Configurations

A point of contact typically experiences a normal and a frictional force. If the line of action of the frictional force is known the combined force at the contact point is within a wedge of a planar pencil. (This model is valid for hard friction contact as defined in [4] and [5].) If the plane of action of the frictional force is known, the combined force at the point of contact is within a cone of a line bundle.

1) A Planar Pencil of Forces: A planar pencil of lines through a point is shown in Fig. (2). If $\mathbf{e}_1$ and $\mathbf{e}_2$ are the bounds of the force direction, any force in the wedge can be written as $\mathbf{e}_f = a\mathbf{e}_1 + (1-a)\mathbf{e}_2$, where $0 \leq a \leq 1$. The direction of the axis of rotation is

$$\mathbf{e}_h = a[J_G]^{-1}\hat{\mathbf{e}}_r \times \hat{\mathbf{e}}_1 + (1-a)[J_G]^{-1}\hat{\mathbf{e}}_r \times \hat{\mathbf{e}}_2. \tag{10}$$

Thus, the direction of the rotation axis varies proportionally with a.

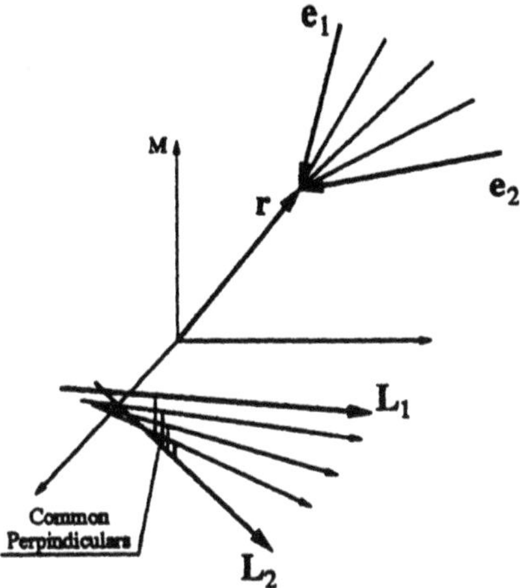

Fig. 2: A wedge of friction applied at a point is a section of a planar pencil of lines.

A position vector from the origin to a point on the axis of rotation is given by

$$\mathbf{d}_h = \frac{\hat{\mathbf{e}}_h \times (a\hat{\mathbf{e}}_1 + (1-a)\hat{\mathbf{e}}_2)}{\hat{\mathbf{e}}_h \bullet \hat{\mathbf{e}}_h} \tag{11}$$

The displacement vector to a point on each axis traces a quadratic curve in a plane perpindicular to the directions of the acceleration axes. A portion of a line congruence can be used to estimate the boundaries of the convex hull that includes this ruled surface.

An interesting feature of this line congruence is that the common perpendiculars between all the lines in this line congruence are parallel. This result is found by taking the cross product of the Plucker vectors formed from Eqs. (10, 11). The direction of the common perpindiculars is $\hat{\mathbf{e}}_p = ([J_G]^{-1}\hat{\mathbf{e}}_r \times \hat{\mathbf{e}}_1) \times ([J_G]^{-1}\hat{\mathbf{e}}_r \times \hat{\mathbf{e}}_2)$.

2) A Bundle of Lines of Force: The result for a bundle of force directions is similar, though slightly more complicated than the result for a planar pencil of force directions. The form of the rotation axes is obtainable from the line congruence found for the planar pencil. The mapping takes the form of a continuous set of quadratics connected by a line. The convex hull is within a line complex. Select five linearly independent lines to determine a conservative representation of the bounds on the convex hull. This representation is the same in principal as that of Salisbury and Roth [6] who approximate the friction cone at a point contact as an n-sided pyramid.

3) A Force Directed through the Center of Mass: An important point force is a force line that passes through the center of mass. Gravity has this line of force so it applies to most problems. The axis of rotation is at the plane at infinity.

Planar Results	Spatial Results
Use instantaneous acceleration center	Use instantaneous rotation axis
Force at center of percussion	Force along axis of percussion [7]
Convex hull of points in 2D	Convex hull of lines in 6D
Result is force direction	Result is force direction

TABLE 1: COMPARISON OF PLANAR AND SPATIAL TECHNIQUES.

B. *Comparison of Planar and Spatial Results*

As noted earlier this technique extends the planar graphical dynamic analysis of Brost and Mason to spatial systems. Tb. (1) lists comparable features from each technique.

IV. Convex-Hull of a Set of Lines

The convex hull of a set of lines is needed to determine possible forces that can produce the desired motion. The convex hull of a set of points in a plane is the smallest polygon that contains all the points. Algebraically, the convex hull satisfies the equation $\mathbf{y} = \sum_{i=1}^{n} \alpha_i \mathbf{x}_i$, where $\mathbf{x}_i$, $i = 1, \cdots, n$ are the vertices of the polygon. The scalars, α_i, $i = 1, \cdots, n$ satisfy the equation $\sum_{i=1}^{n} \alpha_i = 1$, and the inequalities $\alpha_i \geq 0$, for all i.

A similar definition can be used for lines. The convex hull of a set of lines can be defined algebraically as composing those lines that satisfy the equation

$$\hat{\mathbf{p}} = \sum_{i=1}^{n} \alpha_i \hat{\mathbf{L}}_i, \tag{12}$$

where $\hat{\mathbf{L}}_i$, $i = 1, \cdots, n$, are the lines that define the boundary of the convex hull. Because lines are defined with homogeneous coordinates, the only conditions on the scalars are the inequalities $\alpha_i \geq 0$, for all i.

A. *Determination of the Convex Hull*

In plane geometry the boundary of the convex hull can be determined by marching counterclockwise so that all are left of or on the boundary.

A similar technique can be used to define the convex hull of a set of lines. To determine the convex hull of a group of lines, however, choose a linear complex as the boundary and a line as the vertex. A linear complex is determined from all the linear combinations of five linearly independent lines in space. *Any set of five linearly independent lines belong to only one linear complex.*[2]

The linear complex can be written with the coordinates of the polar line in the tangent space to the linear complex. Define this space as $\hat{\mathbf{a}} = \{a_1, a_2, a_3, a_4, a_5, a_6\}$, where for five lines $\hat{\mathbf{L}}_i$, the equations $\hat{\mathbf{a}} \bullet \hat{\mathbf{L}}_i = 0$ for $i = 1, \cdots, 5$, are satisfied.

Given a linear complex, $\hat{\mathbf{a}}$, each line, $\hat{\mathbf{L}}_j$ in the space can be classified as: 1) *above* the linear complex, if it satisfies the inequality $\hat{\mathbf{a}} \bullet \hat{\mathbf{L}}_j > 0$, 2) *below* the linear complex, if it satisfies the inequality

$$\hat{\mathbf{a}} \bullet \hat{\mathbf{L}}_j < 0, \tag{13}$$

or 3) ***dependent*** on the linear complex, if it satisfies the equation $\hat{\mathbf{a}} \bullet \hat{\mathbf{L}}_j = 0$.

The convex hull is found by an ordered march such that all lines in the set remain below each linear complex defining the boundary. To determine the convex hull:

- Find a linear complex such that all lines not on it are below it.
- For each of the linear congruences of the linear complex, construct a second linear complex that is above all the other lines in the set.
- Continue until all the linear complexes that form boundaries of the convex hull are determined.

B. Lower Order Spaces

The checks for the boundaries for lower order spaces use distance measures. The convex hull of a linear ruled surface and a regulus are considered here.

1) The Convex Hull within a Linear Ruled Surface: Consider three lines, $\hat{\mathbf{A}}$, $\hat{\mathbf{B}}$, and $\hat{\mathbf{C}} = b_1\hat{\mathbf{A}} + b_2\hat{\mathbf{B}}$. A two part procedure can be used to determine if $\hat{\mathbf{C}}$ is in the convex hull of $\hat{\mathbf{A}}$ and $\hat{\mathbf{B}}$. First determine four directoricies to $\hat{\mathbf{A}}$ and $\hat{\mathbf{B}}$: $\hat{\mathbf{a}}_i$, $i = 1, \cdots 4$.[2] $\hat{\mathbf{C}}$ is linearly dependent on $\hat{\mathbf{A}}$ and $\hat{\mathbf{B}}$, if $\hat{\mathbf{a}}_i \bullet \hat{\mathbf{C}} = 0$, for all i. Second, note that if the lines are linearly dependent, the common perpindicular between each of the three pairs is the same. If $\hat{\mathbf{C}}$ is within the convex hull of $\hat{\mathbf{A}}$ and $\hat{\mathbf{B}}$, it must lie between them. This occurs if the elliptical distance of the normalized lines [3] satisfy the inequalities

$$Real(\hat{\mathbf{A}} \bullet \hat{\mathbf{B}}) \;\leq\; Real(\hat{\mathbf{A}} \bullet \hat{\mathbf{C}}), \quad Real(\hat{\mathbf{A}} \bullet \hat{\mathbf{B}}) \;\leq\; Real(\hat{\mathbf{B}} \bullet \hat{\mathbf{C}}). \tag{14}$$

2) The Convex Hull within a Regulus: Consider four lines, $\hat{\mathbf{A}}$, $\hat{\mathbf{B}}$, $\hat{\mathbf{C}}$, and $\hat{\mathbf{D}}$. To determine if $\hat{\mathbf{D}}$ is in the convex hull of $\hat{\mathbf{A}}$, $\hat{\mathbf{B}}$, and $\hat{\mathbf{C}}$, first determine three directoricies, $\hat{\mathbf{a}}_i$, $i = 1, \cdots 3$, for the regulus that includes $\hat{\mathbf{A}}$, $\hat{\mathbf{B}}$, and $\hat{\mathbf{C}}$. $\hat{\mathbf{D}}$ is linearly dependent on $\hat{\mathbf{A}}$, $\hat{\mathbf{B}}$, and $\hat{\mathbf{C}}$, if $\hat{\mathbf{a}}_i \bullet \hat{\mathbf{D}} = 0$, for all i.

Second, note that if $\hat{\mathbf{D}} = b_1\hat{\mathbf{A}} + b_2\hat{\mathbf{B}} + b_3\hat{\mathbf{C}}$, the common perpindiculars, $\hat{\mathbf{A}} \times \hat{\mathbf{B}}$, $\hat{\mathbf{A}} \times \hat{\mathbf{C}}$, and $\hat{\mathbf{A}} \times \hat{\mathbf{D}}$ have the same common perpindicular. Similarly $\hat{\mathbf{A}} \times \hat{\mathbf{B}}$, $\hat{\mathbf{B}} \times \hat{\mathbf{C}}$, and $\hat{\mathbf{B}} \times \hat{\mathbf{D}}$ have the same common perpindicular, and $\hat{\mathbf{A}} \times \hat{\mathbf{C}}$, $\hat{\mathbf{B}} \times \hat{\mathbf{C}}$, and $\hat{\mathbf{C}} \times \hat{\mathbf{D}}$ have the same common perpindicular. If $\hat{\mathbf{D}}$ is in the convex hull of $\hat{\mathbf{A}}$, $\hat{\mathbf{B}}$, and $\hat{\mathbf{C}}$, the perpindiculars formed with $\hat{\mathbf{D}}$ must for all cases be inside the other two common perpindiculars. Thus, $\hat{\mathbf{D}}$ is in the convex hull of $\hat{\mathbf{A}}$, $\hat{\mathbf{B}}$, and $\hat{\mathbf{C}}$ if

$$\begin{aligned}
Real[(\hat{\mathbf{A}} \times \hat{\mathbf{B}}) \bullet (\hat{\mathbf{A}} \times \hat{\mathbf{C}})] &\leq Real[(\hat{\mathbf{A}} \times \hat{\mathbf{D}}) \bullet (\hat{\mathbf{A}} \times \hat{\mathbf{C}})], \\
Real[(\hat{\mathbf{A}} \times \hat{\mathbf{B}}) \bullet (\hat{\mathbf{A}} \times \hat{\mathbf{C}})] &\leq Real[(\hat{\mathbf{A}} \times \hat{\mathbf{B}}) \bullet (\hat{\mathbf{A}} \times \hat{\mathbf{D}})], \\
Real[(\hat{\mathbf{A}} \times \hat{\mathbf{B}}) \bullet (\hat{\mathbf{B}} \times \hat{\mathbf{C}})] &\leq Real[(\hat{\mathbf{B}} \times \hat{\mathbf{D}}) \bullet (\hat{\mathbf{B}} \times \hat{\mathbf{C}})], \\
Real[(\hat{\mathbf{A}} \times \hat{\mathbf{B}}) \bullet (\hat{\mathbf{B}} \times \hat{\mathbf{C}})] &\leq Real[(\hat{\mathbf{A}} \times \hat{\mathbf{B}}) \bullet (\hat{\mathbf{B}} \times \hat{\mathbf{D}})], \\
Real[(\hat{\mathbf{A}} \times \hat{\mathbf{C}}) \bullet (\hat{\mathbf{B}} \times \hat{\mathbf{C}})] &\leq Real[(\hat{\mathbf{A}} \times \hat{\mathbf{C}}) \bullet (\hat{\mathbf{C}} \times \hat{\mathbf{D}})], \quad \text{and} \\
Real[(\hat{\mathbf{A}} \times \hat{\mathbf{C}}) \bullet (\hat{\mathbf{B}} \times \hat{\mathbf{C}})] &\leq Real[(\hat{\mathbf{C}} \times \hat{\mathbf{D}}) \bullet (\hat{\mathbf{B}} \times \hat{\mathbf{C}})].
\end{aligned} \tag{15}$$

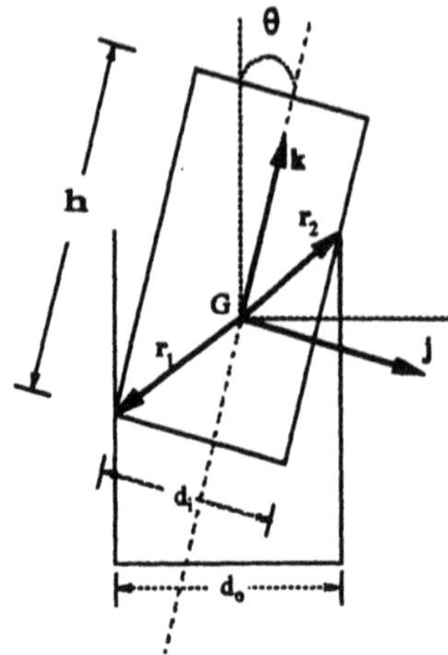

Fig. 3: Insertion of a Cylindrical Peg into a Cylindrical Hole

V. Case Study: Cylindrical Part Insertion

Fig. (3) shows a peg with diameter d_i inserted into a cylindrical hole with diameter d_o. The part is inserted at angle θ about the x-axis, and jams between the sides of the hole. A force needs to be applied to unjam the part. A range of acceptable forces can be determined using acceleration axes.

The two points of contact, $\mathbf{r}_1$ and $\mathbf{r}_2$, are determined from the geometry of the problem. With a body-fixed coordinate frame at the center of the object, they are

$$\mathbf{r}_1 = -\tfrac{d_i}{2}\mathbf{j} - \tfrac{h}{2}\mathbf{k}, \quad \mathbf{r}_2 = \tfrac{d_i}{2}\mathbf{j} + s\mathbf{k}, \tag{16}$$

where h is the height of the object and $s = \frac{d_o - d_i \cos\theta}{\sin\theta}$.

The five limiting lines of force at each contact point are determined from the normal surface of contact and the coefficient of dry friction μ. They are

$$\begin{aligned}
\mathbf{f}_{11} &= cos\theta\mathbf{j} - sin\theta\mathbf{k}, \quad \mathbf{f}_{12} = \mathbf{f}_{11} - \mu\sin\theta\mathbf{j} + \mu\cos\theta\mathbf{k},\\
\mathbf{f}_{13} &= \mathbf{f}_{11} + \mu\sin\theta\mathbf{j} - \mu\cos\theta\mathbf{k}, \quad \mathbf{f}_{14} = \mathbf{f}_{11} + \mu\mathbf{i}, \quad \mathbf{f}_{15} = \mathbf{f}_{11} - \mu\mathbf{i},\\
\mathbf{f}_{21} &= -\mathbf{j}, \quad \mathbf{f}_{22} = \mathbf{f}_{21} + \mu\mathbf{i}, \quad \mathbf{f}_{23} = \mathbf{f}_{21} - \mu\mathbf{i},\\
\mathbf{f}_{24} &= \mathbf{f}_{21} + \mu\mathbf{k}, \quad \mathbf{f}_{25} = \mathbf{f}_{21} - \mu\mathbf{k}.
\end{aligned} \tag{17}$$

The gravitational force through the center of mass acts in the direction $\mathbf{f}_3 = sin\theta\mathbf{j} - cos\theta\mathbf{k}$.

Set $\mu = 0.25$, $\theta = 5^o$, $d_i = 3.98$ cm, d_o= 4.0 cm, $h = 12.0$ cm, and use a mass of 1.2kg. The acceleration axes for the negative gravitational and frictional contact forces are found from Eqs. (6, 7):

$$
\begin{array}{rllllllll}
\hat{\mathbf{L}}_{11} = \{ & 1.00 & 0.00 & 0.00; & 0.00 & -26.66 & 2.33 & \} \\
\hat{\mathbf{L}}_{12} = \{ & 1.00 & 0.00 & 0.00; & 0.00 & -35.11 & -5.83 & \} \\
\hat{\mathbf{L}}_{13} = \{ & 1.00 & 0.00 & 0.00; & 0.00 & -23.41 & 7.73 & \} \\
\hat{\mathbf{L}}_{14} = \{ & 0.94 & -0.34 & 0.11; & -8.32 & -21.97 & 1.69 & \} \\
\hat{\mathbf{L}}_{15} = \{ & 0.94 & 0.34 & -0.11; & 8.32 & -21.97 & 1.69 & \} \\
\hat{\mathbf{L}}_{21} = \{ & -1.00 & 0.00 & 0.00; & 0.00 & -26.66 & 2.33 & \} \\
\hat{\mathbf{L}}_{22} = \{ & -0.92 & -0.35 & -0.13; & -12.32 & 32.71 & -0.51 & \} \\
\hat{\mathbf{L}}_{23} = \{ & -0.92 & 0.35 & 0.13; & 12.32 & 32.71 & -0.51 & \} \\
\hat{\mathbf{L}}_{24} = \{ & -1.00 & 0.00 & 0.00; & 0.00 & 56.51 & -14.1 & \} \\
\hat{\mathbf{L}}_{25} = \{ & -1.00 & 0.00 & 0.00; & 0.00 & 33.11 & 8.28 & \} \\
\hat{\mathbf{L}}_{3} = \{ & 0.00 & 0.00 & 0.00; & 0.00 & 0.09 & -0.99 & \}.
\end{array}
\tag{18}
$$

One logical motion that would unjam the peg is a positive rotation about the x-axis at $\mathbf{r}_1$. The acceleration axis for this motion is $\hat{\mathbf{L}}_{41} = \{1.00, 0.00, 0.00; 0.00, 6.00, -1.99\}$. Use Equation 13 to find that a force producing the desired motion must lie within the convex hull bounded by $\hat{\mathbf{L}}_{11}$, $\hat{\mathbf{L}}_{12}$, $\hat{\mathbf{L}}_{13}$, $\hat{\mathbf{L}}_{21}$, $\hat{\mathbf{L}}_{23}$, $\hat{\mathbf{L}}_{24}$, $\hat{\mathbf{L}}_{25}$, $\hat{\mathbf{L}}_{3}$ and $\hat{\mathbf{L}}_{41}$. Note that when the opposite motion is specified the peg jams into the side of the hole. The convex hull defining such a set of forces is bounded by $\hat{\mathbf{L}}_{11}$, $\hat{\mathbf{L}}_{12}$, $\hat{\mathbf{L}}_{13}$, $\hat{\mathbf{L}}_{21}$, $\hat{\mathbf{L}}_{22}$, $\hat{\mathbf{L}}_{24}$, $\hat{\mathbf{L}}_{25}$, $\hat{\mathbf{L}}_{3}$ and $\hat{\mathbf{L}}_{42}$, where $\hat{\mathbf{L}}_{42} = \{-1.00, 0.00, 0.00; 0.00, -6.00, 1.99\}$.

VI. Case Study: Force Closure Grasp

A force closure grasp is a combination of forces that can resist any externally applied force. Four contact forces with hard friction are sufficient to achieve force closure. [8] [9] For some object geometries, three are sufficient. Force closure occurs when a line of force from each of three [four] friction cones intersect. [10]

The transformation of forces into acceleration axes can be applied to this problem. A minimum of seven lines are required to span the space of acceleration axes. The convex-hull of the acceleration axes span the space if at least six lines are linearly independent and

$$\sum_{i=1}^{N} \alpha_i \mathbf{L}_{hi} = 0, \tag{19}$$

where $\alpha_i \geq 0$ for all i.[11]

To obtain a force closure for a three (or four) finger grasp, select two (or three) grasp positions. Use Eq. (19) to find a third (or fourth) grasp position from an acceleration axis that spans the space of lines in three dimensions.

For instance, consider a cube with two inch sides. This object can obtain a force closure with three fingers. Choose two finger locations at opposite sides of the cube, $\mathbf{R} = \{\pm 1, 0.5, 0\}$. The normal forces are $\mathbf{F}_n = \{\mp 1, 0, 0\}$. The frictional forces are in the plane of the cube surface. Ten acceleration axes are determined from the normal force and the four extremes of the frictional force. These ten axes meet the closure condition in Eq. (19). Only five, however, are linearly independent. Therefore, a third grasp point must be found. Choose a frictional force with an acceleration axis orthogonal to five of the acceleration axes determined by the other two grasp points and tangent to and at a surface of the cube. The contact point for this force yields a full force-closure. Any frictional contact force applied at $\mathbf{R}_3 = \{0, -1, r_z\}$, where $-1.0 < r_z < 1.0$, satisfies the requirements and yields force closure.

Conclusion

This paper has presented a technique to determine a necessary condition an applied force must satisfy to perform a desired motion among multiple frictional contact forces with unknown magnitudes and known direction bounds. This condition occurs frequently in robotics and assembly problems. The technique extends to spatial problems a graphical solution Brost and Mason develop for planar problems. The technique is limited in that it provides a necessary, but not sufficient, condition. This limit results from the uncertainties inherent in the problem. Force closure grasps are determined with this technique by noting that the convex hull of the acceleration axes determined from the frictional contact forces at the grasp points must span all the lines of R^3.

References

[1] R. C. Brost and M. T Mason. Graphical Analysis of Planar Rigid-body Dynamics with Multiple Frictional Contacts. *Robotics Research, Fifth International Symposium*, 293–300 (1988).

[2] O. Veblen and J. W. Young. *Projective Geometry.* Ginn and Company, Boston (1910).

[3] O. Bottema and B. Roth. *Theoretical Kinematics.* North Holland Publ., Amsterdam (1979).

[4] J. Kerr and B. Roth. Analysis of Multifingered Hands. *Int. J. of Robotics Research*, 4(4):3–17 (1985).

[5] V.-D. Nguyen. Constructing Force Closure Grasps. *Int. J. of Robotics Research*, 7(3):3–16 (1988).

[6] J. K. Salisbury and B. Roth. Kinematics and Analysis of Articulated Mechanical Hands. *ASME J. of Mech., Trans., and Auto. in Des.*, ASME-105:35–41 (1983).

[7] W. D. MacMillan. *Dynamics of Rigid Bodies.* Dover Publications, New York, USA (1936).

[8] F. Reuleaux. *The Kinematics of Machinery.* Dover Press, New York, USA (1875).

[9] X. Markenscoff, L. Ni, and C. H. Papadimitiou. The Geometry of Grasping. *Int. J. of Robotics Research*, 9(1):61–74 (1990).

[10] J. Ponce, S. Sullivan, J.-D. Boissonnat, and J.-P. Merlet. On Characterizing and Computing Three- and Four-Finger Force-Closure Grasps on Polyhedral Objects. *Proc. IEEE Robotics and Automation Conf.*, 2:821–827 (1993). Atlanta, USA.

[11] A. J. Goldman and A. W. Tucker. Polyhedral Convex Cones. *Linear Inequalities and Related Systems.* (H. W. Kuhn and A. W. Tucker, Eds.) Princeton Univ. Press (1956).

An Inverse Design Algorithm for a G^2 Interpolating Spline Motion

Q. Jeffrey Ge
Department of Mechanical Engineering
State University of New York at Stony Brook
Stony Brook, NY 11794-2300, USA
E-mail: ge@design.eng.sunysb.edu

Abstract This paper deals with smooth motion interpolation. Recently, a direct construction algorithm was developed for generating rigid body motions with second order geometric continuity (G^2). The present paper shows how the G^2 spline motion can be used to fulfill the task of motion interpolation by solving the problem of inverse design for the G^2 spline motion. The results are useful for computer graphics, mechanical systems animation, and Cartesian trajectory generation for robot manipulators.

Introduction

This paper deals with the problem of motion interpolation and approximation for Computer Aided Animation. The general problem of motion approximation is to find a smooth motion of an object that approximates a given set of configurations or displacements of the object. The motion is said to interpolate the configurations (called *key configurations*) if it allows the object to pass through them.

There are two basic issues in synthesizing motions of a rigid object for computer animation. One is kinematic in nature and is concerned with representation of object configurations. The other is computational geometric in nature and involves interpolation and approximation in the space of configurations. Various representations of configuration space have been explored by many researchers in areas such as Kinematics, Mechanics, and Robotics (see, for example, Arnold (1979), Bottema and Roth (1990), and Lotombe (1991)). This paper follows a series of three papers by Ge and Ravani (1994a, 1994b, and 1993) and uses an *Image Space* representation (see Ravani and Roth 1984) of the configuration space for motion approximation. The Image Space differs from other representations of configuration space in that it captures the group structure of Euclidean displacements. It was made orientable by Ge and Ravani (1994a) to include topological considerations and was used to transform the motion approximation problem into a curve design problem so that ideas and techniques in CAGD may be applied. Ge and Ravani (1994b) used a deCasteljau-type method in conjunction with screw motion interpolant to construct Bézier motions. It also provided answers to several questions raised by Shoemake (1985) on the characteristics of the resulting "spherical Bézier curves". Ge and Ravani (1993) developed geometric conditions for piecing two motion segments smoothly and obtained a direct algorithm for constructing smooth composite Bézier motions that have similarities to Farin-Boehm construction of direct G^2 splines (see Farin 1993). These works by Ge and Ravani were inspired by the idea of using quaternions for animating rotations (see Shoemake 1985 and Pletinckx

A. J. Lenarčič and B. B. Ravani (eds.), Advances in Robot Kinematics and Computationed Geometry, 81–90.

1989).

The geometric algorithm developed by Ge and Ravani (1993) is most suitable for interactive design of a G^2 spline motion, manipulating the control configurations until a desired motion is obtained. The resulting motion, however, only approximates the given set of control configurations. In some applications, however, it may be desirable to obtain spline motions from an interpolation process rather than from an approximation process. The present paper shows how the G^2 spline motion can be used to fulfill the task of motion interpolation by solving the problem of inverse design for the G^2 spline motion. In addition, it provides a simplified G^2 condition for piecing two Bézier motion segments that speeds up the computation required by the G^2 spline motion.

The organization of the paper is as follows. First a review of the Orientable Image Space is given. Section 2 summarizes a direct algorithm for constructing a G^2 spline motion and presents a simplified condition for G^2 continuity. Section 3 shows how the G^2 spline motion can be used for motion interpolation. The material in section 3 and the simplified G^2 condition in Section 2 are the new contributions of this paper.

1 The Image Space

The Image Space of a spatial displacement introduced by Ravani and Roth (1984) is a projective space with three dual dimensions. The coordinates of the Image Space are defined by a set of four dual numbers $\hat{X}_i = X_i + \epsilon X_i^0$ $(i = 1, 2, 3, 4)$, where ϵ is the dual number unit with the property $\epsilon^2 = 0$. The eight real parameters X_i, X_i^0 $(i = 1, 2, 3, 4)$ are homogeneous coordinates of a screw displacement. Since every spatial displacement or configuration can be reduced to a screw displacement, the Image Space is a representation of the configuration space of a rigid object.

There are two different but equivalent ways to define the Image Space coordinates. When introducing the Image Space, Ravani and Roth (1984) used a set of eight Study parameters where the first four, X_i $(i = 1, 2, 3, 4)$ are the Euler parameters of rotation and the second four X_i^0 $(i = 1, 2, 3, 4)$ are defined in terms of the vector of translation $\mathbf{d} = (d_1, d_2, d_3)$ as

$$\begin{bmatrix} X_1^0 \\ X_2^0 \\ X_3^0 \\ X_4^0 \end{bmatrix} = \frac{1}{2} \begin{bmatrix} 0 & -d_3 & d_2 & d_1 \\ d_3 & 0 & -d_1 & d_2 \\ -d_2 & d_1 & 0 & d_3 \\ -d_1 & -d_2 & -d_3 & 0 \end{bmatrix} \begin{bmatrix} X_1 \\ X_2 \\ X_3 \\ X_4 \end{bmatrix}. \tag{1}$$

Only six of these eight components are independent, for they satisfy the relations

$$X_1^2 + X_2^2 + X_3^2 + X_4^2 = 1, \tag{2}$$

$$X_1 X_1^0 + X_2 X_2^0 + X_3 X_3^0 + X_4 X_4^0 = 0. \tag{3}$$

The Image Space coordinates can be alternatively defined using the so-called dual Euler parameters:

$$\begin{aligned} \hat{X}_1 &= \hat{s}_1 \sin(\hat{\theta}/2), \quad \hat{X}_2 = \hat{s}_2 \sin(\hat{\theta}/2), \\ \hat{X}_3 &= \hat{s}_3 \sin(\hat{\theta}/2), \quad \hat{X}_4 = \cos(\hat{\theta}/2), \end{aligned} \tag{4}$$

where $\hat{\mathbf{s}} = \mathbf{s} + \epsilon \mathbf{s}^0 = (\hat{s}_x, \hat{s}_y, \hat{s}_z)$ is a unit dual vector of Plücker coordinates representing the axis of a screw displacement, and $\hat{\theta} = \theta + \epsilon l$ is a dual angle representing the magnitude of the screw displacement. Only three out of the four dual Euler parameters are independent, for they satisfy the relation

$$\hat{X}_1^2 + \hat{X}_2^2 + \hat{X}_3^2 + \hat{X}_4^2 = 1. \tag{5}$$

Details on Study parameters and dual Euler parameters can be found in Bottema and Roth (1990) and McCarthy (1990).

A spatial displacement may be reduced to a pair of "oppositely oriented" screw displacement whose screw axes are directed lines that occupy the same position in space but with opposite sense of direction (see Bottema and Roth 1990). In order to obtain topologically correct interpolating motions, it is important to keep track of the senses of orientation for a series of screw displacements. For this reason, Ge and Ravani (1994a, 1994b) made the Image Space orientable by associating each point in the Image Space with a sense of orientation.

Geometry of the Orientable Image Space (denoted by Σ) is equivalent to that of an oriented dual unit 3-sphere in a space of four dual dimensions. The dual unit 3-sphere can be viewed as an oriented real unit 3-sphere together with the union of all of its tangent 3-spaces[1]. The oriented unit 3-sphere is a representation of the space of rotations that includes the topological consideration needed for motion interpolation[2]. The tangent 3-space at a point on the 3-sphere represents the space of translations of an object at a fixed orientation.

Two fundamental results concerning the Image Space are as follows: (a) all Image-Space algorithms that satisfy the constraint (5) reduce to Euclidean algorithms (or flat algorithms) if all object configurations share the same orientation; (b) all 3-spherical algorithms can be dualized to obtain Image-Space algorithms. The latter conclusion is based on the more general "transfer principle" developed by A.P. Kotel'nikov and E. Study (see Dimentberg 1965).

The distance between two points, $\hat{\mathbf{X}}$ and $\hat{\mathbf{Y}}$, in Σ is a dual angle, $\hat{\phi} = \phi + \epsilon h$, which can be symbolically given by (see Ge and Ravani 1994a):

$$\hat{\phi} = \cos^{-1}(\hat{\mathbf{X}} \cdot \hat{\mathbf{Y}}) = \cos^{-1}(\hat{X}_1\hat{Y}_1 + \hat{X}_2\hat{Y}_2 + \hat{X}_3\hat{Y}_3 + \hat{X}_4\hat{Y}_4). \tag{6}$$

It is uniquely defined, provided that ϕ is restricted to the range $[0, \pi]$. When $\phi < \pi/2$, the two points $\hat{\mathbf{X}}$ and $\hat{\mathbf{Y}}$ are said to be *similarly oriented.*

A dual 1-sphere in Σ, or a dual great circle, is the Image Space representation of a two-degree-of-freedom screw motion which consists of two independent simple motions about the same axis, a rotation and a translation. A one-degree-of-freedom screw motion maps into a special curve in Σ, which may be called a *unifold 1-sphere.* Of special value

---

[1]McCarthy and Ravani (1986) used the unoriented dual unit 3-sphere and developed the differential kinematics for spatial motions.

[2]The unoriented real 3-sphere is the space of unit quaternions used by Shoemake (1985) for animating rotations.

is the following unifold dual spherical interpolation (see Ge and Ravani 1994b):

$$\hat{\mathbf{b}}(t) = \hat{\mathbf{L}}(\hat{\mathbf{b}}_0, \hat{\mathbf{b}}_1; t) = \frac{\sin((1-t)\hat{\phi})}{\sin\hat{\phi}}\hat{\mathbf{b}}_0 + \frac{\sin(t\hat{\phi})}{\sin\hat{\phi}}\hat{\mathbf{b}}_1, \quad t \in [0,1] \tag{7}$$

where $\hat{\mathbf{b}}_0$ and $\hat{\mathbf{b}}_1$ denote two similarly oriented image points for two configurations of an object and $\hat{\phi}$ is the dual angle between them. Let $\mathbf{d}_0, \mathbf{d}_1, \mathbf{d}(t)$ denote the vectors of translation associated with configurations $\hat{\mathbf{b}}_0, \hat{\mathbf{b}}_1, \hat{\mathbf{b}}(t)$, respectively. When $\phi = 0$, the configurations $\hat{\mathbf{b}}_0$ and $\hat{\mathbf{b}}_1$ differ from each other only by a translation and Eq.(7) can be reduced to the following linear interpolation:

$$\mathbf{d}(t) = (1-t)\mathbf{d}_0 + t\mathbf{d}_1. \tag{8}$$

The unifold spherical interpolation corresponds to a constant-speed screw motion and was used by Ge and Ravani (1994b) in conjunction with a deCasteljau-type algorithm to construct one-degree-of-freedom Bézier motions.

A dual 2-sphere in Σ, or simply a dual sphere, is the Image Space representation of a four-degree-of-freedom line-symmetric motion (see Ge and Ravani 1994a).

2 G^2 Composite Bézier Motions

This section summarizes a geometric algorithm for interactive design of a G^2 composite Bézier motion developed by Ge and Ravani (1993). Also provided is a simplied condition for G^2 continuity of two Bézier motion segments that would speed up the computation required by the G^2 algorithm.

For constructing an open loop motion, we are given a set of $N+3$ configurations $\hat{\mathbf{q}}_{-1}, \hat{\mathbf{q}}_0, \ldots \hat{\mathbf{q}}_{N+1}$. The composite screw motion joining these configurations corresponds to a dual-spherical polygon in the Image Space. The G^2 algorithm first computes a series of $3N+1$ Bézier control configurations $\hat{\mathbf{b}}_0, \hat{\mathbf{b}}_1, \ldots \hat{\mathbf{b}}_{3N}$ and then generates all Bézier motion segments for a given knot sequence. To generate the Bézier control configurations, the algorithm proceeds as follows:

1. Select a set of parameters x_i, y_i, z_i $(i = 0, 1, \ldots, N-1)$ such that

$$x_i + y_i + z_i = 1, \quad \text{and } x_i, y_i, z_i > 0, \quad \text{except } x_0 = z_{N-1} = 0. \tag{9}$$

2. First set

$$\hat{\mathbf{b}}_0 = \hat{\mathbf{q}}_{-1}, \quad \hat{\mathbf{b}}_1 = \hat{\mathbf{q}}_0, \quad \hat{\mathbf{b}}_{3N-1} = \hat{\mathbf{q}}_N, \quad \hat{\mathbf{b}}_{3N} = \hat{\mathbf{q}}_{N+1}. \tag{10}$$

   Next, compute Bézier configurations $\hat{\mathbf{b}}_2, \hat{\mathbf{b}}_{3N-2}$ on the screw motions from $\hat{\mathbf{q}}_0$ to $\hat{\mathbf{q}}_1$ and from $\hat{\mathbf{q}}_{N-1}$ to $\hat{\mathbf{q}}_N$, respectively:

$$\hat{\mathbf{b}}_2 = \hat{\mathbf{L}}(\hat{\mathbf{q}}_0, \hat{\mathbf{q}}_1; y_0), \quad \hat{\mathbf{b}}_{3N-2} = \hat{\mathbf{L}}(\hat{\mathbf{q}}_{N-1}, \hat{\mathbf{q}}_N; x_{N-1}). \tag{11}$$

3. Then compute Bézier configurations on the screw motion from $\hat{\mathbf{q}}_{i-1}$ to $\hat{\mathbf{q}}_i$:

$$\hat{\mathbf{b}}_{3i-2} = \hat{\mathbf{L}}(\hat{\mathbf{q}}_{i-1}, \hat{\mathbf{q}}_i; x_{i-1}), \quad \hat{\mathbf{b}}_{3i-1} = \hat{\mathbf{L}}(\hat{\mathbf{q}}_{i-1}, \hat{\mathbf{q}}_i; x_{i-1} + y_{i-1}). \tag{12}$$

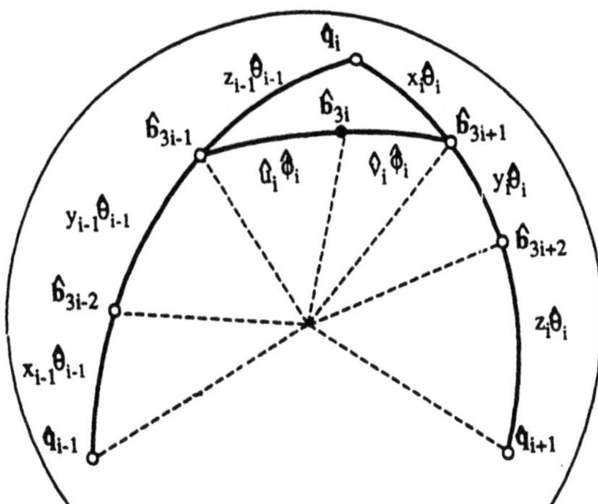

Figure 1: Conditions for G^2 continuity.

4. Finally, find Bézier junction configurations $\hat{\mathbf{b}}_{3i}$ on the screw motion from $\hat{\mathbf{b}}_{3i-1}$ to $\hat{\mathbf{b}}_{3i+1}$ such that G^2 continuity conditions are satisfied.

Essential to the algorithm above are G^2 conditions from which the Bézier junction configurations $\hat{\mathbf{b}}_{3i}$ are determined. These conditions are summarized as follows.

G^1 continuity This requires that the junction configuration $\hat{\mathbf{b}}_{3i}$ to be on the two-degree-of-freedom screw motion from $\hat{\mathbf{b}}_{3i-1}$ to $\hat{\mathbf{b}}_{3i+1}$. In the Image Space Σ, this means that the three corresponding image points lie on the same dual 1-sphere. Let $\hat{\phi}_i = \cos^{-1}(\hat{\mathbf{b}}_{3i-1} \cdot \hat{\mathbf{b}}_{3i+1})$, $\hat{u}_i\hat{\phi}_i = \cos^{-1}(\hat{\mathbf{b}}_{3i-1} \cdot \hat{\mathbf{b}}_{3i})$, and $\hat{v}_i\hat{\phi}_i = \cos^{-1}(\hat{\mathbf{b}}_{3i} \cdot \hat{\mathbf{b}}_{3i+1})$, where $\hat{u}_i + \hat{v}_i = 1$. Then the G^1 condition is given by

$$\sin\hat{\phi}_i\hat{\mathbf{b}}_{3i} = \sin(\hat{v}_i\hat{\phi}_i)\hat{\mathbf{b}}_{3i-1} + \sin(\hat{u}_i\hat{\phi}_i)\hat{\mathbf{b}}_{3i+1}. \tag{13}$$

G^2 continuity This requires two additional conditions. First, five neighboring configurations, $\hat{\mathbf{b}}_{3i-2}, \hat{\mathbf{b}}_{3i-1}, \hat{\mathbf{b}}_{3i}, \hat{\mathbf{b}}_{3i+1}, \hat{\mathbf{b}}_{3i+2}$ must belong to the same four-degree-of-freedom line-symmetric motion. In Σ, this means that the five corresponding image points lie on the same dual 2-sphere (see Figure 1). Secondly, at the junction configuration, the dual numbers $\hat{u}_i, \hat{v}_i$ must satisfy the equation:

$$\frac{\hat{v}_i \sin(\hat{v}_i\hat{\phi}_i)}{\hat{u}_i \sin(\hat{u}_i\hat{\phi}_i)} = \frac{y_{i-1}\hat{\theta}_{i-1}\sin(x_i\hat{\theta}_i)}{y_i\hat{\theta}_i\sin(z_{i-1}\hat{\theta}_{i-1})}. \tag{14}$$

Eq.(14) is a simplification of the original condition for G^2 continuity presented in Ge and Ravani (1993). The dual-number parameters $\hat{u}_i, \hat{v}_i$ can be solved from (14) by obtaining their real and dual parts separately (see Ge and Ravani 1993).

We note that when $\theta_i, \phi_i \to 0$, Eq.(14) reduces to

$$\left(\frac{\hat{v}_i}{\hat{u}_i}\right)^2 = \left(\frac{x_i}{y_i}\right)\left(\frac{y_{i-1}}{z_{i-1}}\right). \tag{15}$$

Eq.(15) is equivalent to the Boehm's curvature continuity condition for G^2 cubic spline curves (see Boehm 1985). Let $\mathbf{d}_{-1}, \mathbf{d}_0, \ldots, \mathbf{d}_{N+1}$ denote the translation vectors associated with the control configurations $\hat{\mathbf{q}}_{-1}, \hat{\mathbf{q}}_0, \ldots \hat{\mathbf{q}}_{N+1}$. In this special case, the geometric algorithm for designing G^2 spline motions reduces to the Farin-Boehm construction (see Farin 1993) of direct G^2 splines with $\mathbf{d}_{-1}, \mathbf{d}_0, \ldots, \mathbf{d}_{N+1}$ as control points.

3 G^2 Spline Motion Interpolation

This section shows how the G^2 algorithm summarized in the previous section can be inverted to fulfill the task of motion interpolation by solving the following inverse design problem:

Given: A set of $N+1$ key configurations $\hat{\mathbf{p}}_i$ $(i = 0, 1, \ldots, N)$ and a set of parameters x_i, y_i, z_i that satisfy conditions (9).

Find: A set of $N+3$ control configurations $\hat{\mathbf{q}}_{-1}, \hat{\mathbf{q}}_0, \ldots \hat{\mathbf{q}}_{N+1}$ such that the resulting G^2 composite Bézier motion interpolates through all the key configurations.

The formulation of the interpolation problem is straightforward. First, we set

$$\hat{\mathbf{b}}_{3i} = \hat{\mathbf{p}}_i; \quad i = 0, \ldots, N. \tag{16}$$

Recall that the Bézier configurations $\hat{\mathbf{b}}_{3i\pm1}$ are related to the key configurations $\hat{\mathbf{p}}_i$ and the control configurations $\hat{\mathbf{q}}_i$ by (see Figure 1):

$$\sin\hat{\phi}_i \hat{\mathbf{p}}_i = \sin(\hat{v}_i\hat{\phi}_i)\hat{\mathbf{b}}_{3i-1} + \sin(\hat{u}_i\hat{\phi}_i)\hat{\mathbf{b}}_{3i+1}, \quad i = 1, \ldots, N-1, \tag{17}$$

$$\sin\hat{\theta}_{i-1}\hat{\mathbf{b}}_{3i-1} = \sin(z_{i-1}\hat{\theta}_{i-1})\hat{\mathbf{q}}_{i-1} + \sin((x_{i-1}+y_{i-1})\hat{\theta}_{i-1})\hat{\mathbf{q}}_i, \quad i = 2, \ldots, N-1, \tag{18}$$

$$\sin\hat{\theta}_i\hat{\mathbf{b}}_{3i+1} = \sin((y_i+z_i)\hat{\theta}_i)\hat{\mathbf{q}}_i + \sin(x_i\hat{\theta}_i)\hat{\mathbf{q}}_{i+1}, \quad i = 1, \ldots, N-2. \tag{19}$$

We obtain the relationships between the unknown $\hat{\mathbf{q}}_i$ and the known $\hat{\mathbf{p}}_i$ by substituting (18) and (19) into (17):

$$\hat{A}_i\hat{\mathbf{q}}_{i-1} + \hat{B}_i\hat{\mathbf{q}}_i + \hat{C}_i\hat{\mathbf{q}}_{i+1} = \hat{\mathbf{p}}_i, \quad i = 1, \ldots, N-1, \tag{20}$$

where

$$\begin{aligned}
\hat{A}_i &= \frac{\sin(z_{i-1}\hat{\theta}_{i-1})\sin(\hat{v}_i\hat{\phi}_i)}{\sin\hat{\theta}_{i-1}\sin\hat{\phi}_i}, \\
\hat{B}_i &= \frac{\sin((x_{i-1}+y_{i-1})\hat{\theta}_{i-1})\sin(\hat{v}_i\hat{\phi}_i)}{\sin\hat{\theta}_{i-1}\sin\hat{\phi}_i} + \frac{\sin((y_i+z_i)\hat{\theta}_i)\sin(\hat{u}_i\hat{\phi}_i)}{\sin\hat{\theta}_i\sin\hat{\phi}_i}, \\
\hat{C}_i &= \frac{\sin(x_i\hat{\theta}_i)\sin(\hat{u}_i\hat{\phi}_i)}{\sin\hat{\theta}_i\sin\hat{\phi}_i}.
\end{aligned} \tag{21}$$

For open-loop motions, special treatment is needed near the end-configurations:

$$\begin{array}{ll} \hat{\mathbf{q}}_{-1} = \hat{\mathbf{p}}_0, & \hat{\mathbf{q}}_0 = \hat{\mathbf{b}}_1 = \text{arbitrary}, \\ \hat{\mathbf{q}}_{N+1} = \hat{\mathbf{p}}_N, & \hat{\mathbf{q}}_N = \hat{\mathbf{b}}_{3N-1} = \text{arbitrary}. \end{array} \tag{22}$$

Assemble all equations in (20) and (22) into the following matrix equation:

$$[M(\hat{\mathbf{Q}})]\hat{\mathbf{Q}} = \hat{\mathbf{P}}, \tag{23}$$

where $\hat{\mathbf{Q}} = [\ \hat{\mathbf{q}}_0 \ \ \hat{\mathbf{q}}_1 \ \ \dots \ \ \hat{\mathbf{q}}_{N-1} \ \ \hat{\mathbf{q}}_N\ ]^T$, $\hat{\mathbf{P}} = [\ \hat{\mathbf{b}}_1 \ \ \hat{\mathbf{p}}_1 \ \ \dots \ \ \hat{\mathbf{p}}_{N-1} \ \ \hat{\mathbf{b}}_{3N-1}\ ]^T$, and

$$[M(\hat{\mathbf{Q}})] = \begin{bmatrix} 1 & & & & & \\ \hat{A}_1 & \hat{B}_1 & \hat{C}_1 & & & \\ & & \ddots & & & \\ & & & \hat{A}_{N-1} & \hat{B}_{N-1} & \hat{C}_{N-1} \\ & & & & & 1 \end{bmatrix}. \tag{24}$$

Eq.(23) is not a linear system since the coefficient matrix $[M(\hat{\mathbf{Q}})]$ depend on the dual angles $\hat{\theta}_i, \hat{\phi}_i$, which in turn depend on the unknowns $\hat{\mathbf{Q}}$. In addition, all the unknown configurations $\hat{\mathbf{q}}_i$ in $\hat{\mathbf{Q}}$ must satisfy the fundamental condition (5).

We propose to solve the set of nonlinear equations (23) using the following iterative method. Assume that the kth iterate for $\hat{\mathbf{Q}}$ is given and is denoted by $\hat{\mathbf{Q}}^k$. We first use the direct G^2 algorithm to estimate the dual angles $\hat{\theta}_i, \hat{\phi}_i$ and dual numbers $\hat{u}_i, \hat{v}_i$. These results are then used to obtain the kth estimate for $[M(\hat{\mathbf{Q}})]$, which is denoted by $[M(\hat{\mathbf{Q}}^k)]$. We obtain the next iterate $\hat{\mathbf{Q}}^{k+1}$ by solving the set of *linear* equations

$$[M(\hat{\mathbf{Q}}^k)]\hat{\mathbf{Q}}^{k+1} = \hat{\mathbf{P}}. \tag{25}$$

The set of equations (25) is a trigiagonal linear system and has a unique solution (after the prescription of two end-configurations $\hat{\mathbf{b}}_1, \hat{\mathbf{b}}_{3N-1}$ that are consistent), if $[M(\hat{\mathbf{Q}})]$ is a *diagonally dominant* matrix, i.e., if

$$\hat{B}_i > \hat{A}_i + \hat{C}_i, \quad i = 1, \dots, N-1. \tag{26}$$

Assuming that the composite screw motion joining all key configurations $\hat{\mathbf{p}}_i$ ($i = 0, \dots, N$) has non-negative pitch, we observe that, in addition to (9), it seems sufficient to require that:

$$x_i \le 1/2, \ \ z_i < 1/2, \text{ or } x_i < 1/2, \ \ z_i \le 1/2, \quad i = 0, \dots, N-1, \tag{27}$$

for (26) to hold. In this case, the unique solution of (25) can be obtained by Gauss elimination without pivoting. Solution methods for tridiagonal systems can be found in Ahlberg et al (1969) and deBoor (1978). The generalization of these methods for dual-number equations (25) is straightforward.

We note that when all key configurations $\hat{\mathbf{p}}_i$ share the same orientation, the coefficients (21) reduce to:

$$\hat{A}_i = z_{i-1}\hat{v}_i, \quad \hat{B}_i = (x_{i-1} + y_{i-1})\hat{v}_i + (y_i + z_i)\hat{u}_i, \quad \hat{C}_i = x_i\hat{u}_i, \tag{28}$$

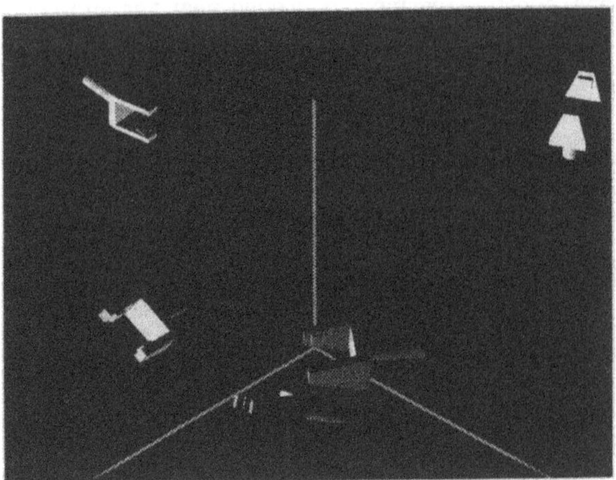

Figure 2: A set of five configurations of an object.

where $\hat{u}_i, \hat{v}_i$ are real parameters which can be obtained from (15) and the relation $\hat{u}_i + \hat{v}_i = 1$. In this case, the problem of G^2 spline motion interpolation reduces to that of G^2 spline interpolation of points in the Euclidean space and can be readily solved. Thus we may use (28) to obtain an initial estimate of $[M(\hat{\mathbf{Q}})]$.

We now outline an algorithm for the inverse design of a G^2 interpolating composite Bézier motion for a given set of parameters x_i, y_i, z_i that satisfy the conditions (9) and (27).

1. Select $\hat{\mathbf{b}}_1, \hat{\mathbf{b}}_{3N-1}$. Use (28) to obtain an initial estimate of $[M(\hat{\mathbf{Q}})]$.

2. Solve equation (25) by a forward substitution and a backward substitution to obtain an estimate of the control configurations $\hat{\mathbf{q}}_i$.

3. Normalize all solutions, $\hat{\mathbf{q}}_i/|\hat{\mathbf{q}}_i| \to \hat{\mathbf{q}}_i$, and then compute the dual angles $\hat{\theta}_i$.

4. Determine all Bézier control configurations $\hat{\mathbf{b}}_i$ from $\hat{\mathbf{q}}_i$ using the direct G^2 algorithm. In the process, compute the dual angles $\hat{\phi}_i$ and obtain $\hat{u}_i, \hat{v}_i$.

5. Compute new coefficients $\hat{A}_i, \hat{B}_i, \hat{C}_i$ using (21) and update equation (25).

6. Compute the dual angle $\hat{\psi}_i = \psi_i + \epsilon\psi_i^0 = \cos^{-1}(\hat{\mathbf{b}}_{3i} \cdot \hat{\mathbf{p}}_i)$. Go back to step 2 until the solution is within an allowable tolerance δ:

$$\sum_{i=0}^{N}(w_i\psi_i^2 + (\psi_i^0)^2) < \delta, \tag{29}$$

where $w_i > 0$ $(i = 0, \ldots, N)$ are positive weight factors.

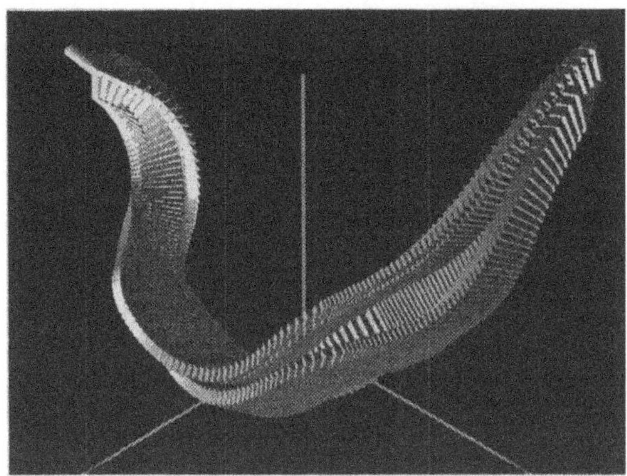

Figure 3: A G^2 composite Bézier motion that interpolates through the five configurations.

For all the test runs that we have conducted, the above algorithm converges rapidly to the correct solution with less than ten iterations. Figure 2 shows a set of five key configurations $\hat{\mathbf{p}}_i$. Figure 3 shows a G^2 composite Bézier motion that interpolates through these key configurations.

Conclusions

In this paper, we have shown how the G^2 spline motion can be used to fulfill the task of motion interpolation by providing an iterative method for solving the problem of inverse design for the G^2 spline motion. The results are useful for computer graphics and CAD/CAM.

Acknowledgment

This work was supported by NSF Research Initiation Award MSS-9396265 to the State University of New York at Stony Brook. Also acknowledged is the assistance of Mr. Donglai Kang in implementing the motion interpolation algorithm.

References

[1] Ahlberg, J., E. Nilson, and J. Walsh, 1967. *The Theory of Splines and their Applications.* Academic Press.

[2] Arnold, V.I., 1979. *Mathematical Methods of Classical Mechanics*, Springer-Verlag.

[3] Boehm, W., 1985. Curvature continuous curves and surfaces, *Computer Aided Geometric Design*, 2:313-323.

[4] Bottema, O. and B. Roth, 1990. *Theoretical Kinematics.* Reprinted Dover Pub.

[5] deBoor, C., 1978. *A Practical Guide to Splines.* Springer-Verlag.

[6] Dimentberg, F.M., 1965, *The Screw Calculus and its Application in Mechanics.* Translated and reproduced by the Clearinghouse for Federal Scientific and Techinical Information, Springfield, VA. Document No. FTD-HT-23-1632-67.

[7] Farin, G., 1993. *Curves and Surfaces for Computer Aided Geometric Design.* 3rd ed., Academic Press, San Diego, 334pp.

[8] Ge, Q.J., and B. Ravani, 1994a. Computer aided geometric design of motion interpolants, to appear in *ASME J. of Mechanical Design.*

[9] Ge, Q.J., and B. Ravani, 1994b. Geometric construction of Bézier Type Motions, to appear in *ASME J. of Mechanical Design.*

[10] Ge, Q.J., and B. Ravani, 1993. Computational geometry and motion approximation. In *Computational Kinematics*, J. Angeles (eds.), Kluwer Academic Publ., Boston, p 229-238.

[11] Latombe, J.-C., 1991. *Robot Motion Planning*, Kluwer Academic Publ., Boston.

[12] McCarthy, J.M., and B. Ravani. 1986, Differential kinematics of spherical and spatial motions using kinematic mapping. *ASME J. of Appl. Mech.*, 53:15-22.

[13] McCarthy, J.M., 1990. *Introduction to Theoretical Kinematics*, MIT Press.

[14] Pletinckx, D., 1989. Quaternion calculus as a basic tool in computer graphics. *The Visual Computer*, 5:2-13.

[15] Ravani, B., and B. Roth, 1984. Mappings of spatial kinematics. *ASME J. of Mech., Transmissions., and Auto. in Design.* 106(3):341-347.

[16] Shoemake, K., 1985. Animating rotation with quaternion curves. *ACM Siggraph*, 19(3):245-254.

Geometry of Parallel-Jaw Gripper Grasps in the Plane

Anil S. Rao[†]
Utrecht University, The Netherlands.

Abstract - Canny and Goldberg [1] observe that in structured robot environments such as a factory, parts manipulation with general purpose robots equipped with complex sensing and control features is often unnecessary. Instead they propose a reduced intricacy in sensing and control (RISC) approach using simple, inexpensive, robust and easily reconfigurable systems. Such systems often have the additional theoretical advantage that planning and control algorithms are simple and *provably* correct.

The parallel-jaw gripper is probably *the* RISC manipulator. This paper considers the parallel-jaw gripper for grasping objects in the plane. Under the right set of assumptions, all of which are realizable in the real world, the problem of grasping an object can be viewed in geometric terms alone. We give an algorithm to compute the "grasp function," a map from the initial pre-grasp orientation of a part to its final orientation after the grasp, from the geometric description of the part boundary. The grasp function is used to plan open-loop grasps of the object in a particular orientation.

1 Introduction

A familiar problem in industrial automation is that of feeding parts: getting parts into a fixed position and orientation. In this paper, we consider planar parts parts which exhibit only planar motion when manipulated. To reduce costs and errors as mentioned in the abstract, it is desirable to perform manipulation without the use of sensors and with a low-complexity manipulator. We assume the parallel-jaw gripper which is a simple two degree-of-freedom gripper. See Fig. 1. A grasp action parameterized by an angle α consists of the following: orienting the gripper at angle α wrt a fixed world frame, closing the jaws as far as possible over the part without deforming it, and then opening the jaws. Note that no sensing is required.

Goldberg and Rao [2, 3] present algorithms that generate optimal open-loop plans (sequence of grasp actions) to orient a planar part. An example of such a plan is shown in Fig. 2.

These papers assume as input the grasp function: a function mapping the initial pre-grasp orientation of the part to its final orientation (Section 2). Except for simple part shapes (polygonal or generalized-polygonal parts, parts with linear-circular edges), they do not give

*This research was supported in part by ESPRIT Basic Research Action No. 6546 (Project PRoMotion)
[†]Department of Computer Science, Email: anil@cs.ruu.nl, Tel: 31-30-535093, FAX: 31-30-513791.

A. J. Lenarčič and B. B. Ravani (eds.), Advances in Robot Kinematics and Computationed Geometry, 91–100.

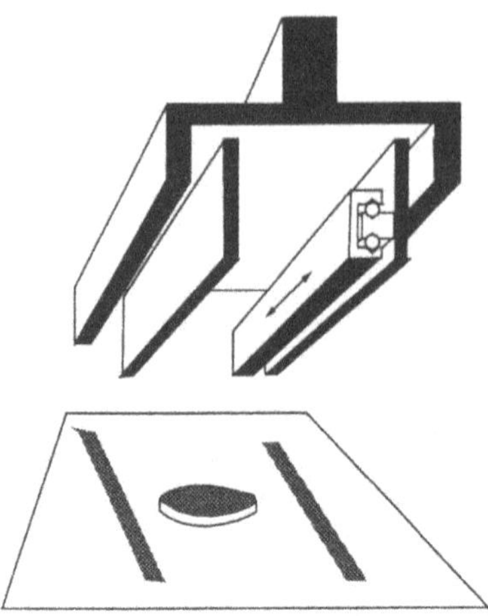

Figure 1: Schematic of parallel-jaw gripper above a part.

algorithms to compute grasp functions from shape descriptions. In [4], we show how to compute the grasp function of an algebraic part – a part whose boundary is composed of algebraic curve segments – given the shape of its boundary in implicit or parametric form.

In this paper we attempt to summarize the results from [2, 5, 4, 3] while focussing on grasp function computations which are purely geometric in nature under appropriate and justifiable mechanical assumptions. We omit details due to space limitations. After describing these assumptions in Section 1.1, we end this section with related work in Section 1.2. Grasp function are introduced in Section 2 and their computation in Section 3. We end with some conclusions in Section 5.

1.1 Geometric and Mechanical Assumptions with Justifications

Part Assumptions The part $\mathcal{P}$ is rigid and of known fixed cross section P which is made up of a sequence of n algebraic (parametrically or implicitly known) arcs. We believe this to be a reasonable restriction since most industrial part boundaries are modelled as piecewise algebraic curves (and surfaces) [6]. If P is not convex, let P refer to its convex hull which may be computed in $O(n)$ time using the algorithm in [7] (given the arcs in order).

Grasp Assumptions We assume a grasping process as in [2, 8]. Briefly, the gripper is based on the ubiquitous parallel-jaw gripper modified to reduce friction to negligible levels [9]. The part is always flat on a horizontal table between the jaws of the gripper. The gripper has two degrees of freedom: in addition to the opening/closing motion of the jaws, they can also together be oriented arbitrarily with respect to the table. As described before, a (squeeze) grasp action begins with the jaws widely separated and oriented at some angle with the table. See Fig. 3. Then they move towards each other and make opposing and simultaneous contact (pure

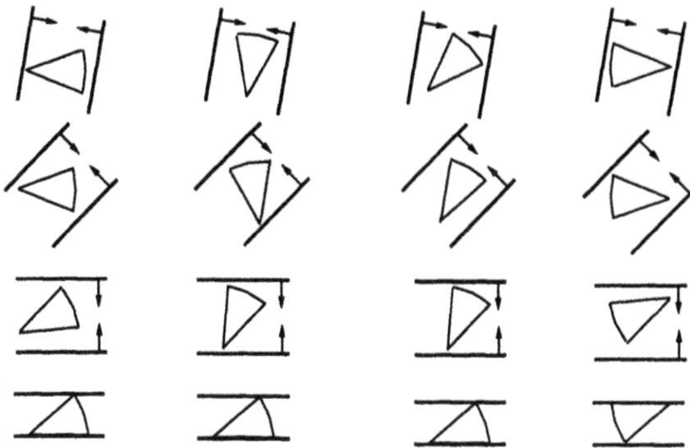

Figure 2: Top view of a three-stage plan for orienting a pie-shaped part with half angle 20° into one of two orientations. Four traces of the plan (running from top to bottom) are shown. Note that the part's initial orientation is different in each trace (top of each column). The gripper orientation at each stage, indicated by two parallel-lines, is the same for each trace; yet the same desired grasp configuration is achieved in all cases.To obtain a single final orientation, push-grasp actions may be used. See Fig. 7.

squeezing) with the part at points at which the jaws are clearly tangent to the part. As the jaws continue closing motion, the part exhibits mechanical compliance changing its orientation. The part never wedges because of negligible friction between part and gripper. Gripper motion stops when further motion will violate part rigidity. The jaws then move apart. Gripper and part motion occurs in the plane and is slow enough that inertial forces are negligible (quasi-static motion as discussed in [10, 11]).

The assumption of simultaneous contact, almost never achieved in practice (Especially without sensors!) is made only for simplicity of presentation. In practice, push-grasp actions [12, 3] would be used. This is briefly discussed in Section 5.

1.2 Related Work

There is a substantial literature on the subject of grasping. See [13] and [14] for reviews.

For parallel-jaw grippers with polygonal parts, Jameson [15] defined a stability condition similar to that of [16]. Building on the work of Mason [17], Brost [12] distinguished between stable and wedged configurations for polygonal parts and gave an algorithm for achieving a stable grasp with a parallel-jaw gripper when the part's initial orientation can be described by a tolerance interval.

The idea of using a sequence of pushing or grasping motions to reduce uncertainty in part orientation was addressed by [18, 19, 20, 21]. Although each of these planning algorithms use realistic models of mechanics, each used heuristics to search for plan strategy, none of which are guaranteed to find a plan in polynomial time. Natarajan [22] ignored the mechanics of parts feeders and focused on the computational problem of planning with a given set of transfer

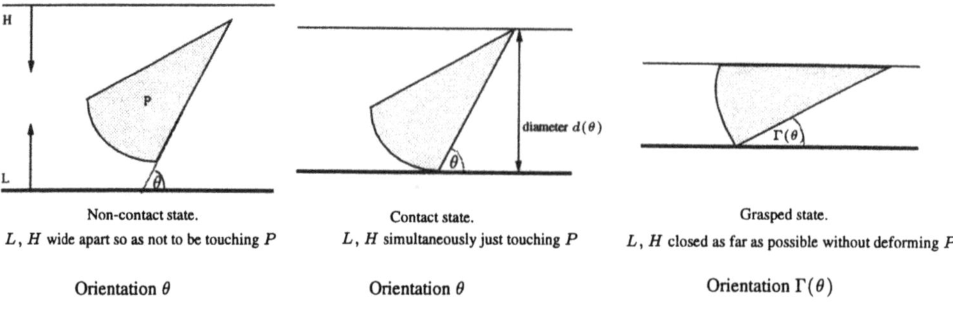

Figure 3: Configurations of the part P wrt the gripper G during the grasping process. In the non-contact state **(top)**, the jaws L,H are not touching P. In the contact state **(middle)**, both jaws just touch P but do not effect its orientation θ. Between the contact and grasped state **(bottom)**, the orientation of P changes to $\Gamma(\theta)$, Γ being the part's grasp function.

functions. For polygonal objects with a finite number of stable orientations, he presented planning algorithms that are polynomial in the the number of edges and the number of transfer functions. The complexities of his algorithms were improved by Eppstein [23].

Finally, our deployment of the parallel-jaw gripper is in tune with Canny's and Goldberg's Reduced Intricacy in Sensing and Control, or RISC, robotics [1]. They observe that for certain industrial tasks, complex (high degree of freedom) manipulators and sensors are unnecessary and that simple, robust, and inexpensive hardware is called for.

2 Diameter and Grasp Functions

The mapping from initial pre-grasp orientations to final orientations $\Gamma : S^1 \to S^1$ is the <u>*grasp function*</u> of the part. Grasp functions may be computed from diameter functions (also called width functions [24]) defined below.

Let S^1 denote the space of planar orientations and $\Re_+$ the set of positive reals. An *interval* is a continuous subset of S^1. Consider two parallel lines making an orientation θ wrt the part, minimally separated and enclosing the part. The distance between the lines is the diameter of the part in orientation θ, denoted as $d(\theta)$. Thus, the <u>*diameter function*</u> d is a $S^1 \to \Re_+$ function. See Fig. 4. The diameter function is continuous, and is oblivious to part shape concavaties. The diameter function has period π because the parallel lines are interchangeable. Additional symmetry in the part can reduce the periodicity in diameter function to a fraction of π [2].

Under the assumptions described in Section 1.1, the chief among them being that of negligible friction and simultaneous contact between the jaws and gripper, it follows (this is shown in [2, 3]) that

Fact 1 *(Squeeze) Grasping minimizes diameter.*

In other words, the diameter function plays the role of a potential function when grasped

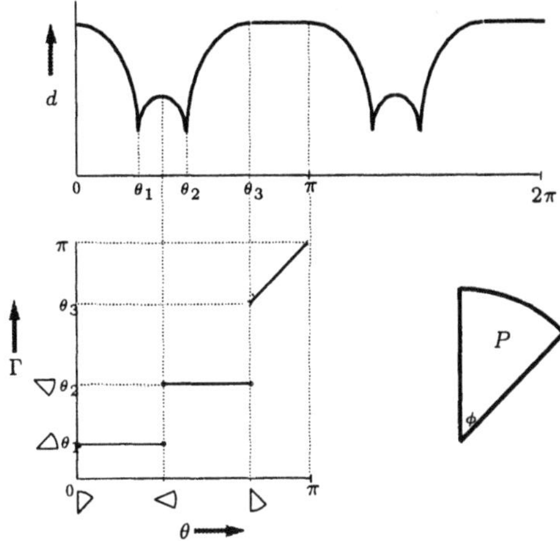

Figure 4: **(top)** The diameter function for the pie-shaped part (shown at right). During a squeeze action, the part rotates so as to reduce the diameter, terminating when the diameter reaches a local minimum. **(bottom)** The grasp function is shown for the portion $[0, \pi)$.

at an orientation θ, the part descends to a valley (local minima orientation) in the diameter function. The grasp function $\Gamma : S^1 \to S^1$ is defined so that this final orientation is $\Gamma(\theta)$. The interval between two consecutive local maxima in d leads to a *step* in Γ with fixed-point equal to the unique local minimum between them. If θ is in the midst of a interval of constant diameter orientations, $\Gamma(\theta) = \theta$. Such intervals are called *ramps* in Γ. Thus the grasp function is a *step-ramp*-function. See Fig. 4. For algebraic parts, [4] shows that the grasp function has a finite number of steps and ramps.

3 Computing Grasp Functions

From Section 2, it follows that to compute the grasp function Γ, we need to compute the extrema in the diameter function d. Any contact configuration between the jaws and the part is defined by the two (opposing) points of contact. Let us refer to an arc S_i or a vertex V_j on the convex boundary (convex hull of the boundary, if not convex) of the part as a *feature*. The features that contain contact points are called *contact features*. Further, if the contact points define an extremum configuration, let the features that contain them be called *extremum contact features*.

In the following, we consider testing if a given pair of features is an extremum contact feature, and if so, to obtain the extremum configuration. This leads to a naive $O(n^2)$ algorithm by considering every feature pair. However, this can be improved to $O(n)$ time, given the order of arcs making up the convex hull of the part. This is discussed in [4]. The idea is to compute all contact feature pairs first: it can be shown that there are at most $O(n)$ of these and they can all be computed in as much time. Each contact feature pair can be tested for extremality.

Computing Extrema Given Feature Pairs

There are three types of feature pairs: arc-arc (a-a), arc-vertex (a-v), and vertex-vertex (v-v) in decreasing order of complexity. A v-v feature pair obviously defines a unique contact configuration which is treated as potentially extremal. Pairs of the a-a and a-v type may result in uncountably many contact configurations. Algebraic tests (discussed below) are applied to these pairs to prune out a finite number (for algebraic curves, at most $(2k-1)k^2$, where k is the maximum polynomial degree in the algebraic representation of the curves) of potentially extremal configurations. A geometric grasp test is applied to each potentially extremal configuration as the final test for extremality.

Algebraic Conditions We consider a-a pairs only. A-v pairs are easier and are not discussed due to the space limitation.
Arc-arc pairs. Let $S(u), T(v)$ be the two arcs under consideration represented parametrically so that u, v in the range $(0, 1)$ sweeps out the two curve segments. Let $X_S(v)$ be the x-coordinate of $S(u)$. Y_S and functions for T are similarly defined. To check if these two features define a extremum contact configuration, we need to solve the following equations:

$$(X_S(u) - X_T(v))\,\frac{dX_S}{du} + (Y_S(u) - Y_T(v))\,\frac{dY_S}{du} = 0, \text{ and}$$

$$(X_S(u) - X_T(v))\,\frac{dX_T}{dv} + (Y_S(u) - Y_T(v))\,\frac{dY_T}{dv} = 0.$$

If the equations have a solution (u_0, v_0) both in the range $0, 1$, the line joining $S(u_0)$ with $T(v_0)$ is the normal at both points and is potentially an extremal contact configuration. We then consider this pair for the grasp test. If there is no solution, we ignore this pair.

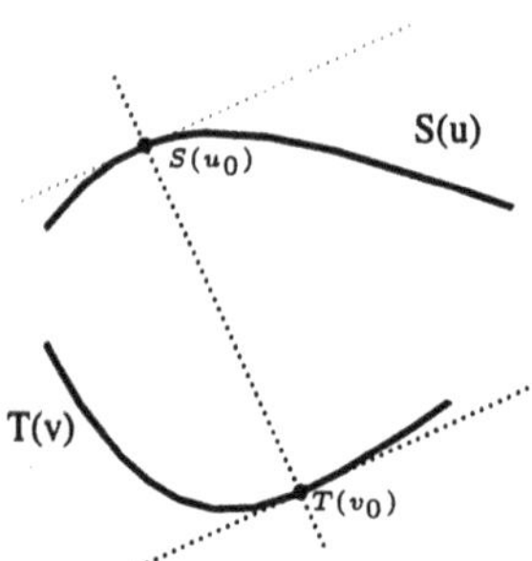

Figure 5: Pairs of curve segments $S(u), T(v)$. If $S(u_0), T(v_0)$ define an extremum orientation, then they satisfy the system of equations given in the text (but not vice versa, which is why grasp tests are necessary). This implies that the following three lines coincide: the normal to $S(u)$ at u_0; the normal to $T(v)$ at v_0; the line joining u_0 and v_0.

Grasp Conditions. Solving the equations presented give configurations that are potentially extremal. Also, all v-v pairs are potentially extremal. All potentially extremal configurations $S(u_0), T(v_0)$ have to satisfy the jaw-contact condition to be extremal: that is, the jaws need to be able touch points $S(u_0), T(v_0)$. This is so if and only if the two lines at $S(u_0), T(v_0)$ perpendicular to the line joining the two points intersect no other point on the part boundary (only intersections with the same or adjacent features need be checked due to convexity and therefore this can be implemented in constant time).

As an example, see Fig. 6.

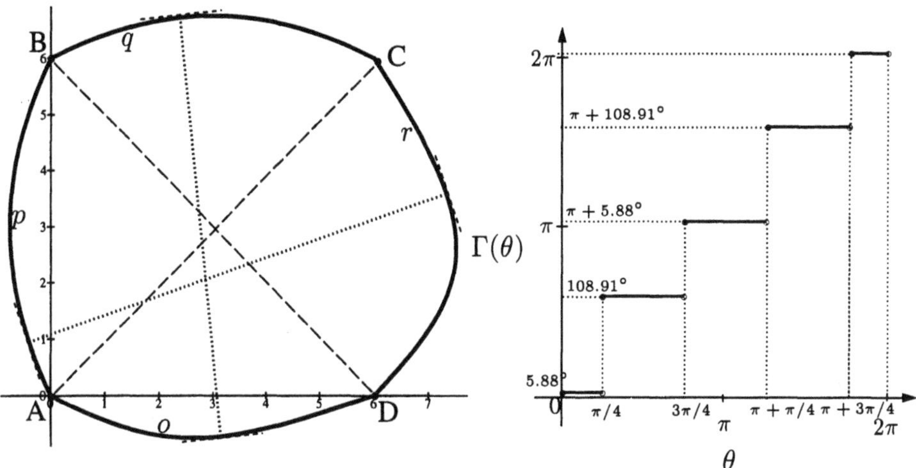

Figure 6: On the left is an algebraic part having two maxima and minima in its diameter function. Its convex boundary is made up of four arcs: o, p, q, r and four vertices A,B,C,D. The arcs are cubic parametric curves as before. Both the maxima ($\pi/4, 3\pi/4$) are vertex-vertex type and indicated by dashed lines joining BD and AC. The minima (5.88°, 108.91°) are arc-arc type: dotted lines joining points in q, o and p, r. The grasp function is plotted on the right. If θ is the initial orientation of the part wrt the gripper, $\Gamma(\theta)$ denotes its orientation after the grasp. The grasp function has a period of symmetry π, *i.e.* $\Gamma(\pi + \theta) = \pi + \Gamma(\theta)$. Within a period of symmetry of width π, the grasp function has two steps corresponding to the two minima (and maxima) in diameter function.

4 Orienting Parts using Grasp Functions

The importance of the grasp function lies in the planning algorithms presented in [2] for polygonal parts and [3] for curved parts. These algorithms, when presented with the grasp function, plan the optimal length sequence of grasp actions to orient the part up to symmetry in its diameter function starting from any arbitrary orientation. Given n steps and ramps as input, the algorithm runs in $O(\min(nN, n^2 \log n + N))$ time, where N is the length of the plan produced. Although the plan length N is optimal, there are no bounds on N in terms of n. We skip the details in this paper which is devoted to computing grasp functions. As an example, when presented with the grasp function at the bottom of Fig. 4 as input, the planning algorithm comes up with the angles $75°, 42°, 0$ which are the grasp actions shown in Fig. 2.

5 Conclusion

In this paper we discussed an algorithm for computing the (squeeze) grasp function of an part bounded by algebraic curves. Together with the algorithm in [3], this results in a complete algorithm to orient algebraic parts, given their boundary representation, up to symmetry in their diameter functions.

Radius Function and Push Grasping Diameter functions have period at most π which implies that the best we can achieve by a sequence of squeeze grasps is one of two final orientations in S^1. To obtain a unique final orientations, push grasp actions may be used. [1] As identified by Brost [12]) a push grasp action consists of a push phase by one jaw towards the other prior to squeezing. Thus, in addition to having periods 2π (generally), push grasps are easer to justify mechanically than squeeze grasps which require simultaneous contact of the two jaws with the part.

A push grasp action requires a sufficient pushing distance [11] and the location of the part's center of mass (com). The mechanics of the push phase is captured by the *push function* defined below [17]. The distance from the part's c.o.m to a line in orientation θ tangent to the part varies with θ. Define the *radius function*, $r : S^1 \rightarrow \Re_+$, to record this variation, *i.e.* $r(\theta)$ equals the distance from the c.o.m. to a tangent line of orientation θ. From Mason [17] and Brost [12], analogous to Fact 1, we get

Fact 2 *Pushing minimizes radius.*

To compute the push function, we need the extrema in the radius function. This may be done in a manner similar to finding extrema in diameter function (in fact, it is simpler).

The push-grasp function, which maps an initial orientation to the corresponding final orientation after the push and squeeze phases, is simply the composition of the squeeze grasp function and the push function because:

Fact 3 *Push-grasping first minimizes radius, and subsequently diameter.*

The push grasp plan for the pie-shaped part is shown in Fig. 7.

---

[1] This is because push grasp functions usually have period 2π rather than π. However in pathological parts (very symmetrical parts with perfectly symmetrically positioned center of mass), push grasp functions can have period less than 2π.

Figure 7: Four traces of 3-step push-grasp plan. The push angles are $0, -56, -128$ degrees. Line with arrow indicates pushing jaw. The initial orientations for the four traces are same as those in Fig. 2. Notice the final orientations in both figures. Initial orientations that are 180° apart (see the first and fourth columns in Fig. 2), end up 180° apart in Fig. 2 (squeeze-grasp actions). However, in this figure, they end up in the identical orientation. Push-grasping resolves the ambiguity.

References

[1] J. F. Canny and K. Y. Goldberg. "risc" for industrial robotics: Recent results and open problems. In *International Conference on Robotics and Automation*. IEEE, May 1994. submitted.

[2] K. Y. Goldberg. Orienting polygonal parts without sensors. *Algorithmica*, 10(2):201–225, Aug 1993. (Special issue on Computational Robotics).

[3] A. S. Rao and K. Y. Goldberg. Manipulating algebraic parts in the plane. Technical Report RUU-CS-93-43, Utrecht University, Department of Computer Science, December 1993. Submitted to the IEEE *Transactions on Robotics and Automation*. A compressed postscript version available by anonymous ftp from `ftp.cs.ruu.nl`: file `/pub/RUU/CS/techreps/CS-1993/1993-43.ps.gz` A preliminary version appears in Proc. ICRA 1992 under the title "Orienting generalized polygonal parts.".

[4] A. S. Rao and K. Y. Goldberg. Computing grasp functions. Submitted to the journal: *Computational Geometry: Theory and Applications*. Preliminary version appears in Proceedings of the IASTED Conference on Robotics and Manufacturing, Oxford, Sep. 1993.

[5] A. S. Rao. *Algorithmic Plans for Robotic Manipulation*. PhD thesis, University of Southern California, Department of Electrical Engineering–Systems, December 1992.

[6] C. M. Hoffmann. *Geometric and Solid Modeling: An Introduction*. Computer Graphics and Geometric Modeling. Morgan Kaufmann, San Mateo, CA 94403., 1989.

[7] A. A. Schaeffer and C. J. Van Wyk. Convex hulls of piecewise-smooth Jordan curves. *Journal of Algorithms*, 8(1):66–94, 1987.

[8] A. S. Rao and K. Y. Goldberg. Shape from diameter: recognizing polygonal parts with a parallel-jaw gripper. *International Journal of Robotics Research*, 13(1):16–37, February 1994.

[9] K. Y. Goldberg and M. Furst. Low friction gripper. U.S. Patent # 5,098,145, March 1992.

[10] M. T. Mason. On the scope of quasi-static pushing. In O. Faugeras and G. Giralt, editors, *The Third International Symposium on Robotics Research.* MIT Press, 1986.

[11] M. A. Peshkin. *Planning Robotic Manipulation Strategies for Sliding Objects.* PhD thesis, Carnegie-Mellon University, Department of Physics, Pittsburgh, Pennylvania, Nov 1986. Also published as a book: *Robotic Manipulation Strategies*, Prentice Hall, 1990, New Jersey.

[12] R. C. Brost. Automatic grasp planning in the presence of uncertainty. *The International Journal of Robotics Research*, 8(1), February 1988.

[13] R. A. Grupen, T. C. Henderson, and I. D. McCammon. A survey of general purpose manipulation. *International Journal of Robotics Research*, 8(1), 1989.

[14] J. Pertin-Troccaz. Grasping: A state of the art. In *The Robotics Review I*, pages 71–98. MIT Press, 1989. edited by O. Khatib, J. J. Craig, and T. Lozano-Perez.

[15] J. Jameson. *Analytic Techniques for Automated Grasp.* PhD thesis, Department of Mechanical Engineering, Stanford University, June 1985.

[16] H. Hanafusa and H. Asada. Stable prehension by a robot hand with elastic fingers. *Trans. of Soc. of Inst. and Control Engineers*, 13(4), 1977. Also see the section by the authors in *Robot Motion: Planning and Control*, edited by Brady *et. al*, MIT Press, 1983.

[17] M. T. Mason. *Manipulator Grasping and Pushing Operations.* PhD thesis, MIT, June 1982. published in *Robot Hands and the Mechanics of Manipulation*, MIT Press, 1985.

[18] M. Mani and R. D. W. Wilson. A programmable orienting system for flat parts. In *Proc: North American Mfg. Research Inst. Conf XIII*, 1985.

[19] M. T. Mason, K. Y. Goldberg, and R. H. Taylor. Planning sequences of squeeze-grasps to orient and grasp polygonal objects. Technical Report CMU-CS-88-127, Carnegie Mellon University, Computer Science Dept., Pittsburgh, PA 15213, April 1988.

[20] M. A. Peshkin and A. C. Sanderson. Planning robotic manipulation strategies for workpieces that slide. *IEEE Journal of Robotics and Automation*, 4(5), October 1988.

[21] R. H. Taylor, M. T. Mason, and K. Y. Goldberg. Sensor-based manipulation planning as a game with nature. In *Fourth International Symposium on Robotics Research*, August 1987.

[22] B. K. Natarajan. Some paradigms for the automated design of parts feeders. *International Journal of Robotics Research*, 8(6):98–109, December 1989. Also appeared in IEEE FOCS, 1986.

[23] D. Eppstein. Reset sequences for monotonic automata. *SIAM Journal of Computing*, 19(3), 1990.

[24] I. M. Yaglom and V.G. Boltyanskii. *Convex Figures.* Holt, Rinehart and Winston, New York, 1951.

Computational Geometry in the Synthesis of Skew Gear Teeth

J. Phillips

Basser Department of Computer Science
University of Sydney, NSW 2006, Australia

Abstract - In skew gears of whatever kind, in general (there being special cases), the normal to the tangent plane at the point contact between teeth, the contact normal, may reside instantaneously at any shortest distance *d* from the pitch line and be inclined at any angle ϕ to it. The parameters of *d* and ϕ may both vary as the point of contact moves from start to finish along its path (and that path need not be straight), but, for constancy of angular velocity transmission as we move from tooth to tooth across teeth, the value of *d* and ϕ must remain a constant *p*, namely the pitch *p* (mm/rad) of the screwing at the pitch line. That, briefly, is the first law of gearing [3]. After a short review of earlier material from [2,3], which we need here, further confirmation of the veracity of the law is provided. This is then used in further discussion of the iterative methods that might be employed to exploit the law. Paper thus begins to explore the kinds of computational geometry that might be used (a) to synthesise the shapes of teeth that are kinematically correct, these to be cut by NC machine tools if necessary, and (b) to analyse the shapes of teeth that are not correct, these having already been designed and cut by approximate methods.

I. Introduction

Although self standing, this paper might best be read in conjunction with an earlier, companion paper [3]. Author gives an historical survey in [3] of the kinematic aspects of the theory of skew gearing and presents and proves the first law of gearing. The law is briefly stated again in the Abstract above. Unless gears of whatever kind (planar, spherical or skew) obey this law, constancy of angular velocity transmission across teeth cannot be achieved. Paper [3] collects from [2] the formulae needed for the location of, and the pitch at, the pitch line and other details required for an exact statement of the law, the law being based, it being clear, entirely upon the found location of the pitch line; paper [3] presents charts to show the influence of the variables which determine the location of and the pitch at the pitch line in any gear set; and paper [3] discusses a few first obvious implications of the law for the kinematic design and analysis of the shapes of skew gear teeth. Needless to say the first law is about to be re-enunciated here for readers who, at the time of reading, are either not as yet, or never to be, in receipt of [3]. The following short review of the mentioned material will provide in any event all that is needed for a first reading of the present work.

II. Short review of earlier material

Three parameters define the configuration of a gear set, (1) the shortest distance C between the shaft axes, (2) the angle Σ between those axes, and (3) the angular velocity

A. J. Lenarčič and B. B. Ravani (eds.), Advances in Robot Kinematics and Computationed Geometry, 101–108.

ratio k. There is a single point of contact Q; several Q are arranged to occur in tendon however; reversibility of rotation obtains (non reversibility due to friction is irrelevant here); rigidity and zero clearances obtain. Parameters C, Σ and k alone determine the location of and the pitch at the pitch line. Pitch line intersects the centre distance line perpendicularly. The theorem of the three axes [1,2,3] is the key: radii r_2 and r_3 from shaft axes to pitch line sum to C, but the ration r_2/r_3 is a function of both k and Σ [2,3]. The relevant angles at the location are similarly determined. The pitch p at the pitch line is that of the relative screwing between the wheels. Measured mm/rad, it is independent of the shapes of the teeth, provided those shapes maintain a constant k of course. This p , the pitch of the gear set, may be right handed positive or left handed negative, but for ordinary spur or helical gears, and bevel gears, pitch is zero. In [2] we took for numerical example C = 40 mm, $\Sigma = 80°$, and $k = \pm 0.6$; p was found to be -15.07 mm/rad (LH) or +20.52 mm/rad (RH), there being two pitch lines and two sets of wheels for a given pair of shafts (these two matching the two different directions of rotation available for the driven wheel the direction of the driver being given). Details of all such matters and more may be found in [2] at chapter 22.

III. The hitherto known body of law

For the shapes of teeth in planar gearing where the shafts are parallel, a collection exists of tight kinematical rules that must obtain if we wish to ensure a constant k as the wheels roll from tooth to tooth. These rules for planar gearing, stated upon the reference plane, are called the fundamental law or laws. They originated with Euler but were first exploited by Willis [4]. They reduce to two. The first relates to the direction of the contact normal at the point contact, the law stating that that line must always pass through a fixed point namely the pitch point P uniquely located for a given k upon the centre distance line. The second relates to the curvatures (the mere circular curvatures) of the profiles mating in mesh at the point of contact [5]. New versions of these laws written for wholly spatial, skew gearing, have not been formulated hitheroto. Putting the new version of the second law aside for comment later (this second law for skew gears is as yet unknown), the new version of the first law may now be states as follows [3].

IV. The first law of gearing

The normal to the tangent plane at the point of contact Q, the contact normal, may be at any shortest distance d from the pitch line and be inclined at any angle ϕ to it; the parameters d and ϕ may both vary as each Q moves from start to finish along its path; but the value of $d \tan \phi$ must at all stages remain a constant namely p. The first law requires, in other words, that the equation $d \tan \phi = p$ obtains at all stages of the meshing. In both cases in gears (a) of intersecting axes where C is zero and the gears are for example bevel gears, and (b) of parallel axes where Σ is zero and the gears are for example spur gears, p is zero, and the first law reduces to the known simple statement for such simple gears that the distance d can only be zero. Fig. 1 shows pitch-line axis P-x , centre-distance axis P-z, a third orthogonal axis P-y, a right handed pitch p, and a legitimate contact normal n-n generally disposed. Line n-n is legitimate because (and only because) $d \tan \phi = p$.

Accordingly *n-n* is drawn binormal to one only of the helices of the system of helices of pitch *p* which coaxially surrounds the pitch line. These helices are mentioned by Litvin [6], but there is the context of rubbing velocities. If *d* is put to zero, that is if *n-n* is chosen ever to cut the pitch line, ϕ can only be $\pi/2$. The law is merely being stated and further investigated here, not proven; see [3]. Notice incidentally that my ϕ here (ϕ comes from [2]) is not the so-called pressure angle; that angle, say δ, might be best seen here to be the angle between the shortest distance *d* and *P-z*. Angle δ is not indicated in Fig. 1 but may be seen by looking there orthogonally along *P-x*.

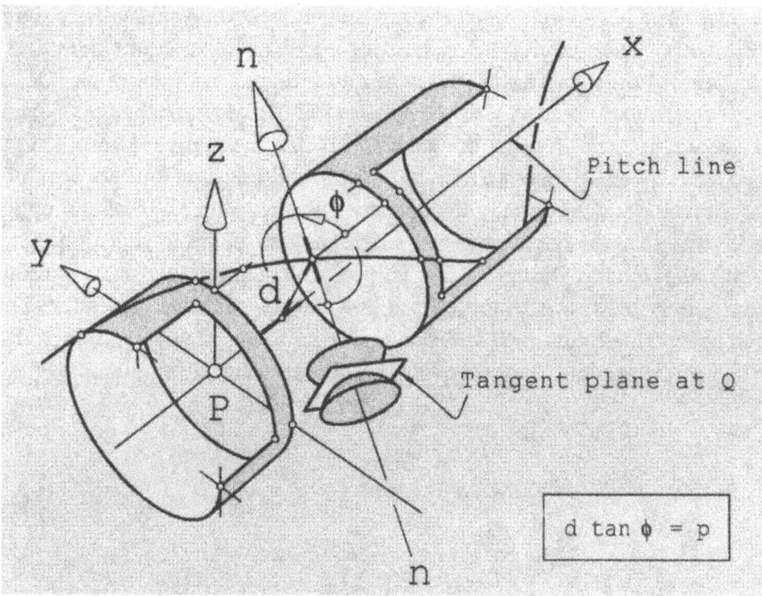

Fig. 1

V. An equivalent mechanism for skew gear set

The discussion under this sub-heading is intended to be one of a number of contributions towards the establishment of proper iterative methods for the design and analysis of the shapes of gear teeth, the said methods being proper in the sense of their being based, not upon approximate rules of thumb with limited ranges of application, but upon exact, universal, kinematic fundamentals, the words exact and universal being used here in the sense that the methods will apply exactly, wherever the regions of meshing are, and in all gears of whatever kind. Fig. 2 shows two hinged contacting gear-bodies 2 and 3 along with their revolutes *R* at frame link 1. This 3-link mechanism represents in general a selected pair of teeth in contact at *Q* (*Q* is eclipsed in the figure) along with the relevant shaft axes. The contacting bodies may be replaced, equivalently for all small displacements and their first derivatives with respect to time namely velocities, by the *RSSR* as shown. The rod *SS* (link 4) is collinear with the contact normal; it may be of any length; but without loss of generality here it has been cut to such a length that the transmission angles at the *S*-joints are both 90° at the instant. Links 2 and 3 of the *RSSR* and the shown cylinders tangential with the rod have radii a_2 and a_3; see also the corresponding angles α_2 and α_2 between the shown velocity vectors and the rod. In a design process aimed at finding properly shaped teeth for a gear set of given *k*, a proposed *Q* may be chosen, at a first choosing, to be anywhere. Next a proposed contact normal at *Q* (a rod *SS*) may be drawn to be anywhere through *Q*, provided it is tangent to two circular cylinders (see Fig. 2) in such a way that the ratio $(a_1 \cos \alpha_1)/(a_2 \cos \alpha_2)$ equates with *k*. In practice this *Q* will be chosen within the as yet vaguely defined yet limited region of the meshing wherein the shapes of the teeth are as yet unknown. The direction for the proposed contact normal will be within some reasonable limits too, due mainly to problems of interference and considerations of force. The region chosen for the meshing of course be quite remote from the pitch line, as it is for example in worms and hypoids, see [2]. There follows now an analysis by means of enumerative geometry, the results of which will be instructive later. Take any two cylinders coaxial respectively with their shaft axes, and consider the number of common tangents that may be drawn. We may choose any point on either one of the cylinders (say *S*) and observe that there is containing *S* a single tangent plane to that cylinder, that that plane intersects the other cylinder in an ellipse, and accordingly that, through *S*, two distinct tangents may be drawn to the ellipse and thus to the other cylinder. This duality reflects the double valued nature of the skew gear problem; having chosen a direction of rotation for one of the wheels, the other may be chosen to rotate in either of the available directions;

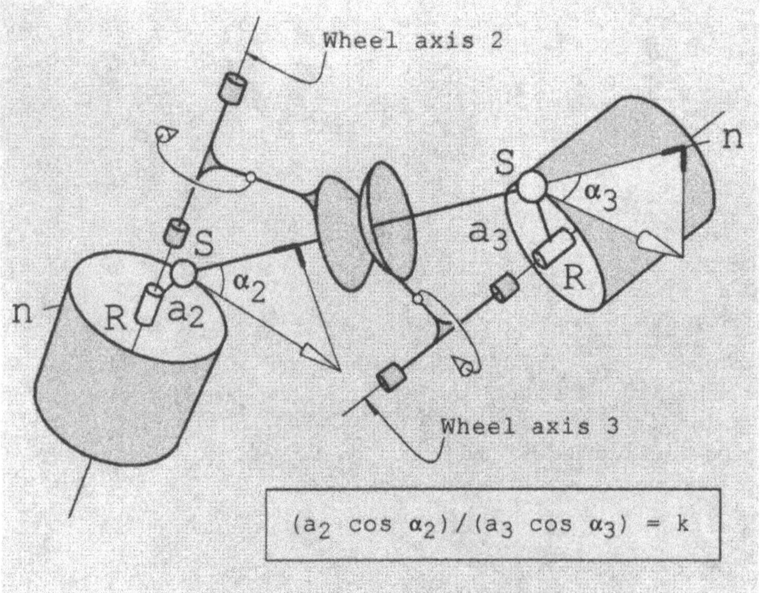

Fig. 2

thus we become aware of the acute and the obtuse angles Σ; see the photo in [2] at chapter 22. But on each cylinder there is an ∞^2 of points S and so, without counting each line an ∞ of times too often, and disregarding the number 2 because twice ∞ is simply ∞, there is an ∞^2 of common tangents SS that may be drawn to the pair of cylinders. Now let each of the cylinders be of any radius, and thereby see that the number of common tangents that may be drawn to all possible pairs of cylinders is ∞^4, namely the number of all of the lines in space. We shall see however that only a tiny fraction of these, namely only an ∞^3, is the somewhat limited number of common tangents that may be taken as legitimate rods SS and thus legitimate contact normals among the actual teeth of the gear set, for given k. All this suggests a certain layout of the legitimate contact normals, namely a linear complex [2,3], and, suspecting that, we might try to show next that, through any chosen Q, for a given k of course, only a single planar pencil of contact normals may be drawn. One may show by computer graphics, as the author has done, (a) that the perpendicular dropped from Q onto the pitch line is a possible rod SS and thus a legitimate contact normal, (b) that a second line through Q, perpendicular at Q to the line just mentioned, but inclined to the pitch line at such an angle ϕ that $\phi = \arctan (p/d)$, is similarly a legitimate contact normal, (c) that all other lines through Q coplanar with the two just found are legitimate contact normals, but (d) that no other line through Q is a legitimate contact normal. The reader might also show to his or her satisfaction that, upon a generally disposed plane, a single infinity of legitimate contact normals lie which are concurrent at a single point.

VI. On the severalty of points of contact Q

Unless there is an overlap between each new pair of teeth coming into mesh and its immediately preceding pair going out of mesh, there will be breaks in the smooth motion. Independently however of the shape of the path of Q, this severalty of Q will not be a trouble, provided of course (a) we have the smooth motion established in the first place, and (b) the first law obtains at all points Q. The severalty Q introduces nothing more than mere overconstraint. It dose mean though that smooth motion will depend not only on proper kinematic design but also on accuracy of manufacture [2,3].

VII. Iterative opportunities for the contact normal

If in design the meshing is put remote from the pitch line (which is often), n-n may never come near to cutting the pitch line. We may, with reservations, freely choose the path of Q; it could be straight or curved; it need not cut P-x or P-z; but the tangent plane at Q must contain the helitangent there, namely the line of the vector of the local rubbing velocity [6]. It may be useful to observe that, whereas all the binormals to all of the helices form a linear complex, all the tangents to all of the helices (the helitangents) form a quadratic complex; refer [2] §21.22; and whereas at each point in a linear complex there is a polar plane of lines, at each point in a quadratic complex there is an elliptical cone of lines; see figure 21.05 in [2]. If reduction of vibration is important we should choose the path of Q and the directions of the contacts normal distributed along it to minimise changes in direction and magnitude of the transmission force at Q, which, in the absence of friction, will be acting along the contact normal there, but otherwise, of course, not. For the sake of symmetry (which is however theoretically speaking unnecessary), we might wish to put the obverse path of Q (namely that other path employed for driving the wheels in the reverse directions) line-symmetrically arranged with respect to P-z. Other considerations of symmetry might also be at issue. Refer [2] §12.08-13, §12.25-27, and distinguish between the following velocities: $\mathbf{v}_{10}$, that relative to frame 1 of the changing and moving point of contact Q, which point follows its path in 1 while remaining fixed in neither of the teeth; $\mathbf{v}_{203}$, that relative to 2 of the point Q_3 in 3, and $\mathbf{v}_{302}$ that relative to 3 of the point Q_2 in 2, which two are equal and opposite, both of which are in the tangent plane at Q, and both equally measures of the rubbing velocity. Look also at $\mathbf{v}_{102}$ and $\mathbf{v}_{103}$, neither of which are in the tangent plane, but whose vector differences $\mathbf{v}_{0203}$ and $\mathbf{v}_{0302}$ are both parallel with the tangent plane and equal and opposite measures of the rubbing velocity. Distinguish also between the following paths: those of Q_2 and of Q_3 in 1, which are non-coplanar circles intersecting at Q; and those of Q in tooth 2 and of Q in tooth 3, which latter (shown up by their respective skid-marks on the real surfaces of the teeth) are spatial curves also intersecting at Q. These curves of course are those we seek in synthesis. In terms say of measured crank angle at input, we might calibrate the chosen path of Q in 1 in some chosen way, thus determining the shapes of the paths of Q in 2 and Q in 3, thus the corresponding 'twisted ribbons' upon the teeth, there being (of necessity) a smoothly twisting thin ribbon of the elementary tangent planes distributed along the paths Q on 2 and Q on 3. To get these shapes we might engage in step-by-step inversion of the mechanism to reconstruct the shapes of the paths across those infinitesimally small, neighbouring neighbourhoods where Q has been or will be. But continuos smooth motion must obtain; the tangent plane of the button pair [2,7] upon the contact normal must at all stages of the sliding reside already embedded, as it were, tangentially in the surfaces of the teeth. When in contact at Q, moreover, the tangent planes must (as was said before) contain the line of the vector of the local rubbing velocity. In all of this there lies a difficulty; we are blundering here into the realm of the unknown second law; whatever it is, this law must surely involve such matters as the circular and spherical curvatures of the surfaces at the successive points of contact, the torsion of twisted curves, and spatial envelopes.

VIII. On the overall body of law

Going to Baxter in [5], we find his figure 1-10 and his algebraic formulation of the law for the curvatures of profiles mating at Q in planar gearing. Remember that his ϕ is not my ϕ. We see here a double application of the Euler-Savary geometry of the instantaneous relative motion of two laminae (circa 1750); refer [8,9]. Baxter takes two scenes, those (a) of wheel 2 and the basic rack, and (b) of wheel 3 and the basic rack, erects two inflexion circles, not shown in [5], writes two equations and adds them to arrive at his laws. Baxter's laws do not fix the radii of curvature ρ_2 and ρ_3 of the mating profiles, but gives an algebraic relationship that must exist between them. He draws attention to relative curvature $1/\rho_0$, a measure of the likely Hertzian area of contact in gears that are non-rigid, namely actual gears. It is easy to see in the doubly special case of involute gearing that Baxter's laws are obeyed and that the laws moreover fix, for a given k and pressure angle, not only the relationships between, but the actual curvatures themselves. The radii of curvature ρ_2 and ρ_3 in planar involute gearing must sum (by definition) to the length of the common tangent between the base circles, and, with Q at the pitch point they relate inversely to one another in the ratio k. But what can be said about the full version of this simple law, the (as yet unknown) second law of gearing? It may be said that a full version or the spatial analogue of the Euler-Savary geometry was established and effectively studied by Skreiner [10] in 1967. Therein must surely lie important clues.

IX. Iterative opportunities for smooth motion

Let us imagine ourselves in Baxter's figure 1-10 in ignorance of the involute shape and its fortuitous, beneficial properties. Let us have to hand some curved and calibrated path for Q (chosen with the first law available but otherwise with nothing other than cunning); and let us begin to imagine how we would synthesise the shapes of the profiles of some conjugate teeth. Read Mac Cord [11] who wrote so engagingly about such matters in 1883. At least we could do the following: we could deal first with some initial trial location of the moving Q in 1, thus automatically plotting the first two coincident locations (with their tangent planes at the instant coplanar) of the moving Q on 2 and the moving Q on 3. Next we could either (a) rotate both wheels to the next Q, paying attention to the angular velocity ratio, or better (b) invert the mechanism by fixing one of the wheels, rotate the frame and thus roll the other wheel relative to the fixed one, until the next Q. We would take successive infinitesimal steps while carefully watching the successively intersecting tangent planes at the button pairs. But how would we do this latter exactly? Our next problem of course is to deal with such step-by-step inversions and the layout of teeth in three dimensions. How much overlap should we give at the said intersections? Given conditions at the tops and bottoms of teeth, is there room for bounded, finite element analyses here? We can in any event, with the new first law and some intelligent clues about the second, check the faults in old designs. How bad is the expected tooth-ripple of transmission error to be found at the output of all machine-cut skew gears?

X. Conclusions

This paper has not shown much in the way of actual computational technique, but it has begun to explore the possibilities. It has set down some fundamental facts, and reflected the author's own recent efforts in computational geometry at the computer screen. There is much to learn. It is clear however that the first law is a strong law; for upon its basis we can make progressively better synthesis and incisive analyses, for example, analyses, even, of already-cut skew gears not performing properly. Noise and vibration in gears is primarily caused by teeth whose shapes are wrong [12]. Design of the hypoid set is as yet a black art in the eyes of many, worms are correspondingly mysterious, crossed helicals can be seen to obey the first law but are in themselves a kind of mild absurdity, and there is a host of difficulties. One for example is the metrology; how does one actually locate an existing contact normal accurately? By means somehow of spherical geometry and spheres? In the arena of synthesis, however, we have the emergence now of (a) mechanical circumstances that might well require slow-mowing, skew gear-sets displaying continuously accurate, constant *k*, and (b) rapidly improving NC machine tools with which we might be able to manufacture newly synthesised teeth accurately and cheaply.

XI. References

[1] J.R. Phillips and K.H. Hunt, "On the Theorem of Three Axes in the Spatial Motion of Three Bodies", *Aust. J. App. Sci.*, *15*(4), 267-287, CSIRO, Melbourne (1989).

[2] J. Phillips, *Freedom in Machinery*, in two volumes, Cambridge University Press, UK, (1984-1990).

[3] J. Phillips, "The First Law of Gearing", paper submitted for publication, *Mechanism and Machine Theory*.

[4] R. Willis, "On the Teeth of Wheels", *Trans. I. Civ. E.*, *2*, London (1838). see also same author, *Principles of Mechanism*, 2nd ed. Longmans Green, London (1870).

[5] M.L. Baxter, "Basic Theory of Gear-Tooth Action and Generation", *Gear H'book,* (Dudley), McG Hill, (1962).

[6] F.L. Litvin, *Theory of Gearing*, NASA reference publication 1212, US Govt. Printing Office, (1989).

[7] J.R. Phillips, "Button Pair Synthesis of Some Simple f3 Joints", *Proc. 6th World Congress IFToMM, New Delhi, 2*, 1260-62, (1983).

[8] N. Rosenauer and A.H. Willis, *Kinematics of Mechanisms*, Assoc. Gen. Publications, Sydney (1953), Dover (1967).

[9] A.S. Hall, Jr., *Kinematics and Linkage Design*, Prentice Hall (1961).

[10] M. Skreiner, "A Study of the Geometry and Kinematics of Instantaneous Spatial Motion", *Journal of Mechanisms, 1*, 115-143, Pergamon (1966).

[11] C.W. Mac Cord, *Kinematics or Mechanical Movements*, John Wiley, New York, (1889).

[12] J. Derek Smith, *Gears and Their Vibration*, Marcel Dekker, The Macmillan Press Ltd. New York (1983).

3. Kinematic Errors and Calibration

A CAD Tool for Remote Calibration of Space Station Robotic Test Bed

Salvatore Illiano Giovanni Iodice Bruno Siciliano

Dipartimento di Informatica e Sistemistica
Università degli Studi di Napoli Federico II
Via Claudio 21, 80125 Napoli, Italy

Abstract — This paper describes the evolution of a CAD programming tool for remote video calibration of the Internal Automation & Robotics Technology Test Bed (IARTTB) available in the WKR Section at ESTEC. First the correspondence between the CAD model of the workcell in the off-line programming system and the real world is created. Then the robot is remotely calibrated with respect to the two racks present in the workcell by using a video camera on the end effector. A software package with a user-friendly interface has been developed to carry out the calibration procedure in a systematic way.

I. Introduction

In the latest years the interest towards space robotics applications is increasing, thanks to the foreseen construction of space stations. To carry out duties in internal and external operations it may become convenient to use robot manipulators, especially if those have to work in a known structured environment.

The launch stress and altered gravity conditions may modify some physical parameters of the manipulator, which then requires a calibration procedure to find an accurate relationship between the joint variables and the end-effector location to be used by the controller in planning and execution tasks [1],[2],[3].

Generally the calibration operations, needing the presence of an expert operator close to the robot, are not always possible (unmanned missions) or they have drawbacks such as increase of crew workload, cost, and robot utilization time.

This study investigates the possibility of performing local robot calibration remotely via the development of a standard working procedure. The approach is to compute for each location of a manipulation task the displacement error vector between the real and calculated locations of the robot end effector, and to use this vector to get an accurate execution of the task. The calibration method relies on the use of a video camera mounted on the robot end effector, and the prior knowledge-based algorithm [4] is adopted to recognize three-dimensional objects from single two-dimensional images.

The method has been implemented on the working environment available in the WKR Robotics Section of the European Space Agency establishment ESTEC. The workcell, called Internal Automation & Robotics Technology Test Bed (IARTTB), has

A. J. Lenarčič and B. B. Ravani (eds.), Advances in Robot Kinematics and Computationed Geometry, 111–118.

been conceived to develop technological solutions to the problems related to the use of a robot for Intra-Vehicular Activities (IVA) servicing in a space station.

To generate application programs which are not affected by manipulator positioning errors a unique correspondence between the CAD model of the workcell and the real world has to be created. The calibration of the off-line programming system is performed in two steps. In the first step three-dimensional measures of all the objects in the workcell are obtained to derive representative CAD models. In the second step the best values of the parameters are computed to update the models in the off-line programming system in order to perform an accurate inverse kinematics procedure.

A software package has been developed to handle communication between the robot controller and a workstation, to integrate the video images of the camera with those of the CAD system, and to compute the displacement between the actual and the expected locations of the robot end effector. A user-friendly interface has been realized to simplify and speed up the calibration procedure.

The performance of the calibration system is demonstrated by means of numerical examples on the above robotic test bed.

II. Video Calibration

The video calibration procedure has been implemented by using the prior knowledge-based approach in [4] which is effective when the vision system operates in a completely structured environment, i.e. when the types of objects are known and it is desired to compute their location in space.

The following equations describe the projection of a 3D object model point p into a 2D image point of coordinates (u, v):

$$p' = Rp \tag{1}$$

$$\begin{bmatrix} u \\ v \end{bmatrix} = \begin{bmatrix} \dfrac{f_u p'_x}{(p'_z + d_z)} + d_x \\ \dfrac{f_v p'_y}{(p'_z + d_z)} + d_y \end{bmatrix} \tag{2}$$

where R is the rotation matrix transforming p in the original object frame into a point $p' = (p'_x, p'_y, p'_z)$ in the camera frame. These coordinates are then combined by the parameters f_u, f_v which are proportional to the camera focal lengths along the two axes of the image plane to perform the perspective projection into a point of coordinates (u, v); the parameters d_x and d_y specify the coordinates of the origin of the object frame on the image plane whereas d_z represents the distance between the origins of the two frames.

The rotation matrix R can be expressed as the initial rotation matrix of the object with respect to the camera frame premultiplied by three rotation matrices describing incremental rotations by (ϕ_x, ϕ_y, ϕ_z) about the axes of the current frame; such rotations are to be considered nearly independent for small increments.

Therefore the goal is to solve (1),(2) for $h = [\, d_x \quad d_y \quad d_z \quad \phi_x \quad \phi_y \quad \phi_z \,]^{\mathrm{T}}$, given a number of object model points and their corresponding locations into an image available by the camera. This can be achieved by applying Newton's method which requires computation of partial derivatives of u and v with respect to the unknown parameters. It is not difficult to show that:

$$\frac{\partial u}{\partial d_x} = 1 \quad \frac{\partial u}{\partial d_y} = 0 \quad \frac{\partial u}{\partial d_z} = -f_u c^2 p'_x \tag{3}$$

$$\frac{\partial u}{\partial \phi_x} = -f_u c^2 p'_x p'_y \quad \frac{\partial u}{\partial \phi_y} = f_u c(p'_z + c p'^2_x) \quad \frac{\partial u}{\partial \phi_z} = -f_u c p'_y \tag{4}$$

and

$$\frac{\partial v}{\partial d_x} = 0 \quad \frac{\partial v}{\partial d_y} = 1 \quad \frac{\partial v}{\partial d_z} = -f_v c^2 p'_x \tag{5}$$

$$\frac{\partial v}{\partial \phi_x} = -f_v c(p'_z + c p'^2_y) \quad \frac{\partial v}{\partial \phi_y} = f_v c^2 p'_x p'_y \quad \frac{\partial v}{\partial \phi_z} = f_v c p'_x \tag{6}$$

where $c = 1/(p'_z + d_z)$.

For each point in the model that should match against some corresponding point in the image, the model point is projected into the image using the current parameter estimates and the errors e_u and e_v are measured with respect to the given image point. Then the *two* equations:

$$\frac{\partial u}{\partial h}\Delta h = e_u \qquad \frac{\partial v}{\partial h}\Delta h = e_v \tag{7}$$

are used for *three* different points to form a linear system of *six* equations to be solved for the vector of corrections Δh at each iteration. If the *two* focal lengths are not exactly known, they can be easily embedded into the algorithm by computing the partial derivatives of u and v with respect to f_u and f_v; in that case a *fourth* point is needed to generate the corrections Δf_u and Δf_v at each iteration.

III. Space Station Robotic Test Bed

The Internal Automation & Robotics Technology Test Bed (IARTTB) is a technology workcell aimed at developing solutions to the problems related to the use of a robot for Intra-Vehicular Activities (IVA) servicing in a space station. The test bed is composed of:

- A 6-DOF industrial manipulator COMAU SMART 10, equipped with a passive Remote Center of Compliance (RCC), a tool exchange device that can handle a two-finger gripper Shunk SEG 10 or a screwdriver. Special brackets, fixed to the gripper, allow placing of two video cameras on it.
- Two racks, equipped with several objects and their corresponding storage locations, one drawer and two locations to host the two robot tools.

All the objects or moving parts in the two racks are equipped with special fixtures designed to optimize the contact with the gripper fingers during the grappling operations. The robot may perform the classical operations of picking objects from one rack and placing them in the other rack so as to simulate tasks to be executed automatically in a space station.

The robot is controlled by the COMAU C3G controller and an Off-line Programming System (OPS) is available for programming, operating and monitoring the robot in an easy and fast way. The OPS is composed of a Silicon Graphics IRIS workstation with a CAD system especially developed by Tecnomatix for robotics applications, named RobCAD. This software package includes the robot model and allows designing a complete model of the workcell and of the grappling fixture (Fig. 1) as well as planning and simulating tasks by simple mouse operations. Once task feasibility has been ascertained, the corresponding procedure is implemented automatically in PDL2 (Program Definition Language) on the controller.

IV. CAD Tool

The calibration procedure normally used to calibrate the OPS consists in:

- measuring the robot workcell and updating the CAD models inside the OPS,
- calculating extended robot parameters and creating an accurate inverse kinematics solution inside the OPS.

Whereas the first step does not present any technical difficulties, the second one requires several extensive and accurate measurements which have to be performed by an expert operator using special tools, such as theodolites. This is not useful for a test bed conceived for IVA, because the time required to perform the required measurements makes inefficient use of the manipulator in a space environment.

An alternative approach is based on the video calibration technique previously described. Once the camera location is known with high accuracy, the end-effector location can be computed and calibration of the OPS becomes simple, systematic and remotely executable. The main features of the CAD tool developed to perform the calibration procedure are described in the following.

While the robot model is included in RobCAD, the model of the workcell has to be created by the user. This operation can be achieved in three steps:

- measurement of the real location for each object in the workcell,
- design of their CAD models,
- graphic construction of the workcell model.

The measurements of real object locations were performed by ESTEC Metrology Office using the 3D theodolite-based measuring system WILD TMS with a measuring accuracy of ± 0.1 mm; all measurements were expressed in the robot base frame.

The design of each component of the workcell has been implemented in a special environment of RobCAD, named `component`, by referring to the original drawings used for their constrcution.

The graphic assembly has been carried out in a different environment of RobCAD, named `workcell`. The accuracy in the location of the rack models has the same order as that reached in the measurement process.

During the implementation of the automatic video calibration procedure it is necessary to build a list of correspondences between the locations and the associated correction vectors to be used by the robot controller. This can be achieved according to the following steps:

- encoder measurements of real joint variables and their storage in the controller memory,
- serial communication between the controller and the workstation,
- graphic animation of the task executed by the manipulator with the transmitted data.

The first step is performed by a PDL2 procedure named **measure**. The serial communication is suitably synchronized by a C procedure which cyclically sends a new data request and recognizes a possible string of joint values transmitted from the controller. The last step is carried out by a screen refresh after each set of data which, in view of the typically slow operational speeds in space-oriented robotics applications, allows continuous animation of the CAD model.

The implementation of the video calibration procedure has been developed by devising a simple user-friendly interface. It comprises a command window for the routines to activate and three graphic windows: the first one displays the real object image shot by the camera attached to the gripper, the second one displays the virtual image of the same object obtained by the CAD model of the camera, and the third one displays the CAD model of the whole workcell.

The command window structure is realized by using the RobCAD User Interface Language, whereas a suitably modified version of the software package VideoLab has been used to show the real camera shots. To minimize errors due to image distortion, which are not taken into account in the projection equations used in the calibration procedure, only the central part of the frame is shown. The possibility that the operator selects a point on the real image that is slightly different from its corresponding point on the CAD image has been reduced by implementing a cross-shaped cursor which allows selection of the single pixel on the screen.

The solution of the linear system equations required at each step of the calibration technique is performed by a Gaussian elimination routine developed in PASCAL to transform the coefficient matrix in triangular form.

The RobCAD communication system with the database is a serial pipe and does not allow communication with other processes running outside the RobCAD environment. The linear system solving routine has to receive 2D coordinates from VideoLab and 3D coordinates from the RobCAD database. To overcome the above drawback VideoLab could have been treated as a child process launched inside the RobCAD environment from the parent calibration procedure. In this way both processes could have had access to the pipe communication system, transferring data from one to the other through the database. Unfortunately the child process does not inherit the graphic features from the parent process, so that no real image could have been shown and thus no selection of 2D coordinates would have been possible.

The interprocessing communication has been efficiently implemented by creating a shared memory segment in the RobCAD environment where all the 2D coordinates of the selected image points are stored in order to be used by the calibration procedure.

The main menu displayed by RobCAD has four main procedures: **programming**, **simulation**, **calibration**, **execution**. The last two procedures are of concern for the actual calibration procedure and their operation is described below.

To start off calibration, take the robot to the desired location by driving the joint variables obtained by kinematic inversion; an error displacement will result in view of model uncertainties.

Activate the procedure **calibration**; a new menu appears on the command window and, at the same time, the procedure **monitoring** drives the CAD model of the manipulator to the corresponding location. A shared memeory segment in the RobCAD environment is also initialized so that VideoLab can access it and get the data used to form the video image.

The user can now select on the command window the function **target object**; a form will appear on the screen and the user has to select a workcell component. After this, two functions are selectable: **point** and **accept**. Clicking on the first one causes a new form to be displayed. The user has to select a point on the object model and its coordinates are obtained from the RobCAD database; then the same point has to be selected on the video image to get its 2D coordinates to be stored in the shared memory through VideoLab. If there is a mistake in the selection, the user can repeat the operation until the correct correspondence is created. Once the selections are correct, the second function has to be clicked to get the 3D coordinates which are stored as well in the assigned shared memory segment. This sequence of operations is repeated as many times as required by the calibration technique. Three couple of points are to be selected if the focal lengths are known, while four are needed if they have to be computed too.

When the point acquisition phase is over, the function **calibrate** becomes active on the command window. All the steps in the Newton's algorithm are executed, including proper transformation of 3D data from the base frame to the object frame; the initial data of h is the current one available in the RobCAD database.

Monitoring of the calibration task can be obtained by selecting the function **execution** in the main menu of the command window. A new menu appears with two functions. The first one, named **monitoring**, causes the CAD model to show the movements and the operations executed by the manipulator in real time. On the other hand the second one, named **live** shows the video camera shots generated by VideoLab.

V. Results

The video calibration procedure acquires from the RobCAD database the 3D coordinates of a set of three points chosen on the grappling fixture referred to the robot base frame and from VideoLab the corresponding 2D coordinates referred to the image plane. The obtained values are compared and the Newton's algorithm is used to obtain matching of the computed and real values of video camera location.

To simplify the debugging of the procedure, data acquisition from VideoLab has been simulated first. It is assumed that the real camera is placed in a number of different locations and the projection equations (1),(2) are used to compute the corresponding 2D points. The resulting errors between the real camera locations and the locations computed through the camera model are used to apply the calibration procedure. The results are reported in Tab. 1 for a number of 10 tests, where the two rows for each test are respectively the initial and final displacements referred to the camera model location used as starting value by the procedure. It can be recognized that both the position and orientation final errors are very limited.

Other tests have been performed by considering errors also on the focal lengths and thus including other two points. The results, not reported here for brevity, have demonstrated that the procedure is able to estimate the focal values with a precision of the order of 10^{-4}, but the accuracy on the location parameters is decreased by an order of magnitude on the average. Also the starting displacement error had to be less than 40 mm in translation and 15 deg in rotation, otherwise the procedure did not converge [5]. From a physical viewpoint this means that the chosen points have to be projected on the linear part of the image where the distortion effect can be considered negligible.

Having checked the numerical robustness of the calibration procedure, the real camera data can be used from the various image shots. At this stage it is crucial that the vision system itself be accurately calibrated (focal lengths and origin of the image plane) because this will affect the entire robot calibration procedure. The difficulties concerned with this kind of calibration are discussed in [5].

Finally a local calibration test has been carried out. Usually the displacement between the actual robot end effector location and that assumed by the model is on the order of a few millimeters in translation and a few degrees in rotation. Therefore the accuracy requested after calibration is typically of 0.1 mm in translation and 0.1 deg in rotation.

The real manipulator end effector location measured with a 3D tool was (1146.24, -428.85, 1367.70, 0.0, 0.0, 0.2), while the result of the calibration gave (1145.87, -429.92, 1367.34, 0.64, -0.03, 0.0), that is a final displacement error of (0.37 1.07 0.36 -0.64 0.03 0.2). The obtained accuracy is less than expected and this was imputable mainly to imperfections of the calibration tool; more details can be found in [5]. Current work is in progress at ESTEC to develop a more accurate calibration tool for the vision system so as to obtain a fully satisfactory accuracy of the overall robotic test bed.

Acknowledgements

This work was performed during a six-month stay of the first two authors in the WKR Section of ESTEC, Noordwijk, The Netherlands, under the direction of Wim De Peuter. Useful tutoring by Gianfranco Visentin is gratefully acknowledged.

References

[1] Z.S. Roth, B.W. Moring, B. Ravani, "An overview of robot calibration," *IEEE J. of Robotics and Automation*, 1987.

[2] S. Hayati, K. Tso, G. Roston, "Robot geometry calibration," *IEEE J. of Robotics and Automation*, 1988.

[3] J.M. Hollerbach, "A survey of kinematics calibration," *The Robotics Review*, 1, MIT Press, 1989.

[4] D.C. Lowe, "Three-dimensional object recognition from single two-dimensional images," *Artificial Intelligence*, 31, Elsevier Science Publishers, 1987.

[5] G. Iodice, S. Illiano, *Video Calibration of Internal A&B Technology Test Bed IARTTB*, ESTEC Working Paper n. 1721, 1993.

Calibration involves making several measurements of the endpoint position and orientation error dx, and determining the optimal $d\theta$ and dp that minimize the error in some least-squares sense, *e.g.*, minimizing $\|dx - \frac{\partial f}{\partial \theta} d\theta - \frac{\partial f}{\partial p} dp\|^2$.

The Denavit-Hartenberg (D-H) parameters are easily the most popular set of parameters used for describing the kinematics of mechanisms. Although they are attractive because of their minimal set property, it is well-known that the D-H parameters are singular when neighboring joint axes are nearly parallel: because the common normal varies wildly with small changes in the axis orientation, these parameters are ill-conditioned. Various modifications have been proposed to handle the parallel axis case. Hayati [3] introduces a modified set of four parameters as well as a five parameter representation that also handles prismatic joints. Stone [11] adds two parameters to the D-H model to allow for arbitrary placement of link frames. Zhuang, Roth, and Hamano [13] also propose a six-parameter kinematic representation, in which four parameters represent the joint axis, and the remaining two parameters allow for arbitrary placement of the link frame. Zhuang *etal* emphasize that their model, unlike the D-H or certain other kinematic models, is continuous and parametrically complete.

While many of the above calibration methods overcome the problem of singularities with the D-H representation, most of these methods are ad hoc, and often unnecessarily complicate the calibration procedure with arbitrarily defined parameters and computational algorithms that depend on whether the joint is prismatic or revolute. Also, the majority of these kinematic models are developed expressly for calibration, and their usefulness for other applications remains unclear. In this article we propose a kinematic calibration method based on representing the forward kinematics as a product of matrix exponentials of the form

$$f(x_1, \ldots, x_n) = e^{A_1 x_1} e^{A_2 x_2} \cdots e^{A_n x_n} M$$

Here $x_1, \ldots, x_n$ represent the joint variables, M is a 4×4 homogeneous transformation representing the tool frame in its home position, and the A_i are 4×4 constant matrices of a special structure that completely specify the kinematics of the mechanism. Unlike the D-H representations, the A_i vary *smoothly* with changes in the joint axes, and form a complete set of parameters for modeling mechanisms with both revolute and prismatic joints. This representation, called the *product-of-exponentials* (POE) formula, is based on the concept of a one-parameter subgroup of a Lie group, and provides a modern geometric interpretation of classical screw theory [1]. The POE formula is a zero reference model that eliminates the need to attach local reference frames to each link, and shares a number of similar features with the kinematic model proposed by Mooring and Tang [5]. By drawing upon well-established principles of classical screw theory and modern differential geometry, the POE formula eliminates many of the ad hoc features found in existing calibration algorithms. More importantly, it offers a general kinematic representation that is useful for applications besides kinematic calibration (Park [8]).

The article is organized as follows. In Section 2 we review the necessary background on SE(3), the Lie group of Euclidean motions, and introduce the POE formula for modeling the kinematics of open chains. In Section 3 we linearize the POE formula with respect to its kinematic parameters, and derive a least-squares algorithm for kinematic calibration. Although we do not assume any prior knowledge of Lie groups and Lie algebras on the part of the reader, we suggest consulting [2] for additional background.

2 Geometric Background

The Lie Group SE(3)

For our purposes it is sufficient to think of SE(3), the Euclidean group of rigid-body motions (or the *Special Euclidean Group,* also known in the robotics literature as the *homogeneous transformations*), as consisting of matrices of the form $\begin{bmatrix} \Theta & b \\ 0 & 1 \end{bmatrix}$, where $\Theta \in SO(3)$ and $b \in \Re^3$. Here SO(3) denotes the group of 3×3 proper rotation matrices. SE(3) has the structure of both a differentiable manifold and an algebraic group, and is an example of a *Lie group.*

Associated with any matrix Lie group is its *Lie algebra,* which for our purposes can be thought of as the set of tangent matrices at the identity element. On SO(3) it can be shown that its Lie algebra, denoted, so(3), consists of 3×3 real skew-symmetric matrices of the following form:

$$\begin{bmatrix} 0 & -\omega_3 & \omega_2 \\ \omega_3 & 0 & -\omega_1 \\ -\omega_2 & \omega_1 & 0 \end{bmatrix} \triangleq [\omega] \tag{1}$$

where $\omega \in \Re^3$. Note that an element $[\omega] \in so(3)$ can also be represented as a vector $\omega \in \Re^3$. It can also be shown that the Lie algebra se(3) consists of 4×4 matrices of the form $\begin{bmatrix} [\omega] & v \\ 0 & 0 \end{bmatrix}$, where $[\omega] \in so(3)$ and $v \in \Re^3$.

An important connection between a Lie group and its Lie algebra is the *exponential mapping*—defined on each Lie algebra is the exponential mapping into the corresponding Lie group. On matrix groups the exponential mapping corresponds to the usual matrix exponential, *i.e.*, if A is an element of the Lie algebra, then $\exp A = I + A + \frac{A^2}{2!} + \ldots$ is an element of the Lie group. On so(3) and se(3) the exponential mapping is *onto*, *i.e.*, for every $X \in SO(3)$ (resp., SE(3)) there exists an $x \in so(3)$ (resp., se(3)) such that $\exp x = X$. The following explicit formula for the exponential mapping on se(3) is derived in Park [10].

Lemma 1 *Let* $[\omega] \in so(3)$ *and* $v \in \Re^3$, *and* $\|\omega\|^2 = w_1^2 + w_2^2 + w_3^2$. *Then*

$$\exp \begin{bmatrix} [\omega] & v \\ 0 & 0 \end{bmatrix} = \begin{bmatrix} \exp[\omega] & Av \\ 0 & 1 \end{bmatrix} \tag{2}$$

is an element of SE(3), *where*

$$\exp[\omega] = I + \frac{\sin\|\omega\|}{\|\omega\|} \cdot [\omega] + \frac{1 - \cos\|\omega\|}{\|\omega\|^2} \cdot [\omega]^2 \tag{3}$$

$$A = I + \frac{1 - \cos\|\omega\|}{\|\omega\|^2} \cdot [\omega] + \frac{\|\omega\| - \sin\|\omega\|}{\|\omega\|^3} \cdot [\omega]^2 \tag{4}$$

Note that if A is an element of some Lie algebra, then $e^{At}, t \in \Re$, itself forms a group, in this case a subgroup of the Lie group. Such groups are called ***one-parameter subgroups*** of a Lie group, and play a special role in the kinematic description of mechanisms.

Explicit formulas for the inverse map log : SE(3) $\rightarrow$ se(3) can also be derived (Park [8]). Although SE(3) is not a vector space, se(3) is: the log formula provides a set of ***canonical coordinates*** for representing neighborhoods of SE(3) as open sets in a vector space.

Lemma 2 *Let* $\Theta \in \mathrm{SO}(3)$ *such that* $\mathrm{Tr}(\Theta) \neq -1$, *and let* ϕ *satisfy* $1 + 2\cos\phi = \mathrm{Tr}(\Theta)$, $|\phi| < \pi$. *Finally, suppose* $b \in \Re^3$. *Then*

$$\log \begin{bmatrix} \Theta & b \\ 0 & 1 \end{bmatrix} = \begin{bmatrix} [\omega] & A^{-1}b \\ 0 & 0 \end{bmatrix} \tag{5}$$

where $[\omega] = \frac{\phi}{2\sin\phi}(\Theta - \Theta^T)$, *and*

$$A^{-1} = I - \frac{1}{2} \cdot [\omega] + \frac{2\sin||\omega|| - ||\omega||(1 + \cos||\omega||)}{2||\omega||^2 \sin||\omega||} \cdot [\omega]^2 \tag{6}$$

Also, $||\log\Theta||^2 = \phi^2$.

The matrix exponential and logarithm show explicitly the connection between SE(3) and the classical screw theory of rigid-body motions: the screw $(\omega, v) \in \Re^6$ defines a rigid-body motion given by the matrix exponential of Lemma 1.

An element of a Lie group can also be identified with a linear mapping between its Lie algebra as follows. Let **G** be a matrix Lie group with Lie algebra **g**. For every $X \in \mathbf{G}$ the mapping $\mathrm{Ad}_X : \mathbf{g} \rightarrow \mathbf{g}$ defined by $\mathrm{Ad}_X(x) = XxX^{-1}$ is linear. Ad_X is known as the ***adjoint map*** (or ***adjoint representation***) of X. If $X = \begin{bmatrix} \Theta & b \\ 0 & 1 \end{bmatrix}$ is an element of SE(3), then its adjoint map acting on an element $x = \begin{bmatrix} [\omega] & v \\ 0 & 0 \end{bmatrix}$ of se(3) is given by

$$\mathrm{Ad}_X(x) = \begin{bmatrix} \Theta & b \\ 0 & 1 \end{bmatrix} \begin{bmatrix} [\omega] & v \\ 0 & 0 \end{bmatrix} \begin{bmatrix} \Theta^T & -\Theta^T b \\ 0 & 1 \end{bmatrix} = \begin{bmatrix} \Theta[\omega]\Theta^T & [b]\Theta\omega + \Theta v \\ 0 & 0 \end{bmatrix} \tag{7}$$

It is not difficult to verify that $\Theta[\omega]\Theta^T = [\Theta\omega]$. Alternatively if x is regarded as a six-dimensional column vector the Ad mapping admits the 6×6 matrix representation

$$\mathrm{Ad}_X(x) = \begin{bmatrix} \Theta & 0 \\ [b]\Theta & \Theta \end{bmatrix} \begin{bmatrix} \omega \\ v \end{bmatrix} \tag{8}$$

It is also easily verified that $\mathrm{Ad}_X^{-1} = \mathrm{Ad}_{X^{-1}}$ and $\mathrm{Ad}_X \mathrm{Ad}_Y = \mathrm{Ad}_{XY}$ for any $X, Y \in$ SE(3). These identities will be particularly useful in linearizing the POE formula.

The Product-of-Exponentials Formula

The product-of-exponentials (POE) formula is a useful theoretical and computational tool for modeling the kinematics of open chains (see Brockett [1], Paden and Sastry [7], Park [8]) in which the the connections between kinematics and Lie theory are shown explicitly. If a right-handed reference frame is fixed at the tip of each link of the chain, then the Euclidean transformation which describes the position and orientation of the ith frame in terms of the $(i-1)^{st}$ frame is $f_{i-1,i} = e^{P_i x_i} M_i$, where $M_i \in \mathrm{SE}(3)$, $P_i \in \mathrm{se}(3)$, and $x_i \in \Re$ is the joint variable, $i = 1, 2, \ldots, n$. The frame fixed at the tip is related to that at the base by the product

$$f(x_1, \ldots, x_n) = e^{P_1 x_1} M_1 e^{P_2 x_2} M_2 \cdots e^{P_n x_n} M_n \tag{9}$$

This equation can be simplified using the fact that $M^{-1} e^P M = e^{M^{-1} P M}$, so that by repeatedly applying the identity $M e^P = e^{M P M^{-1}} M$, f can be written

$$f(x_1, x_2, \ldots, x_n) = e^{A_1 x_1} e^{A_2 x_2} \cdots e^{A_n x_n} M \tag{10}$$

where $A_1 = P_1$, $A_2 = M_1 P_2 M_1^{-1}$, $A_3 = (M_1 M_2) P_3 (M_1 M_2)^{-1}$, *etc.* Similarly, one can apply the identity $e^A M = M e^{M^{-1} A M}$ to rewrite the above as

$$f(x_1, x_2, \ldots, x_n) = M e^{B_1 x_1} e^{B_2 x_2} \cdots e^{B_n x_n} \tag{11}$$

where $B_i = M^{-1} A_i M$, $i = 1, \ldots, n$. The matrix exponentials in this formula can be easily computed from the earlier closed-form formulas.

The POE formula can alternatively be derived independently of the D-H parameters, by appealing to the interpretation of the A_i (B_i) as the screw parameters for joint i's motion (see Paden and Sastry [7]). Specifically, denote the rotational and translational components of A_i (B_i) by ω_i and v_i, respectively. Then ω_i is a unit vector in the direction of joint axis i, expressed in inertial (tip) frame coordinates; v_i is a vector, also described in inertial (tip) frame coordinates, such that the pitch of the screw motion generated by joint i is $\omega_i^T v_i$. Note that for revolute joints the pitch is zero, and $\omega_i \times v_i$ is required to be a point lying on the joint axis; for prismatic joints $\omega_i = 0$ and v_i is in the direction of movement.

Example 1 Let the forward kinematics for the 3R open chain of Figure 1 be of the form $f(x_1, x_2, x_3) = e^{A_1 x_1} e^{A_2 x_2} e^{A_3 x_3} M$. Setting the joint axes to zero, the tip frame $M = (\Theta, b)$ relative to the base frame is given by $\Theta = I$, $b = (L_1 + L_2, 0, 0)$. Joint 1 has a screw axis in the direction $\omega_1 = (0, 0, 1)$; since $\omega_1 \times v_1$ must be a point lying on the joint axis, such that v_1 is normal to ω_1, it follows that $v_1 = (0, 0, 0)$. Similarly, for joints 2 and 3 we have $\omega_2 = (0, -1, 0)$, $v_2 = (0, 0, 0)$, and $\omega_3 = (0, -1, 0)$, $v_3 = (0, 0, -L_1)$.

As the above example illustrates, the POE formula for an open kinematic chain can be obtained directly without resorting to the D-H parameters. The POE formula also has a number of modeling advantages over the D-H representations which make it attractive for calibration. First, it is a type of *zero reference position description* of the kinematics: once a tip and inertial frame have been chosen, and a zero position selected

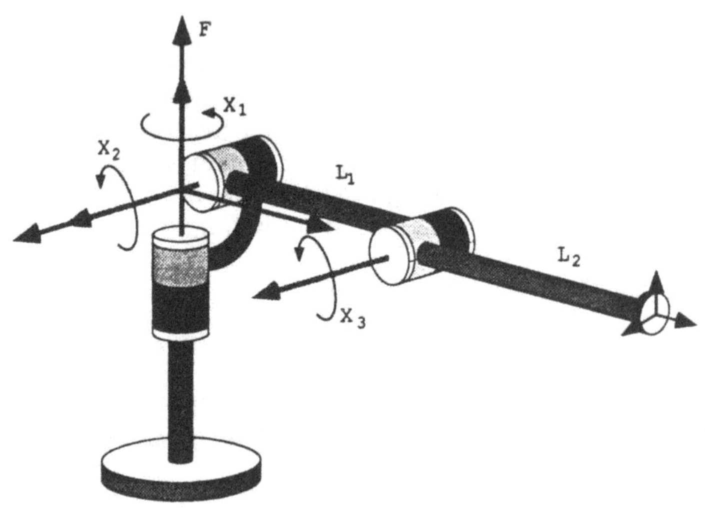

Figure 1: A 3R elbow manipulator.

for each of the joints, there exists a unique set of constant matrices $A_1, A_2, \ldots, A_n \in$ se(3) and $M \in$ SE(3) that describes the forward kinematics of the mechanism. Unlike the D-H representations it is not necessary to attach reference frames to each link. Second, the POE formula treats both revolute and prismatic joints in a uniform way; recall that using the D-H parameters the joint variable can be either θ_i or d_i depending on whether the joint is revolute or prismatic. Third, as discussed earlier, the D-H parameters are extremely sensitive to small kinematic variations when neighboring joint axes are nearly parallel. This sensitivity is what makes kinematic calibration routines based on DH parameters unnecessarily complicated. On the other hand the POE formula is a continuous parametric model, in the sense that the A_i's vary smoothly with variations in the joint axes.

One of the most attractive features of the POE formula is the compact expression for the Jacobian. If f describes the position and orientation of the tip frame relative to the inertial frame, then recall that $\dot{f}f^{-1}$ is an element of se(3) corresponding to the generalized velocity of the tip frame relative to the inertial frame. Using the fact that $(e^{Ax})^{-1} = e^{-Ax}$, a direct calculation shows that

$$\dot{f}f^{-1} = A_1\dot{x}_1 + \mathrm{Ad}_{e^{A_1x_1}}(A_2\dot{x}_2) + + \ldots \mathrm{Ad}_{e^{A_1x_1}\cdots e^{A_{n-1}x_{n-1}}}(A_n\dot{x}_n) \qquad (12)$$

If $\dot{f}f^{-1}$ is rearranged as a six-dimensional vector, then the right-hand side can be expressed as the product of a $6 \times n$ Jacobian matrix $J(x)$ with an n-dimensional joint velocity vector $(\dot{x}_1, \ldots, \dot{x}_n)$, so that each term in the right-hand side can be identified with a column of the Jacobian matrix.

3 Kinematic Calibration Using the POE Formula

3.1 Linearizing the POE Formula

The following result is the key to our POE-based calibration algorithm:

Lemma 3 *Let* $X(t) = e^{A(t)}$ *be a curve in* SE(3). *Then*

$$\dot{X}X^{-1} = \int_0^1 e^{A(t)s}\dot{A}(t)e^{-A(t)s}\,ds \tag{13}$$

In fact, this integral can be evaluated to the following explicit formula. Denoting $A(t)$ by $\begin{bmatrix} [\omega(t)] & v(t) \\ 0 & 0 \end{bmatrix}$ and $e^{A(t)}$ by $\begin{bmatrix} \Theta(t) & b(t) \\ 0 & 1 \end{bmatrix}$, a direct calculation shows that $\dot{\Theta}\Theta^{-1} = [R\dot{\omega}]$, where

$$R = I + \frac{1 - \cos\|\omega\|}{\|\omega\|^2}\cdot[\omega] + \frac{\|\omega\| - \sin\|\omega\|}{\|\omega\|^3}\cdot[\omega]^2$$

Also, $b = Rv$, from which a closed-form expression for $\dot{b} - \dot{\Theta}\Theta^{-1}b$ can be obtained.

We apply this result to linearize the POE formula with respect to its kinematic parameters. As before let the forward kinematics be given by Equation (10), where each $A_i = \begin{bmatrix} [\omega_i] & v_i \\ 0 & 0 \end{bmatrix}$. Note that M can also be expressed using the log formula as $M = e^{\Gamma}$ for some constant $\Gamma \in$ se(3). Then

$$\begin{aligned} df\cdot f^{-1} &= d(e^{A_1x_1})e^{-A_1x_1} + e^{A_1x_1}d(e^{A_2x_2})e^{-A_2x_2}e^{-A_1x_1} \\ &+ \ldots + e^{A_1x_1}\cdots e^{A_{n-1}x_{n-1}}d(e^{A_nx_n})e^{-A_nx_n}\cdots e^{-A_1x_1} \\ &+ e^{A_1x_1}\cdots e^{A_nx_n}(dM)M^{-1}e^{-A_nx_n}\cdots e^{-A_1x_1} \end{aligned} \tag{14}$$

From Lemma 3, each $d(e^{A_ix_i})e^{-A_ix_i}$ in (14) is expanded as

$$d(e^{A_ix_i})e^{-A_ix_i} = x_i\int_0^1 e^{A_ix_is}dA_ie^{-A_ix_is}ds + A_idx_i \tag{15}$$

Similarly, $(dM)M^{-1}$ in (14) becomes

$$(dM)M^{-1} = d(e^{\Gamma})e^{-\Gamma} = \int_0^1 e^{\Gamma s}d\Gamma e^{-\Gamma s}ds \tag{16}$$

In terms of the adjoint map (14) can now be expressed as

$$\begin{aligned} df\cdot f^{-1} &= A_1\,dx_1 + \mathrm{Ad}_{e^{A_1x_1}}(A_2)\,dx_2 + \ldots + \mathrm{Ad}_{e^{A_1x_1}\cdots e^{A_{n-1}x_{n-1}}}(A_n)\,dx_n \\ &+ x_1\int_0^1 \mathrm{Ad}_{e^{A_1x_1s}}(dA_1)\,ds + x_2\mathrm{Ad}_{e^{A_1x_1}}\left(\int_0^1 \mathrm{Ad}_{e^{A_2x_2s}}(dA_2)\,ds\right) \\ &+ \ldots + x_n\mathrm{Ad}_{e^{A_1x_1}\cdots e^{A_{n-1}x_{n-1}}}\left(\int_0^1 \mathrm{Ad}_{e^{A_nx_ns}}(dA_n)\,ds\right) \\ &+ \mathrm{Ad}_{e^{A_1x_1}\cdots e^{A_nx_n}}\left(\int_0^1 \mathrm{Ad}_{e^{\Gamma s}}(d\Gamma)\,ds\right) \end{aligned} \tag{17}$$

The linearized equation above can also be expressed in the usual matrix form $\mathbf{Ap} = \mathbf{y}$, where $\mathbf{p} \in \Re^{7n+6}$ is the vector of kinematic parameters

$$\mathbf{p} = [dx_1 \; dx_2 \cdots dx_n \; d\omega_1^T \; dv_1^T \cdots d\omega_n^T \; dv_n^T \; d\omega_M^T \; dv_M^T] \tag{18}$$

and $\mathbf{y}$ is the six-dimensional representation of $df \cdot f^{-1}$. Let

$$e^{A_i x_i} = \begin{bmatrix} \Theta_i & b_i \\ 0 & 1 \end{bmatrix}, \quad e^{A_i x_i s} = \begin{bmatrix} R_i(s) & d_i(s) \\ 0 & 1 \end{bmatrix} \tag{19}$$

$$M = e^{\Gamma} = \begin{bmatrix} \Theta_M & b_M \\ 0 & 1 \end{bmatrix}, \quad e^{\Gamma s} = \begin{bmatrix} R_M(s) & d_M(s) \\ 0 & 1 \end{bmatrix} \tag{20}$$

with $e^{A_0 x_0}$ defined to be the identity matrix, and

$$r_k = \left(\prod_{i=0}^{k-1} \begin{bmatrix} \Theta_i & 0 \\ [b_i]\Theta_i & \Theta_i \end{bmatrix} \right) \begin{bmatrix} \omega_k \\ v_k \end{bmatrix} \tag{21}$$

$$Q_k = \left(\prod_{i=0}^{k-1} \begin{bmatrix} \Theta_i & 0 \\ [b_i]\Theta_i & \Theta_i \end{bmatrix} \right) x_k \int_0^1 \begin{bmatrix} R_k(s) & 0 \\ [d_k(s)]R_k(s) & R_k(s) \end{bmatrix} ds \tag{22}$$

$$Q_M = \left(\prod_{i=0}^{n} \begin{bmatrix} \Theta_i & 0 \\ [b_i]\Theta_i & \Theta_i \end{bmatrix} \right) \int_0^1 \begin{bmatrix} R_M(s) & 0 \\ [d_M(s)]R_M(s) & R_M(s) \end{bmatrix} ds \tag{23}$$

Then (17) can be expressed as

$$\begin{aligned} \mathbf{y} &= \Big[r_1 \,|\, r_2 \,|\, \cdots \,|\, r_n \,|\, Q_1 \,|\, Q_2 \,|\, \cdots \,|\, Q_n \,|\, Q_M \Big] \mathbf{p} \\ &\triangleq \mathbf{Ap} \end{aligned} \tag{24}$$

Next we consider the left-hand side $df \cdot f^{-1}$. Let T_n and T_a be the nominal and actual end-effector frames, respectively,where T_a is obtained from measurement data and T_n is computed using the nominal parameter values. Then $df \cdot f^{-1} = (T_a - T_n)T_n^{-1} = T_a T_n^{-1} - I$. If $T_a T_n^{-1} = I + \Omega + \Omega^2/2! + \ldots$, where $\Omega = \log(T_a T_n^{-1})$, and T_a and T_n are sufficiently close, then to first order

$$df \cdot f^{-1} = \Omega = \log(T_a T_n^{-1})$$

3.2 A Least-Squares Algorithm for Kinematic Calibration

Calibration proceeds by positioning the mechanism in many points of the workspace and combining the error vectors $\mathbf{y}_i$ and the Jacobians $\mathbf{A}_i$ into a single equation:

$$[\,\mathbf{y}_1 \;\cdots\; \mathbf{y}_m\,]^T = [\,\mathbf{A}_1 \;\cdots\; \mathbf{A}_m\,]^T \mathbf{p} \tag{25}$$

or more compactly, $\mathcal{Y} = \mathcal{A}\mathbf{p}$. The least-squares solution for $\mathbf{p}$ is, as well known, $\mathbf{p} = (\mathcal{A}^T\mathcal{A})^{-1}\mathcal{A}^T\mathcal{Y}$. If $\mathbf{P}$ denotes the vector of original kinematic parameters, the updated kinematic parameter values $\mathbf{P}'$ are obtained by $\mathbf{P}' = \mathbf{P} + \mathbf{p}$. Since this is a nonlinear estimation problem, this procedure is iterated until the variations $\mathbf{p}$ approach zero and the parameters $\mathbf{P}$ have converged to some stable values. At each iteration the $\mathbf{A}_i$'s are evaluated with the current parameters.

Note that each $A_i = \begin{bmatrix} [\omega_i] & v_i \\ 0 & 0 \end{bmatrix}$ in the POE formula (10) models a general helical joint, in which the joint axis is directed along ω_i, and the pitch is $\omega_i^T v_i$. Hence, purely revolute joints must satisfy the condition $\omega_i^T v_i = 0$, while for prismatic joints $\omega_i = 0$. If revolute joints are to be modeled as being perfectly revolute (*i.e.*, without any kinematic imperfections that may result in a slightly helical motion) then this additional constraint can be imposed. Hence, to calibrate an n-revolute joint manipulator we minimize

$$J(\mathbf{p}) = \|\mathcal{A}\mathbf{p} - \mathcal{Y}\|^2$$

subject to the quadratic equality constraint $\mathbf{p}^T\mathbf{Q}\mathbf{p} = 0$, where $\mathbf{Q}$ is a $6n \times 6n$ constant block-diagonal matrix with each diagonal block of the form $\begin{bmatrix} 0 & I \\ I & 0 \end{bmatrix}$. Here $I \in \Re^{3\times3}$ is the identity matrix. For prismatic joints only v_i need be identified.

4 Conclusions

In this article we have formulated a kinematic calibration algorithm based on the product of exponentials formula for open kinematic chains. The POE formula enjoys a number of modeling advantages over other comonly used representations, and calibration can be performed in a straightforward and uniform way, without the ad hoc methods of other calibration algorithms. In particular, no special procedures are required for adjacent joints whose axes are nearly parallel. (Further details as well as simulation examples can be found in Okamura [6]). The POE formula can be viewed as a modern geometric formulation of classical screw theory, and as such is a general tool that is useful for a range of applications besides calibration. For example, using the POE formula the dynamic equations can be formulated elegantly and in a form that admits easy factorization and differentiation (Park *et al*, [9]).

The calibration algorithm discussed in this article determines the least-squares solution, but in principle other estimation algorithms can be used, *e.g.*, Levenberg-Marquardt methods, total least-squares, *etc.* Also, only geometric errors have been considered; for more accurate modeling one may wish to include non-geometric effects such as backlash, gear transmission error, and joint compliance. Many of these issues are discussed in Hollerbach [4] and the references cited, and extending the POE calibration method to include these effects merits further study.

References

[1] R. W. Brockett, "Robotic manipulators and the product of exponentials formula," *Proc. Int. Symp. Math. Theory of Networks and Systems*, Beer Sheba, Israel (1983).

[2] M. L. Curtis, *Matrix Groups*, New York, Springer-Verlag (1984).

[3] S. Hayati, "Robot arm geometric parameter estimation," *Proc. IEEE Int. Conf. Decision and Control* (1983).

[4] J. M. Hollerbach, "A survey of kinematic calibration," *The Robotics Review I* (O. Khatib, J. J. Craig, and T. Lozano-Perez, Eds.), Cambridge, MIT Press (1989).

[5] B. W. Mooring and G. R. Tang, "An improved method for identifying the kinematic parameters in a six axis robot," *Proc. Int. Comput. in Eng. Conf. Exhibit*, 79-84, Las Vegas (1984).

[6] K. Okamura, "Geometric calibration of manipulators via the product of exponentials," UCI Mechanical Systems Technical Report, August (1993).

[7] B. Paden and S. S. Sastry. "Optimal kinematic design of 6R manipulators," *Int. J. Robotics Research*, Vol. 7 (1988).

[8] F. C. Park, "Computational aspects of the product-of-exponentials formula for robot kinematics," *IEEE Trans. Auto. Contr.* 39(3), 643-647 (1994).

[9] F. C. Park, J. E. Bobrow, and S. R. Ploen, "A Lie group formulation of robot dynamics," *Proc. 3rd Int. Workshop on Advanced Motion Control*, Berkeley, CA (1994).

[10] F. C. Park, *The Optimal Kinematic Design of Mechanisms*, Ph.D. thesis, Harvard University(1991).

[11] H. W. Stone, *Kinematic Modeling, Identification, and Control of Robot Manipulators*, Boston, Kluwer (1987).

[12] D. E. Whitney, C. A. Lozinski, and J. M. Rourke, "Industrial robot forward calibration method and results," *J. Dynamic Systems, Meas., Control*, vol. 108, 1-8 (1986).

[13] H. Zhuang, Z. S. Roth, and F. Hamano, "A complete and parametrically continuous kinematic model for robot manipulators," *IEEE Trans. Robotics Autom.* vol. 8, 451-463 (1992).

A Systematic Approach to Model Arbitrary Non Geometric Kinematic Errors

M. Vincze, K.M. Filz, H. Gander, J.P. Prenninger, and G. Zeichen

Institute of Flexible Automation
Technical University of Vienna
Gusshausstr. 27-29/361, 1040 Vienna, Austria

Abstract - We have developed the SYNE-model (SYstematic Non-redundant and Extandible) which increases modelling accuracy by adding a minimum number of non-geometric model parameters. The framework of the model is the geometric model. It contains a complete and minimal set of parameters that are chosen depending on the axes configuration of the manipulator. To increase accuracy non-geometric parameters are included without generating redundancy or reducing the systematic approach of the model[1].

I. Introduction

The motion of manipulators can be described with a model of the manipulator's joints and links. The input to the model are axes encoder readings and the output is the pose in three dimensional space. The capability of the model is of the utmost importance concerning the accuracy of the description. Another significant item is to find the actual parameter values. A generally accepted method to yield an accurate set of parameters is to model the mechanism and through measurements determine the model parameter. This process is referred to as calibration. A model for calibration should exhibit the following five properties.

- The model must be complete, i.e., describe any possible motion.
- The model must be proportional [1].
- The model should contain terms for non geometric errors, since motion in no actual mechanism will be ideal (section III.).
- The model should be non-redundant, i.e. there should be no parameter that can be expressed with a combination of other parameters in the model.

[1]This work is supported by the Austrian Science Foundation grant number P9275.

A. J. Lenarčič and B. B. Ravani (eds.), Advances in Robot Kinematics and Computationed Geometry, 129–138.

- And the model should allow parameter identification independent of the measurement method.

Complete models are the zero reference model [2] and the Sheth-Uicker model as used in [3]. In both models redundant parameters must be eliminated before calibration, which is not always a straightforward task. The Complete and Parametrically Continuous (CPC) model [4] is based on a singularity free line description, but is in its application more computationally expensive than the conventional DH-model [5]. Redundancies are handled by deriving explicit equations for the redundant parameters. An extension of the CPC model towards non geometric parameters has not been presented yet. Duelen and Schröer [6] distinguish several cases of axes configurations, but their model also contains redundant parameters that are suppressed for identification.

Whitney et. al. [7] and Ahmad [8] first addressed the issue of non geometric parameters. Their approach investigates errors often found with gear driven robots. Additional link deflections are included in the non geometric model in [6]. Everett [3] proposed a Fourier series for each model parameter, which is theoretically capable of describing any error source, but requires an infinite number of parameters.

We propose a model that utilises the well known homogeneous matrix description. Four cases are extracted that enable to describe any configuration of mechanical axes with a non-redundant geometric model (section II.). Section III. summarises non geometric errors and their models and section IV. lists the rules to include the models into the geometric description.

II. Systematic Geometric Model

There are two assumptions at the level of geometric modelling. First, the motion accomplishes a lower-pair joint, and second, the links connecting the joints are rigid. To meet the aforementioned properties it is necessary to consider four different cases of geometric axes configurations.

Last Joint Axis: The transformation $\mathbf{A}_{gen}$ between a coordinate system (CS) on a joint axis and a CS on the link that moves around this axis is described using six parameters for the six possible degrees of freedom (DOF) in 3D space. Applying the standard homogeneous notation [5] this transformation is given by

$$\mathbf{A}_{gen} = \mathbf{Rot}(z,\theta)\mathbf{Rot}(y,\beta)\mathbf{Rot}(x,\alpha)\mathbf{Trans}(a,b,d), \tag{1}$$

where $\theta, \beta, \alpha, a, b$ and d are the actual parameters each composed of the nominal parameter plus an error (e.g. $a = a_{no} + \Delta a$). The angle θ also contains the joint variable q. This means that for the identification only the summed (i.e. actual) parameter can be identified (e.g. $a_{no} + \Delta a$).

Two Orthogonal Joint Axes: If two axes are orthogonal the well-known Denavit and Hartenberg model [9] best suits the requirements mentioned. This model is applied in all cases where the angle between the two axes is greater or equal to 45 degrees.

Two Parallel Joint Axes: If the configuration of two consecutive axes does not belong to the case of orthogonal axes, the parallel model is applied. The parallel model used has been introduced by Hayati [10], and contains four parameters, just as the orthogonal model (this is sufficient as geometrical considerations confirm [11, 12]).

Transformation between World CS and Base CS: In most cases of robot placement the z axis of the world CS and the robot base CS coincide. Then the transformation between the world or measurement CS and the robot base CS is given by

$$\mathbf{A}_w = \mathbf{Rot}(y, \beta)\mathbf{Rot}(x, \alpha)\mathbf{Trans}(a, b, 0). \quad (2)$$

This transformation holds four parameters (defined as before) and no joint variable. A translation as well as a rotation around the z axis is redundant to the respective parameter in the models of the other three cases (for a thorough treatment see [11]). If the axis of the base CS points along the x or y coordinate axis, the x respectively y parameters of the general transformation in eq. (1) are omitted.

III. Non Geometric Models

The aim of the non geometric terms is to increase model accuracy towards the limit set by the axis repeatability. Since accuracy is a static characteristic, dynamic errors are not regarded.

Besides proportionality and non-redundancy, a model for non geometric errors should satisfy the following properties.

- The model gives a good (in the sense of sufficiently accurate for the application at hand) description with a minimum number of measurements, since these are generally tedious.
- The model displays a unique relation between a model parameter and the physical source of error.

The first requirement is fulfilled by replacing the theoretically perfect Fourier series [3] with a function assembled of terms that more significantly describe an error (e.g., a hysteresis function instead of several orders of a Fourier series).

Tb. 1 shows quite a comprehensive list of errors often found with mechanisms constructed of rotary joints (theodolites, manipulators). It also summarises the effect of each error on accuracy. The effect depends largely on the specific mechanism. Thus the categorising expresses the present view of the authors, based on experiments with a 6 DOF direct-drive (DD) robot and a 6 DOF transmission-drive (TD) robot. Besides the local effect on accuracy the distance of a joint to the end-effector of the mechanism must be taken into account. A model is proposed for errors with strong effect on accuracy. Subsequently, the factors and models are outlined shortly. Since space is limited only the main features are given.

errors at	type of error	effect on DD	effect on TD	model	apply to
joint axis	tumbling	l	s	equation (3)	$a, b, d, \alpha, \beta, \theta$
	axis compliance	s	l	$\theta_c = p_c M$	θ
	clearance in bearing	s	vs	stochastic	—
	backlash	-	l	$\theta_b = p_b \,\text{sign}(M)$	θ
	tooth, tooth spacing	-	s	higher order FS necessary	θ
	gear axes alignment	-	s	1st, 2nd, n_{tr}th order terms	θ
	transmision ration	-	s	polynom, higher order FS terms	θ
link	deflection	e-vl	s-l	$\mathbf{Rot}(z, d\theta) \cdot \mathbf{Rot}(y, d\beta) \cdot \mathbf{Rot}(x, d\alpha) \cdot \mathbf{Trans}(dx, dy, dz)$	after a, b, d
	temperature sensitivity	e-l	s	$a(T) = a(T_0)(1 + a_t \cdot \mu \cdot (T - T_0))$	a, b, d
encoder	non-linear output	s	vs	table, higher order FS terms	θ
	coupling	s	vs	equation (3)	θ
	hysteresis	s	vs	equation (4)	θ
envi-ronment	temperature; draft; humidity; electrostatic or magnetic fields; etc.	nk	nk	stochastic	—
computation	path computation; digital round off; steady state control error;	vs	vs	neglegible, otherwise stochastic	—

symbol	explanation	symbol	explanation
r	repeatability	θ_c	angular axis compliance
vl	very large ($\approx$ 100r)	p_c	axis compliance parameter
l	large ($\approx$ 10r)	θ_b	angular backlash
e	equal ($\approx$ 1r)	p_b	half the total backlach
s	small ($\approx$ 0.1r)	n_{tr}	transmission ratio
vs	very small ($\approx$ 0.01r)	$d\alpha, d\beta, d\theta$	link inclinations
nk	not known	dx, dy, dz	link deflections
DD	direct-drive axis	$a(T)$	link length at temperature T
TD	transmission-drive axis	T_0	reference temperature
FS	Fourier series	a_t	temperature parameter
sign	signum function	μ	heat extension coefficient
M	moment active at joint		

Table 1: List of errors that decrease robot accuracy.

A. Errors at the Joint Axis

Tumbling: The geometric model assumes that the joint axis rotates in itself. Imperfections in the roundness of the axis, the bearings, and the shells will introduce a motion on the joint axis, which is called tumbling. The absolute tumbling error at each bearing is small. Since the distance between two bearings is small compared to the length of a robot arm, the error is enlarged and tumbling cannot be neglected, especially for DD axes.

Tumbling affects each parameter in the geometric model independent of the configuration. It will most probably be cyclic and a Fourier series (FS) therefore offers itself to replace the constant parameters, e.g., to replace a with

$$a = a_{no} + \Delta a + \sum_{j=1}^{n_{sc}} (a_{js} \sin(j \cdot q) + a_{jc} \cos(j \cdot q)) , \quad (3)$$

where n_{sc} is the number of orders applied. Experiences from Fourier analysis and measurements on an axis confirm that the first three orders describe tumbling with an accuracy in the range of axis repeatability [11].

Axis Compliance: Axis compliance for DD-robots can be neglected since the axis is very short and the encoder is mounted close to the motor. TD mechanisms exhibit larger errors due to joint and transmission axes torsion. Additionally the encoder measures the motor and not the axis location.

Backlash: Backlash, results from most types of transmissions, and the subsequent errors at joint level can be found only with TD robots. With gears it can be assumed that the last transmission affects accuracy most. Backlash is best modelled with a hysteresis function around the desired joint location (Tb. 1). It changes direction when the motor torque changes direction and is above a certain threshold [6].

Gear Axis Alignment: Errors from gear axis alignment, eccentric gears or tilted gear axes can be modelled with transcendental functions. Geometrical considerations show that the first, second and n_{tr}th (transmission ratio) order terms are most significant to modelling accuracy [13].

Transmission Ratio: A non constant transmission ratio results from gear eccentricities, bad axis alignment, or tooth errors. These models are already outlined. Principally cyclic or polynomial models offer themselves. Coupling terms must also be considered [6].

B. Errors at the Link

The geometric model assumes the links to be rigid. In reality, deflections and temperature changes influence this relationship.

Link Deflections: Gravity is the only static force acting on a robot. In principal any model of link deflections can be applied. A simple model containing any possible deflection or inclination is proposed in [11]. This model is based on the assumption that gravity acts in the minus z direction of the world or measurement system. It then utilises the theory of the strength of materials to find the deflection terms. The six components yield deflections and inclinations that are introduced into the geometric model after the link extension. The deflection model is similar to the general model of eq. (1) and given in Tb. 1. The six parameters describe the actual behaviour of the link, i.e., the actual cross section, Young's modulus, shear modulus, the constant moments of inertia, and the constant torsional moment of inertia.

The six deflection components compose no redundancies to the other model parameters, since they are proportional to the forces and moments at each joint.

Though the parameters are in principle free of redundancies, the model must be applied carefully. Due to the special configuration of many available robots not all forces act at each link. Even if a force acts it must be secured that the value changes. For example, if the load on the robot is not changed and the first link is parallel to the z axis of the world coordinate system (the common set-up), then a constant force acts along the z direction. A constant elongation therefore results, which can be identified only together with the geometric parameter. A similar consideration holds for a moment around a first vertical robot axis.

Temperature: Temperature changes can result from the environment, or, especially for DD motors, heat developing while the motor is in motion or must hold a load.

The influence of temperature sensitivity on accuracy can be modelled with the application of sensors. A relation between a temperature change and a change in the model parameter is established for each sensor mounted to the link. Each length (small offsets that are in the ideal model equal or close to zero can be neglected) is modelled with such an expression. This expression is a first estimate of the behaviour of the length of an axis due to temperature variations. By adding more temperature sensors or additional (polynomial) terms, this expression can be made arbitrarily complex.

C. Errors at the Encoder

The limit in accuracy of an encoder is its resolution. The errors vary with the type of encoder used. The following list concentrates on the most commonly applied disc encoders.

Non-linear Output: Non-linear encoder characteristics result from spacing errors of the disc marks. They can be tabulated or a higher order FS (Fourier series) employed. Other non-linear errors, such as wrong encoder readings or errors in the signal transmission, are stochastic in nature and therefore difficult to model.

Coupling: The coupling between encoder and joint axis causes inaccuracies, when the axes are misaligned or bracing introduces forces and moments on the encoder axis. These forces or moments will incline the encoder axis and alter the encoder reading. Couplings in the form of cardan joints transform the angle of rotation when the two axes are not aligned. In both cases the error is small. Periodic errors (axis tilt, eccentricities, bearing or roundness errors, and others) overlap with the tumbling model of the same joint parameter, but the amplitude is at least one magnitude smaller.

Hysteresis Contrary to joint axes, an encoder axis carries no load - it is compensated by the motor. The only load effective on the encoder axis is the encoder inertia and friction. In a static consideration, the resulting error is an angular offset θ_{hy} caused by stiction and the spring constant of the coupling, formulated as

$$\theta_{hy} = \begin{cases} p_{hy} < (\theta - \theta_{ch}) : p_{hy} \cdot \text{sign}(\theta - \theta_{ch}) \\ otherwise : 0 \end{cases} , \tag{4}$$

where p_{hy} is half the total hysteresis and θ_{ch} the joint location at the point of change in rotational direction.

D. Environmental and Computational Errors

In general, environmental errors are stochastic and evade a modelling. Factors such as draft or room temperature influence the link temperature and are treated together with the link errors. Other factors, such as, humidity, electrostatic, or magnetic fields, are very difficult to detect and cannot be modelled. It is best to avoid them in robot environments.

Errors resulting from robot path computation or digital round off can be neglected with the currently available computers. Steady state control errors must be reduced with appropriate control algorithms. For today's robots this can be presumed.

IV. The SYNE-Model

Any chain of axes can be separated in the above four cases of axis configurations. If the transformations are strictly combined in the sense of the classification, a geometrically complete, proportional, and non-redundant geometric model results. Furthermore, the number of parameters is minimal, i.e., any elimination of a parameter renders the model incomplete, and any additional parameter renders the geometric model redundant.

The first step in the application of the SYNE-model is to find the geometric model. With this basis the accuracy can be evaluated and the effect of non geometric parameters investigated.

A. Significant Parameters

If there are ample measurements available, many orders in the FS can be identified. From the magnitude of the identified parameters conclusions can be drawn on which parameters to select [3]. If few measurements are available, tests on certain factors can help to find the errors with high effect on accuracy.

The deflection of the robot is easily tested by applying additional loads to the end-effector with fully extended arm. The amount of the resulting deflection is detected with a simple repeatability measurement system.

Similarly, the temperature sensitivity can be tested during the start up phase or by artificial heating of the arms by applying high loads or fast movements between measurements.

With a disc mounted on the end-effector axis tumbling can be measured [11]. Usually the first two or three orders are sufficient to yield accuracy in the range of repeatability.

B. Redundancy of Non Geometric Parameters

The investigation of significant non-geometric errors tells us, which models to include into the geometric model. Tb. 1 gives an overview of models for various non-geometric errors. The combination of several errors has to obey the following rules:

1. the geometric parameter is replaced by a function (Tb. 1 gives some examples),
2. add the models that are functions of sensor readings or other inputs (e.g, the deflection model), and
3. similar terms in the function are combined (only one parameter for each function term can be identified).

The first and second rule establish the full model including both geometric and non-geometric terms. The third rule secures that each parameter function contains only mathematically orthogonal terms.

For example, constant offsets, each order in a FS, or polynomial terms should show up only once, since all these terms depend on the joint angle. The terms or matrices for models depending on sensors or other inputs (e.g., forces and moments on the link) do not interfere with these functions. It must be also checked that the functions on sensor or other inputs contain only orthogonal terms.

The third rule is a simple method to generate a non-redundant model. If the parameter functions for each parameter are non-redundant, then **the entire SYNE-model is non-redundant.** The replacement of the constant geometric parameters with parameter functions does not disturb the properties of the geometric model.

The physical interpretation of the parameter functions is straight forward. Each parameter indicates the error to a particular model and thus a special form of error. Merely in the case of overlapping non-geometric models the attribution is difficult. For example, tumbling around the joint axis and encoder coupling errors both use the same terms of a FS. It is therefore not possible to differentiate the physical magnitude of the tumbling error and the magnitude of the encoder coupling error. Only the combined error can be modelled. To detect the separate errors special measurements at the axis are necessary, but the result on the end-effector will be the same.

An advantage of the SYNE-model formulation is that the limitation to the minimum of necessary parameters yields low computational complexity. The composite homogeneous transformations are less computationally expensive than other models [2, 3, 4]. The main advantage of the SYNE-model is the simplicity in its application and the possibility to include arbitrary functions to yield the requested model accuracy.

C. Parameter Identification

After modelling, a transformation matrix describes each joint and link. The composite transformation is then used to identify the parameters that best fit the measurements of the end-effector pose. This approach is the most consistent one since the desired relationship between end-effector pose and joint angles is used directly.

The number of parameters in the SYNE-model might be large; e.g., the first PUMA robot axis, which is followed by an orthogonal axis, could be described with a geometric model, with tumbling (first 3 orders), axis compliance, backlash, gear axis alignment, and deflection terms contains 35 (4;24;1;1;3-2;6-2) parameters. Four parameters are redundant and eliminated: the 1st and 2nd order terms in the gear axis alignment model are redundant to the tumbling model, and the parameters for deflection and inclination around the z axis are constant and therefore already modelled with the geometry.

Though the number of parameters might be large, the basic Gauß-Newton algorithm can be utilised for identification. It can be used since the strict geometric modelling and the three rules formulated above for non-geometric terms form a non-redundant model. When applying this algorithm, the identification procedure converges fast and, the more measurements are available, the more reliable are the results.

Fig. 1 shows a typical convergence for a three axis model with 129 parameters containing terms for tumbling, encoder hysteresis, temperature sensors, and link deflection [11]. The slower convergence in the second iteration results from the deflection parameters. The relation between the deflection parameters and accuracy is different to the same relation of other model parameters. The reason is that the deflection model uses the original constant parameters of a model exploiting beam theory. The magnitude of these parameters could be adjusted to yield a relation in the same magnitude as the other parameters, but identification convergences nonetheless.

In Fig. 1 the adjustment of magnitude between the parameters is adjusted in iteration step 2. Afterwards convergence is unperturbed. The convergence criteria used is the euclidian distance between the poses of the simulated measurements and the poses reached with the new set of parameters.

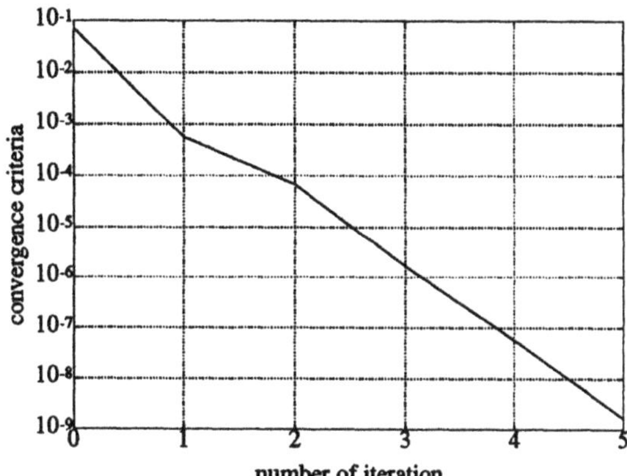

Figure 1: Convergence of the parameter identification procedure.

Measurements of actual robots contain a certain amount of measurement error. The remaining error between measurements and model cannot be reduced as indicated in Fig. 1. Tests show that after two iterations the optimal parameters are already found with a precision of less than a micrometer [11]. This is another confirmation of the reliable and smooth convergence of the identification procedure.

V. Conclusion and Perspective

The SYNE-model presented describes the motion of open loop mechanisms with arbitrary accuracy. Since the model is systematic it is simple to apply. The non geometric errors can be added depending on the specific mechanism or robot. Since the model is non-redundant the parameter identification is simple and a relation to physical error sources can be established.

Two further extensions of the SYNE-model comprise prismatic joints and closed loop mechanisms. Both cases will be (hopefully) achieved by devising appropriate rules to define the geometric models.

The accuracy of existing measurement systems sets presently the limit on accuracy improvements with the application of then SYNE-model. Measurements better than $0.1mm$ are hardly achievable. A precise pose measurement systems with an accuracy in the range of a robot's repeatability is presently in development at our Institute [14].

VI. References

[1] L.J. Everett, M. Driels, B.W. Mooring, "Kinematic Modelling for Robot Calibration", *Conf. on Robotics and Automation Vol.1*, 183-189 (1987).

[2] B.W. Mooring, Z.S. Roth, M.R. Driels, "Fundamentals of Manipulator Calibration", *John Wiley & Sons*, (1991).

[3] L.J. Everett, "Models for Diagnosing Robot Error Sources", *Conf. on Robotics and Automation Vol.2*, 155-159 (1993).

[4] H. Zhuang, Z.S. Roth, F. Hamano, "A Complete and Parametrically Continuous Kinematic Model for Robot Manipulators", *IEEE Transactions on Robotics and Automation 8*(4), 451-463 (1992).

[5] R.P. Paul, "Robot Manipulators - Mathematics, Programming and Control", *MIT Press*, (1981).

[6] G. Duelen, K. Schöer, "Robot Calibration - Method and Results", *Robotics & Computer Intergrated Manifactoring 8*(4), 223-231 (1991).

[7] D.E. Whitney, C.A. Lozinsky, J.M. Rourke, "Indutrial Robot Calibration Method and Results", *Proceedings International Computers and Engineering Conference*, 92-100 (1984).

[8] S. Ahmad, "Second Order Nonlinear Kinematic Effects, and their Compensation", *Conf. on Robotics and Automation*, 307-314 (1985).

[9] J. Denavit, R.S. Hartenberg, "A Kinematic Notation for Lower-Pair Mechanisms based on Matrices", *Trans. of ASME J. Applied Mechanics 22*, 215-221 (1955).

[10] S.A. Hayati, "Robot Arm Geometric Link Parameter Estimation", *Conference on Decision and Control*, 1477-1483 (1983).

[11] M. Vincze, "A Model of Direct-Drive Axes to Improve Robot Accuracy", *Ph.D. thesis*, Institute of Flexible Automation, Technical University of Vienna, (1993).

[12] W. Khalil, M. Gautier, C. Enguehard, "Identifiable Parameters and Optimum Configurations for Robots Calibration", *Robotica 9*(1), 63-70 (1991).

[13] R.P. Judd, A.B. Knasinski, "A Technique to Calibrate IR with Experimental Verification", *IEEE Trans. on Robotics and Automation 6*(1), 20-30 (1990).

[14] J.P. Prenninger, M. Vincze, H. Gander, "Contactless Position and Orientation Measurement of Robot End-Effectors", *IEEE Conference on Robotics and Automation, Vol.1*, 180-185 (1993).

4. Kinematics of Mobile Robots

The Wheeled Actively Articulated Vehicle (WAAV): An Advanced Off-Road Mobility Concept

S. V. Sreenivasan, P. K. Dutta, and K. J. Waldron

Advanced Robotics Laboratory
Department of Mechanical Engineering
The Ohio State University
Columbus, OH 43210, U. S. A.

Abstract

This manuscript contains descriptions of the kinematic configuration, the mechanical systems, and the sensing, control and actuation sub-systems of the Wheeled Actively Articulated Vehicle (WAAV). It includes a discussion of the current capabilities of the WAAV. System enhancements that are being considered for future experiments have also been addressed.

1 Introduction

The Wheeled Actively Articulated Vehicle (WAAV) shown in Figure 1 is an experimental system that has been developed to demonstrate the use of *active coordination* in terrain adaptive wheeled vehicles. Actively coordinated vehicle systems are desired in fields which require off-road mobility, and these fields include military applications, agriculture, particularly forestry, mining and planetary exploration. Actively coordinated vehicles refer to vehicles that possess independently controlled actuators for the suspension and the locomotion degrees of freedom. The contact force vectors at the vehicle-terrain contact locations can be directly influenced by using these actuators. These vehicles also have the capability to vary their geometry to accommodate to terrain obstacles. The coordination of these vehicles requires digital integration of all the actuators and associated sensors, and the control strategy requires an understanding of the kinematics and dynamics of spatial hybrid series-parallel chains. Actively coordinated vehicle systems should possess superior mobility characteristics in unstructured terrains when compared to traditional, passively suspended vehicle systems. Traditional vehicles differ from actively coordinated vehicles since they emphasize the use of systems with a single prime mover, remotely actuating the vehicle degrees of freedom, with the help of a mechanical transmission unit. In addition, they do not possess the capability of varying their configuration to surmount obstacles.

Actively coordinated vehicle systems such as the WAAV are different from road vehicles with active suspensions, such as the one discussed by Milliken [1988]. In that work, only the three degrees of freedom which are accommodated by passive suspensions in conventional vehicles, are under integrated active control. These systems essentially function as active vibration dampers. The WAAV is also distinct from vehicles that possess complex kinematic configurations, but only a subset of their degrees of freedom are controlled actively. Examples of such systems include the CARD [Wilcox and Gennery, 1987], and the Rocker-Bogie [Bickler, 1990; Chottiner, 1992] vehicle configurations. While these machines have relatively large numbers of independently controllable degrees of freedom, they lack the ability to control the distribution of load among the wheels or to configure themselves to optimally attack obstacles or changing terrain conditions.

Actively coordinated vehicles include walking machines and wheeled locomotion systems. Walking machines have been studied extensively in recent years [McGhee and

A. J. Lenarčič and B. B. Ravani (eds.), Advances in Robot Kinematics and Computationed Geometry, 141–150.

Ishwandhi, 1979; Hirose and Umetani, 1980; Pugh *et al.*, 1990; Raibert, 1986; Song and Waldron, 1988; Kumar and Waldron, 1990]. Actively coordinated wheeled vehicles have not been studied in as much detail. Some of the coordination issues of active wheeled systems operating in complex environments can be found in the literature [Waldron *et al.*, 1991a, 1991b, 1987; Kumar and Waldron, 1989; Waldron, 1989]. The mechanical design of the WAAV was described by Yu and Waldron [1991], and the WAAV mechanical hardware was fabricated almost entirely based on this design. A vehicle configuration similar to the WAAV is the Articulated Transporter/Manipulator System (ATSM) described by Chiang *et al.* [1992]. This vehicle is not fully actively coordinated since the articulations joining adjacent vehicle modules are actively actuated only about two axes (the pitch and the yaw). Of course the ATSM is intended only for operation in a 2D environment with 2D obstacles. Therefore, even though its geometry is similar to that of the WAAV, its coordination and control issues are quite different. The WAAV is almost identical in its geometry to a vehicle that was studied by Martin-Marietta (see Figure 2) [Spiessbach and Woodis, 1988]. Martin-Marietta only studied the concept and did not pursue the configuration any further.

The kinematic configuration, and the mobility features of the WAAV are discussed in Section 2. The mechanical systems and the sensing, control and actuation sub-systems of the WAAV are described in Section 3. A discussion of the current capabilities of the WAAV system and future WAAV experiments that are being planned is included in Section 4.

2 Kinematic Configuration

The kinematic configuration of the WAAV is shown in Figure 3. Each module consists of a body and two wheels and the consecutive modules are connected by three degree of freedom articulations consisting of three successively orthogonal revolute joints. The six wheels and the three degrees of freedom of the articulations are independently actuated. The axle of the middle module is mounted on a guide way so that it can translate along this guide way from left to right (in the fore-aft direction in Figure 3). This translatory motion allows the center of mass of the system to be located on either side of the axle of the middle module as appropriate to maintain stability. This movement of the center of mass is very useful in mobility maneuvers such as step climbing and self-recovery from an accident [Waldron *et al.*, 1987]. The WAAV kinematic configuration has many attractive features that include surmounting large obstacles, crossing wide ditches, and self-recovery from an overturn.

One of the primary objectives of constructing a fully actively coordinated wheeled vehicle is to fully optimize the contact conditions at each of the wheels. This means, effectively, minimizing the maximum ratio of tangential to normal contact force over all the wheels. This has the effect both of maximizing traction and of minimizing power consumption [Waldron *et al.*, 1991a]. Passive distribution of power to the wheels via differentials, as in conventional automotive technology, works well only on surfaces which are close to level. Even then, there is no possibility of optimizing the load distribution among the wheels.

Another capability which comes with an actively controllable configuration, is a capability for anticipatory changes in configuration when approaching large obstacles. Actively coordinated vehicle geometries also possess the capability for self-recovery from roll-overs or traction failures. This capability can be very useful if an autonomous vehicle mission of high reliability is desired [Waldron *et al.*, 1987]. A summary of the mobility capabilities of the WAAV is listed in the table below. The mobility capabilities of the WAAV are discussed in greater detail in Yu and Waldron [1991].

Figure 1 The Wheeled Actively Articulated Vehicle (WAAV)

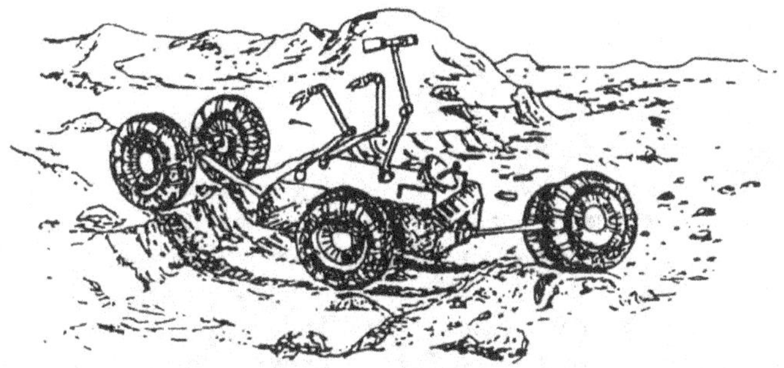

Figure 2 Martin Marietta's Actively Articulated Six Wheeled Vehicle Concept

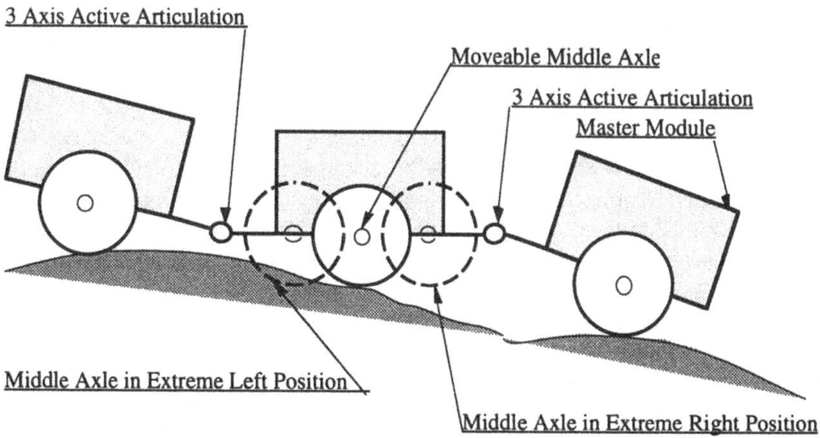

Figure 3 Kinematic Configuration of the WAAV

Table 1 Mobility Capabilities of the WAAV

1	Nominal Dimension	6'2" × 2'5" × 1'6"
2	System Weight	275 lb.
3	Anticipated Maximum Speed	1 mile/hour
4	Gradability	35% loose sand gradient (speed ≈ 0.5 mile/hr) 60% hard surface gradient (speed ≈ 0.5 mile/hr)
5	Articulation joint limits	Roll: ± 180°, Yaw: ± 40°, Pitch: +85° to -120°
6	Middle Axle Travel Stroke	± 4 5/8"
7	Average Anticipated Power Consumption	0.29 Hp.
8	Obstacle Crossing Ability	Vertical: 16", Horizontal: 25"

3 Description of the Sub-Systems

The WAAV has 12 twelve rotary brushless DC motors, one for each of the six wheels, and three for each of the two articulations. These 12 actuators are under continuous microcomputer control. A rotary actuator (a DC motor) and a power screw are used to obtain the fore-aft translatory motion of the middle axle. This actuator is not to be operated under continuous control. It is to be used in a three position mode: middle axle in the extreme left position, middle axle in the extreme right position, or middle axle in the central position. The middle axle in the central position is suited for normal operations involving motion on all six wheels since, on even terrain, this will lead to low loads on the articulation actuators. The middle axle in the extreme positions (right or left) are required during special mobility maneuvers such as step climbing or self-recovery. The middle module will house the onboard computing equipment and some centralized hardware to process the sensor information from the position encoders. The WAAV does not possess a vision system at this time. If a vision system is included, it will be mounted on the 'master' or the first module. The desired kinematic quantities of the master module include the direction of the gravity vector, and the six components of linear and angular velocity. An inertial sensing package that was used in the Adaptive Suspension Vehicle [Pugh *et al.*, 1990] consisted of a vertical gyroscope, rate gyroscopes directed along three orthogonal axes, and accelerometers on the same three axes. A similar inertial package is to be used to obtain the kinematic quantities of the master module. Inertial navigation using 'dead-reckoning' may not be adequate when the vehicle traverses significant distances on the terrain, due to drift problems associated with the gyroscopes and integration errors. During preliminary experiments this is not expected to be a problem, since the vehicle will traverse small distances. Inertial sensing systems involve drift errors which can be significant when the vehicle traverses relatively large distances. A *drift-free* star sensor based navigation scheme such as the one investigated by Sreenivasan and Waldron [1994] can be used for long range experiments.

The mechanical design details of the WAAV are shown in Figure 4. The vehicle structure is made of Aluminum Alloy 6061-T6. This material is relatively light weight, possesses superior welding characteristics, and has almost temperature invariant mechanical properties. The last property allows the use of data generated at terrestrial ambient temperatures for design purposes over a large range of temperatures. ASTM 4140, a typical high strength alloy steel, is used for high-stress components such as drive shafts in order to keep their dimensions small. Due to the relatively high torque and low speed operational requirements of the WAAV, the DC motors are equipped with appropriate speed reduction units. The reduction units account for a significant part of the vehicle weight. Each of the wheel motors are provided with a commercially available harmonic drive made by the Emhart

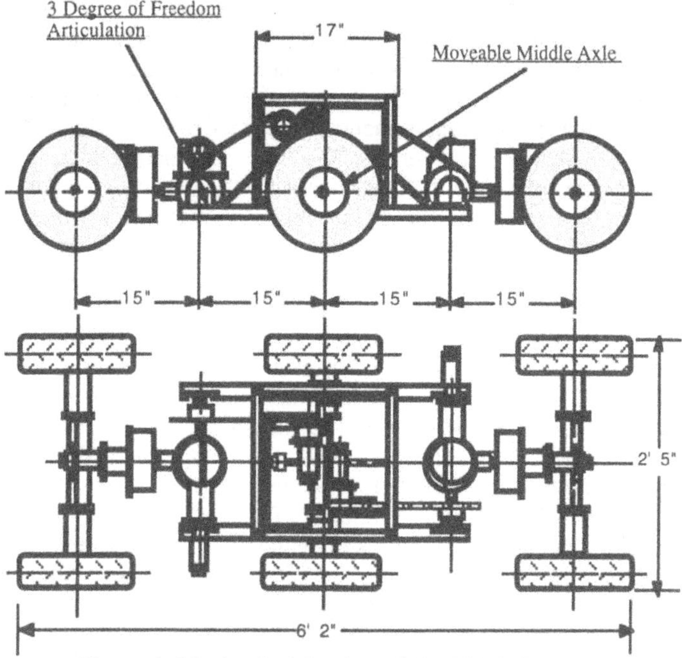

Figure 4 Mechanical Design of the WAAV

Machinery Group. These harmonic drives provide a reduction ratio of 160 to 1, and they have a transmission efficiency of 75%. All six articulation actuators also use the same harmonic drive as part of multi-stage reduction units. The pitch actuators of the two articulations are mounted inside the middle module frame as shown in Figure 4. The motor shaft is connected to a harmonic drive followed by a single stage spur gear pair that provides a second stage reduction ratio of 3 to 1. Finally, the output of the spur gear pair is connected to the pitch axis by a roller chain and sprocket drive that provides a reduction ratio of 2 to 1 leading to an overall reduction ratio of 960 to 1, and an overall transmission efficiency of about 57.4%. The yaw actuators are mounted outboard of the pitch actuators. The yaw actuators are provided with a compact planetary gearhead unit of ratio 4 to 1 followed by a harmonic drive. Finally, the output of the harmonic drive is connected to the yaw axis by a spiral bevel gear pair of ratio 2 to 1. The yaw motors are mounted above the pitch axes such that the axes of these motors are parallel to the pitch axes (Figure 4). The choice of the transmission units and the locations of the yaw motors were based on geometric interference studies that ensured the desired motion ranges of the articulation axes. The overall reduction ratio of the two yaw actuation sub-systems is 1280 to 1, and the transmission efficiency is about 57%. The roll actuators are mounted outboard of the yaw actuators on the front and the rear modules. The roll motor is connected to a harmonic drive followed by a single stage spur gear pair that provides a second stage reduction ratio of 3.43 to 1. The overall reduction ratio is about 549 to 1, and the overall transmission efficiency is about 67.5%.

At present, it is not planned to have an onboard power supply unit. Invertors that work from the 120V AC supply are being used for laboratory testing. Power transformers for the main motors are located off board, and the power supply to the vehicle is *via* an umbilical cable. Four invertor lines rated at 50V DC @ 10A each are required. Each motor is rated at 50 V DC with a maximum current of 5A. Each line supports a set of three motors. The three motors belonging to a set are chosen so that a current of 10 A will be sufficient to drive the three motors at any given time. This is because all three of these motors are not expected to operate at their peak torque requirements at the same time. A block diagram of the power distribution system is shown in Figure 5. The two wheel motors of the front module and the front roll actuator are connected to power line # 1. The left wheel of the middle module and the pitch and the yaw actuators of the front module are connected to power line # 2. The right wheel of the middle module and the pitch and the yaw actuators of the rear module are connected to power line # 3, and the two wheel motors of the rear module and the rear roll actuator are connected to power line # 4. The four motor sets contain motors that are located close to each other. This reduces the problem of shielding required to isolate the power lines. The power circuit includes a fuse box that ensures that the current drawn by each of the twelve motors does not exceed the designed maximum value. The power system described here is primarily suited for laboratory experiments. A trailer equipped with gasoline powered generators, which was used during the Adaptive Suspension Vehicle project [Pugh *et al.*,1990], is available for use during field experiments.

The emphasis of the experimental study of the WAAV is towards understanding the mobility, coordination and control issues of such systems. If a similar vehicle is to be used in an autonomous mode of operation, the system design may have to be changed to accommodate an on board power supply unit. An on board power supply unit may employ batteries or, in the case of a extra-terrestrial missions, radio-isotope thermoelectric generators. In either case, the mechanical system has to be fabricated using low weight to strength ratio materials. Further, use of high torque to weight ratio rare-earth motors can reduce the required speed reduction ratios to less than 150 to 1 for each of the twelve motors. This will decrease the overall system weight significantly. It is also desirable to minimize the transmission losses as far as possible to conserve energy. A vehicle designed in this manner will have a smaller overall volume, which leads to larger ranges of motion for the articulation joints, and hence to superior mobility. Of course, a system that is designed taking into account the above considerations is bound to be more expensive.

One of the reasons for choosing actively coordinated vehicles instead of conventional vehicle systems is the fact that active coordination possesses the ability to optimize the load distribution among the actuators. In passively suspended systems, poor force distribution leads to large, fluctuating structural loads, especially when these systems operate on unstructured terrains. This requires such vehicles to be relatively bulky in order to avoid fatigue failures of the structural elements. This in turn leads to increased power consumption. Of course, actively coordinated vehicles have a large number of actuators and sensors which have overheads associated with them. Therefore, the need for active coordination is not justified unless the terrain is 'sufficiently' rugged. Further, if the actively coordinated system must use bulky transmission and actuation components, its ability to conserve power will not be properly exploited. As the performance of available actuation and transmission systems gets better, and as more efficient systems design techniques are developed, the option of using active coordination should become a more viable one.

The WAAV's actuator assortment consists of twelve brushless DC motors (Manufactured by Pittman Industries, Harleysville, PA), and a small, high-torque DC motor. The twelve main motors are on four isolated circuits, each serving three motors as shown in Figure 5. The thirteenth motor, also known as the carriage-slider motor, is a small, high-torque motor used for moving the middle axle back and forth. This motor is powered by a small switching supply that also supplies power to the drive electronics for controller boards of the main motors. There is a 120V line powering this switching supply *via* the umbilical cable.

The control architecture is implemented on a layered, networked system of computers. An on board 80386 computer is responsible for the majority of the localized motor and sensor functions. An off board 80386 computer, equipped with a monitor and keyboard for user interaction, is responsible for generating the desired position, force, and velocity commands. Information exchange between the two computers is made possible *via* an asynchronous serial link achieved by a null modem connection.

Twelve high resolution incremental optical encoders (Hewlett-Packard series HEDS-5000), each attached to a brushless DC motor, are used to read position information on the motor shaft. Decoding is performed on the middle module of the WAAV in the on board computer. To decode the encoder position, there is a high speed, 16-bit encoder card utilizing the Advanced Micro Devices 9513 programmable IC (available from Omega Electronics). The card is configured for 20 channels of pulse counting, but with the use of a programmable array logic for quadrature decoding, there are 8 quadrature encoder inputs and four additional 16-bit counters. The quadrature inputs coming into the card are converted to "up" and "down" channels. A motor position is the down index subtracted from the up index, which results in a 16-bit position index. This information, combined with the positions of the other motors, leads to an absolute position vector stored in an integer array. This angular position information is sufficient for motion planning laboratory experiments that are to be performed. A discussion of the motion planning experiments is included in the next section.

In order to test the special mobility capabilities of the WAAV, and its performance in unstructured terrain conditions, the vehicle should be operated in a force control mode. The coordination scheme is similar to that of the Adaptive Suspension Vehicle, and is described by Kumar and Waldron [1989]. Control of the force distribution in the vehicle requires the sensing of the contact forces at the wheel-terrain interface. Contact force information can also be used to estimate the location of the contact center with respect to a vehicle fixed frame [Kumar and Waldron, 1989]. The WAAV design includes a provision for the sensing of the contact force components in the plane of the wheel. The lateral force component normal to the wheel plane cannot be sensed accurately for the wheel-axle design of the WAAV. It may be possible to estimate the lateral force component using other sensed information. The contact force components in the plane of the wheels are sensed using a combination of radial force sensors and wheel torque sensors. One set of radial force sensors and wheel torque sensors have been calibrated for sensing purposes. These sensors will be incorporated into all six wheels. The reaction at each wheel center (in the plane of the wheel) due to the respective contact force is sensed by an array of strain gages mounted on the bearing housing of the wheel axle. The moment applied to the wheels is sensed by an LVDT mounted on the wheel rim that measures the angular deflection of a pointer mounted on the wheel hub. Using the sensed force and moment, the contact forces in the plane of the wheel can be estimated. Linear calibration curves have been obtained for both the radial and torque wheel sensors in static tests. Dynamic testing followed by integration of these sensors with the data acquisition system of the WAAV has to be performed. The design of a torque sensing wheel is shown in Figure 6. In order to obtain linear torque to deflection calibration curves, this sensor had to be fabricated such that the five spokes, the hub and the rim were machined from one piece. The rim of the sensor was then keyed to the wheel. Otherwise hysterisis in the welded joints at the ends of the spokes led to nonlinear calibration curves.

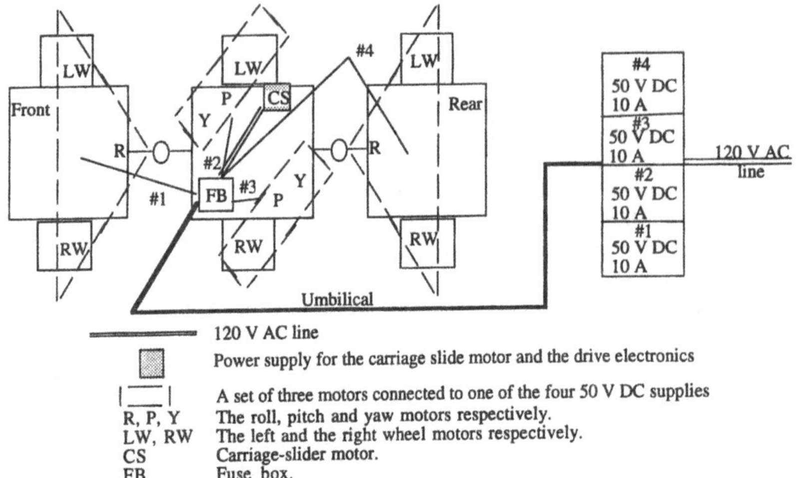

Figure 5 A block diagram of the power distribution system

The twelve DC motors are controlled by onboard drive electronics which include PWM torque control and high-current field effect transistor three phase motor drivers. The controller boards can track reference current values, and they have a current feedback loop that accounts for the actuator dynamics. The reference motor current values are set by an analog voltage input of ± 10 volts to the PWM board. Motor torque is a linear function of this controller input voltage. The full scale range is -10V to 10V and is available as the output of a 12-bit (plus sign) D to A converter located in the onboard 80386 computer. The thirteenth motor is not controlled directly by the computer. Instead, it is controlled by a small Motorola 6811 based micro-controller board called the MiniBoard 2.0. (Designed by Fred Martin of the MIT Media Lab. Information is available in PostScript/UNIX compressed form *via* anonymous ftp to: cher.media.mit.edu in the /pub/miniboard directory.) The MiniBoard 2.0 has on board H-bridge motor driver IC for driving the carriage-slider motor. It has several analog and digital inputs that can be used for control purposes in the future. The control architecture of the WAAV that includes the position feedback loop from the optical encoders is shown in Figure 7. The incorporation of the force feedback loop is a topic of future research.

This concludes the discussion of the various sub-systems of the WAAV.

4 Current Capabilities and Future Experiments

The mechanical fabrication and assembly of the WAAV has been completed. Testing of the actuation sub-systems and the control architecture has also been performed. The D to A card has been interfaced with the motors allowing for computer control of the actuators. The carriage-slider motor is controlled by a local circuit using the Miniboard that allows for the positioning of the carriage-slider in one of three desired positions. The choice of the location of the carriage slider can be done from the offboard computer through the umbilical. The WAAV has been tested in an open-loop fashion to ensure that it can perform some basic maneuvers that may be required during mobility experiments. The encoders on the motors have been connected to a data acquisition board and the motor position feedback is available.

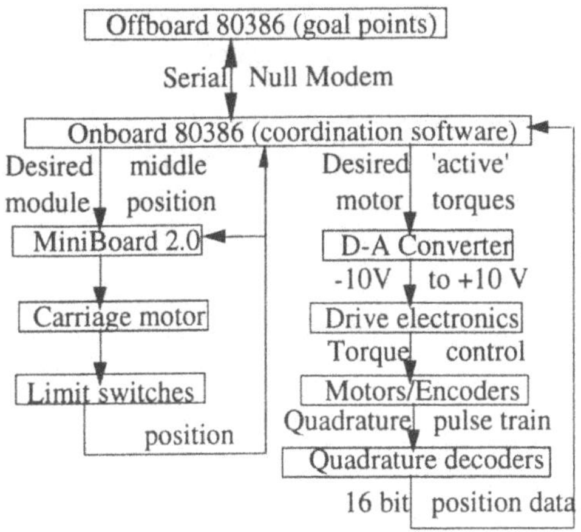

Figure 6 The control architecture

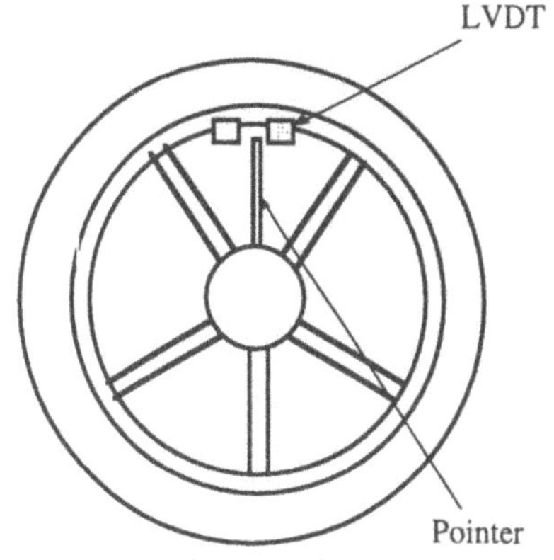

Figure 7 Wheel design for torque sensing

The vehicle system is first expected to be used for position control experiments to test basic motion planning algorithms of the WAAV on even terrain in a laboratory setting. The presence of rolling contacts at the wheel ground interface that do not allow for controlled lateral motion makes the motion planning problem a nonholonomic one. The coordination problem involved in achieving a desired lateral motion for a conventional wheeled system is a surprisingly difficult one as noted by Murray and Sastry [1992]. The nonholonomic motion planning problems result in paths requiring several iterative loops before the vehicle can reach a desired position. The presence of active articulations in the WAAV makes it distinct from the systems that have been studied in the literature. For instance, the WAAV can lift the middle module off the ground and move laterally using a series of actuator maneuvers. Position control experiments should identify the fine motion capabilities of the WAAV.

The second set of experiments are the force control experiments. These will include the investigation of the WAAV performance with respect to geometric objects in a laboratory setting, and comparison of this performance with numerical simulation results. All the twelve active actuators can be tested, for instance, by requiring the vehicle to track an arc of a circle on an inclined plane. The desired kinematics and contact forces will be spatial, and an active control of all the twelve actuators will be required. The variable configuration capability can be tested by investigating the system performance while attempting various mobility maneuvers such as obstacle climbing, ditch crossing, self-recovery etc.

The final set of experiments involves field testing. The vehicle will be made to traverse in-situ soils and rocks in the fields and required to track desired paths. The vehicle will also be made to perform mobility maneuvers requiring the variable configuration capability. The coordination and control schemes will have to be tuned in order to make them robust with respect to unmodelled uncertainties introduced by the unstructured environment. Once the WAAV performs satisfactorily in these experiments, the experimental data that is obtained can be used to compare the performance of the WAAV with other vehicles. The experimental data can also be used, along with simulations, to predict the performance of other vehicles similar to the WAAV, and to predict the system behavior in various terrain conditions.

The force control experiments will require the contact force sensing systems to be integrated with the control architecture. This has not yet been done on the WAAV. In addition, the field experiments will require an inertial sensing package as described in Section

3. The integration of the inertial sensing system into the control scheme of the WAAV will require special hardware and software design, and is a topic of future research.

5 Conclusions

A technical description of the electro-mechanical systems of the WAAV has been provided in this manuscript. A discussion of the current capabilities of the WAAV system, and the mobility experiments that are planned for the future is also included.

6 References

1. Bickler, D. B., 1990, "Computing True Traction Forces on an Eccentrically Loaded Vehicle," *SAE Off-Highway Conference.*
2. Chiang, S., Crane, C. D., Duffy, J., Sept., 1992, "Path Planning for an Articulated Transporter/Manipulator System," *Proc., ASME Mech. Conf.*, Arizona, DE-Vol. 45, pp. 405-412.
3. Chottiner, J., "Simulation of a Six Wheeled Martian Rover Called The Rocker Bogie," M.S. Thesis, Mechanical Engineering, Ohio State University, Spring, 1992.
4. Hirose, S., and Umetani, Y., "The Basic Motion Regulation System for a Quadruped Walking Machine," ASME Paper No. 80-DET-34, September 1980.
5. Kumar, V., and Waldron, K. J., "Force Distribution in Walking Vehicles," *ASME Journal of Mechanical Design*, Vol. 112, pp. 90-99, March 1990.
6. Kumar, V. and Waldron, K. J., 1989, "Actively Coordinated Vehicle Systems," *Journal of Mechanisms, Transmissions and Automation in Design*, March 1989.
7. McGhee, R. B., and Iswandhi, G. I., 1979, "Adaptive Locomotion of a Multi-Legged Robot Over Rough Terrain," *IEEE Trans. on Systems, Man, and Cybernetics*, Vol. 9, No. 4, pp. 176-182.
8. Milliken, W. F. Jr., "Active Suspension," *SAE Transactions* Paper No. 880799, 1988.
9. Murray, R., and Sastry, S., May, 1992, "Nonholonomic Motion Planning Using Sinusoidal Steering Inputs," *Proceedings of the IEEE Robotics and Automation Conference*, Nice, France.
10. Pugh, R. D., Ribble, E., Vohnout, V., Bihari, T. E., Patterson, M. R., Waldron, K. J., "Tech. Description of the Adaptive Suspension Vehicle," 1990, *Int. J. of Rob. Res.*, Vol.9, pp. 24-42.
11. Raibert, M. H., "Legged Robots That Balance," M.I.T. Press: Cambridge, MA, 1986.
12. Song, S. M., and Waldron, K. J., "Machines that walk: The Adaptive Suspension Vehicle," M.I.T Press, October 1988.
13. Spiessbach, A.J., and Woodis, W.R., 1988, Mars Rover/Sample Return, (MRSR) Rover Mobility and Surface Rendezvous Studies, Martin Marietta Space Systems, Denver, Final Report.
14. Sreenivasan, S. V., and Waldron, K. J., 1994, "A Drift-Free Navigation System for A Mobile Robot Operating on Unstructured Terrain," To appear in *The ASME Journal of Mechanical Design.*
15. Waldron, K. J., Sreenivasan, S. V., and Varadhan, V., November, 1991a, "Mobility Enhancement Using Active Coordination," *Proceedings of the Second Conference on Applied Mechanisms and Robotics*, Cincinnati, OH.
16. Waldron, K. J., Sreenivasan, S. V., Chottiner, J., Sept., 1991b "Coordination of Complex Mechanisms in Unstructured Environments," *Proc. of Int. Advanced Robotics Programme, Sensor Fusion and Environmental Modelling*, Oxford, U.K.
17. Waldron, K. J., "A Simulation Study of a Baseline, Actively Coordinated Locomotion System for a Mars Rover," Final Report, JPL, Cal. Institute of Tech., Pasadena, California, July, 1989b.
18. Waldron, K. J., Kumar, V., Burkat, A., "An Actively Coordinated Mobility System for a Planetary Rover," *Proc. , Int. Conf. on Advanced Robotics*, Versailles, France, Oct. 13-15, 1987.
19. Wilcox, B. H., and Gennery, D. B., October 1990, "A Mars Rover for the 1990's," *Journal of the British Interplanetary Society.*
20. Yu, J., and Waldron, K. J., "Design of Wheeled Actively Articulated Vehicle," *Proc. of the Second National Applied Mechanisms and Robotics Conf., Cincinnati*, Ohio, 3-6 Nov., 1991.

Limitations of Kinematic Models for Wheeled Mobile Robots

F. Demick Boyden and Steven A. Velinsky

Advanced Highway Maintenance & Construction Technology Center
Department of Mechanical & Aeronautical Engineering
University of California, Davis
Davis, CA 95616-5294

Abstract - Wheeled Mobile Robots (WMRs) have been discussed extensively in the literature, and most of the reported work considers low speed, low acceleration, and lightly loaded applications. In such cases, kinematic models provide reasonably accurate results. As WMRs are designed to perform more demanding, practical applications with high speeds and/or high loads, kinematic models are no longer valid representations. This paper examines the limitations of WMR kinematic models through comparison to simulation results of dynamic models for both differentially and conventionally steered WMRs. An important feature of the dynamic models employed, which have been discussed in Boyden and Velinsky [1], is the use of a complex tire representation which is necessary to accurately account for the tire/ground interaction.

I. Introduction

Over the last several years, numerous prototype Wheeled Mobile Robots (WMRs) have been developed and a wide variety of research related to WMRs has been reported. The prototype WMRs have been built primarily as research vehicles with emphasis on such aspects as perception and control (e.g., [2]) and machine reasoning and intelligence [3]. A fundamental aspect concerning the path generation and control of a WMR involves the manner in which the WMR's position and orientation is tracked. Most researchers have used kinematic models to accomplish this task [4,5] arguing that because of the low speeds, low accelerations, and lightly loaded conditions under which WMRs operate, these kinematic models are valid. However, as WMRs are designed to perform heavy duty work and travel at higher speeds, dynamic modeling of these vehicles becomes increasingly important. A few researchers have derived dynamic models for wheeled mobile robots [6,7], but for large, high load vehicles these models may not be valid due to imposed restrictions and potentially inaccurate tire models. Accordingly, Boyden and Velinsky [1] have developed dynamic models for both differentially steered and conventionally steered WMRs which incorporate complex tire models thus alleviating the restrictions of the earlier work. The tire model is based on the vast amount of work in the automotive field and accounts for longitudinal slip, slip angle, and the trade-off between the forces generated in the longitudinal and lateral directions.

The impetus behind the aforementioned dynamic models is the current development of a high load WMR for highway maintenance and construction tasks at the University of California, Davis, in conjunction with the California Department of Transportation, as part of the Advanced Highway Maintenance and Construction Technology Center [8]. The WMR is a differentially steered, self-propelled robot which works in close proximity to a support

A. J. Lenarčič and B. B. Ravani (eds.), Advances in Robot Kinematics and Computationed Geometry, 151–160.

vehicle. Initial applications of the WMR involve the use of high load highway equipment such as pavement routers which develop channels for better crack sealant penetration, drilling devices for slab mud jacking, etc. Based on these heavy duty tasks, the WMR is expected to weigh in excess of 2225 N (500 lbs). Such tasks exert very large external forces on the WMR and they have the potential to significantly influence its path and direction. Furthermore, future anticipated WMR uses include tasks to be performed at relatively high speeds in order to minimize the necessity for lane closures during maintenance.

Generally, kinematic models are significantly easier to use with much reduced computational requirements than dynamic models, and thus, within their limits of accuracy, the use of a kinematic model is normally desirable. Accordingly, this paper is aimed at understanding the limitations of kinematic models for WMRs. This is accomplished through comparison to the dynamic models with particular attention to WMR speed and load effects.

II. Analysis

An overhead view of a typical WMR with the forces which are exerted upon it is depicted in Fig. 1. This figure shows the planar version of a standard vehicle body centered reference frame (SAE vehicle axis system) in which x_2 denotes the longitudinal direction (positive forward), y_2 denotes the lateral direction, and ψ denotes the in-plane rotation (about the z_2 axis which is positive downward and is not shown). Furthermore, u and v denote velocities in the x_2 and y_2 directions, respectively, and r denotes the in-plane rotational velocity also referred to as the yaw rate. The unit vectors in the x_2 and y_2 directions are $\mathbf{i}_2$ and $\mathbf{j}_2$, respectively. For a conventionally steered WMR, the front wheel steered angle δ is used to control the motion of the vehicle. An alternative configuration is the differentially steered WMR which includes a caster at the front as opposed to an actively steered wheel, and the motion is controlled through the velocities of the rear drive wheels.

A. Kinematic Models

Conventionally Steered WMR: The primary assumption in the existing WMR kinematic models is that no tire slippage occurs. As such, simple rigid body kinematics can be employed to describe the WMR's motion. For the conventionally steered WMR, the front wheel steered angle δ, and the angular velocity of any one wheel must be known (the right rear wheel angular velocity ω_r is used herein). The characteristic kinematic equations are then

$$u = \frac{\omega_r R_t}{1 - \dfrac{T_r \tan\delta}{2(a+b)}} \tag{1}$$

$$v = \frac{bu \tan\delta}{(a+b)} \tag{2}$$

and

$$r = \frac{u \tan\delta}{(a+b)} \tag{3}$$

where u and v denote the forward and lateral velocities of the center of mass, respectively, r denotes the yaw velocity, R_t denotes the tire radius, T_r is the wheel track, and b is the distance from the rear axle to the center of mass.

Figure 1 shows two points on the WMR that are of particular significance, points P and O. Point P designates an arbitrary location where the WMR's tooling interacts with the surrounding environment (e.g., router location for the highway maintenance application). Thus, P_x and P_y are corresponding tooling forces. Generally, a WMR's position is tracked by observing a specific location upon the WMR. Herein, point O designates the point on the WMR which is tracked (physically or optically, etc.). That is, P is the point on the WMR that must follow a prescribed path as dictated by the application, and the path is followed by controlling both the location of point O and the orientation of the WMR. As such, in the analysis to follow, we will assume that external forces (other than tire/road forces) act at point P, but we will use point O to understand the efficacy of the models for tracking the WMR.

In the world coordinate system, the velocity of the tracked point $\mathbf{V}_o$ of the WMR is

$$\mathbf{V}_o = U\hat{i}_1 + V\hat{j}_1 \tag{4}$$

where

$$U = u\cos(\psi) - v\sin(\psi) - er\sin(\psi) \tag{5}$$

$$V = u\sin(\psi) + v\cos(\psi) + er\cos(\psi) \tag{6}$$

and the yaw angle, ψ, is the integral of the yaw velocity, r.

The position of the tracked point $\mathbf{R}_o$ in the world coordinate system is expressed as

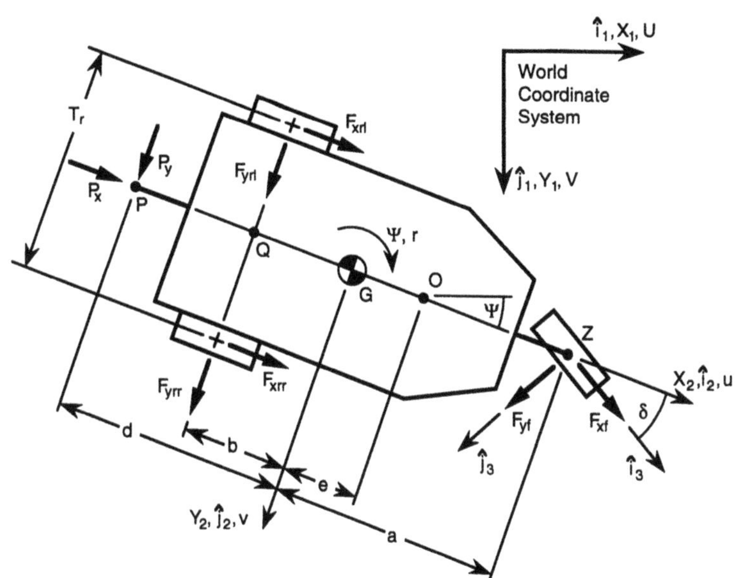

Fig. 1: Overhead view of a typical WMR

$$\mathbf{R_o} = X_1\hat{i}_1 + Y_1\hat{j}_1 \tag{7}$$

where X_1 and Y_1 are the integrals of U and V, respectively.

Differentially Steered Wheeled Mobile Robot: For the differentially steered WMR, the steered wheel at the front is replaced by a caster as noted above. Since the caster wheel does not provide any lateral forces (assuming it is free to pivot), the steered angle of the caster is dictated by the velocity of its pivot. That is, δ is a function of the WMR kinematics which can be written in terms of the left and right driving wheel angular velocities, ω_l and ω_r, respectively, as follows:

$$u = (\omega_r + \omega_l)\frac{R_t}{2} \tag{8}$$

$$v = \frac{R_t b}{T_r}(\omega_l - \omega_r) \tag{9}$$

$$r = \frac{R_t}{T_r}(\omega_l - \omega_r). \tag{10}$$

Again, the velocity and position of the tracked point on the WMR are found by (4) - (7).

B. Dynamic Model

With the model limited to a 3 degree-of-freedom planar model with corresponding independent variables u, v, and r, summation of the applicable forces and moments with respect to the body centered reference frame yields [1]:

$$\sum F_x = m(\dot{u} - vr) = F_{xrl} + F_{xrr} + F_{xf}\cos\delta - F_{yf}\sin\delta + P_x \tag{11}$$

$$\sum F_y = m(\dot{v} + ur) = F_{yrl} + F_{yrr} + F_{yf}\cos\delta + F_{xf}\sin\delta + P_y \tag{12}$$

$$\sum M_z = I_z\dot{r} = \frac{T_r}{2}(F_{xrl} - F_{xrr}) - b(F_{yrl} + F_{yrr}) + a(F_{yf}\cos\delta + F_{xf}\sin\delta) - dP_y \tag{13}$$

where m is the WMR mass, I_z is the WMR moment of inertia about the vertical axis, a and d are the distances from the center of mass to the front axle and tool location, respectively, F_{xrr}, F_{yrr}, F_{xrl} and F_{yrl} are the driving wheel tire forces, and F_{xf} and F_{yf} are the front tire forces.

In state variable form, (11) - (13) are written as:

$$\dot{u} = \frac{F_{xrl} + F_{xrr} + F_{xf}\cos\delta - F_{yf}\sin\delta + P_x}{m} + vr \tag{14}$$

$$\dot{v} = \frac{F_{yrl} + F_{yrr} + F_{yf}\cos\delta + F_{xf}\sin\delta + P_y}{m} - ur \tag{15}$$

$$\dot{r} = \frac{\frac{T_r}{2}(F_{xrl} - F_{xrr}) - b(F_{yrl} + F_{yrr}) + a(F_{yf}\cos\delta + F_{xf}\sin\delta) - dP_y}{I_z}. \tag{16}$$

The longitudinal, lateral, and yaw velocities (u, v, r) are simply the integrals of their accelerations ($\dot{u}$, $\dot{v}$, $\dot{r}$) and the velocity and position of the tracked point on the WMR are, in turn, found by equations (4) - (7).

The same equations apply for the differentially steered WMR with it being recognized that a denotes the distance from the center of mass to the caster wheel pivot for this case, and we additionally recognize that $F_{yf}= 0$ since the caster is free to pivot.

Tire Model: In order to make use of the dynamic equations, it is necessary to find the longitudinal and lateral tire forces for each tire on the WMR, and especially for high speed or heavily loaded applications, the tire/ground interface needs to be modeled in an accurate manner. Lateral tire forces are the result of a tire having a slip angle, that is, it being pointed in a direction that differs from its velocity. Longitudinal forces are generated from tire slip, the difference between the radius times the rotational speed and the forward velocity of the tire. Both slip angle and tire slip are easily determined from the WMR velocities (u, v, and r) and the wheel rotational velocities.

Automotive engineers have modeled tires for many years and one of the key aspects is that there is a trade-off between tractive (longitudinal) forces and cornering (lateral) forces. This trade-off has led to the friction circle concept. Most generally, the friction circle concept says that the tire has a finite amount of force that it can produce based on the dynamic coefficient of friction between the surface and the tire, the normal load at the tire/road interface, etc., and in instances in which there is a combination of tractive and cornering forces, there is a trade-off between the amount of force that can be produced in each direction.

The Dugoff tire friction model [9] makes use of this friction circle concept and is used to predict tire forces in this paper. Gunter and Sankar [10] later presented a method of simplifying the calculations required for simulation purposes. However, this simplified method is only valid if the vehicle is restricted to forward motion and slip angles of less than 90 degrees. Outside of this range, this simplified method has many inconsistencies. This method has been extended [11] to overcome some of these inconsistencies by placing limits on slip and friction variables, checking for additional pathological cases, removing non-holonomic conditions, and removing the sign ambiguity associated with the slip angle.

III. Results and Discussion

The kinematic models are compared to the dynamic models to examine the effects of speed and load for the conventionally and differentially steered WMRs. For all simulations, the vehicle parameters presented in Tb. 1, which are representative of the high load highway maintenance WMR under development, are employed.

A. Speed Dependent Maneuver

Conventionally Steered WMR: For the conventionally steered model, steady-state turning at various speeds and steer angles is simulated to examine speed dependence on the resulting path. There are no tool forces exerted on the WMR for this simulation ($P_x = P_y = 0$). Figure 2 shows the results of simulation at a steered input of 0.2 radians. In this figure,

WMR mass	272 kg (18.6 slugs)
Tire radius	0.3048 m (1.0 ft)
Lateral tire stiffness	40 kN/rad (9000 lb/rad)
Longitudinal tire stiffness	40 kN/unit slip (9000 lb/slip)
Road/tire interface coefficient of friction	0.8
Yaw moment of inertia	407 kg-m^2 (300 ft-lb-sec^2)
Steered front wheel moment of inertia	2.71 kg-m^2 (2 ft-lb-sec^2)
Combined wheel, gearbox, and motor rotor moments of inertia	6.78 kg-m^2 (5 ft-lb-sec^2)
Center of mass to front axle distance, a	0.762 m (2.5 ft)
Center of mass to rear axle distance, b	0.6096 m (2.0 ft)
Center of mass to tool location distance, d	0.9144 m (3.0 ft)
Center of mass to point being tracked distance, e	0.2286 m (0.75 ft)
Rear Track, T_r	0.9144 m (3.0 ft)

Tb. 1: Parameter Values for Highway Maintenance WMR

which is typical of all of the steer angles examined, we have normalized the steady-state radius of curvature derived from the dynamic model with that of the kinematic model which is speed independent for a given steer input. This best illustrates that the error due to the kinematic model is highly dependent upon the lateral acceleration and that this error grows excessively at lateral accelerations above 0.5 g. Under such circumstances, the tire forces begin to approach their limits of adhesion (the highly nonlinear region of the tire model) which the kinematic model fails to take into account. Accordingly, for conventionally steered WMR's, kinematic models should be reasonably accurate when lateral accelerations are less than approximately 0.2 g. This can be accomplished by appropriate speed and path planning; i.e., ensuring that turning radii are adequately large and cornering speeds are adequately small.

Differentially Steered WMR: To examine the speed dependence on the path of differentially steered WMRs, we have simulated steady-state turning at various speeds and left-to-right wheel speed ratios. Figure 3 shows the normalized steady-state radius of curvature (radius of curvature of the dynamic model divided by the kinematic model's radius of curvature) as a function of lateral acceleration for three different left-to-right wheel speed ratios. The kinematic model's radius of curvature is independent of speed for a given left-to-right wheel speed ratio.

Of considerable interest is that the errors due to the kinematic model become significant at very low lateral accelerations. Furthermore, the error increases as the wheel speeds become closer to each other; i.e., for very similar wheel speeds, the error of the kinematic model will be quite large at minimal lateral acceleration. While this result may oppose intuition, it

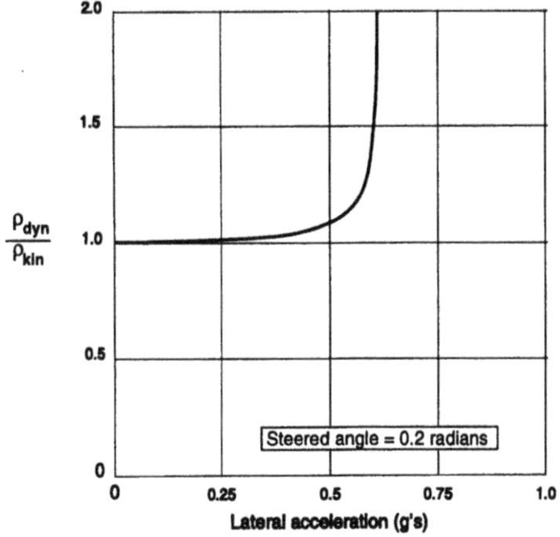

Fig. 2: Normalized radius of curvature vs. lateral acceleration for a conventionally steered WMR

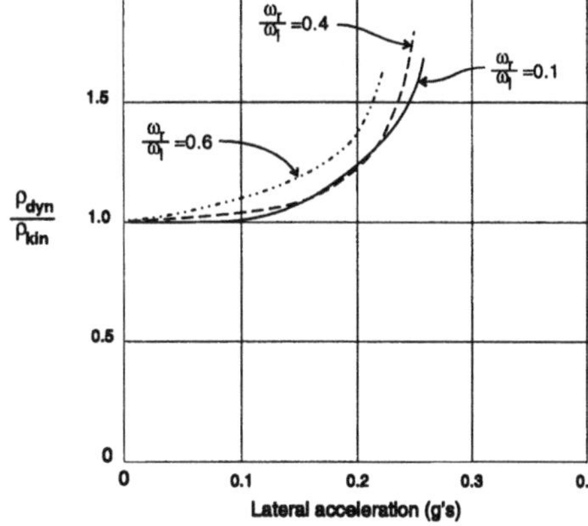

Fig. 3: Normalized radius of curvature vs. lateral acceleration for a differentially steered WMR

comes from the very different manner in which differentially steered vehicles turn as opposed to conventionally steered WMRs as well as other vehicles with which we are accustomed.

The rotation of a vehicle or WMR during a turn comes from a moment about the center of mass. In a conventionally steered WMR, this moment comes from lateral forces due to tire slip angle. On the other hand, a differentially steered WMR employs differences in the longitudinal forces on either side of the vehicle to develop the moment about the center of mass, and the longitudinal forces arise from tire slip. Through examination of the conventionally steered WMR (Figure 2), we can conclude that slip angle related error associated with the kinematic model does not become excessive until about 0.5 g. However, the kinematic model is much more sensitive to tire slip related error. Accordingly, a kinematic model is inadequate for differentially steered WMRs with any substantial mass.

B. Load Dependent Maneuver

Conventionally Steered WMR: To examine load dependence of the models, a transient turning maneuver with various tool loads is simulated. For this simulation, equal torques are applied to the rear driving wheels throughout the maneuver and the steered angle is specified. The WMR starts with an initial velocity of 0.3048 m/sec (1 ft/sec) with a steered angle of zero degrees. From time t=2 seconds until time t=4 seconds, the front wheel is steered to a maximum angle of -10 degrees (a negative steering angle causes a left hand turn) as a quarter sine wave. The steered angle is held at -10 degrees for 2 seconds (from time t = 4 seconds until t = 6 seconds) and then steered back to zero degrees over a period of 2 seconds (from time t = 6 seconds until t = 8 seconds) as a quarter sine wave. The vehicle is then allowed to travel with a steering angle of zero degrees for an additional 2 seconds; the total time simulated is 10 seconds. The lateral tool force, P_y, is assumed zero, and the longitudinal force, P_x, is varied. To arrive at the kinematic model's path, the steered angle and the wheel speeds as determined by the dynamic model are used in conjunction with equations (1) - (7). Such an approach is comparable to the dead reckoning measurement method.

Boyden and Velinsky [1] have previously simulated this maneuver with no tool load. The earlier work has found a kinematic model error of approximately 1-2% for 40.8 N-m (30 ft-lb) driving torque on each wheel (which causes forward acceleration ≈ 0.12 g). This error arises with the tires well below saturation (≈ 30%) with the noted loads.

Figure 4 depicts results for a tool load P_x = -533 N (-120 lb) and driving torques of 122 N-m (90 ft-lb), and Fig. 5 depict results for a tool load P_x = -890 N (-200 lb) and driving torques of 176 N-m (130 ft-lb). The tool loads in these cases are indicative of that expected in the actual highway maintenance machine. For the lower load case, the kinematic model error grows to approximately 7%, and here the tire is at approximately 2/3 of saturation. For the higher load case, errors become quite excessive due to the tires being loaded very near to saturation.

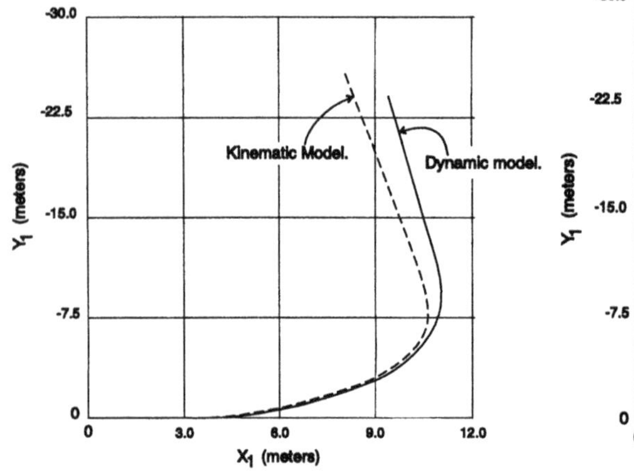

Fig. 4: Path of conventionally steered WMR with 533 N (120 lb) load. Applied wheel torque = 122 Nm (90 ft-lb). External force P_x = -533 N (-120 lb)

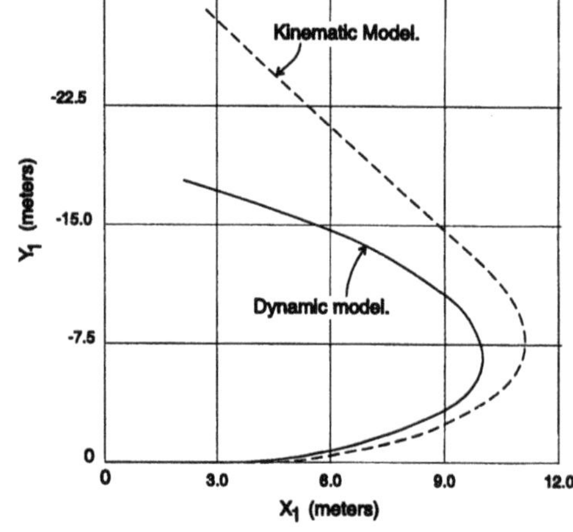

Fig. 5: Path of conventionally steered WMR with 890 N (200 lb) load. Applied wheel torque = 176 Nm (130 ft-lb). External force P_x = -890 N (-200 lb)

Differentially Steered WMR: The simulated maneuver for this comparison is again, a left hand turn. In this simulation, the WMR begins with an initial velocity of 0.3048 m/sec (1 ft/sec) in the forward direction (both the right and left wheels are spinning accordingly) and equal torques are applied to the right and left driving wheels. At time t = 2 seconds, the torque applied to the left wheel is reduced and applied for 2 seconds (until t = 4 seconds), and then resumed to its original value. The right wheel torque remains constant throughout. The total time simulated for each case is identical (6 seconds). As above, it is again assumed that there are no lateral tooling forces. Also, the kinematic model results are determined in the same manner.

Boyden and Velinsky [1] have previously simulated this maneuver with no tool load, initial driving torques of 27.2 N-m (20 ft-lb), and the left wheel torque reduced by (54.4 N-m) 40.0 ft-lb for the period t = 2 through 4; e.g., -27.2 N-m (-20 ft-lb) over that time. The earlier work has found a kinematic model error of approximately 5% although the tires are well below saturation (≈ 16%) with the noted loads.

In this paper, we employ the same torque reduction over the period from t = 2 to 4, and additionally apply a longitudinal tool load. The following two cases are thus simulated: 1) initial drive torques = 81.3 N-m (60 ft-lb) and P_x = -356 N (-80 lb), and 2) initial drive torques = 122 N-m (90 ft-lb) and P_x = -979 N (-220 lb). Results are depicted in Figs. 6 and 7, respectively. For the lower load case, the kinematic model error has risen to about 10% while the tires are at about 50% of saturation, whereas for the higher load case, the tires are saturated and the error excessive. Again, it is apparent that the kinematic model is inadequate for this type of WMR with any significant load or mass.

IV. Conclusions

This paper has attempted to examine the speed and load dependence of errors arising from the use of kinematic models for moderately sized wheeled mobile robots. The kinematic models were compared to dynamic models for both conventionally steered and differentially steered WMRs through computer simulation. The dynamic model has employed a complex tire model to realistically represent the tire/ground interface. Based on the results of the simulations, it can be concluded that lightly loaded conventionally steered WMRs can be reasonably represented with kinematic models when lateral accelerations are kept below 0.2 g. For conventionally steered WMRs with any moderate load, the kinematic models produce significant errors even well within the saturation limits of the tires. Due to the manner in which they generate forces, kinematic models are inaccurate for differentially steered WMRs with any substantial mass and/or load. In conclusion, kinematic models must be limited to lightweight WMRs which operate under very low accelerations and lightly loaded conditions, and thus, dynamic modeling is necessary for virtually all WMRs for practical applications.

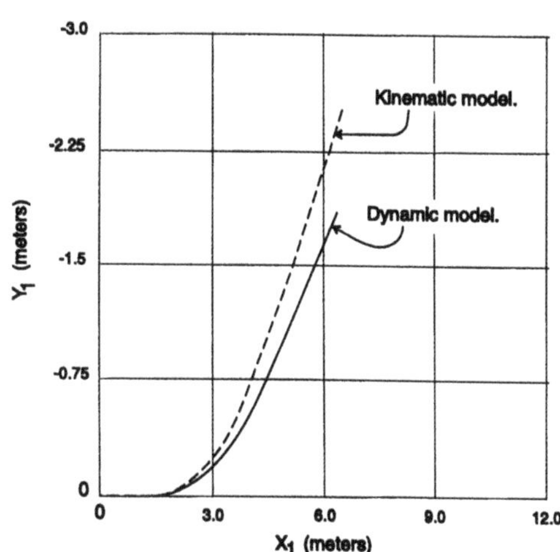

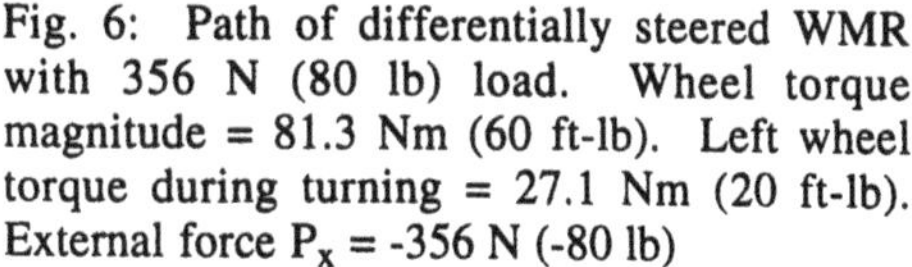
Fig. 6: Path of differentially steered WMR with 356 N (80 lb) load. Wheel torque magnitude = 81.3 Nm (60 ft-lb). Left wheel torque during turning = 27.1 Nm (20 ft-lb). External force P_x = -356 N (-80 lb)

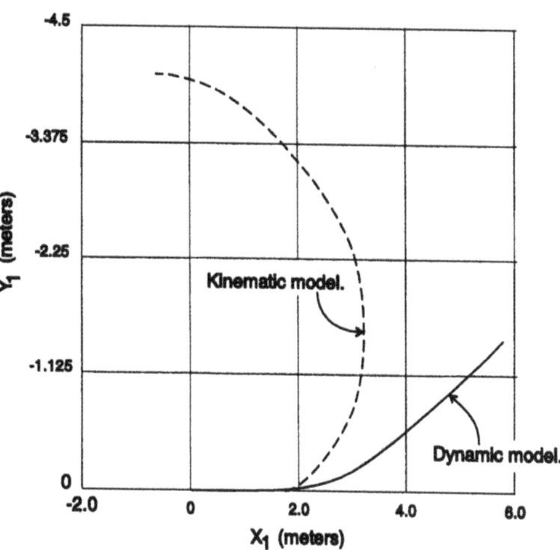

Fig. 7: Path of differentially steered WMR with 979 N (220 lb) load. Wheel torque magnitude = 176 Nm (130 ft-lb). Left wheel torque during turning = 122 Nm (90 ft-lb). External force P_x = -979 N (-220 lb)

V. Acknowledgment

The authors gratefully acknowledge the Office of New Technology, Materials and Research of the California Department of Transportation for the support of this work.

VI. References

[1] F. D. Boyden, S.A. Velinsky, "Dynamic Modeling of Wheeled Mobile Robots for High Load Applications", *Proc. IEEE Intl. Conf. on Robotics and Automation*, to appear, San Diego, USA (1994).

[2] H.P. Moravec, "The Stanford Cart and the CMU Rover", *Proc. of the IEEE*, 71(7), 872-884 (1983).

[3] J.D. Steele, N.D.Ebrahimi, "Control of Mobile Robots", *Intl. J. Robotics and Automation*, 1(2), 40-45 (1986).

[4] A. Segovia, M. Rombaut, A. Preciado, D. Meizel, "Comparative Study of the Different Methods of Path Generation for a Mobile Robot in a Free Environment", *Fifth Intl. Conf. on Advanced Robotics*, 1667-1670 (1991).

[5] J.C. Alexander, J.H. Maddocks, "On the Kinematics of Wheeled Mobile Robots", *Intl. J. Robotics Research*, 8(5), 15-27 (1989).

[6] A. Hemami, M.G. Mehrabi, R.M.H. Cheng, "A New Control Strategy for Tracking in Mobile Robots and AGV's," *Proc. IEEE Conf. on Robotics and Automation*, 1122-1127, Cincinnati, USA (1990).

[7] A. Hamdy, E. Badreddin, "Dynamic Modelling of a Wheeled Mobile Robot for Identification, Navigation, and Control", *Robotics and Flexible Manufacturing Systems* (J.C. Gentina and S.G. Tzafestas, Eds.), Elsevier Science, North-Holland, 119-128 (1992).

[8] S. E. Winters, D. Hong, S.A. Velinsky, K. Yamazaki, "A New Robotic System Concept for Automating Highway Maintenance Tasks," *Proc. ASCE Conf. on Robotics for Challenging Environments*, 374-382, Albuquerque, USA (1994).

[9] H. Dugoff, P.S. Fancher, L. Segal, "An Analysis of Tire Traction Properties and Their Influence on Vehicle Dynamic Performance", *SAE Transactions*, Paper #700377, 1219-1243 (1970).

[10] R. Guntur, S. Sankar, "A Friction Circle Concept for Dugoff's Tyre Friction Model", *Intl. J. Vehicle Design*, 1(4), 373-377 (1980).

[11] F.D. Boyden, "Dynamic Modeling of Wheeled Mobile Robots," M.S. Thesis, University of California, Davis (1993).

A Spatial Leg Mechanism With Anthropomorphic Properties for Ambulatory Robots

A. Kecskeméthy

Fachgebiet Mechatronik
University of Duisburg
47048 Duisburg
Germany

Abstract - A novel anthropomorphic design of a robotic leg device featuring a spherical hip joint and linear actuators is presented. The typical large swivel motions at hip and knee arising in the case of flexion and extension of the human leg are implemented through appropriate two-loop transmission mechanisms, while the limited range of lateral and longitudinal rotations at the hip is used for furnishing the mechanism with low-cost self-aligning joints in lieu of ideal spherical pairs. The paper describes the closed-form, recursive solution of the direct kinematics and the statics of the mechanism, as well as some results of computer simulations.

I. Introduction

Legged undercarriages are playing an increasingly important role in the realization of mobile robots. As a locomotion element, the leg features several advantages over the traditional wheeled ambulatory devices, among which are the better adaptability to unkown and rough terrain, the increased aptitude for the conquest of unforeseen obstacles, the mitigated threat of environmental damaging, and its better traction properties particularly in cases where low gravity, steep escarpments or high lift effects reduce the normal forces. For this reason, legged ambulatory robotic devices are viewed by several researchers as the ideal candidate for applications in highly unstructured terrain and with limited to impossible human access, as e. g. contaminated nuclear reactors, interior of pipelines, unknown surfaces of planets and comets, deep-sea soils, etc. Another field of application currently investigated is the use of exo-skeleton devices for assistance of dismounted infantry-men.

The design and manufacture of a walking machine is a complex and extensive task requiring intelligent solutions for kinematics, sensorics, actuatorics, and control-related issues. In this setting, the conception of an appropriate leg mechanism plays a fundamental role, as it affects all subsequent stages of the devising process. Correspondingly, a large amount of scientific interest has been devoted to this topic in the past years. The resulting types of striding gears can be roughly categorized into two groups [1]: (a) insect-type legs with essentially horizontal structure, and (b) mammal-type legs with basically vertical structure. While mammal-type legs exhibit better efficiency characteristics than insect-type legs, the latter have become more widespread due to their better understanding and easier implementation. Examples of such devices are the vehicles *Ambler* [2] and *ODEX I* [3], as well as the '*walking stick*', named after the *Carausius morosus*, a six-legged grashopper species [4]. However, the legs of mammal type have led to more flexible and dexterous devices, such as the *Adaptive Suspension Vehicle* [1] and 'hoppers' featuring dynamical equilibrium properties [5]. Other

A. J. Lenarčič and B. B. Ravani (eds.), *Advances in Robot Kinematics and Computationed Geometry*, 161–170.

important aspects for leg design, besides similarities with living beings, concern their intrinsic kinematical properties. Recently, some new devices where conceived featuring a high degree of 'kinematical isotropy', which is a measure for the well-posedness of the kinematical transmission equations mapping joint coordinates to spatial motion [6].

Surprisingly, to the knowledge of the author no leg mechanism has been designed so far which is explicitly targeted to emulate human leg behaviour. Particularly, while the human hip supplies three degrees of freedom, existing leg mechanisms sustain at most two degrees of freedom, which are arranged in a gimbal-like fashion. Anthropomorphic properties of stride gear are however of paramount importance in many application fields. An example is teleoperated locomotion control, which becomes necessary when the terrain is too unstructured for an automatic control scheme to function. In this case, the desired motion of the machine legs is measured out from corresponding mechanisms attached to the operator legs, and discrepancies between these two systems tampers the operator's commands. For instance, we experiment with the teleoperation of the combined wheeled and legged robot 'roboTRAC', a device currently under investigation between several researchers ([7]). In an experimental setup (Fig. 1) the operator prescribes the motion of the legs of a scaled model of the roboTRAC through a sensing mechanism attached to his legs. While it is possible, after few hours of training, to maneuver the vehicle through a testbed with several obstacles, the discrepancy between the structure and proportions of human and machine legs makes the operation tedious and error-prone.

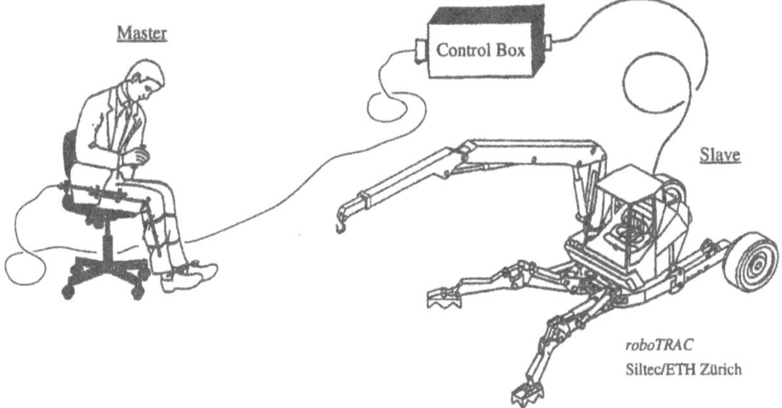

Figure 1: Master-slave control of walking machine 'roboTRAC'

In this paper, we investigate the design and analysis of an anthropomorphic leg mechanism featuring a spherical hip joint and linear actuators. In Section **II**, the addressed anthropomorphic motion characteristics of the mechanism are described, and an overview of the leg device is given. In Section **III**, the closed-form, recursive solution for the direct kinematics and statics problems is detailed, which is important for the future design of real-time control schemes. Some results of computations carried out with our object-oriented modeling package 'MOBILE' in Section **IV** illustrate the feasibility of the concepts.

II. Anthropomorphic properties of the leg design

Human leg motion comprises a total of four degrees of freedom. The motion at the knee involves a pure rotation realizing the flexion and extension of the (lower) leg relative to the thigh. At the hip, the femur articulates with the pelvic bone through a ball-and-socket joint. The corresponding three degrees of freedom can be decomposed in (1) a vertical rotation implementing the flexion and extension of the thigh, (2) a lateral rotation moving the thigh outwardly (*abduction*) or inwardly (*adduction*), and (3) a longitudinal rotation realizing the inward and outward rolling motion of the leg aroung the femur axis (Fig. 2). Associated to these degrees of freedom are muscle groups which

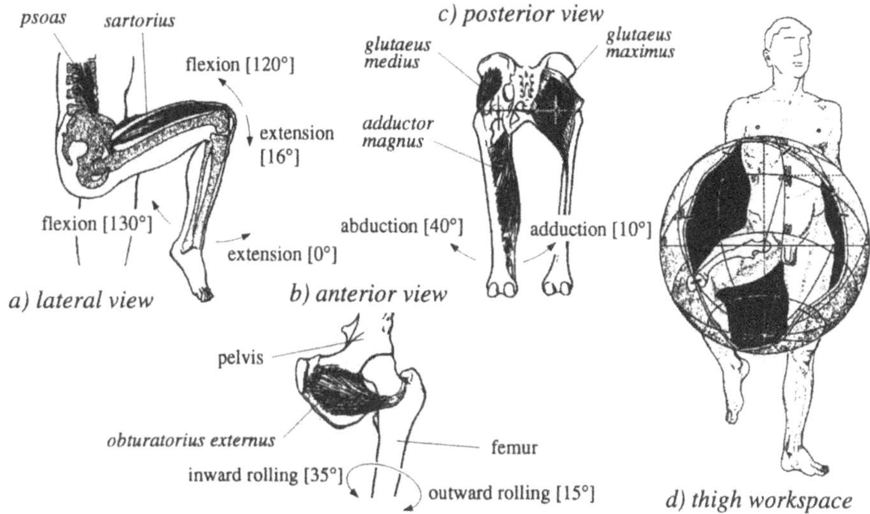

Figure 2: Basic characteristics of human leg motion (from [8])

act to a large degree independently of each other. Some representatives of these muscle groups are the *psoas* and *rectus femoris*, which perform the flexion of the thigh; the *glutaeus maximus*, which implements the extension; the *glutaeus medius* and *adductor magnus*, which implement the abduction and adduction motions, respectively; and the *obturatorius internus*, which carries out the outward rolling motion ([9]). Because of several tendons and ligaments which tie up the bones at the articulations, the corresponding degrees of freedom are unilaterally constrainted by non-symmetric, irregular boundary functions. The resulting 'workspace' of reachable points is characterized by a large range for vertical rotation and smaller ranges for lateral and longitudinal rotations (Fig. 2d).

A mechanical device for emulating these basic human leg motion properties is shown in Fig. 3. It features a spherical joint H at the hip, a revolute pair K at the knee, and further revolute joints $R_1, \ldots, R_5$ and spherical joints $S_1, \ldots, S_7$. By making explicit use of the restricted workspace of the thigh, the spherical pairs can be realized as standard self-aligning joints with tilting motion limited by approx. ± 15 degrees. Furthermore, the leg mechanism is furnished with four hydraulic cylinders $P_1, \ldots, P_4$, one for each degree of freedom, which are arranged so as to correspond largely to the muscle groups discussed above. Moreover, two additional two-loop transmission mechanisms are incorporated which enable to implement ranges of vertical motion at the hip and the knee of approx. 170°. The design is such that the direct kinematics are solvable in a closed-form, as described in the next section.

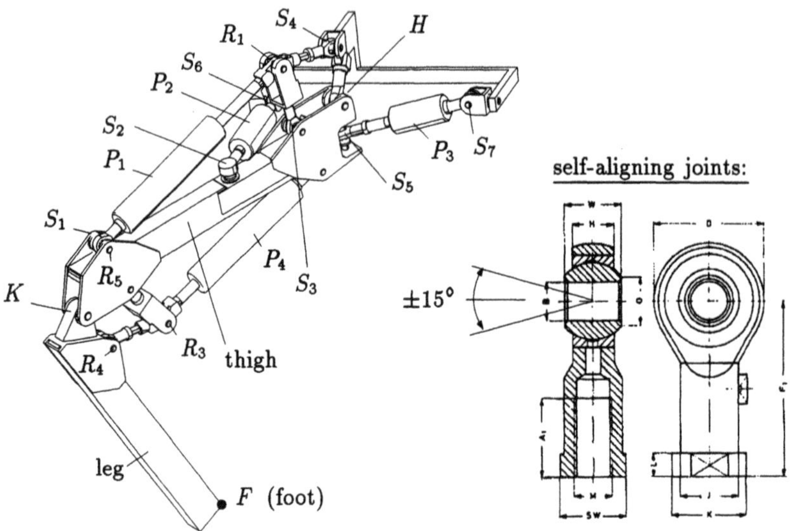

Figure 3: Overview of the leg mechanism

III. Kinematic and static analysis

A. Geometrical parameters and basic notations

The leg mechanism can be regarded as a concatenation of a spherical parallel manipulator implementing the orientation of the thigh and a planar subsystem realizing the motion at the knee (Fig. 4). For spherical parallel manipulators, generic procedures for obtaining closed form solutions where previously derived [10, 11]. However, in the present case it is necessary to establish a dedicated solution scheme because additional transmission mechanisms are interweaved into to overall structure. Characteristic to the design are the following geometric properties: (i) Points (S_1, S_2, S_3, H), (K, R_5, R_2, S_5) and (S_6, H, S_7) are, respectively, colinear; (ii) Points S_1, S_3, S_4 are coplanar; (iii) the axis of R_1 is normal to the plane defined by S_1, S_3, S_4; (iv) the axes of joints R_2, R_3, R_4, R_5 and K are parallel.

We define the displacements $s_1, \ldots, s_4$ to be the total distances between the endpoints of the four cylinders, respectively. Furthermore, we introduce an inertial and a thigh-fixed coordinate system, xyz and $x'y'z'$, respectively, as follows; both origins coincide with the center of the hip joint H; the z-axis of xyz passes through S_4, and its y-axis through S_7; the x'-axis of $x'y'z'$ is aligned along point S_1 and the y'-axis is parallel to the axis of joint K. In the depicted reference configuration both reference

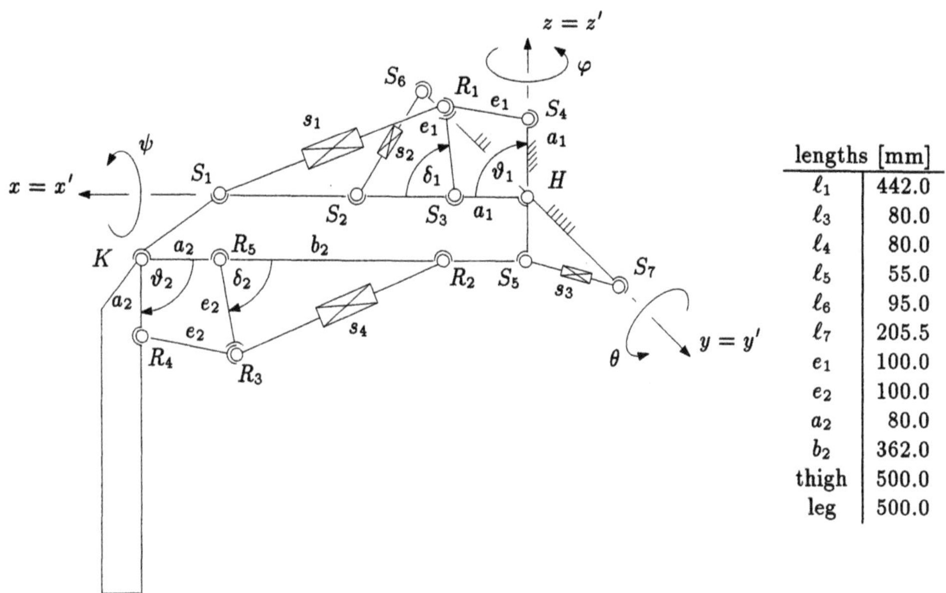

lengths [mm]	
ℓ_1	442.0
ℓ_3	80.0
ℓ_4	80.0
ℓ_5	55.0
ℓ_6	95.0
ℓ_7	205.5
e_1	100.0
e_2	100.0
a_2	80.0
b_2	362.0
thigh	500.0
leg	500.0

Figure 4: Kinematic structure of the leg in reference configuration

systems are congruent and the lower leg is orthogonal to the thigh. Further, we denote by $\underline{\ell}_i$ the radius vectors connecting the point H with the joints S_i, decomposed in the global frame; by $\underline{\ell}_i'$ the corresponding vectors decomposed in the thigh-fixed coordinate system; and by ℓ_i their corresponding lengths. Finally, we introduce the net rotation angles at hip and knee as $\Theta = \vartheta_1 - \pi/2$ and $\zeta = \vartheta_2 - \pi/2$, respectively.

B. Position analysis

The general solution of the kinematics of multibody systems can be regarded as a two-stage process ([12]); in the *relative kinematics*, the relationship between the independent kinematical inputs of the system, $\underline{q}$, and the dependent joint variables describing the relative motion at the joints, $\underline{\beta}$, is established by solving a system of constraint equations

$$\underline{f}(\underline{\beta};\underline{q}) = 0 \ ; \tag{1}$$

then, the ***absolute kinematics*** determines the absolute position of the interesting bodies as a function of the joint variables $\underline{\beta}$. One can solve (1) iteratively using well-known numerical methods. However, it is computationally more efficient and reliable to use closed-form solutions, if possible. This is the case when, possibly after a reordering of equations or unknowns, Eq. (1) is recursively solvable, i.e. its ith equation depends only on the unknowns $\beta_1, \ldots, \beta_i$ and each equation by itself is solvable for this unknown.

The present system represents such a case, even if at first glance, e.g. after applying the topological criteria described in [13] or [14], this is not obvious. We proceed by establishing this system of equations, and leave out the absolute kinematics, which is trivial for the present case. First, we set the input vector as $\underline{q} = [s_1, s_2, s_3, s_4]$ and collect the joint variables in $\underline{\beta} = [\delta_1, \Theta, \varphi, \psi, \delta_2, \zeta]$. A system of constraint equations which features the targeted structure is:

$$
\begin{aligned}
f_1(\delta_1; s_1) &\equiv b_1 + e_1^2 - 2b_1 e_1 \cos\delta_1 - s_1^2 && (2)\\
f_2(\delta_1, \Theta) &\equiv \ell_3(\ell_3 + e_1\cos\delta_1)(1+\sin\Theta) - \ell_3 e_1 \sin\delta_1 \cos\Theta && (3)\\
f_3(\Theta, \varphi; s_2) &\equiv \ell_2^2 + \ell_6^2 + 2\,\ell_2\,\ell_6 \sin\varphi \cos\Theta - s_2^2 && (4)\\
f_4(\Theta, \varphi, \psi; s_3) &\equiv \ell_5^2 + \ell_7^2 - 2\,\ell_5\,\ell_7 \cos\varphi \sin\psi + 2\,\ell_5\,\ell_7 \sin\varphi \sin\Theta \cos\psi - s_3^2 && (5)\\
f_5(\delta_2; s_4) &\equiv b_2 + e_2^2 - 2b_2 e_2 \cos\delta_2 - s_4^2 && (6)\\
f_6(\delta_2, \zeta) &\equiv a_2(a_2 + e_2\cos\delta_2)(1+\sin\zeta) - a_2 e_2 \sin\delta_2 \cos\zeta && (7)
\end{aligned}
$$

where $b_1 = \ell_1 - \ell_2$. This system of equations is obtained as follows. Eq. (2) results from the cosine law applied to $\Delta(S_1, S_2, R_1)$, while Eq. (3) is a consequence of the condition of constant length of the coupler R_1–S_4,

$$e_1^2 = [e_1 \cos\delta_1 + a_1(1-\cos\vartheta_1)]^2 + [1 \sin\vartheta_1 - e_1 \sin\delta_1]^2 ,$$

and the substitutions $\vartheta_1 = \Theta + \pi/2$ and $a_1 = \ell_3$. The equations (4) and (5) result from considerations involving the orientation matrix of the thigh, $\boldsymbol{R}$, describing the transformation of vector components from the thigh-fixed frame $x'y'z'$ to the fixed frame xyz. We describe this matrix as three concatenated elementary rotations about axes $z \to y \to x$ of the moving frame, respectively,

$$
\begin{aligned}
\boldsymbol{R} &= \mathtt{Rot}[z,\varphi] \circ \mathtt{Rot}[y,\Theta] \circ \mathtt{Rot}[x,\psi] \\
&= \begin{pmatrix} \cos\varphi\cos\Theta & -\cos\psi\sin\varphi + \sin\Theta\sin\psi\cos\varphi & \sin\psi\sin\varphi + \sin\Theta\cos\psi\cos\varphi \\ \sin\varphi\cos\Theta & \cos\psi\cos\varphi + \sin\Theta\sin\psi\sin\varphi & -\sin\psi\cos\varphi + \sin\Theta\cos\psi\sin\varphi \\ -\sin\Theta & \sin\psi\cos\Theta & \cos\psi\cos\Theta \end{pmatrix} \quad (8.)
\end{aligned}
$$

Next, we formulate the following conditions ([10, 11])

$$\sigma_i^2 = \|\, \boldsymbol{R}\,\underline{\ell}'_{t(i)} - \underline{\ell}_{p(i)} \|^2 = \ell'^2_{t(i)} + \ell^2_{p(i)} - 2\,\underline{\ell}^{\mathrm{T}}_{p(i)}\,\boldsymbol{R}\,\underline{\ell}'_{t(i)} \quad i = 1,2,3 \;, \qquad (9)$$

where $t(i)$ and $p(i)$ are the points of attachment of the ith actuation unit at the thigh and pelvis, respectively, and σ_i represents the corresponding distance between these attachment points, $\sigma_1 \equiv \overline{S_1\,S_4}$, $\sigma_2 \equiv s_2$, $\sigma_3 \equiv s_3$. Then, it follows

$$
\begin{aligned}
\sigma_1^2 &= \ell_1^2 + \ell_4^2 - 2\,\ell_1\,\ell_4 \sin\Theta \;, && (10)\\
s_2^2 &= \ell_2^2 + \ell_6^2 + 2\,\ell_2\,\ell_6 \sin\varphi \cos\Theta \;, && (11)\\
s_3^2 &= \ell_5^2 + \ell_7^2 - 2\,\ell_5\,\ell_7 \cos\varphi \sin\psi + 2\,\ell_5\,\ell_7 \sin\varphi \sin\Theta \cos\psi \;. && (12)
\end{aligned}
$$

Eq. (10) just corresponds to the cosine law for $\Delta(S_1\,H\,S_4)$ so it represents no additional independent condition besides Eqs. (2) and (3). The other two equations yield Eqs. (4) and (5). Note that this simple relationship would not hold if another sequence of elementary rotations, say $y \to z \to x$, would have been taken as a basis in Eq. (8). Thus this decomposition is material. Eqs. (6) and (7) are derived in a similarly to Eqs. (2) and (3).

Equations (2)to (7) all have the structure

$$A \cos\beta + B \sin\beta + C = 0 \quad , \tag{13}$$

where β represents the unknown related to the actual equation, and the coefficients A, B and C may depend on the input variables $\underline{q}$ as well as on already determined unkowns. Such an equation can be easily resolved for β, giving two solutions (see e. g. [15]). Thus, the complete system of constraint equations $f_1, \ldots, f_6$ can be resolved in closed form, yielding $2^6 = 64$ possible solutions.

C. Velocity analysis

The velocity relationships for the leg are determined from the time differentiation of the constraint equations (1). It follows

$$\boldsymbol{J}_\beta \dot{\underline{\beta}} + \boldsymbol{J}_q \dot{\underline{q}} = 0 \quad , \tag{14}$$

where

$$\boldsymbol{J}_\beta = \frac{\partial \underline{f}}{\partial \underline{\beta}} = \begin{pmatrix} j_{11} & 0 & 0 & 0 & 0 & 0 \\ j_{21} & j_{22} & 0 & 0 & 0 & 0 \\ 0 & j_{32} & j_{33} & 0 & 0 & 0 \\ 0 & j_{42} & j_{43} & j_{44} & 0 & 0 \\ 0 & 0 & 0 & 0 & j_{55} & 0 \\ 0 & 0 & 0 & 0 & j_{65} & j_{66} \end{pmatrix} \quad , \quad \boldsymbol{J}_q = \frac{\partial \underline{f}}{\partial \underline{q}} = \begin{pmatrix} s_1 & 0 & 0 & 0 \\ 0 & 0 & 0 & 0 \\ 0 & s_2 & 0 & 0 \\ 0 & 0 & s_3 & 0 \\ 0 & 0 & s_4 & 0 \\ 0 & 0 & 0 & 0 \end{pmatrix} \tag{15}$$

and

$$\begin{aligned} j_{11} &= 2b_1 e_1 \sin\delta_1 \quad , \quad j_{21} = -e_1\ell_3[\cos(\delta_1 - \Theta) + 1] \quad , \quad j_{22} = e_1\ell_3 \cos(\delta_1 - \Theta) + \ell_3 \cos\Theta \\ j_{32} &= -2\ell_2\ell_6 \sin\varphi \sin\Theta \quad , \quad j_{33} = 2\ell_2\ell_6 \cos\varphi\cos\Theta \quad , \quad j_{42} = 2\ell_5\ell_7 \sin\varphi\cos\Theta\cos\psi \\ j_{43} &= 2\,\ell_5\,\ell_7\,(\sin\varphi\,\sin\psi + \cos\varphi\,\sin\Theta\,\cos\psi) \\ j_{44} &= -2\,\ell_5\,\ell_7\,(\cos\varphi\,\cos\psi + \cos\varphi\,\sin\Theta\,\sin\psi) \quad , \quad j_{55} = 2b_2 e_2 \sin\delta_2 \\ j_{65} &= -e_2 a_2[\cos(\delta_2 - \zeta) + 1] \quad , \quad j_{66} = e_2 a_2 \cos(\delta_2 - \zeta) + a_2 \cos\zeta \end{aligned}$$

Note that $\boldsymbol{J}_\beta$ has triangular structure, so the resolution of Eq. (14) is particularly simple. Note also that, in the reference configuration, all entries of $\boldsymbol{J}_\beta$ outside the diagonal vanish, so that, by appropriate length specification, the mechanism can be made isotropic ([6]).Finally, the velocity twist $\underline{t}_F = [\boldsymbol{\omega}_F, \boldsymbol{v}_F]^{\mathrm{T}}$ of the foot point F, where $\boldsymbol{\omega}_F$ and $\boldsymbol{v}_F$ are, respectively, the corresponding angular and linear velocity ([16]), is easily determined as

$$\underline{t}_F = \boldsymbol{J}_F \dot{\underline{\beta}} \quad , \tag{16}$$

where $\boldsymbol{J}_F$ is the usual Jacobian encountered in robotics.

D. Force analysis

Besides pure kinematics, some applications, like force-feedback control, also require the knowledge of the resulting forces within the mechanism, both at the cylinders and at the joints. Because of the existence of closed loops, the determination of these forces is not as straight-forward as it is in a tree-type system. We present here a simple and efficient method for solving this problem.

Each of the constraint equations defined above introduces a scalar constraint force which can be identified as a Lagrange multiplier [17]. These Lagrange multipliers can be given the mechanical interpretation of 'tensions' acting in direction of the measurements from which the constraint functions result. In the present case all constraint equations result from the measurement of the squared distance between two points. Thus, the Lagrange multipliers correspond to the actual tensions arising within the couplers or cylinders (Fig. 5). However, their magnitude has to by divided by twice the value of the measurement to give the physical tension.

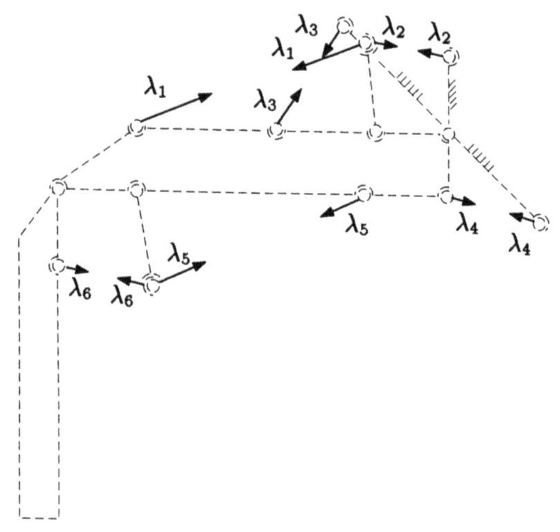

Figure 5: Lagrange multipliers and their associated lines of action

We now associate to each unknown joint variable β_i a corresponding generalized force Q_{β_i} , obtaining in the present case a vector $\underline{Q}_\beta = [Q_{\beta_1}, \ldots, Q_{\beta_6}]^{\mathrm{T}}$ of corresponding torques. By linearity, these torques depend on the constraint forces as

$$\underline{Q}_\beta = \boldsymbol{J}_\beta^{\mathrm{T}} \underline{\lambda} + \underline{\widehat{Q}}_\beta \ , \tag{17}$$

where $\underline{\lambda} = [\lambda_1, \ldots, \lambda_6]^{\mathrm{T}}$ is the vector of Lagrange multipliers, and $\underline{\widehat{Q}}_\beta$ is the vector of 'remaining' applied forces. This vector is easily established by computing the equilibrium conditions for the 'opened' system, i. e. the one for which all Lagrange multipliers are artificially set equal to zero. In the present case, for instance, we obtain

$$\underline{\widehat{Q}}_\beta = \boldsymbol{J}_F^{\mathrm{T}} \underline{\mathrm{w}}_F \ , \tag{18}$$

where $\underline{\mathrm{w}}_F = [\boldsymbol{\tau}_F, \boldsymbol{f}_F]^{\mathrm{T}}$ is the wrench operating at the foot and containing the moment $\boldsymbol{\tau}_F$ and force $\boldsymbol{f}_F$ at that point. Now, the global condition of equilibrium demands that the resulting torques in Eq. (17) identically vanish ([18]), from which it follows

$$\underline{\lambda} = \boldsymbol{J}_\beta^{-\mathrm{T}} \boldsymbol{J}_F^{\mathrm{T}} \underline{\mathrm{w}}_F \ . \tag{19}$$

After this, all other forces can be determined easily as in tree-type systems.

IV. Results

The design discussed above was modeled and simulated with an object-oriented multibody modeling package 'M∞BILE' [19], exhibiting good mobility and force transmission properties. The similarity between the motion of the actual design and that of a human leg is illustrated in Fig. 6, which also shows some results of the kinematics and statics, the latter computed for a constant load of 200N applied to the foot in negative inertial z-direction.

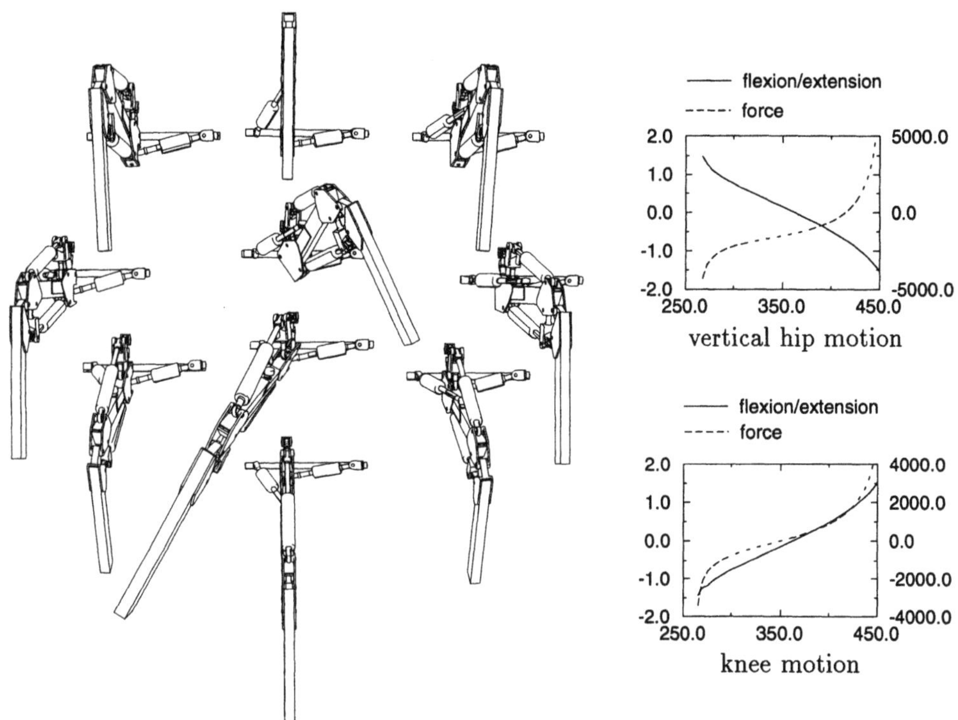

Figure 6: Mobility study of the leg mechanism

V. Conclusions

The discussed leg mechanism offers several analogies to human leg motion, which are favourable for legged robots being teleoperated by humans through directly connected master-slave devices. The kinematics of the device are solvable in closed-form, thus reducing the necessary computational effort for feed-back control. The use of standard self-aligning joints for the spherical pairs makes the construction technologically feasible and inexpensive, while application of hydraulic cylinders guarantees a high energy density and light-weight construction. The simulation confirms the feasibility of the approach and the intended anthropomorphic properties of the mechanism. A prototype is currently being manufactured which is aimed to weight 6–8kg for a net load of 20kg.

References

[1] S.-M. Song and K. J. Waldron. *Machines That Walk: The Adaptive Suspension Vehicle.* The MIT-Press, Cambridge, Massachusetts, 1989.

[2] S.N. Dwivedi and S. Mahalingham. Terrain adaptive gaits for the ambler. *Advanced Robotics*, 5(2):109–132, 1991.

[3] J.S. Byrd and K.R. DeVries. A six-legged telerobot for nuclear applications. *The International Journal of Robotics Research*, 9(2), 1990.

[4] H.-J. Weidemann. *Dynamik und Regelung von sechsbeinigen Robotern und natürlichen Hexapoden.* Fortschritt-Berichte VDI, Reihe 8 Nr. 362. VDI-Verlag, 1993.

[5] M.H. Raibert. *Legged Robots that Balance.* The MIT-Press, Boston, 1985.

[6] J. Angeles, F. Pfeiffer, and H.-J. Weidemann. The kinematic design of walking machines under isotropy criteria. In V. Parenti-Castelli and J. Lenarcic, editors, *3rd International Workshop on Advances in Robot Kinematics*, pages 230–235, Ferrara, Italy, 1992.

[7] M. Hiller, G. Schweitzer, and C. Woernle. Kinematical control of the combined wheeled and legged vehicle “RoboTRAC”. In *Proceedings of the 8th CISM-IFToMM Symposium Ro.Man.Sy*, 1990.

[8] A. Benninghof and K. Goerttler. *Lehrbuch der Anatomie des Menschen*, volume 1. Urban & Schwarzenberg, München, Wien, Baltimore, 1980.

[9] Encyclopedia of human biology. Academic Press, 1991.

[10] C. Gosselin and E. Lavoie. Spherical parallel manipulators: Dexterity and isotropy. In V. Parenti-Castelli and J. Lenarcic, editors, *3rd International Workshop on Advances in Robot Kinematics*, pages 143–149, Ferrara, Italy, 1992.

[11] C. Innocenti and V. Parenti-Castelli. Direct kinematics of the 6-4 fully parallel manipulator with position and orientation uncoupled. In *European Robotics and Intelligent Systems Conference*, Corfù, 1991.

[12] M. Hiller and A. Kecskeméthy. Equations of motion of complex multibody systems using kinematical differentials. *Transactions of the Canadian Society of Mechanical Engineers*, 13(4):113–121, 1989.

[13] P. Fanghella and C. Galletti. A modular approach for computational kinematics. In J. Angeles, G. Hommel, and P. Kovacs, editors, *Computational Kinematics.* Kluwer Academic Publishers, 1993.

[14] A. Kecskeméthy. On closed form solutions of multiple-loop mechanisms. In J. Angeles, G. Hommel, and P. Kovacs, editors, *Computational Kinematics.* Kluwer Academic Publishers, 1993.

[15] J. Duffy. *Analysis of Mechanisms and Robot Manipulators.* Edward Arnold, London, 1980.

[16] J. Angeles. *Rational Kinematics.* Springer Tracts in Natural Philosophy 34. Springer-Verlag, New York, 1988.

[17] P.E. Nikravesh. *Computer-Aided Analysis of Mechanical Systems.* Prentice-Hall, 1988.

[18] A. Kecskeméthy and M. Hiller. An object-oriented approach for an effective formulation of multibody dynamics. *Computer Methods in Applied Mechanics and Engineering*, 1994. To appear.

[19] A. Kecskeméthy. MOBILE - An object-oriented tool-set for the efficient modeling of mechatronic systems. In *Proceedings of the Second Conference on Mechatronics and Robotics, September 27–29*, Moers/Duisburg, 1993.

An Investigation of the Kinematic Control of a Six-legged Walking Robot

André Preumont
Université Libre de Bruxelles

Abstract - This paper describes a semi autonomous six-legged walking machine with hexagonal architecture that has been developed for research on gait control. The machine weights 13 kg; each leg has 3 d.o.f. with closed-loop kinematics, actuated by d.c. electric drives. The kinematics is designed to achieve gravitational decoupling, which simplifies considerably the velocity control of the legs. The control architecture is decentralized, each leg being controlled by a separate microcontroller. A central computer coordinates the motion of the various legs (gait) and controls the attitude of the vehicle. The gait can be programmed arbitrarily.

I. Introduction

Wheeled mobile robots have a limited mobility which prevents their use in a strongly unstructured environment. They must be replaced by tracked, legged, or articulated body vehicles. The superior terrain adaptability of legged robots makes them particularly attractive for space exploration and disaster prevention vehicles. The interest of the engineering community in walking robots started 10 to 15 years ago and has been growing steadily ever since (e.g.[11,15]). A wide variety of prototypes have been constructed with various sizes and architectures (e.g.[1,9,14,16]). Note that the addition of suction cups or electromagnetic feet can transform a walking machine into a *climbing* machine [3].

So far, walking robots are considerably less elegant and less effective than insects, particularly on rough ground. This results from the following facts:

- Animal limbs have a larger workspace and some degree of redundancy which gives them a superior agility, allowing them to climb or avoid obstacles which are by far higher than themselves, and to recover stability when it has been lost because of changing ground conditions.
- Animal limbs possess distributed tactile sensors which are used effectively in obstacle avoidance, foothold selection and gait adaption to terrain [4].
- Wider variety of gaits available and continuous transition from one to another. Dynamic gaits offer even wider possibilities for stability recovery, high speed and obstacle avoidance [14].
- Combined use of terrain preview and tactile information for gait control.

The control architecture of walking machines is often hierarchical, with the following levels:

- *Level A*: navigation & planning
- *Level B*: gait control
- *Level C*: leg trajectory

A. J. Lenarčič and B. B. Ravani (eds.), Advances in Robot Kinematics and Computationed Geometry, 171–178.

Level A is not significantly different from what it is for a wheeled robot. Level B coordinates the motion of the legs; three different strategies can be used:

- central supervision with fixed rules (e.g. regular gaits);
- leg coordination based on control influences exchanged between the legs, duplicating the behaviour observed on insects (*neurobiological coordination*) [4,19];
- rule-based control within a finite set of gait states (*free gait*) [7]. Besides its great flexibility with respect to changes of direction, the free gait can accomodate a damaged leg.

Level C (trajectory and servo control) can typically be distributed at the leg level (decentralize control architecture), but some coupling with level B is necessary to account for the tactile information (obstacle avoidance) and for attitude control.

The locomotion on rough ground of a blind vehicle implies some tactile information. In the neurobiological coordination, this information is part of the control influences processed by the leg coordination modules. If force sensors are placed in the feet, force feedback can be used to correct the kinematic input and produce an *active compliance* [5,6]. This allows to follow contours and contributes to produce a more uniform force distribution between the supporting legs.

This project was partly financed by the EEC in the program *TELEMAN*. The first phase consisted of an investigation of the regular, symmetric gaits [13], which were demonstrated on a small hexapod (1.3 kg) with 2 d.o.f. per leg [12]. The second phase, described in this paper, started in 1991 with the objective of demonstrating the kinematic control (level B & C) of a research machine representative of future, larger size vehicles.

II. Electromechanical architecture

The vehicle consists of an hexapod with an hexagonal architecture (Fig.1) similar to that of the ODEX-1 [1,2]. Because of the 60° symmetry, this configuration does not require a change of orientation if the direction of motion is changed (*crab walking*). The weight of the robot is 13 kg and its typical size is about 60cm (these specifications have been selected with no special mission in mind, but rather to be compatible with a University environment). Each leg consists of a closed loop structure with 3 d.o.f., actuated by d.c. motors of 6 watts connected to planetary gears and encoders. The two rotary d.o.f. (q_2 and q_3 in Fig.2) are provided with potentiometers for calibration; the vertical d.o.f. q_1 is actuated by a ball screw. A set of contact switches are installed at various places in the mechanism for calibration and safety purposes. Each foot is provided with a contact switch for ground detection. The computer boards are located in the central body of the vehicle.

III. Kinematics

Each leg can rotate about a vertical axis (coordinate q_3) by an amount of $\pm 35°$; for a

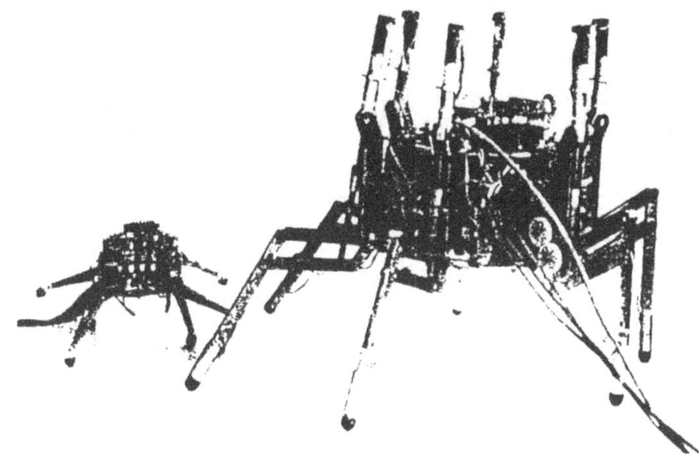

Fig. 1: General view of the robot. The small robot developed in the first phase of the project is also shown.

fixed value of q_3, the leg moves in a plane as represented in Fig.2. The geometry of the mechanism has been selected to achieve a reasonable compromise between the following objectives:

- gravitational decoupling,
- constant stroke R within the workspace,
- maximum mobility (i.e. maximum Δz within the workspace).

Gravitational decoupling is desirable to reduce the energy loss associated with the vertical cycling of the center of mass of the vehicle during the motion (energy cannot be recovered during deceleration); it also simplifies the control considerably by reducing by one the order of the inverse Jacobian equation to be solved in real time for trajectory generation. Perfect gravitational decoupling is not possible, but a small error can be easily compensated by the compliance of the leg.

Unduly small strokes R in large parts of the workspace would put heavy restrictions on the velocity V of the vehicle, because it can be shown [18] that

$$V = \frac{R}{\tau}\frac{(1-\delta)}{\delta} \tag{1}$$

where δ is the duty factor and τ is the return or transfer time [the time necessary to return the leg from the posterior extreme position (*PEP*) to the anterior extreme position (*AEP*)].

τ includes the lifting and the acceleration, the transfer at maximum velocity and the descent until the foot touches the ground; it is not reduced significantly for smaller strokes. Maximum mobility is, of course, the prime objective of the legged vehicle.

Because the attitude control keeps the vertical axis of the robot parallel to the gravity vector and the gravitational decoupling makes the foot trajectory in the horizontal plane

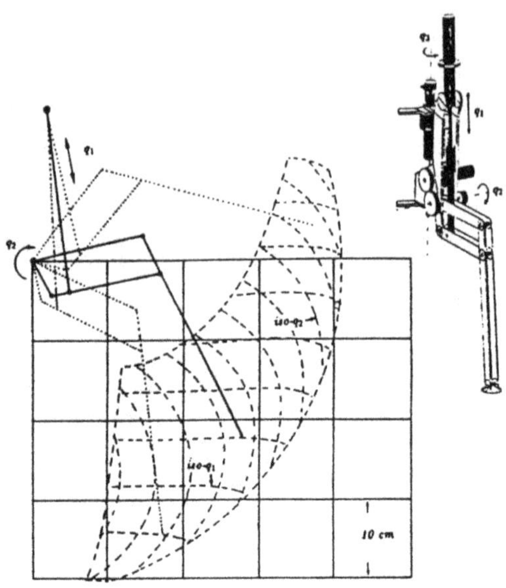

Fig. 2: Workspace of the leg. The trajectories for constant values of q_1 show a good gravitational decoupling.

independent of the vertical coordinate q_1, the joint trajectories can be computed by solving the inverse kinematics of the leg in the plane normal to the gravity vector at the tip of the leg.

A motion of the vehicle with velocity v will be achieved if all the legs in support phase move with the same velocity in the reverse direction (Fig.3). Using a local system of coordinates with x along the velocity vector, the joint velocities of leg i can be obtained in two steps: First, the velocity in polar coordinates is calculated by

$$\begin{pmatrix} \dot{r}_i \\ \dot{\theta}_i \end{pmatrix} = J_i^{-1} \begin{pmatrix} -V \\ 0 \end{pmatrix} \tag{2}$$

$$J_i = \begin{pmatrix} \cos\psi_i & -r_i \sin\psi_i \\ \sin\psi_i & -r_i \cos\psi_i \end{pmatrix} \tag{3}$$

where $\psi_i = (2i\text{-}1)\pi / 6 - \alpha + \theta_i$, α is crab angle and i is the leg index. Next, we convert the velocity in polar coordinates into joint velocities according to

$$\dot{q}_2^i = f^{-1}\left(q_1^i, q_2^i\right)\dot{r}_i \tag{4}$$

$$\dot{q}_3^i = \dot{\theta}_i \tag{5}$$

where $f(q_1^i, q_2^i)$ describes the kinematics of the leg. Observe that, because of the gravitational decoupling, the vertical coordinate appears only as a parameter in Equ.(4).

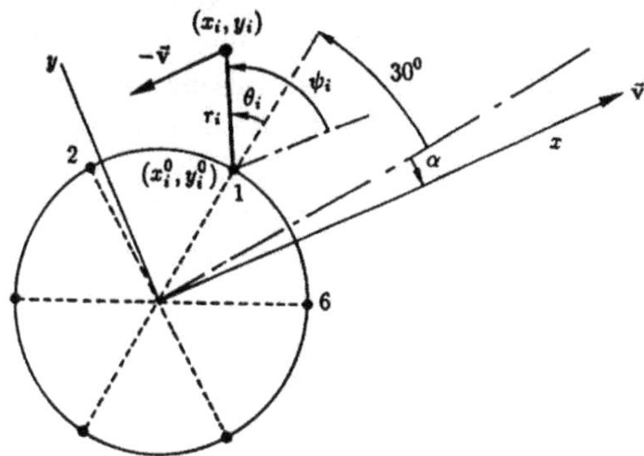

Fig. 3: Trajectory generation

IV. Control architecture

The control architecture is described in Fig.4. Each leg is connected to its own control board based on a microcontroller *Intel 87C196KC* (level C) the main task of the microcontroller is to solve in real time the inverse kinematic equations. To speed up the process, all the trigonometric functions appearing in the Jacobian and the function f describing the leg kinematics in Equ.(4) are tabulated. The microcontroller supplies the reference angular velocities of the various motors to dedicated circuits which control the d.c. motors with *PWM* via built-in *PID*. A serial link connects all the level C control boards to a *PC* which acts as gait generator and attitude controller (level B).

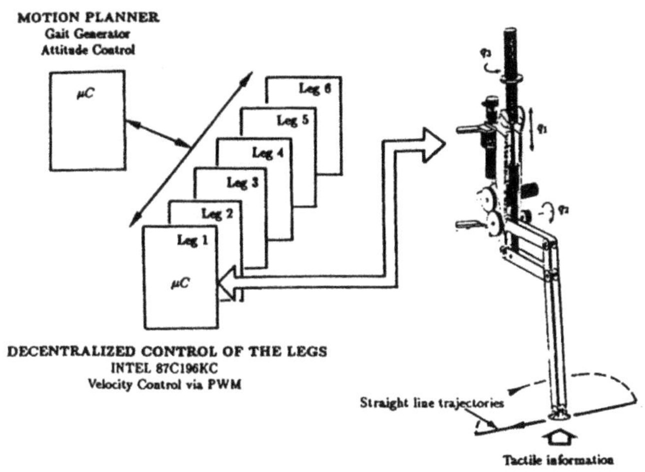

Fig. 4: Control architecture.

This control architecture allows a very flexible programming of the leg motion and is very well suited to regular gaits as well as free gaits. So far, we have only implemented regular gaits which are appropriate for walking in straight line on moderately rough ground. In this case, the information transmitted to the level C control consists of the velocity V, the crab angle α, the cycle time T, the duty factor δ, the phase relation to the first leg, ϕ_i, and the starting position (x_o, y_o). Note that the control strategy is purely kinematic; subtantial errors in the leg trajectories would produce internal forces in the system. Fortunately, the build up of internal forces is impossible because of the relaxation at regular intervals, during the transfer phase. In fact, minor errors can be tolerated because of the combined effect of the flexibility and the backlash in the gears. In our machine, the effect of backlash corresponds to an uncertainty circle of approximate diameter 20mm, which is larger than the integrated error in one stroke.

V. Attitude control

The robot is provided with a dual channel inclinometer measuring the *roll* (γ) and *pitch* (β) angles. The corresponding rotation matrix is

$$R = \begin{pmatrix} \cos\beta & 0 & \sin\beta \\ 0 & 1 & 0 \\ -\sin\beta & 0 & \cos\beta \end{pmatrix} \begin{pmatrix} 1 & 0 & 0 \\ 0 & \cos\gamma & -\sin\gamma \\ 0 & \sin\gamma & \cos\gamma \end{pmatrix}$$

or

$$R = \begin{pmatrix} \cos\beta & \sin\beta\sin\gamma & \sin\beta\cos\gamma \\ 0 & \cos\gamma & -\sin\gamma \\ -\sin\beta & \cos\beta\sin\gamma & \cos\beta\cos\gamma \end{pmatrix} \tag{6}$$

The axis $\boldsymbol{n}$ and angle of rotation θ can be computed from

$$1 + 2\cos\theta = \mathrm{Trace}(R) = \cos\beta + \cos\gamma + \cos\beta\cos\gamma \tag{7}$$

and

$$n_x = \frac{r_{32} - r_{23}}{2\sin\theta} = \frac{\sin\gamma(\cos\beta + 1)}{2\sin\theta} \tag{8}$$

$$n_y = \frac{r_{13} - r_{31}}{2\sin\theta} = \frac{\sin\beta(\cos\gamma + 1)}{2\sin\theta} \tag{9}$$

$$n_z = \frac{r_{21} - r_{12}}{2\sin\theta} = -\frac{\sin\gamma\sin\beta}{2\sin\theta} \cong 0 \tag{10}$$

For small angles, the axis of rotation is approximately contained in the horizontal plane; the attitude of the vehicle can be corrected by moving vertically all the legs in support phase according to

$$\Delta z_i = -\theta h_i \tag{11}$$

where h_i is the distance between the current position of leg i with respect to the axis of rotation (Fig.5).

VI. Concluding remark

A six-legged walking machine has been constructed for research on gait control. A decentralized control architecture has been chosen to allow a flexible programming of the gait. So far, only regular gaits have been tested, but further research is under way to develop a free gait algorithm based on maximizing the static stability within a set of geometric constraints. Such a free gait has been demonstrated successfully by A.Halme and his coworkers [7,8].

So far, the only tactile information which is used by our walking machine comes from a on-off switch placed in each foot. It is clear that a more elaborate tactile information could be extremely helpful in improving the obstacle avoidance capability of the robot, allowing to climb stairs, etc... If a force sensor is available in each leg, force feedback can be added to the kinematic loop to produce active compliance. We plan to implement active compliance following the ideas developed by E.Devjanin, A.Shneider and coworkers [5,6].

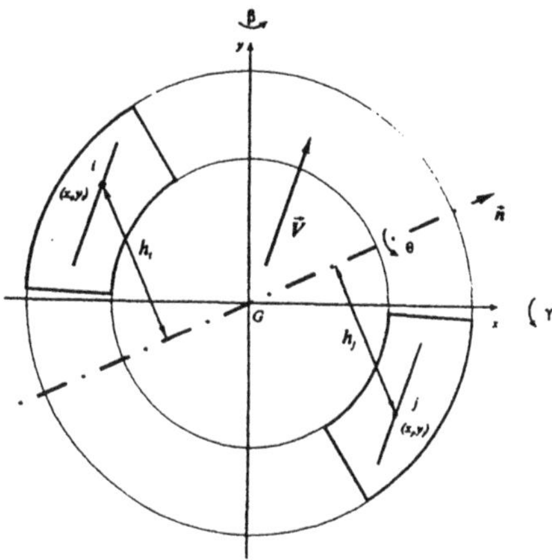

Fig. 5: Principle of attitude control.

VII. References

[1] T.G.Bartholet, The first functionoid developed by ODETICS Inc., Proc. Int. Conf. on Advanced Robotics , 293-298, 1983.

[2] J.S.Byrd & K.R.De Vries, A six legged telerobot for nuclear applications development, *Int. Journal of Robotics Research, Vol.9, No 2,43-52, April 1990.*

[3] A.Collie, Unusual robots, *Industrial Robot, Vol.19,No 4, 13-16, 1992.*

[4] H.Cruse, J.Dean, U.Muller & J.S.Schmitz, The stick insect as a walking robot, Int. Conf. on Advanced Robotics, Pisa, 936-940, June 1991.

[5] E.A.Devjanin et al., The six-legged walking robot capable of terrain adaptation, *Mechanism and Machine Theory, Vol.18, No 4, 257-260, 1983.*

[6] D.M.Gorinevsky & A.Yu.Shneider, Force control in locomotion of legged vehicles over rigid and soft surfaces, *Int. Journal of Robotics Research, Vol.9, No 2, 4-23, April 1990.*

[7] A.Halme, K.Hartikainen & K.Karkkainen, Terrain adaptive motion and free gait of a six-legged walking machine, *IFAC* Workshop on Intelligent Autonomous Vehicles, Southampton, April 1993.

[8] K.K.Hartikainen, A.J.Halme, H.Lehtinen & K.O.Koskinen, *MECANT 1*: A six legged walking machine for research purposes in outdoor environment, *IEEE,* Int. Conf. on Robotics and Automation, Nice, 157-163, May 1992.

[9] S.Hirose, A study of design and control of a quadruped walking vehicle, *Int. Journal of Robotics Research, Vol.3,No 2, 1984.*

[10] S.Hirose & O.Kunieda, Generalized standard foot trajectory for a quadruped walking vehicle, *Int. Journal of Robotics Research, Vol.10, No 1, 3-13, February 1991.*

[11] *The International Journal of Robotics Research*, Special Issue on Legged Locomotion, *Vol.9, No 2, April 1990.*

[12] A.Preumont, P.Alexandre & D.Ghuys, Gait Analysis and implementation of a six leg walking machine, Int. Conf. on Advanced Robotics, Pisa, 941-945, June 1991.

[13] A.Preumont, D.Ghuys & C.Malékian, On the static stability of hexapods, 3rd Int. Workshop on Advances in Robot Kinematics, Ferrara, 126-133, June 1992.

[14] M.H. Raibert, *Legged Robots that Balance*, MIT Press, 1986.

[15] *Robotics in Nuclear Facilities*, Special Issue, *SMIRT-11*, Tokyo, August 1991.

[16] S.M.Song & K.J.Waldron, *Machine that Walk: The Adaptive Suspension Vehicle*, MIT Press, 1989.

[17] K.J.Waldron, Force and motion management in legged locomotion, *IEEE Transactions on Robotics and Automation, Vol.RA-2, No 4, December 1986.*

[18] K.J.Waldron,V.J.Vohnout, A.Pery & R.B.McGhee, Configuration design of the adaptive suspension vehicle, *Int. Journal of Robotics Research, Vol.3, No2, 37-48, Summer 1984.*

[19] H.-J.Weidemann, J.Eltze & F.Pfeiffer, Leg design based on biological principles, *IEEE* Int. Conf. on Robotics and Automation, Atlanta, 352-358, May 1993.

5. Kinematic Performance

Kinematic Isotropy in 3R Positioning Manipulators

M. L. Husty J. Angeles

Department of Mechanical Engineering &
McGill Centre for Intelligent Machines
McGill University
817 Sherbrooke St. West
Montreal, Canada, H3A 2K6
husty@cim.mcgill.ca angeles@cim.mcgill.ca

Abstract - Discussed in this paper is the existence of *isotropy points* in the end effector of 3-revolute manipulators with a given architecture at a given posture. We show that, in general, such points are not possible. Moreover, when these points occur, they are not unique. The results of this discussion should find applications in manipulator design, manipulator control, and in trajectory planning.

I Introduction

With the aim of enhancing the performance of manipulators from an accuracy point of view, *kinematic isotropy* has been proposed as a design criterion [1, 2, 3, 4]. Briefly stated, a manipulator for either positioning or orienting tasks only, has a *dimensionally-homogeneous* Jacobian matrix, which allows for a straightforward definition of the associated condition number [5]. Manipulators for both positioning and orienting tasks have Jacobian matrices that are dimensionally inhomogeneous, which makes such a definition impossible. However, as proposed by [6], the introduction of a *characteristic length* allows one to extend the concept of manipulator condition number to the latter. In any case, accuracy is enhanced when the minimum value of unity for the condition number of a manipulator can be attained, which occurs at postures whereby the Jacobian matrix becomes *isotropic*. For completeness, we recall here that a matrix is isotropic if all its singular values are identical and different from zero. Moreover, manipulators with such a property, termed *kinematic isotropy*, are called *isotropic*.

Discussed here is the concept of kinematic isotropy as pertaining to 3-revolute manipulators for positioning tasks only, and hence, we will be able to dispense with character-

A. J. Lenarčič and B. B. Ravani (eds.), Advances in Robot Kinematics and Computationed Geometry, 181–190.

istic lengths, for the associated Jacobian matrices have all entries with units of length. Moreover, we focus on manipulators with a *given design*, at *given postures* and look for points in their end-effector (EE) that, if used as *operation points*, will render the Jacobian matrix isotropic at that particular posture. Note that these points may lie within the physical limits of the given EE or in an extension thereof. The relevance of this investigation lies in that most of six-axis industrial manipulators are of the decoupled type, i.e., have spherical wrists, with the last three revolute axes intersecting at a point that plays the role of the center of the sphere. The inverse kinematics of these manipulators is greatly simplified because the two tasks, positioning and orienting, can be decoupled. Hence, when designing manipulators of this kind, we should be aiming at wrist centers that can render the positioning 3×3 Jacobian matrix isotropic.

II Conditions for Kinematic Isotropy

As stated in [4], kinematic isotropy occurs in a manipulator when its Jacobian matrix can be rendered isotropic at certain configurations or postures. The manipulator of interest to us has the general architecture shown in Fig. 1. In this figure, the first three revolutes serve to position point C, the center of the spherical wrist. Thus, C lies on the axis of the fourth revolute of Figure 1, which is the first revolute of the wrist.

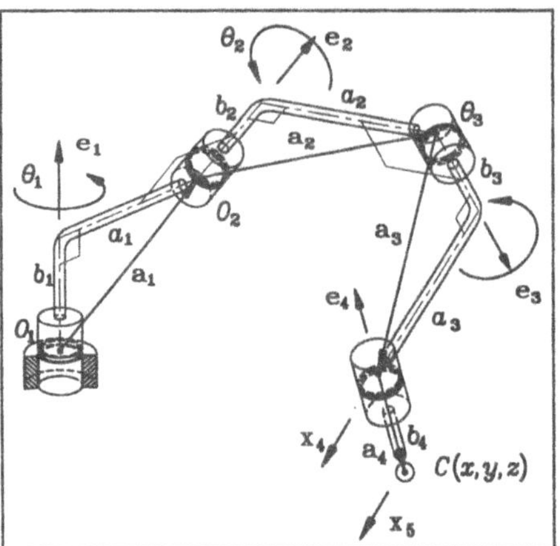

Figure 1: General 3R Positioning Robot

For quick reference, we recall here the form of the Jacobian matrix of that manipulator, namely,

$$\mathbf{J} = [\mathbf{n}_1 \quad \mathbf{n}_2 \quad \mathbf{n}_3] \tag{1}$$

where $\mathbf{n}_k$, for $k = 1, 2, 3$, denote the *moments* of the three revolute axes, i.e.,

$$\mathbf{n}_k \equiv \mathbf{e}_k \times \mathbf{r}_k \tag{2}$$

with $\mathbf{e}_k$ and $\mathbf{r}_k$ denoting the unit vector parallel to the kth axis and the vector directed from any point, say O_k on the same axis, as shown in Fig. 1, to the *operation point* C of the EE.

In the discussion below we investigate the existence of kinematic isotropy in manipulators with given links at given postures, upon changes in the operation point. To do this, we recall from [4] that the manipulator at hand is in an isotropic posture if all moments $\{\mathbf{n}_k\}_1^3$ are both mutually orthogonal and of equal magnitude, namely,

$$\mathbf{n}_1 \cdot \mathbf{n}_2 = \mathbf{n}_2 \cdot \mathbf{n}_3 = \mathbf{n}_1 \cdot \mathbf{n}_3 = 0, \qquad \|\mathbf{n}_1\| = \|\mathbf{n}_2\| = \|\mathbf{n}_3\| = \sigma \tag{3}$$

Furthermore, as shown in [7], all points equidistant from two lines lie in a hyperboloid, called the *equidistance hyperboloid.*

A coordinate frame $\mathcal{F}$ is now chosen with origin at the midpoint of the common normal of the two lines. Further, the common normal is defined as the Z'-axis, the X' and Y' axes being chosen so that they make equal angles with the two lines. If we now assume that these two lines make an angle of ϕ and are a distance $a_1 \neq 0$ apart, then the equidistance hyperboloid $\mathcal{H}$ of these two lines has the equation

$$z = \frac{\sin\phi}{a_1} xy \tag{4}$$

The special case in which the two lines intersect will be discussed later. Now we study the equidistant points of three lines $\{\mathcal{L}_k\}_1^3$. Let $\mathcal{H}_{12}$ and $\mathcal{H}_{23}$ be the equidistance hyperboloids associated with the pairs $(\mathcal{L}_1, \mathcal{L}_2)$ and $(\mathcal{L}_2, \mathcal{L}_3)$. The intersection curve of the two equidistance hyperboloids is a quartic space curve $\mathcal{C}$. Clearly, all points of $\mathcal{C}$ are equidistant to all three lines. The projection of $\mathcal{C}$ onto the X-Y plane has the equation

$$\begin{aligned} &x^2\cos^2\frac{\phi}{2} + y^2\sin^2\frac{\phi}{2} + r_{31}^2 + r_{32}^2 + r_{33}^2 - \frac{1}{4}a_1^2 - 2xr_{31} - 2yr_{32} \\ &-2\frac{\sin\phi}{a_1}xy - (xe_{31} + ye_{32} + e_{33}xy\frac{\sin\phi}{a_1})^2 = 0 \end{aligned} \tag{5}$$

where r_{3i} and e_{3i} are the components in $\mathcal{F}$ of vectors $\mathbf{r}_3$ and $\mathbf{e}_3$, respectively. With this equation we have the first condition for the operation point C to render the Jacobian matrix isotropic. For brevity, we call such points *isotropy points.* Thus, for an EE point to be an isotropy point, *it has to lie on the intersection curve* $\mathcal{C}$. An arbitrary point P on the first equidistance hyperboloid has the coordinates $x, y, (xy/a_1)\sin\phi$. To calculate the moments of the three axes with respect to this point, we take point O_k, for $k = 1, 2, 3$, on each axis $\mathcal{L}_k$. The three moments are given below:

$$\mathbf{n}_1 = \begin{bmatrix} [xy\sin(\phi/2)\sin\phi]/a_1 - (a_1/2)\sin(\phi/2) \\ [xy\cos(\phi/2)\sin\phi]/a_1 - (a_1/2)\cos(\phi/2) \\ -y\cos(\phi/2) - x\sin(\phi/2) \end{bmatrix}$$

$$\mathbf{n}_2 = \begin{bmatrix} -(a_1/2)\sin(\phi/2) - [xy\sin(\phi/2)\sin\phi]/a_1 \\ (a_1/2)\cos(\phi/2) + [xy\cos(\phi/2)\sin\phi]/a_1 \\ -y\cos(\phi/2) + x\sin(\phi/2) \end{bmatrix}$$

$$\mathbf{n}_3 = \begin{bmatrix} r_{33}e_{32} - r_{32}e_{33} + e_{33}y - e_{32}(xy/a_1)\sin\phi \\ -r_{33}e_{31} + r_{31}e_{33} - e_{33}x + e_{31}(xy/a_1)\sin\phi \\ r_{32}e_{31} - r_{31}e_{32} + e_{32}x - e_{31}y \end{bmatrix}$$

Now, from the three normality conditions, $\mathbf{n}_i \cdot \mathbf{n}_j = 0$, for i, $j = 1,2,3$ and $i \neq j$, we derive three equations that define three curves $\mathcal{C}_i$, for $i = 1,2,3$, in the X-Y plane, of the form

$$\mathcal{C}_1 \; : \; \frac{-x^2}{2} + \frac{y^2}{2} + \left(-\frac{a_1^2}{4} + \frac{x^2}{2} + \frac{y^2}{2} + \frac{x^2y^2}{4a_1^2}\right)\cos\phi - \frac{x^2y^2}{4a_1^2}\cos 3\phi = 0 \tag{6}$$

$$\begin{aligned} \mathcal{C}_2 \; : \; & (r_{31}e_{32} - r_{32}e_{31} - e_{32}x + e_{31}y)(y\cos\frac{\phi}{2} + x\sin\frac{\phi}{2}) + \\ & (r_{33}e_{32} - r_{32}e_{33} + e_{33}y - e_{32}\frac{xy}{a_1}\sin\frac{\phi}{2})\sin\frac{\phi}{2}(\frac{xy}{a_1}\sin\phi - \frac{a_1}{2}) + \\ & (r_{31}e_{33} - r_{33}e_{31} - e_{33}x + e_{31}\frac{xy}{a_1}\sin\frac{\phi}{2})\cos\frac{\phi}{2}(\frac{xy}{a_1}\sin\phi - \frac{a_1}{2}) = 0 \end{aligned} \tag{7}$$

$$\begin{aligned} \mathcal{C}_3 \; : \; & (r_{31}e_{32} - r_{32}e_{31} - e_{32}x + e_{31}y)(y\cos\frac{\phi}{2} + x\sin\frac{\phi}{2}) + \\ & (r_{33}e_{32} - r_{32}e_{33} + e_{33}y - e_{32}\frac{xy}{a_1}\sin\frac{\phi}{2})\sin\frac{\phi}{2}(\frac{a_1}{2} + \frac{xy}{a_1}\sin\phi) + \\ & (r_{31}e_{33} - r_{33}e_{31} - e_{33}x + e_{31}\frac{xy}{a_1}\sin\frac{\phi}{2})\cos\frac{\phi}{2}(\frac{xy}{a_1}\sin\phi + \frac{a_1}{2}) = 0 \end{aligned} \tag{8}$$

If $\phi \neq k\pi/2$, for $k = 1,2,...$, and neither e_{31} nor e_{32} vanishes, then these three curves are quartic like eq. (5). Their most interesting property is that they have common points at the line at infinity of the X-Y plane. By homogenizing the equations one realizes immediately that these points are $(1:0:0)$ and $(0:1:0)$[1]. This is an important issue because we have to know how many intersection points of these curves have finite locations. Moreover, each curve passes twice through each point at infinity, so that two of them have eight points in common with the line at infinity, and at most eight finite intersection points. To fulfill the normality conditions of eq.(3), the three curves $\mathcal{C}_1$, $\mathcal{C}_2$ and $\mathcal{C}_3$, as well as the equidistance curve $\mathcal{C}$, have to have common points. In general such points will not exist, but we can, for example, obtain eight finite solutions for any two of them. This means that, by intersecting, for example, $\mathcal{C}$ with $\mathcal{C}_1$, we would obtain eight positions of the operation point C, where C is equidistant from all three axes and the moments with respect to the first and the second axes are orthogonal. However, note that this property does not have a special meaning for the performance of this point at the given posture. Especially, a point could even be the intersection of two curves and render the Jacobian singular, if that point is taken as the operation

[1] In special cases the order of the curves can be lower and the points at infinity can have a different behaviour.

point. This is so because the third moment vector could be linearly dependent with the other two. It also can happen that all three moments are orthogonal, and that $\mathcal{C}_1$, $\mathcal{C}_2$ and $\mathcal{C}_3$ have a common intersection point, but this intersection point is not on $\mathcal{C}$. This means that we cannot expect, in general, the three curves to have a finite common intersection point. We have, then, in summary:

Theorem 1: *A general 3R positioning manipulator has no isotropy points on its EE for arbitrary postures.*

Furthermore, note that $\mathcal{C}_1$ is independent of the location of the third axis. This is so because the relative layout of the first two axes does not change during any motion of the manipulator. Therefore, $\mathcal{C}_1$ and its equidistance hyperboloid $\mathcal{H}$ do not change either; they undergo rigid-body motions only. Recall that the geometric meaning of $\mathcal{C}_1$ is orthogonality of the first two moments, which leads us to:

Conjecture 1: *A general 3R manipulator can have isotropic postures only if its operation point lies on* $\mathcal{C}_1 \subset \mathcal{H}$.

III Special Axis Layouts

If the first two axes intersect, $a_1 = 0$, and the foregoing derivations are no longer valid, because the equidistance hyperboloid decomposes into two planes, namely,

$$x = 0 \qquad y = 0 \tag{9}$$

Hence, instead of eq.(5) we obtain the two equations below:

$$\begin{aligned}
y^2\sin(\phi/2) + r_{31}^2 + r_{32}^2 + r_{33}^2 - 2(yr_{32} + zr_{33}) - (ye_{32} + ze_{33})^2 &= 0 \\
x^2\cos(\phi/2) + r_{31}^2 + r_{32}^2 + r_{33}^2 - 2(xr_{31} + zr_{33}) - (xe_{31} + ze_{33})^2 &= 0
\end{aligned}$$

which represent two quadratic curves, one lying in the Y-Z plane and the other in the X-Z plane. For the points on these two curves we calculate the three moments analogous to the general case and find, for example, for the points on the first curve:

$$\mathbf{n}_1 = \begin{bmatrix} z\sin(\phi/2) \\ z\cos(\phi/2) \\ -y\cos(\phi/2) \end{bmatrix}, \quad \mathbf{n}_2 = \begin{bmatrix} -z\sin(\phi/2) \\ z\cos(\phi/2) \\ -y\cos(\phi/2) \end{bmatrix}, \quad \mathbf{n}_3 = \begin{bmatrix} r_{33}e_{32} - r_{32}e_{33} + e_{33}y - e_{32}z \\ -r_{33}e_{31} + r_{31}e_{33} + e_{31}z \\ r_{32}e_{31} - r_{31}e_{32} - e_{31}y \end{bmatrix}$$

The normality conditions yield again three equations in z and y, which yield in turn three curves in the Y-Z plane, namely,

$$
\begin{aligned}
\mathcal{C}_1 &: \frac{y^2}{2}(1+\cos^2\phi) + z^2\cos^2\phi = 0 \\
\mathcal{C}_2 &: (r_{31}e_{32} - r_{32}e_{31} + e_{31}y)y\cos\frac{\phi}{2} + (r_{33}e_{32} - r_{32}e_{33} + e_{33}y - e_{32}z)\sin\frac{\phi}{2} + \\
&\quad +(-r_{33}e_{31} + r_{31}e_{33} + e_{31})z\cos\frac{\phi}{2} = 0 \qquad (10) \\
\mathcal{C}_3 &: (r_{31}e_{32} - r_{32}e_{31} + e_{31}y)y\cos\frac{\phi}{2} + (r_{33}e_{32} - r_{32}e_{33} + e_{33}y - e_{32}z)\sin\frac{\phi}{2} + \\
&\quad +(r_{31}e_{33} - r_{33}e_{31} + e_{31}z)z\cos\frac{\phi}{2} = 0
\end{aligned}
$$

with similar equations for the $y = 0$ plane.

The special case where the first two axes are parallel can be handled with eqs.(5) – (8) by setting $\phi = 0$. The equidistance surface is now a plane through the midpoint of one of the common normals and perpendicular to the same normal.

IV Example

For an example we choose a 3R manipulator with the HD parameters given in Table 1. It should be pointed out that the last offset and the last distance between axes are not

Link	a	b	α
1	1	1	$\pi/2$
2	1	1	$-\pi/2$
3	x	y	—

Table 1: HD-Parameters of Example

determined, for they are dependent on the location of the operation point, which we want to determine so as to render it an isotropy point. Shown in Figs. 2–6 are the four curves $\mathcal{C}$, $\mathcal{C}_1$, $\mathcal{C}_2$ and $\mathcal{C}_3$ in different locations of the third axis of this manipulator. We start at the configuration whereby $\theta_2 = 0$ and then, from figure to figure, change θ_2. In all figures, $\mathcal{C}$ is plotted as a thick curve, $\mathcal{C}_1$ as a thinner curve, $\mathcal{C}_2$ as a long-dashed curve and $\mathcal{C}_3$ as a short-dashed curve. Moreover, Fig. 2 pertains to $\theta_2 = 0$, where we have two points P_1 and P_2 that are common to all four curves. This gives, then, two possibilities for C to be an isotropy point. The coordinates of these two points in the $\mathcal{F}$ frame are seen in the plot and can be computed as $P_1(\sqrt{2}/2, \sqrt{2}/2, 1/2)$ and $P_2(\sqrt{2}/2, -\sqrt{2}/2, -1/2)$. The z-coordinate was derived from eq.(4), which states the condition that P_1 and P_2 lie on the equidistance hyperboloid of $\mathcal{L}_1$ and $\mathcal{L}_2$. The base coordinates of these points are $P_1(0,1,0)$ and $P_2(1,0,-1)$.

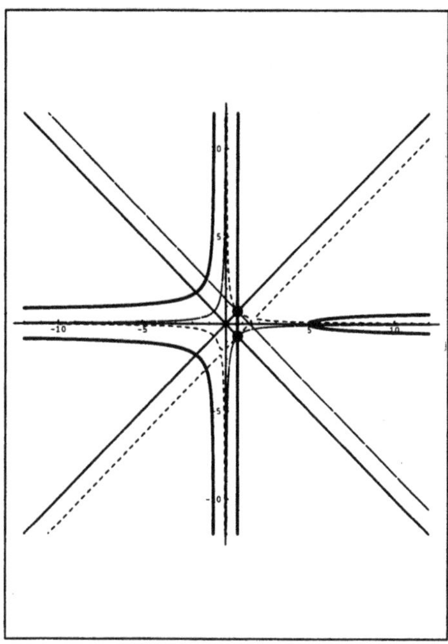

Figure 2: Isotropy Points P_1 and P_2

Furthermore, we observe that, for this special example, curve C_1 breaks down into a pair of lines bisecting the coordinate axes. In Cartesian space the corresponding points on the equidistance hyperboloid are on two parabolas. This means, according to Conjecture 1, that, whenever isotropy points are possible on the EE, they lie on these two parabolas. Because the second and the third axes are always normal to each other, the curve C_3 decomposes in this case into a fixed line, with the equation $x+y=\sqrt{2}$, and a hyperbola whose center moves on the line $x+y=0$. For $\theta_2=0$, the same happens with C_2 decomposing into the line $x-y=\sqrt{2}$ and a hyperbola. This layout gives us the symmetric configuration of Fig. 2. If the second joint starts moving, while the third one is locked, we loose immediately this symmetric configuration, as shown in Fig. 3, and recover symmetry only when the same joint is rotated through an angle of $\pi/2$, Fig 4. Here the first and the third axes are parallel, which is a source of symmetry. However, there are no points common to all three curves, so that no isotropy point is possible for that layout. The next symmetry arises after rotating the same joint through an angle of π, Fig. 5. Here we find again two points, P_3 and P_4, that are common to all four curves. The coordinates of the two points in the $\mathcal{F}$ frame are $P_3(\sqrt{2}/2, \sqrt{2}/2, 1/2)$ and $P_4(-\sqrt{2}/2, \sqrt{2}/2, -1/2)$. The base coordinates of these points are $P_3(0,1,0)$ and $P_4(1,0,1)$. Plotting these four curves for different angles between $\pi \leq \theta_2 \leq 2\pi$ (e., g., Fig. 6) we can see that no more isotropy points are possible until we return to the configuration where $\theta_2 = 2\pi$. In Figs. 7 – 10 we show the four isotropic configurations found. Figures 7 and 8 pertain to $\theta_2=0$, while Figs. 9 and 10 to the configuration at which $\theta_2 = \pi$. The most interesting conclusion that we can draw from these figures is that we do not have four different manipulators, because, apparently, Figs. 8 and 10 show the same manipulator in two different postures.

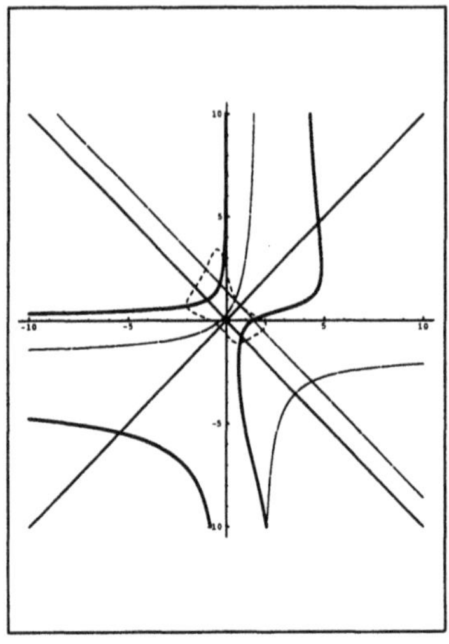

Figure 3: $\theta_2 = 3\pi/8$

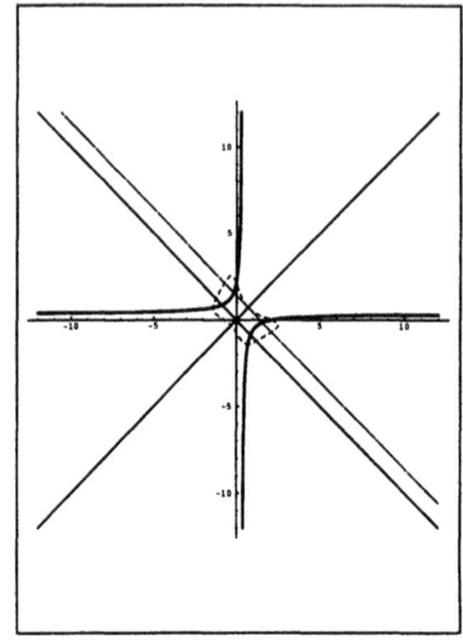

Figure 4: $\theta_2 = \pi/2$

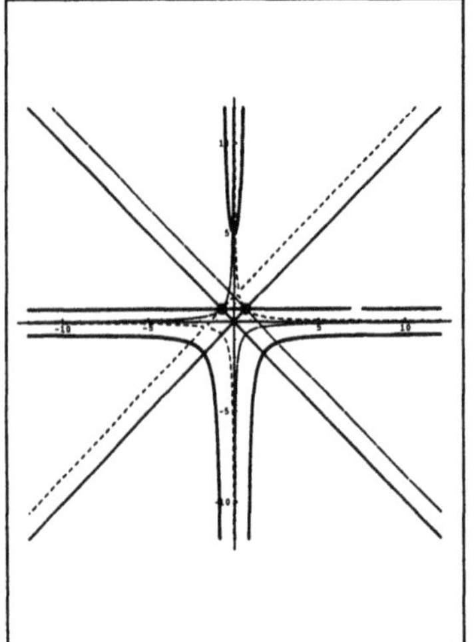

Figure 5: $\theta_2 = \pi$

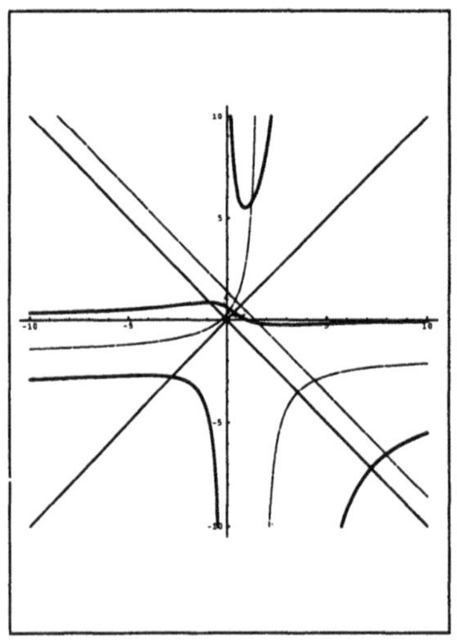

Figure 6: $\theta_2 = 11\pi/8$

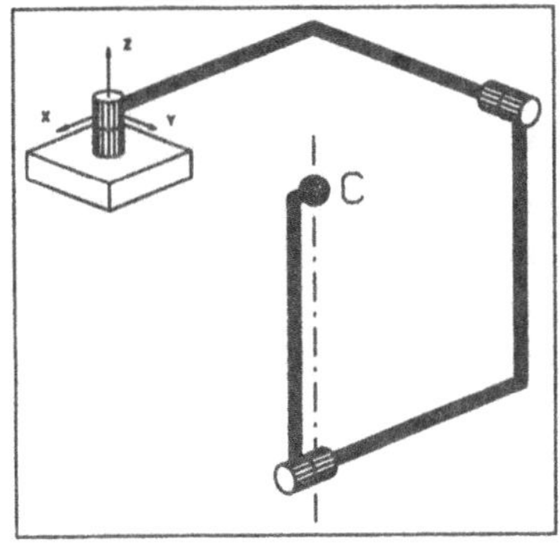

Figure 7: configuration for $P_1(0,1,0)$

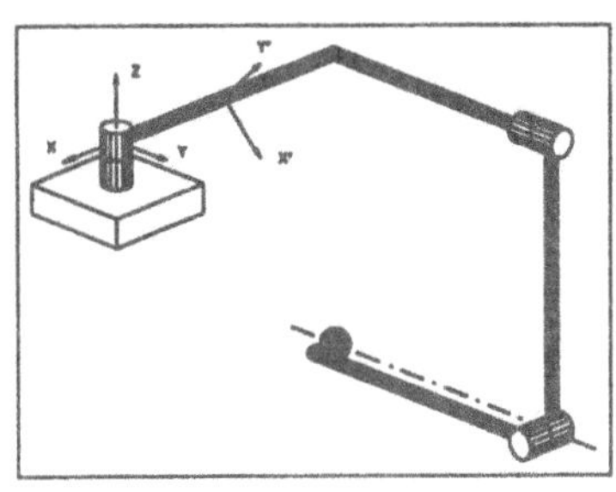

Figure 8: configuration for $P_2(1,0,-1)$

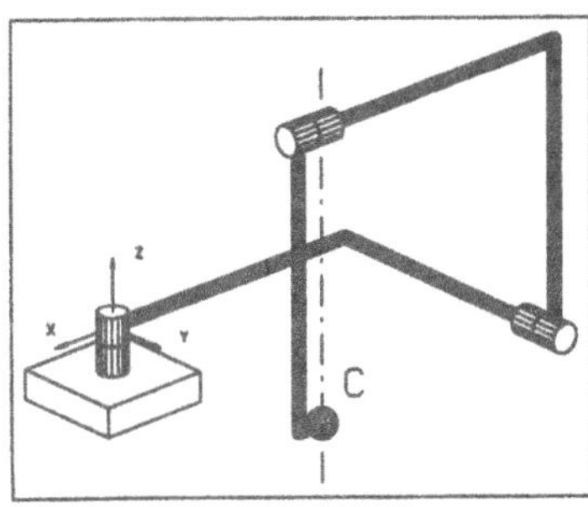

Figure 9: configuration for $P_3(0,1,0), \theta_2 = \pi$

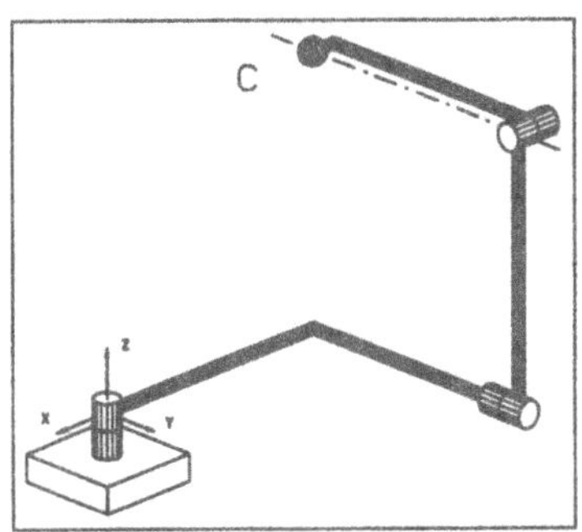

Figure 10: configuration for $P_4(1,0,1)$

V Conclusions

We defined isotropy points of 3R positioning manipulators in terms of the concept of kinematic isotropy. Moreover, we showed that, in general, arbitrary manipulators in arbitrary postures do not have isotropy points in their EE. For cases where isotropy points are possible, we derived a procedure to determine all of them.

The results appear applicable to the design and control of manipulators of the type discussed here for purposes of enhancing their positioning accuracy.

other similar optimal synthesis and performance analysis tasks. The proposed measures are based on a new class of norms to quantify the "distance" in space [1]. In this paper, the theoretical basis of the proposed measures together with examples of their application are presented.

An Object Shape Dependent Concept of "Distance"

Most problems in kinematics and dynamics analysis and control of mechanical systems require estimation, analysis and manipulation of quantifiable measures of position and orientation of rigid bodies in a plane or space. A rigid body has three degrees of freedom in plane, and six degrees of freedom in space. These degrees of freedom define both the orientation and position of the object. Kinematic and dynamic analysis and control of mechanical systems often involve guiding a rigid body (mathematically or physically) from one position (current position) to another (desired position). The term position hereafter also implies orientation. The distance between two finitely separated positions is not, however, intuitively understood. This "distance" or "norm" is an abstraction of our usual concept of length and its definition and formulation directly affects the mathematical modeling of the system and its physical consequences. In this section, the class of norms proposed by the authors in [1] for quantifying a measure for the "distance" between two positions of a rigid body in plane or in space is reviewed. Based on this measure, object shape dependent kinematic manipulability measures are then developed in the following section.

The generalized coordinates required to define the position of the object in the Euclidean two dimensional plane or three dimensional space involve both angular and linear displacements. Over the past two hundred years, many different mathematical representations and methods have been developed to deal with finite and instantaneous kinematics and then dynamics of rigid bodies. For an extensive discussion of the subject, the reader is referred to [11-13]. To locate or move a rigid body in space using some type of mechanical system, it is possible to define the generalized coordinates corresponding to appropriate relative motions within the mechanical systems. Such positioning problems lead to mapping in the form of a system of nonlinear equations. This system, once differentiated with respect to an independent parameter such as time, yields a linearized relationship where a Jacobian matrix maps the rates of the space coordinates onto the system generalized coordinates. Since the original equations are usually highly coupled and nonlinear, the latter relationships often become central to the computational kinematics and dynamics and control of these mechanical systems [14-16]. Although the displacement of a rigid body in space is a physical phenomenon and independent of the choice of the coordinate system and the metrics (i.e., scales) used, the above relation and the algorithms developed based on them in different fields are very sensitive to the choice of metrics and coordinate system. This problem is pointed out in detail for hybrid control strategies in robotic applications in [17].

The six independent coordinates in space (3 in the plane) may be folded to define a measure of the "distance" between two specified positions of a rigid body. This norm can be used as the objective function in the design of mechanisms or a convergence criterion in the computational kinematics and dynamics or the control of mechanical systems. Serious concerns have, however, been expressed about the physical interpretation, non-homogeneity, and coordinate dependence of the distance norm and the linearized velocity relations (e.g., [17-18]). Poor choices of the coordinate systems and reference points can result in ill conditioned or highly sensitive systems. In short, these norms or the above linearized relations are highly sensitive to: 1)- the choice of the coordinate system; and 2)- the choice of metrics (scale.)

In a recent article, [1], the authors propose a new class of norms for quantifying a measure for the "distance" between two positions of a rigid body in plane or in space as:

$$\delta = \frac{1}{V}\int_v \omega d^2 dv$$

in which v is the volume, w a weight factor (fixed or variable over the volume) and d^2 the square of the distance between the two positions of a point in the rigid body. This norm is shown to be: 1)- coordinate independent; 2)- homogeneous in its definition; 3)- dependent on the volume and shape of the object; and 4)- emphasizes (with a varying intensity) on the volume(s) of interest.

Object Shape Dependent Manipulability Measures

Consider planar motion of an object in the fixed Cartesian coordinate system XY, Fig. 1. In Fig. 1, the object is shown in its original position A and after a small motion to a new position A'. For infinitesimally separated positions of the object, the motion from the position A to the position A' can be described as a pure rotation $\Delta\alpha$ about the instantaneous center of rotation I_c.

Let dS represent a differential area of the planar object, Fig. 1. ρ is the radius of the instantaneous rotation of the element dS. During the motion, dS translates ΔX and ΔY in the directions of the X and Y coordinates, while traveling along its instantaneous circular path. Using a similar but more general definition than that proposed in [1], the "displacement" of the object in the X and Y directions, and the total "distance" traveled by the surface elements of the object along their circular paths may be indicated by the following integrals

$$\Delta D_x = \left[\frac{1}{S}\int\int_S w_x|\Delta x|^n dS\right]^{1/n} \tag{1a}$$

$$\Delta D_y = \left[\frac{1}{S}\int\int_S w_y|\Delta y|^n dS\right]^{1/n} \tag{1b}$$

$$\Delta D_\alpha = \left[\frac{1}{S}\int\int_S w_\alpha|\rho\Delta\alpha|^n dS\right]^{1/n} \tag{1c}$$

where w_x, w_y and w_α are some weights, S is the total area of the object, and n is a nonzero integer. The weights may be assigned according to the importance of the various region to the task being performed, such as the accuracy of motion in the process of part mating in assembly operations. The displacements are obviously dependent on the shape of the object and the instantaneous (local) path of motion. Depending on the value of the integer n, the displacements become dependent on the different aspects of the "distribution of the area over the surface of the object", i.e., on the shape of the object and its various features, and the distribution of the corresponding weighting functions. For example, when $n = 1$ and with unity weights, ΔD_x and ΔD_y are the "average length of travel" of the object in the x and y directions, respectively. ΔD_α is the "average length of rotational travel". When $n = 2$, the sum of the displacements ΔD_x and ΔD_y gives an indication of the total "distance" between the two positions of the object as defined in [1], in which case the latter sum becomes ΔD_α.

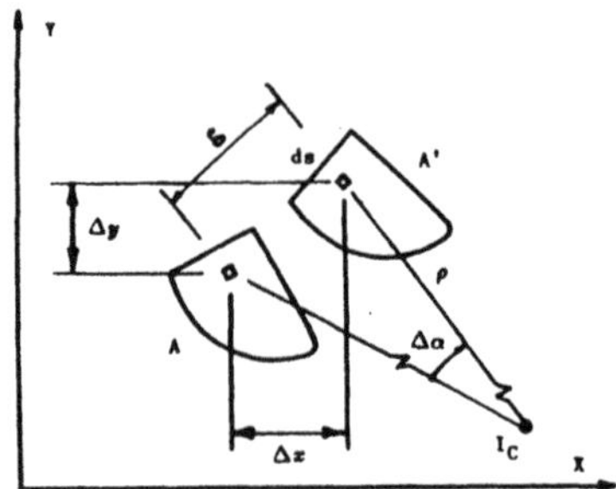

Figure 1: An infinitesimal motion of a plane object.

Equations (1a) and (1b) also hold for finite displacements of an object. For finite displacements, however, the displacement ΔD_α can not be defined unless a corresponding center of rotation is defined. For general motions, ΔD_α is dependent on the path of motion. In the form of equation (1c), ΔD_α is intended to be used primarily for infinitesimal motions. The sum of the displacements ΔD_x and ΔD_y, or when $n \neq 2$, the sum of the three displacements, or other combinations of the displacements may be used when appropriate as measures of the total distance traveled by the object.

Note that the displacements (1) have definite physical meaning. They indicate the corresponding differential lengths of travel of the individual "surface elements" or "particles" that comprise the object. The displacements are therefore very general. No rigid body constraints are imposed. Therefore the object may be deforming, stretching or shrinking.

In this section, the proposed shape dependent object manipulability measures are introduced by their application to the planar object shown in Fig. 1. To simplify analytical derivations and since the objective of this paper in to introduce the basic concepts being developed, the case of $n = 2$ alone is examined.

Consider the plane 3R manipulator shown in Fig. 2. The end-effector is holding the object at the point H. The coordinate system UV is fixed to the object. In the following formulations, the coordinates of the origin of the UV coordinate system in the fixed XY coordinate system are indicated by (x_o, y_o). (x_H, y_H) and (u_H, v_H) are the coordinates of the holding point in the XY and UV coordinate systems, respectively.

Let (x, y) indicate the coordinates of the differential area dS in the XY coordinate system. From Fig. 2, x and y can be written in terms of the coordinates (u, v) of the differential area in the UV coordinate system as

$$x = x_0 + u \cos\alpha - v \sin\alpha \tag{2a}$$

$$y = y_0 + u \sin\alpha + v \cos\alpha \tag{2b}$$

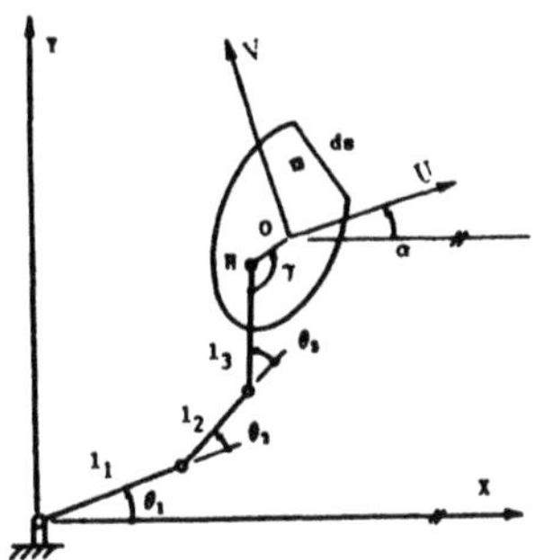

Figure 2: A plane object and a planar 3R manipulator.

With a small motion of the object, the coordinates of the differential area is displaced Δx and Δy in the directions of the X and Y coordinates and the angle α is changes by $\Delta\alpha$ given by

$$\Delta x = \Delta x_o - u\Delta\alpha \sin\alpha - v\Delta\alpha \cos\alpha \tag{3a}$$

$$\Delta y = \Delta y_o + u\Delta\alpha \cos\alpha - v\Delta\alpha \sin\alpha \tag{3b}$$

$$\Delta\alpha = \Delta\theta_1 + \Delta\theta_2 + \Delta\theta_3 \tag{3c}$$

Substituting (3) in (1), the displacements ΔD_x, ΔD_y and ΔD_α are found as

$$\Delta D_x = \frac{1}{S}\int\int_s \omega \Delta x^2 dxdy = \frac{1}{S}\int\int_s \omega(\Delta x_o - u\Delta\alpha\sin\alpha - v\Delta\alpha\cos\alpha)^2 dudv \tag{4a}$$

$$\Delta D_y = \frac{1}{S}\int\int_s \omega \Delta y^2 dxdy = \frac{1}{S}\int\int_s \omega(\Delta y_o + u\Delta\alpha\cos\alpha - v\Delta\alpha\sin\alpha)^2 dudv \tag{4b}$$

$$\Delta D_\alpha = \frac{1}{S}\int\int_s \omega(\rho\Delta\alpha)^2 dxdy = \frac{1}{S}(\Delta\alpha)^2\int\int_s \omega(\rho)^2 dudv \tag{4c}$$

where $\rho^2 = (u - u_R)^2 + (v - v_R)^2$, with (u_R, v_R) indicating the coordinates of the instantaneous center of rotation R. Δx_o and Δy_o can be written in terms of changes in the position of the hand Δx_H and Δx_H from equations (3a) and (3b) as

$$\Delta x_o = \Delta x_h + u_H\Delta\alpha\sin\alpha + v_H\Delta\alpha\cos\alpha \tag{5a}$$

$$\Delta y_o = \Delta y_H - u_H\Delta\alpha\cos\alpha + v_H\Delta\alpha\sin\alpha \tag{5b}$$

For the 3R manipulator shown in Fig. 2, the position of the end-effector, i.e., the object holding position H, and the angle α are

$$x_H = l_1\cos\theta_1 + l_2\cos(\theta_1+\theta_2) + l_3\cos(\theta_1+\theta_2+\theta_3) \tag{6a}$$

$$y_H = l_1\sin\theta_1 + l_2\sin(\theta_1+\theta_2) + l_3\sin(\theta_1+\theta_2+\theta_3) \tag{6b}$$

$$\alpha = (\theta_1+\theta_2+\theta_3) + \gamma - tan(v_h/u_h) - \pi \tag{6c}$$

$[M_y]$ and $[M_x] + [M_y]$ are calculated. As an example, the matrix $[M]$ becomes

$$[M] = [M_x] + [M_y] = \begin{bmatrix} 7.976 & 5.495 & -0.093 \\ 5.495 & 7.015 & 3.427 \\ -0.093 & 3.427 & 3.839 \end{bmatrix}$$

The eigenvalues of $[M]$ become $\lambda_1 = 0.023$, $\lambda_2 = 5.243$ and $\lambda_3 = 13.56$. The corresponding semi-major and semi-minor axes of the resulting object displacement (velocity) transmission ellipsoids; i.e., the directions of minimum and maximum object displacement (velocity) transmissions are then obtained. Object manipulability measures, such as one based on the condition number or its inverse can then be calculated for the displacement of interest and a map of its distribution over the workspace of the manipulator constructed. For more detail on the steps involved, the reader is referred to [6-8].

For the above manipulator and for a rectangular object with $p = 0.2$ and $q = 1.0$ held at position $u_H = -0.1$ and $v_H = 0.1$, by setting $(\Delta D_x)^2$, $(\Delta D_y)^2$ and $(\Delta D_x)^2 + (\Delta D_y)^2$ equal to unity, the related object manipulability ellipsoids (elliptic cylinders for $(\Delta D_x)^2$ and $(\Delta D_y)^2$ when the motion of the object is in the direction of the X and Y axes), are obtained. The intersections of the manipulability ellipsoids with the $\theta_1 - \theta_3$ plane for $(\Delta D_x)^2 + (\Delta D_y)^2 = 1$ are shown in Fig. 4.

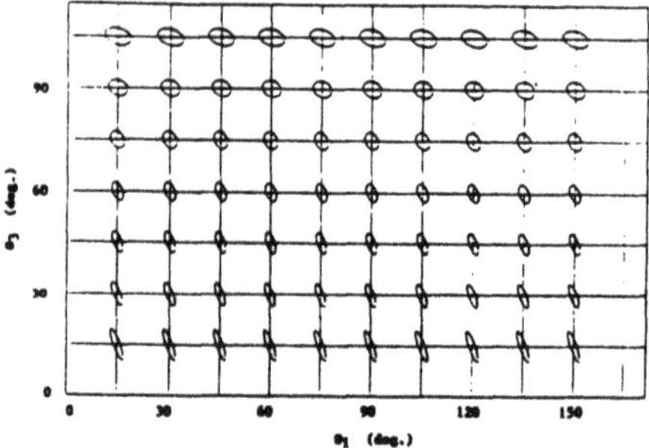

Figure 4: The intersection of the object manipulability ellipsoids with the $\theta_1 - \theta_3$ plane for unity $(\Delta D_x)^2 + (\Delta D_y)^2$.

The manipulator is then placed at various positions within its workspace while holding a rectangular object with $p = 0.4$ and $q = 0.5$, Fig. 3, at the position $u_H = -0.3$ and $v_H = 0.2$. By setting $[\Delta\theta]^T [\Delta\theta] = 1$, the corresponding object velocity transmission ellipsoids are obtained. In Fig. 5, the intersection of planes containing the θ_3 axis and making an angle of 60 and 30 degrees with the $\theta_1 - \theta_3$ plane are shown with solid and dashed lines, respectively. In this illustration, the directions X and Y also indicate the directions of the ΔD_x and ΔD_x. The relative size and directions of the semi-major and semi-minor axes are observed. The locations at which one of the eigenvalues become large, i.e., where the area of the ellipsoid tends to zero, the manipulator is close to its related singular position. Similar plots are shown in Fig. 6 with the object keeping the same surface area but constructed as rectangles with sides $p = 0.6$ and $q = 0.333$, and $p = 0.4$ and $q = 0.5$. The objective is to study the effects of the object shape on the above object manipulability measure. The results are shown in solid and dashed lines for the first and the second rectangles, respectively.

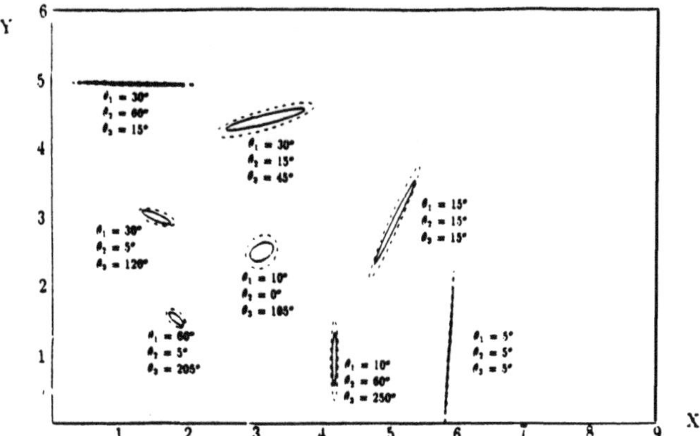

Figure 5: The effects of the various positioning on the object in the task space on the object manipulability.

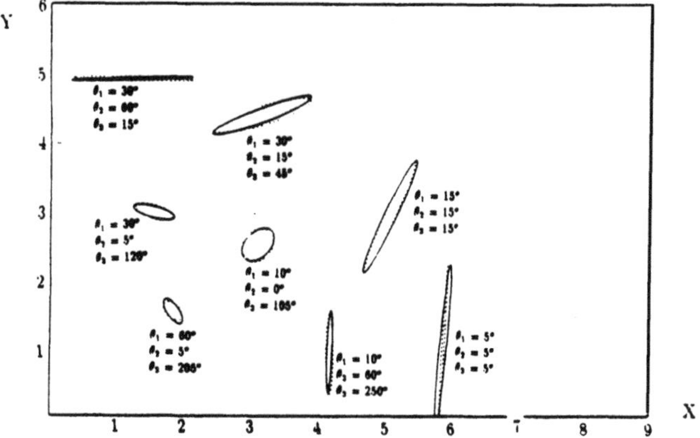

Figure 6: The effects of the object shape on the object manipulability at different positioning of the object within the task space.

Conclusions

In this paper, object manipulability measures are developed that can be used to quantify the ability of a manipulator to displace an object within its workspace. The proposed manipulability measures are shape dependent. They can be used for optimal path planning for maximum or minimum displacement transmission, for optimal positioning of mating parts within the workspace of the manipulator, for shape optimization from the point of view of object manipulability, or for other similar objectives. The primary objective of this article is to present the new concept of object manipulability and its basic formulation. The formulations and the examples are for plane objects and planar motions. However, one only needs to replace surfaces with volumes to obtain similar manipulability formulations for solid objects. The object may be rigid, deforming, stretching or shrinking during its motion.

References

[1] K. Kazerounian, and J. Rastegar, "Object Norms: A Class of Coordinate and Metric Independent Norms for Displacements", *1992 ASME Mechanisms Conference*, pp. 271-275 (1992).

[2] T. Yoshikawa, "Dynamic Manipulability of Robot Manipulators", *J. of Robotic Systems 2(1)*, pp. 113-124 (1985).

[3] T. Yoshikawa, "Manipulability of Robotic Mechanisms", *The Intern. J. of Robotics Research 4(2)*, pp. 3-9 (1985).

[4] S. L. Chiu, "Task Compatibility of Manipulator Postures," *The Intern. J. of Robotics Research 7(5)*, pp. 13-21 (1988).

[5] M. J. Tsai and Y. H. Chiou, "Manipulability of Manipulators", *Mechanism and Machine Theory 25(5)*, pp. 575-585 (1990).

[6] C. Gosselin and J. Angeles, "A New Performance Index for the Kinematic Optimization of Robotic Manipulators," *ASME Trends and Developments in Mechanisms, Machines, and Robotics DE-Vol. 15-3*, pp. 441-447 (1988).

[7] J. Rastegar and R. J. Singh, "Determination of Manipulator Kinematic Characteristics: A Probabilistic Approach," *Intern. J. of Robotics & Automation 7(4)*, pp. 161-170 (1992).

[8] J. Rastegar and R. J. Singh, "A New Probabilistic Method for the Performance Evaluation of Manipulators", *ASME J. of Mechanical Design*, in press.

[9] R. J. Singh and J. Rastegar, "Optimal Synthesis of Robot Manipulators Based on Global Kinematic Parameters," *ASME Mechanisms Conference*, Chicago (1990).

[10] R. J. Singh and J. Rastegar, "Optimum Synthesis of Robot Manipulators Based on Global Dynamic Parameters," *Intern. J. of Robotics Research 11(6)*, (1992).

[11] K. H. Hunt, *Kinematic Geometry of Mechanisms*, Oxford University Press, Oxford, England (1978).

[12] J. Duffy, *Analysis of Mechanisms and Robot Manipulators*, Edward Arnold Publishers Ltd., London and John Wiley and Sons, Inc., New York (1980).

[13] O. Bottema and B. Roth, *Theoretical Kinematics*, North Holland Pub. (1979).

[14] D. E. Whitney, "Resolved Motion Rate Control of Manipulators and Human Prostheses," *IEEE Trans., Man-Machine Systems MMS-10* (1969).

[15] K. C. Gupta and K. Kazerounian, "An Investigation of Time Efficiency and Near Singular Behavior of Numerical Robot Kinematics," *Mechanism and Machine Theory 22(4)* (1987).

[16] K. Kazerounian, "On the Numerical Inverse Kinematics of Robotic Manipulators," *ASME J. of Mechanisms, Trans., and Auto. in Design 109(3)*, (1987).

[17] J. Duffy, "The Fallacy of Modern Hybrid Control Theory That is Based on 'Orthogonal Complements' of Twist and Wrench Spaces," *J. of Robotic Sys.* (1990).

[18] J. Lonacaric, "Normal Forms of Stiffness and Compliance Matrices", *IEEE J. of Robotics and Automation RA-3 (6)* (1987).

Application of Dexterity Indices to Spatial Articulated Hands

Benyoucef Imçaoudene and Clément M. Gosselin
Département de Génie Mécanique
Université Laval
Ste-Foy, Québec, Canada, G1K 7P4

Abstract - The optimum kinematic design of spatial articulated hands can be formulated as an optimization problem. To this end, the concept of kinematic dexterity is applied to such mechanical systems in this work. The kinematic dexterity is based on the condition number of the Jacobian matrix of the mechanical system under study. However, when the position and orientation of the end effector of a manipulator are introduced, the Jacobian matrix contains entries with dimensional inhomogeneities, which renders the condition number physically meaningless. This problem has been addressed recently by a few researchers. The solutions proposed to overcome this problem are applied to articulated hands in this paper. An example is given to illustrate the practical applications in the context of design and trajectory planning of spatial articulated hands.

I. Introduction

The optimum kinematic design of manipulators and articulated hands has attracted the attention of researchers for several years.

Salisbury and Craig [1] first introduced the condition number of the Jacobian matrix as an optimization criterion of a manipulator. Later, Angeles and Rojas [2] applied the condition number of the Jacobian matrix to the design of a three-degree-of-freedom spatial manipulator and a three-degree-of-freedom spherical wrist. Klein and Blaho [3] presented a comprehensive review of the dexterity indices which can be applied to redundant manipulators. Dexterity indices have also been used for the kinematic optimization of parallel manipulators [4, 5]. Moreover, a global performance index for the kinematic optimization of robotic manipulators has been defined in [6]. This quantity characterizes the kinematic accuracy of the manipulator over its whole workspace.

However, when both position and orientation of the end effector are introduced in the kinematic equations, the Jacobian matrix contains entries with different physical units and the condition number of the Jacobian matrix becomes variable under the scaling of the manipulator dimensions. To overcome this problem, Gosselin [7] has suggested two kinematic formulations to eliminate the dimensional inhomogeneities in the Jacobian matrix. Instead of changing the kinematic formulation, Tandirci et al. [8] introduced the concept of the natural link length. Thus, the Jacobian matrix becomes dimensionless, when the translational entries are divided by a natural link length, which is defined as to minimize the condition number of the Jacobian matrix. In this same context, the concept of the natural link length has been applied to redundant

A. J. Lenarčič and B. B. Ravani (eds.), Advances in Robot Kinematics and Computationed Geometry, 201–208.

where the joint velocity vector $\dot{\boldsymbol{\theta}}$, is defined as

$$\dot{\boldsymbol{\theta}} = [\dot{\boldsymbol{\theta}}_a^T, \dot{\boldsymbol{\theta}}_{f_1}^T, \dot{\boldsymbol{\theta}}_{f_2}^T, ..., \dot{\boldsymbol{\theta}}_{f_n}^T]^T \tag{10}$$

in which $\dot{\boldsymbol{\theta}}_a$ is the vector of joint velocities associated with the arm sub-chain, i.e., the kinematic sub-chain supporting the palm and $\dot{\boldsymbol{\theta}}_{f_i}$ is the vector of joint velocities associated with the ith finger.

The generalized twist vector, $\mathbf{t}_H$, contains all the Cartesian coordinates of interest of the hand. They can be linear and/or angular velocities of the fingertips. Hence, the Jacobian matrix will take on the form

$$\mathbf{J} = \begin{bmatrix} \mathbf{J}_a^1 & \mathbf{J}_{f_1} & 0 & 0 & \dots & 0 \\ \mathbf{J}_a^2 & 0 & \mathbf{J}_{f_2} & 0 & \dots & 0 \\ \mathbf{J}_a^3 & 0 & 0 & \mathbf{J}_{f_3} & \dots & 0 \\ \vdots & \vdots & \vdots & \vdots & \vdots & \vdots \\ \mathbf{J}_a^n & 0 & 0 & 0 & \dots & \mathbf{J}_{f_n} \end{bmatrix} \tag{11}$$

where $\mathbf{J}_a^i$ is the Jacobian matrix associated with the kinematic chain supporting the palm when the fingertip of the ith finger is taken as the Cartesian point of interest and $\mathbf{J}_{f_i}$ is the Jacobian matrix associated with ith finger. The kinematic dexterity of the hand is based on the conditioning of the global Jacobian matrix $\mathbf{J}$. However, each of the sub-Jacobian matrices contained in $\mathbf{J}$ can involve positioning and/or orientation terms. To eliminate the problem of mixed dimensions, these submatrices are first treated using the approach developed in [7] and summarized in the last section. Each of the twists is conceptually replaced by the velocity of three points on the end-effector and, consequently, the corresponding Jacobian sub-matrix is premultiplied by a matrix $\mathbf{R}$ which relates twists to the velocities of 3 points on the body. Therefore, the Jacobian matrix $\mathbf{J}$ is modified, to render it homogeneous, as follows

$$\mathbf{J}' = \mathbf{K}\mathbf{J} \tag{12}$$

where

$$\mathbf{K} = \begin{bmatrix} \mathbf{R}_a^1 & \mathbf{R}_a^2 & \mathbf{R}_a^3 & \dots & \dots & \mathbf{R}_a^n \\ \mathbf{R}_{f_1} & 0 & 0 & 0 & \dots & 0 \\ 0 & \mathbf{R}_{f_2} & 0 & 0 & \dots & 0 \\ \vdots & \vdots & \vdots & \vdots & \vdots & \vdots \\ 0 & 0 & 0 & 0 & \dots & \mathbf{R}_{f_n} \end{bmatrix} \tag{13}$$

in which matrices $\mathbf{R}$ are equal the to identity matrix if the corresponding sub-Jacobian matrix involves linear velocities only and are equal to the matrix defined in [7] if the corresponding sub-Jacobian matrix involves both angular and linear velocities.

IV. Example

A spatial twelve-degree-of-freedom multifingered robot hand is shown in Fig. (1). If the fingers of a hand have a sufficient number of joints, it is possible to manipulate and to grasp the objects relative to the hand. Motion of the fingers will thus result in motion of the object. In general, the contact between the fingertips and the object is considered to be point contact with friction, and the contact points do not move with respect to the object during the motion. It is assumed here that the Cartesian coordinates of interest are the position of the contact point and the orientation of the last phalanx of finger 1 and the position of the contact point on fingers 2 and 3. The Hartenberg-Denavit parameters of the hand are given in Tb. (1). Hence the Cartesian velocity vector is given by

$$\mathbf{t} = [\mathbf{v}_1^T, \mathbf{v}_2^T, \mathbf{v}_3^T]^T$$

where $\mathbf{v}_i, i = 1, 2, 3$ is the velocity vector of each fingertip, and the components of each of these vectors are written as

$$\mathbf{v}_1 = [\omega_x, \omega_y, \omega_z, v_{O'x}, v_{O'y}, v_{O'z}]^T \tag{14}$$

$$\mathbf{v}_2 = [v_{O''x}, v_{O''y}, v_{O''z}]^T \tag{15}$$

$$\mathbf{v}_3 = [v_{O'''x}, v_{O'''y}, v_{O'''z}]^T \tag{16}$$

The velocity vector of the fingertip number one contains entries with positioning and orientation components but the last two vectors contain only positioning components. The joint velocity vector is defined as

$$\dot{\boldsymbol{\theta}} = [\dot{\theta}_1^0, \dot{\theta}_2^0, \dot{\theta}_3^0, \dot{\theta}_1^1, \dot{\theta}_2^1, \dot{\theta}_3^1, \dot{\theta}_1^2, \dot{\theta}_2^2, \dot{\theta}_3^2, \dot{\theta}_1^3, \dot{\theta}_2^3, \dot{\theta}_3^3]^T \tag{17}$$

where $\dot{\theta}_i^j$ is the joint variable associated with the ith joint of the jth kinematic subchain. The subchains are defined as follows: subchain 0 is the manipulator supporting the palm of the hand and subchains 1,2 and 3 are associated with the first, second and third fingers, respectively. The Jacobian matrix can be defined as,

$$\mathbf{J} = \begin{bmatrix} \mathbf{J}_{6\times6}^{01} & \mathbf{0}_{6\times3} & \mathbf{0}_{6\times3} \\ \mathbf{J}_{3\times3}^{02} & \mathbf{J}_{3\times3}^{2} & \mathbf{0}_{3\times3} \\ \mathbf{J}_{3\times3}^{03} & \mathbf{0}_{3\times3} & \mathbf{J}_{3\times3}^{3} \end{bmatrix} \tag{18}$$

where $\mathbf{J}_{6\times6}^{01}$ is the Jacobian matrix of the six-axis spatial manipulator defined by the supporting manipulator and the first finger, which is given by:

$$\mathbf{J}_{6\times6}^{01} = [\mathbf{A}, \mathbf{B}] \tag{19}$$

with

$$\mathbf{A} = \begin{bmatrix} \mathbf{e}_1^0 & \mathbf{e}_2^0 & \mathbf{e}_3^0 \\ \mathbf{e}_1^0 \times \mathbf{r}_1^{01} & \mathbf{e}_2^0 \times \mathbf{r}_2^{01} & \mathbf{e}_3^0 \times \mathbf{r}_3^{01} \end{bmatrix}, \quad \mathbf{B} = \begin{bmatrix} \mathbf{e}_1^1 & \mathbf{e}_2^1 & \mathbf{e}_3^1 \\ \mathbf{e}_1^1 \times \mathbf{r}_1^1 & \mathbf{e}_2^1 \times \mathbf{r}_2^1 & \mathbf{e}_3^1 \times \mathbf{r}_3^1 \end{bmatrix} \tag{20}$$

and $\mathbf{J}_{3\times3}^{0i}, i = 2, 3$ is the Jacobian matrix associated only with translation for the first three links of the base manipulator ($i = 0$) expressed as:

$$\mathbf{J}_{3\times3}^{0i} = [\mathbf{e}_1^0 \times \mathbf{r}_1^{0i} \quad \mathbf{e}_2^0 \times \mathbf{r}_2^{0i} \quad \mathbf{e}_3^0 \times \mathbf{r}_3^{0i}], \quad i = 2, 3 \tag{21}$$

Moreover, $\mathbf{J}^i_{3\times3}, i = 2,3$ is the Jacobian matrix associated with the two fingers $i = 2,3$ and obtained only from the positioning vector. These matrices are written as:

$$\mathbf{J}^i_{3\times3} = [\mathbf{e}^i_1 \times \mathbf{r}^i_1 \quad \mathbf{e}^i_2 \times \mathbf{r}^i_2 \quad \mathbf{e}^i_3 \times \mathbf{r}^i_3], \quad i = 2,3 \tag{22}$$

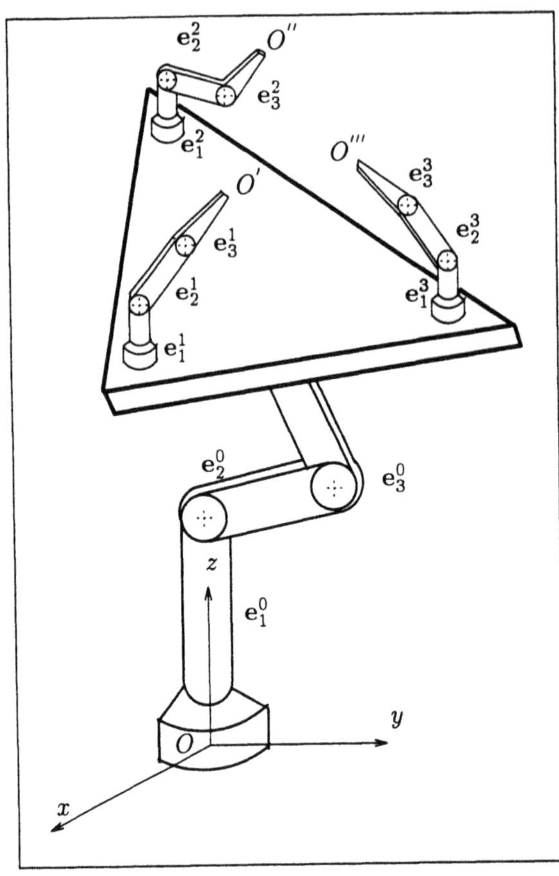

Figure 1: 12 DOF spatial multifingered robot hand.

In the above expressions, vector $\mathbf{e}^j_i$ is a unit vector defined along the axis of the ith joint of the jth subchain while $\mathbf{r}^j_i$ is the vector connecting the origin of the ith coordinate frame of the jth subchain to the tip of the subchain. The exception in this case are vectors $\mathbf{r}^{0j}_i$ which are defined from the origin of frame i of subchain 0 to the tip of subchain j.

Since we address the problem of both positioning and orientation, the Jacobian matrix obtained in Eq. (18) is dimensionally inhomogeneous and the measure of the kinematic accuracy based on the condition number of Eq. (4) is physically meaningless. To overcome this problem, the approach proposed in [7] is used and the velocity vector of the end effector of the first fingertip is replaced by the velocity of three noncolinear points A, B and C on the body.

The new Jacobian matrix is written as

$$\mathbf{J}' = \begin{bmatrix} \mathbf{R}_{6\times6} & 0_{6\times6} \\ 0_{6\times6} & \mathbf{1}_{6\times6} \end{bmatrix} \begin{bmatrix} \mathbf{J}^{01}_{6\times6} & 0_{6\times3} & 0_{6\times3} \\ \mathbf{J}^{02}_{3\times6} & \mathbf{J}^2_{3\times3} & 0_{3\times3} \\ \mathbf{J}^{03}_{3\times6} & 0_{3\times3} & \mathbf{J}^3_{3\times3} \end{bmatrix} \tag{23}$$

where matrix $\mathbf{R}_{6\times6}$ is as described in [7]. Then, we minimize the condition number of new Jacobian matrix. Based on this, we can obtain a configuration in which the corresponding homogeneous Jacobian matrix is well conditioned by minimizing the condition number over the following set of parameters, $\mathbf{x}$, i.e.,

$$\mathbf{x} = [\theta^0_2, \theta^0_3, \mathbf{u}_1, \mathbf{u}_2, \mathbf{u}_3, x_b, x_c, y_c]^T \tag{24}$$

a_1^0	a_2^0	b_1^0	b_2^0	α_1^0	α_2^0
0.0	0.9	0.6	0.0	90	0.0

i	a_3^{0i}	b_3^{0i}	α_1^{0i}	a_1^i	a_2^i	a_3^i	b_1^i	b_2^i	b_3^i	α_1^i	α_2^i	α_3^i
1	0.6	-0.8	90	0.0	0.4	0.2	0.8	0.0	0.0	90	0.0	0.0
2	1.0	0.0	90	0.0	0.4	0.2	0.8	0.0	0.0	90	0.0	0.0
3	0.6	0.8	90	0.0	0.4	0.2	0.8	0.0	0.0	90	0.0	0.0

Table 1: HD parameters of the multifingered robot hand.

where $\mathbf{u}_i, i = 1,2,3$ are the sub-vectors for each finger, these sub-vectors are given by

$$\mathbf{u}_i = [\theta_1^i, \theta_2^i, \theta_3^i, a_3^i, b_3^i]^T, \quad i = 1,2,3 \tag{25}$$

and x_b, x_c, y_c are the components of the position vectors points B and C defined on finger 1.

It is clear that $\alpha_3^i, i = 1,2,3$ and θ_1^0 do not play any role in the condition number, and hence, they are not considered as design variables. The optimization problem has to be solved using a numerical method [11] subject to the following constraints on the geometric parameters, namely,

$$a_3^i \geq 0, \quad x_b > 0, \quad x_c \geq 0, \quad y_c > 0, \; i = 1,2,3 \tag{26}$$

The physical meaning of the constraints is that we would like to guarantee nonnegative architecture dimensions and also avoid singularities of matrix $\mathbf{R}$. The design parameters and joint variables associated with the optimum design are shown in Tb. (2) and the condition number of the Jacobian matrix $\mathbf{J}'$ is very close to unity in this case. The solution of this design problem is now independent from the dimensional scaling factor.

V. Conclusion

The condition number of the Jacobian matrix of a manipulator has been used as a measure of the kinematic accuracy in the past. The dimensional inhomogeneity of the Jacobian matrix can be eliminated by redefining the end-effector velocity vector. The new Jacobian matrix obtained is independent from the dimensional scaling factor of the manipulator. This concept has been applied to articulated hands in this paper. A general formulation of the problem has been derived. The application to a twelve-degree-of-freedom articulated hand has been presented. Moreover, the optimization of the hand has been performed through the minimization of the condition number of the homogeneous Jacobian matrix obtained. An optimum architecture has been found. It is anticipated that the general procedure presented here can be used for the optimization of other complex mechanical systems such as parallel and hybrid manipulators.

$\theta_2^0 = 235.21$	Finger	$i=1$	$i=2$	$i=3$
$\theta_3^0 = 254.71$	θ_1^i	130.07	-3.7980	347.52
$x_b = 0.8384$	θ_2^i	-8.3668	104.24	102.09
$x_c = 0.1841$	θ_3^i	120.56	131.64	135.78
$y_c = 0.6537$	a_3^i	0.4349	0.3805	0.3913
$\kappa = 1.4$	b_3^i	-0.2771	0.0046	-0.0533

Table 2: Optimum design of the articulated multifingered hand.

References

[1] J.K. Salisbury and J.J. Craig, "Articulated hands: force control and kinematic issues", *The International Journal of Robotics Research*, 1(1),4–17 (1982).

[2] J. Angeles and A.A. Rojas, "Manipulator inverse kinematics via condition number minimization and continuation", *The International Journal of Robotics and Automation*, 2(2),61–69 (1987).

[3] C.A. Klein and B.E. Blaho, "Dexterity measures for design and control of kinematically redundant manipulators", *The International Journal of Robotics Research*, 6(2),72–83 (1987).

[4] C. Gosselin and J. Angeles, "The optimum kinematic design of a planar three-degree-of-freedom parallel manipulator", *ASME Journal of Mechanisms, Transmissions, and Automation in Design*, 110(1),35–41 (1988).

[5] R. Stoughton and T. Kokkinis, "Some properties of a new kinematic structure for robot manipulators", *Proceedings of the 13th Design Automation Conference*,73–79, Boston, USA (1987).

[6] C. Gosselin and J. Angeles, "A global performance index for the kinematic optimization of robotic manipulators", *ASME Journal of Mechanical Design*, 113(3),220–226 (1991).

[7] C.M. Gosselin, "The optimum design of robotic manipulators using dexterity indices", *Journal of Robotics and Autonomous Systems*, 9(4),213–226 (1992).

[8] M. Tandirci, J. Angeles and F. Ranjbaran, "The characteristic point and the characteristic length of robotic manipulators", *ASME Mechanisms Conference*, 45,203–208 (1992).

[9] J. Angeles, "The Design of isotropic manipulator architectures in the presence of redundancies", *The International Journal of Robotics Research*, 11(3),196–201 (1992).

[10] G. H. Golub and C. F. Van Loan, *Matrix computations*, The Johns Hopkins University Press, Baltimore (1983).

[11] P.E. Gill, W. Murray W., M.A. Saunders M.A. and M.H. Wright, "User's guide for NPSOL (Version 4.0): A fortran package for nonlinear programming", *Technical Report Sol 86-2*, Miami (1986).

Characterization of Kinematic and Static Performances of Robotic Geared Wrists

N. P. Belfiore and E. Pennestrì

University of Rome *La Sapienza*
Dept. of Mechanics and Aeronautics
via Eudossiana, 18
00184 Rome, Italy

II University of Rome *Tor Vergata*
Dept. of Mechanical Engineering
via della Ricerca Scientifica
00133 Rome, Italy

Abstract - Kinematic and static characterization of remotely actuated bevel-gear spherical wrist is proposed. The input-space to output-space transformation is splitted in two stages: topological and dimensional. An optimization technique can therefore be based on the separation of the two design phases. The first stage is herein analyzed and some algebraic and numerical results are presented.

I. Nomenclature

Φ	3 dimensional generalized velocity vector in the actuators space;
Θ	n dimensional generalized velocity vector in the joints space;
X	3 dimensional generalized velocity vector in the Cartesian space;
τ	n dimensional generalized resultant torque vector in the joints space;
ξ	3 dimensional generalized input torque vector in the actuators space;
A	$n \times 3$ structural matrix;
A^+	$3 \times n$ generalized inverse matrix of the structural matrix;
J	$3 \times n$ Jacobian matrix;
J^+	$n \times 3$ generalized inverse matrix of the Jacobian matrix;
F	3 dimensional generalized output force vector in the Cartesian space;
$\mathbf{W}_x$, $\mathbf{W}_\Phi$, $\mathbf{W}_F$, $\mathbf{W}_\xi$	3×3 weighting matrices;
$\mathbf{W}_\Theta$,$\mathbf{W}_\tau$	$n \times n$ weighting matrix;
$\lvert \mathrm{V} \rvert^2$	$\mathrm{V}^T \mathrm{W_V} \mathrm{V}$, weighted norms of vector V
K_v	ratio of the output and input generalized velocity magnitudes;
T_m	number of teeth of gear m;
$N_{ji} = \pm \frac{T_j}{T_i}$	gear ratio;
η_i^2, μ_i^2, λ_i^2	eigenvalues of transform matrices N, Q and R, respectively.

A. J. Lenarčič and B. B. Ravani (eds.), Advances in Robot Kinematics and Computationed Geometry, 209–218.

II. Introduction

With reference to geared robotic wrists, an extention of the indices of merit introduced by Asada and Cro Granito [1] is proposed. For this purpose, Tsai and Chen's model [2÷4] has been adopted in order to simplify the optimization of the kinematic and static characteristic for a given wrist. In particular, the actuator-space to cartesian-space mapping has been divided in two stages so that an optimization can be separately performed both on gear ratios and on angles between rotoidal joint axes. The *kinematic condition number* KCN and the *velocity capacity* VC have been redefined introducing topological and dimensional KCNs and VCs. The topological stage refers to the transformation from the actuator space to the joint space, which is completely described by the structural matrix $\boldsymbol{A}$. In this matrix, only the gear ratios N_{ij} appear as variables so that an optimal set of N_{ij} are to be found for a given application. The dimensional stage refers to the transformation from the joint space to the cartesian space. In this case, the jacobian matrix $\boldsymbol{J}$ is the transformation matrix.
In this paper only the topological stage has been considered. A general computer code for the algebraic evaluation of the structural matrix was written and tested [5] for each structure listed in the atlas [6-7]. Therefore it has been possible to derive automatically the symbolic expression of the topological VCs for all the structures described in the atlas. As for the topological KCNs, plots are herein presented. They show the influence of two gear ratios versus fixed value of the remaining.
The proposed optimization techinque is general and gives hints on the choice of the set of gear ratios which best suit the design requirements.

III. Revisiting a Kinematic and Static Characterization of Wrist Mechanisms

In 1985 Asada and Cro Granito [1] suggested an original approach to design robotic wrists. In their paper new performance indices were introduced with the aim of improving both kinematic and static characteristics of a wrist. According to the adopted nomenclature their model can be described as follows:

$$\Phi \longrightarrow \boxed{J} \longrightarrow X \tag{1}$$

and

$$\xi \longrightarrow \boxed{(J^T)^+} \longrightarrow F \tag{2}$$

First of all they extended the concept of *velocity ratio* and *mechanical advantage* from the single-input-single-output to the multi-input-multi-output class of mechanisms. In particular, the term*velocity ratio* K_v has been generalized by introducing the instantaneous ratio of the output and input velocity magnitudes:

$$K_v^2 = \frac{|X|^2}{|\Phi|^2} \tag{3}$$

Ratio K_v depends on the input joint angles and velocities. In this paragraph we briefly recall the results obtained in [1] in a form suitable for a spherical not redundant wrist mechanism with n=3.

It can be shown that the following relation holds:

$$K_v^2 = \frac{\Phi^T J^T W_x J \Phi}{\Phi^T W_\Phi \Phi} \tag{4}$$

and a transmission matrix N can be introduced:

$$N = W_\Phi^{-1} J^T W_x J \tag{5}$$

Matrix N has three real and non-negative eigenvalues:

$$\eta_1^2 \leq \eta_2^2 \leq \eta_3^2 \tag{6}$$

such as:

$$\eta_1 \leq K_v \leq \eta_3 \tag{7}$$

The square root of the eigenvalues are so defined as generalized velocity rattios GVRs.

The typical values of K_v can be evaluated by the following measure of the average of GVRs:

$$M_{ave} = \sqrt[3]{\eta_1 \eta_2 \eta_3} \tag{8}$$

or by the product of the square root of the eigenvalues:

$$V.C. = \eta_1 \eta_2 \eta_3 \tag{9}$$

From property (6), an index for measuring how K_v is sensitive to any change of configuration or input velocity values can be defined as follows:

$$M_{iso} = \frac{\eta_3}{\eta_1} \tag{10}$$

and is usually referred to as *isotropic mobility* [1] or *kinematic condition number*, KCN.

As for the static characterization, Asada and Cro Granito [1] introduced the instantaneous ratio of the output torque magnitude to the input torque magnitude:

$$K_m^2 = \frac{|F|^2}{|\xi|^2} = \frac{F^T W_F F}{F^T J W_\xi J^T F} \tag{11}$$

They demonstrated that if

$$W_\xi^{-1} = W_\phi \tag{12}$$

$$W_F = W_X^{-1} \tag{13}$$

then

$$\frac{1}{\eta_3} \leq K_m \leq \frac{1}{\eta_1} \tag{14}$$

defining the inverse of the GVRs as *generalized mechanical advantages* GMAs.

For completeness we will add the concept of *mechanical advantage capacity* MAC as the inverse of the velocity capacity:

$$MAC = (\eta_1 \eta_2 \eta_3)^{-1} = \frac{1}{VC} . \tag{15}$$

The objective function to be minimized has the form:

$$OBJ = (KCN - 1)^2 + w\left(VC^{1/3} - \eta_d\right)^2, \tag{16}$$

where η_d is the desired average magnitude of the GVRs and w is a weighting factor. It is worth noticing that, when η_d is too small, the output velocity has a small magnitude, while when η_d is large, then a poor load capacity is expected.

IV. Structural matrix

When a wrist is actuated through a direct drive, model (1) and (2) describe both kinematic and static behaviour. However, the transformation matrix, from the actuator space to the cartesian space, is not the simple jacobian matrix. In particular, for a bevel-gear remotely actuated wrist mechanism, models (1) and (2) should change as follows:

$$\Phi \;\longrightarrow\; \boxed{(A^T)^+} \;\longrightarrow\; \Theta \;\longrightarrow\; \boxed{J} \;\longrightarrow\; X \tag{17}$$

and

$$\xi \;\longrightarrow\; \boxed{A} \;\longrightarrow\; \tau \;\longrightarrow\; \boxed{(J^T)^+} \;\longrightarrow\; F , \tag{18}$$

where structural matrix A is defined by

$$\tau = A\, \xi \tag{19}$$

as suggested by Tsai [2]. In this case the performance indices proposed in the previous paragraph are still interesting even though their actual evaluation is not straightforward. For example, a possible way to evaluate K_v is

$$K_v^2 = \frac{\Phi^T \left[A^+ J^T W_x J \left(A^T\right)^+ \right] \Phi}{\Phi^T W_\Phi \Phi} \tag{20}$$

and the transmission matrix

$$N = W_\Phi^{-1} A^+ J^T W_x J (A^T)^+ \tag{21}$$

The explicit expression of the eigenvalues of N , as appear in (21), is impractical [4]. Analogously, ratio K_m can be expressed by:

$$K_m^2 = \frac{\xi^T A^T J^+ W_F (J^T) A \xi}{\xi^T W_\xi \xi} \tag{22}$$

or

$$\frac{1}{K_m^2} = \frac{F^T J (A^+)^T W_\xi A^+ J^T F}{F^T W_F F}. \tag{23}$$

Inequalities (14) toghether with conditions (12) and (13) hold.

V. Evaluation of performance indices

The methodology suggested is based on models (17) and (18). The ratio K_v can be written in the following way:

$$\frac{1}{K_m^2} = \frac{F^T J (A^+)^T W_\xi A^+ J^T F}{F^T W_F F}, \tag{24}$$

where a topological factor:

$$K_{va} = \frac{|\Theta|^2}{|\Phi|^2} \tag{25}$$

and a dimensional factor

$$K_{vj}^2 = \frac{|X|^2}{|\Theta|^2}, \tag{26}$$

are defined.

The latter, K_{vj} , is coincident with the K_v as intended by Asada and Cro Granito, who developed in their paper the described optimization technique. A similar optimization procedure can be separately performed on K_{va}. Only the actuator-space to joint-space mapping is considered, which makes easier both symbolic and numerical evaluation of the performance indices. From models (17) and (18) follows:

$$\begin{cases} \tau = A\xi \\ \Theta = (A^T)^+ \Phi, \end{cases} \tag{27}$$

from which we obtain:

$$K_{va}^2 = \frac{|\Theta|^2}{|\Phi|^2} = \frac{\Phi^T A^+ W_\Theta (A^T)^+ \Phi}{\Phi^T W_\Phi \Phi}. \tag{28}$$

The new transmission matrix can be written in the form:

$$Q = W_\Phi^{-1} A^+ W_\Theta (A^T)^+. \tag{29}$$

The square root of the eigenvalues of Q bounds the values of K_{va}:

$$\mu_1 \le K_{va} \le \mu_3 \tag{30}$$

As for K_{vj} we obtain:

$$K_{vj}^2 = \frac{|X|^2}{|\Theta|^2} = \frac{\Theta^T J^T W_x J \Theta}{\Theta^T W_\Theta \Theta} \tag{31}$$

together with the transmission matrix:

$$R = W_\Theta^{-1} J^T W_x J \tag{32}$$

and the properties of the square root of its eigenvalues:

$$\lambda_1 \le K_{vj} \le \lambda_3 \ . \tag{33}$$

VII. Optimization of the Kinematic Condition Number

From equation (24) appears that the two ratios K_{va} and K_{vj} are independent. Let us now define a *topological kinematic condition number*:

$$KCN_a = \frac{\mu_3}{\mu_1} \tag{34}$$

and a *dimensional kinematic condition number:*

$$KCN_j = \frac{\lambda_3}{\lambda_1} \tag{35}$$

Equation (24) together with inequalities (30) and (33) shows that an optimization technique can be separately performed on the two KCNs. In particular, a unit value is required for both in order to achieve an isotropic transformation from the actuator-space to the joint-space, and from the joint-space to the cartesian-space. However, it must be pointed out that this condition is only necessary to achieve isotropy from the actuator space to the cartesian space, which is granted by the global optimization of the KCN. An example of global optimization for a two-D.O.F. planar gear coupled manipulator is offered by Chen [4].

VI. Optimization of the Mechanical Advantage Capacity

The new model allows a separate performance of the optimization of the VC or its inverse *MAC*. Let us define a *topological* average of the velocity ratios:

$$M_{aveA} = \sqrt[3]{\mu_1 \mu_2 \mu_3} \tag{36}$$

and a *dimensional* one:

$$M_{aveJ} = \sqrt[3]{\lambda_1 \lambda_2 \lambda_3} \ . \tag{37}$$

Similarly, topological and dimensional VCs are defined by:

$$VC_a = \mu_1 \mu_2 \mu_3, \tag{38}$$

$$VC_j = \lambda_1 \lambda_2 \lambda_3 . \tag{39}$$

It is possible to show that:

$$VC_a = \sqrt{\det(Q)} = \sqrt{\det\left(W_\Phi^{-1} A^+ W_\Theta \left(A^T\right)^+\right)} \tag{40}$$

and

$$VC_j = \sqrt{\det(R)} = \sqrt{\det\left(W_\Theta^{-1} J^T W_x J\right)} . \tag{41}$$

Adopting the expression of VC proposed by Chen [4], we can write:

$$VC = \frac{\sqrt{\det\left(J^T W_x J\right)}}{\sqrt{\det\left(A W_\Phi A^T\right)}} . \tag{42}$$

From (40), (41), and (42) we obtain:

$$VC = VC_a \, VC_j \tag{43}$$

that is, the global VC is equal to the product of the topological and dimensional VCs. Finally, the global object function becomes:

$$OBJ = \left(KCN_a - 1\right)^2 + w_1\left(KCN_j - 1\right)^2 + w_2\left(VC_j^{1/3} VC_a^{1/3} - \eta_d\right)^2 . \tag{44}$$

VIII. Numerical results

In this investigation an analysis of the topological kinematic condition number and topological velocity capacity has been performed. A MAPLE code has been developed to evaluate structural matrix A for any feasible kinematic structure of a geared wrist. The evaluation of structural matrix has been performed according to the theoretical bases described in [5].

For a given structure, the symbolic expression of MAC must be deduced. For example, with reference to the classification criteria adopted in the atlas [7], when the weighting matrices are all equal to the unit matrix, the topological mechanical advantage capacity of the 7 link wrist mechanism, shown in Fig. 1, is:

$$MAC = N_{75} \, N_{47} \, N_{36} ; \tag{45}$$

while, for wrist No. 9 2 3 23 depicted in Fig. 2, we obtain:

$$MAC = \frac{N_{49} \, N_{95} \, N_{37} \, N_{86}}{N_{82}} \tag{46}$$

In Figs. 1 and 2 a planar representation of the corrisponding graph is also offered.

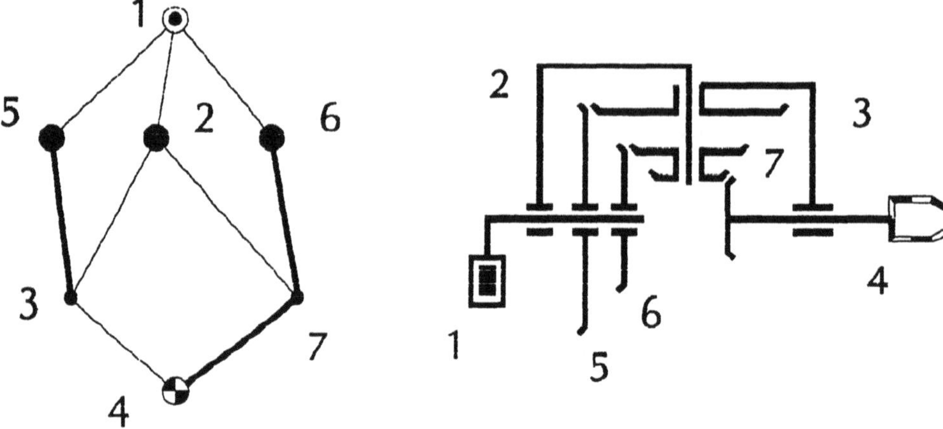

Fig. 1 Graph and functional representation of the 7 link wrist mechanism of the atlas [7]

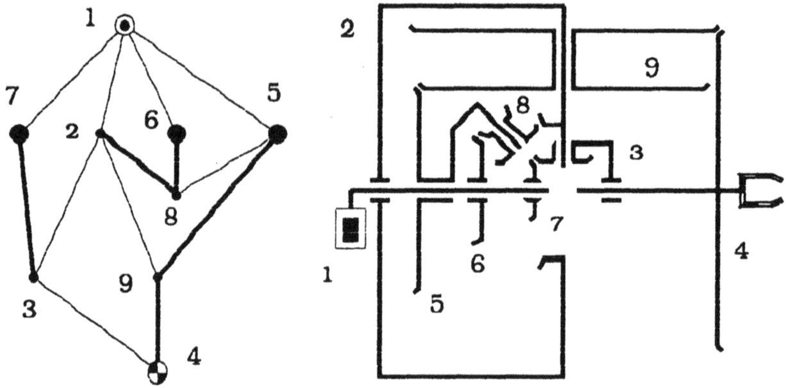

Fig. 2 Graph and functional representation of wrist mechanism No. 9 2 3 23 of the atlas [7]

The symbolic expression of the KCN_a is more difficult to obtain, since we have a third order characteristic polynomial and an explicit ordering of the magnitude of the eigenvalues is required. For this reason a numerical evaluation of topological KCN was performed for prescribed values of the gear ratios. In Figs. 3 - 5 the values of the topological KCNs have been plotted, respectively, for fixed values of N_{36}, N_{47}, and N_{75}, for the 7 link wrist of Fig. 1. For the same structure, the *MAC* is plotted in Fig. 6 with a fixed value of N_{47}. Figures 7 and 8, respectively, represent the plots of the topological KCN and topological *MAC* relative to the structure 9 2 3 23 depicted in Fig. 2, with fixed values of N_{86}, N_{82}, and N_{95}.

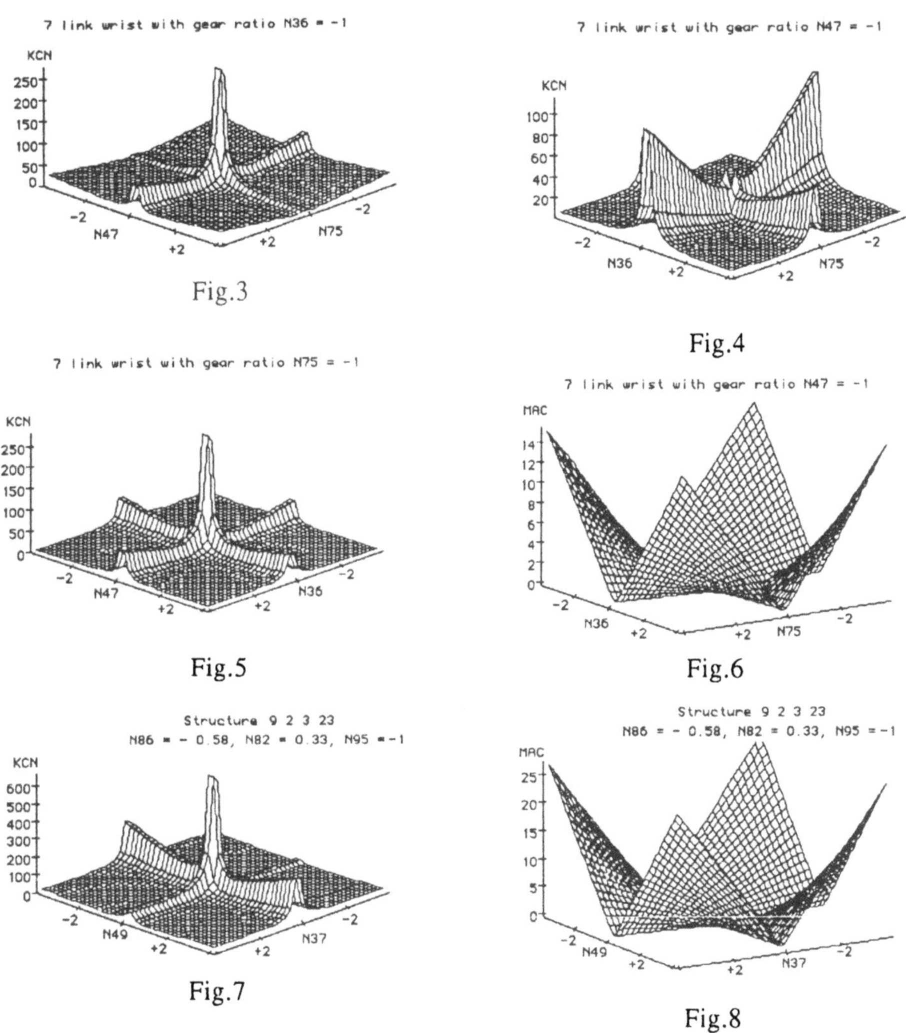

Fig.3

Fig.4

Fig.5

Fig.6

Fig.7

Fig.8

In order to make the plot of the topological KCN clear, the values corresponding to null values of the variable gear ratios has been avoided. Figures 3, 4, 5, and 7, are the plot of 40×40 values of KCN relative to a range from -3.9 to 3.9 to each variable gear ratio. Therefore, the zero value is avoided. It should observed that, from the design point of view, bevel gear ratios equal to +1 are not feasible.

IX. Conclusions

The results herein discussed show how promising appears the proposed approach for the optimization of both kinematic and static characteristics of a given topology. In the first stage, the gear ratios are optimized independently from the other dimensional parameters which will be determined during a second stage. As for the latter stage, results obtained in the previous works, as for example, the one of Asada and Cro Granito [1] fit into the proposed optimization scheme.

X. References

[1] Asada, H., Cro Granito, J.A., "Kinematic and static characterization of wrist joints and their optimal design", *Proc. IEEE Int. Conf. on Robotics and Automation*, March 1985, St. Louis, pp. 244 - 250.

[2] Tsai, L.-W., "The Kinematics of Spatial Robotic Bevel-Gear Trains", *IEEE Journal of Robotics and Automation*, April 1988, Vol. 4, No. 2, pp. 150 - 156.

[3] Chang, S.L, Tsai, L.-W., "Topological Synthesis of Articulated Gear Mechanisms", IEEE Transactions on Robotics and Automation, Feb. 1990, Vol. 6, No. 1, pp. 97 - 103.

[4] Chen, D.-Z., "Dynamic Analysis and Synthesis of Geared Robotic Mechanisms",Doctoral Thesis, Ph.D. 91-6, University of Maryland, College Park, MD USA, 1991.

[5] Belfiore, N.P., Pennestrì, E., "Kinematic and Static Force Analysis of remotely actuated geared wrists", submitted to the IFAC Int. Congress.

[6] Belfiore, N.P., Tsai, L.W., "A new Methodology for Structural Synthesis of Geared Robotic Wrists", *Proc. 2nd Nat. Conf. on Applied Mech. and Robotics*, Cincinnati, Nov. 1991, paper No. VIB.5.

[7] Belfiore, N.P., "An Atlas of Remote Actuated Bevel-Gear Type Wrist Mechanisms of up to Nine Links", *The Int. Journal of Robotic Research*, Vol. 12, No. 5, October 1993, pp. 448 - 459.

Acknowledgements

Financial support from Consiglio Nazionale delle Ricerche - Progetto Finalizzato Robotica is gratefully acknowledged.

6. Kinematics in Control

A Dynamically Consistent Strategy for Manipulator Control at Singularities

Kyong-Sok Chang and Oussama Khatib

Robotics Laboratory
Computer Science Department
Stanford University

Abstract - This paper presents a general strategy for manipulator control at kinematic singularities. The basic idea in this strategy is to treat the manipulator in the neighborhood of singular configurations as a redundant mechanism with respect to the motion of the end-effector in the subspace orthogonal to the singular directions. The control in this subspace is based on operational forces, while null space joint torques are used to deal with the control in the singular directions. The control is based on the *dynamically consistent* force/torque relationship that guarantees decoupled behavior. Two different types of kinematic singularities have been identified and strategies dealing with these singularities have been developed. Experimental results of the implementation of this approach on a PUMA 560 manipulator are presented.

I. Introduction

The difficulty with joint space control techniques lies in the discrepancy between the space where robot tasks are specified and the space in which the control is taking place. By its very nature, joint space control calls for transformations whereby joint space descriptions are obtained from the robot task specifications.

The joint space task transformation problem is exacerbated for mechanisms with redundancy or at kinematic singularities. The typical approach involves the use of pseudo- or generalized inverses to solve an under-constrained or degenerate system of linear equations, while optimizing some given criterion [3,4]. Other inverses with improved performance also have been investigated, e.g., the singularity robust inverse [1,8].

In addition, dealing with dynamics is essential for achieving higher performance. In free motion, the effects of dynamics increase with the range of motion, speed, and acceleration at which a robot is operating. In part mating operations, dynamics effects also increase with the rigidity of the mating object. Furthermore, control of the end-effector contact forces in some direction is affected by the inertial forces caused by end-effector motion in the subspace orthogonal to that direction. These effects must be taken into account to achieve higher performance.

The limitations of joint space control techniques, especially in constrained motion tasks, have motivated alternative approaches for dealing with task-level dynamics and control. The *operational space formulation*, which falls within this line of research, has been

A. J. Lenarčič and B. B. Ravani (eds.), Advances in Robot Kinematics and Computationed Geometry, 221–228.

driven by the need to develop mathematical models for the description, analysis, and control of robot dynamics with respect to task behavior.

In this framework, the control of redundant manipulators relies on two basic models: a *task-level dynamic model* obtained by projecting the manipulator dynamics into the operational space [5], and a *dynamically consistent force/torque relationship* that provides decoupled control of joint motions in the null space associated with the redundant mechanism [6]. These two models are the bases for implementing the control strategy for kinematic singularities. At singular configurations, a manipulator is treated as a redundant mechanism in the subspace orthogonal to the singular direction.

In this paper, we propose a classification of kinematic singularities from control point of view and present a general strategy for manipulator control at kinematic singularities. The effectiveness of this approach is experimentally demonstrated on a PUMA 560 manipulator.

II. Kinematic Singularities

A *singular configuration* is a configuration $\mathbf{q}$ at which the end-effector mobility – defined as the rank of the Jacobian matrix – locally decreases. At a singular configuration, the end-effector locally loses the ability to move along or rotate about some direction of the Cartesian space.

Singularities and mobility are characterized by the determinant of the Jacobian matrix for non-redundant manipulators; or by the determinant of the matrix product of the Jacobian and its transpose for redundant mechanisms. This determinant is a function, $s(\mathbf{q})$, that vanishes at each of the manipulator singularities. This function can be further developed into a product of terms,

$$s(\mathbf{q}) = s_1(\mathbf{q}) \cdot s_2(\mathbf{q}) \cdot s_3(\mathbf{q}) \ldots s_{n_s}(\mathbf{q}); \tag{1}$$

each of which corresponds to one of the different singularities associated with the mechanism. n_s is the number of different singularities. To a singular configuration there corresponds a *singular direction*. It is in or about this direction that the end-effector presents infinite effective mass or effective inertia. The end-effector movements remain free in the subspace orthogonal to this direction. In reality, the difficulty with singularities extends to some neighborhood around the singular configuration. The neighborhood of the ith singularity, $\mathcal{D}_{s_i}$, can be defined as

$$\mathcal{D}_{s_i} = \{\mathbf{q} \mid |s_i(\mathbf{q})| \leq \eta_i\}; \tag{2}$$

where η_i is positive.

III. Control Strategy

The basic concept in our approach to end-effector control at kinematic singularities is described as follows: In the neighborhood $\mathcal{D}_{s_i}$ of a singular configuration $\mathbf{q}$, the manipulator is treated as a redundant system in the subspace[1] orthogonal to the singular direction. End-effector motions in that subspace are controlled using the operational space redundant manipulator control, while null space joint torques are used to deal with the control in the singular direction according to the type of singularity. The use of the dynamically consistent force/torque relationship guarantees decoupled behavior between end-effector control and null space control.

A. Types of Singularities

In previous work [7], singularities have been distinguished in terms of the internal freedom of motion a manipulator has at a singular configuration, while the end-effector remains fixed.

However, for control purposes, we separate singularities in terms of the control characteristics of their null spaces: (Type 1) singularity at which the end-effector can be controlled in the singular directions using null space torques, and (Type 2) singularities where null space torques only affect internal joint motions.

Any projection to the null space associated to Type 1 results only in a finite end-effector motion in the singular direction while any projection to the null space associated to Type 2 results only in a finite change of the singular direction through finite internal joint motions.

B. Dynamic Consistency

In order to achieve dynamically consistent behavior of the redundant manipulator, it is necessary to decouple the motion in the null space and the motion in the operational space. In other words, the selection of null space motion control torques from a null space should not generate any acceleration at the end-effector.

This can be achieved using the dynamically consistent generalized inverse of the Jacobian matrix, $\bar{J}$, which is defined [6] as

$$\bar{J}(\mathbf{q}) = A^{-1}(\mathbf{q})J^T(\mathbf{q})\Lambda(\mathbf{q}); \tag{3}$$

where $\mathbf{q}$ is the vector of joint coordinates, $A(\mathbf{q})$ is the joint space kinetic energy matrix, $J(\mathbf{q})$ is the Jacobian matrix, and $\Lambda(\mathbf{q})$ is the operational space kinetic energy matrix. $\bar{J}$ has been shown to be unique [6].

The dynamically consistent relationship between joint torques and operational forces is

$$\mathbf{\Gamma} = J^T(\mathbf{q})\mathbf{F} + \left[I - J^T(\mathbf{q})\bar{J}^T(\mathbf{q})\right]\mathbf{\Gamma}_0. \tag{4}$$

This relationship provides a decomposition of joint torques into two dynamically decoupled control vectors: joint torques corresponding to forces acting at the end effector

[1] a subspace of the end-effector operational space.

$(J^T\mathbf{F})$; and joint torques that only affect internal joint motions, $\left([I - J^T(\mathbf{q})\overline{J}^T(\mathbf{q})]\mathbf{\Gamma}_0\right)$. $\mathbf{F}$ refers to the operational space control forces acting on the end-effector; $\mathbf{\Gamma}_0$ refers to the joint control torques for desired motions in the null space; and $[I - J^T(\mathbf{q})\overline{J}^T(\mathbf{q})]$ is the dynamically consistent null space.

Using this decomposition, the end effector can be controlled by operational forces, while motions in the null space can be independently controlled by joint torques that are guaranteed not to alter the end effector's dynamic behavior.

This decomposition of a non-redundant mechanism is done as follows: In the neighborhood of singular configurations, first, singular directions and associated singular frames are identified. A *singular frame* is the frame where one of its axes is aligned to the singular direction. Then, the Jacobian matrix is rotated to the singular frame and its rows corresponding to singular directions are eliminated. This redundant Jacobian matrix corresponds to the redundant mechanism with respect to the motion of the end-effector in the subspace orthogonal to the singular directions. The null space generated by the dynamically consistent inverse of this redundant Jacobian matrix is used in the null space motion control.

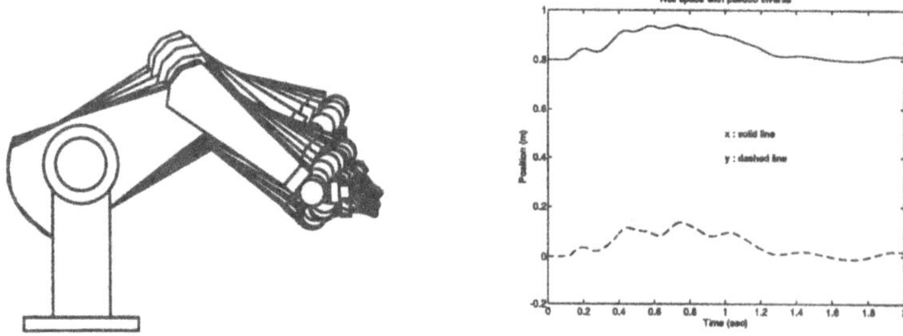

Figure 1: Null Space with Pseudo Inverse Simulation

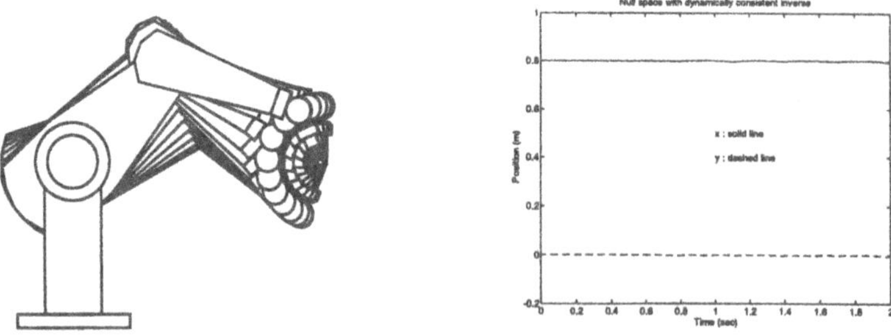

Figure 2: Null Space with Dynamically Consistent Inverse Simulation

The impact of the dynamically consistent control decomposition is illustrated on the 3R-planar manipulator shown in Figure 1 and Figure 2. This manipulator is treated as a redundant mechanism with respect to the task of positioning the end-effector.

Plots in Figure 1 and Figure 2 show the result when a step impulse is added to two different null spaces; one constructed by the dynamically consistent inverse from Equation 3 and the other constructed by the Moore-Penrose (Pseudo) Inverse, $J^{+} = J^{T}(JJ^{T})^{-1}$. The latter null space is dynamically inconsistent since J^{+} is not dynamically weighted [2].

The goal is to maintain the end-effector position, x and y, while the manipulator is moving in the null space. As expected, with the dynamically consistent null space in Figure 2, the manipulator has no difficulty maintaining its end-effector position, while with the other null space in Figure 1, the manipulator is unable to account for the accelerations generated by the null space.

IV. Null Space Control

An additional task to be carried out using the null space can be realized by constructing a potential function, $V_0(\mathbf{q})$, whose minimum corresponds to the desired task. By selecting

$$\mathbf{\Gamma}_0 = -A(\mathbf{q})\nabla V_0; \tag{5}$$

where $A(\mathbf{q})$ is the joint space kinetic energy matrix, one obtains the needed attraction to the desired task. The interference from the additional torques on the end-effector is eliminated by projecting this gradient in the dynamically consistent null space.

Two different strategies are developed according to the two types of singularities defined in this paper. For each strategy, two sub-cases are considered; (Case 1) end-effector motion in the singular direction is required and (Case 2) end-effector motion in the singular direction is not required.

A. Singularity Type 1 : control of the end-effector motion

Case 1: The end-effector motion in the singular direction can be controlled directly through the associated null space by selecting a potential function whose minimum corresponds to the desired configuration. The resulting torques from Equation 5 affects the end-effector motions only along the singular direction.

Case 2: Null space control torques should be applied to *asymptotically* stabilize the mechanism in the null space, e.g., $\mathbf{\Gamma}_0 = -kvq\dot{\mathbf{q}}$.

B. Singularity Type 2 : control of the internal joint motion

Case 1: The configuration of the manipulator itself is controlled to change the singular direction until the singular direction is orthogonal to the operational force vector. By constructing a potential function such that its minimum corresponds to the configuration where the singular direction is orthogonal to the operational force vector, the singular direction can be changed in the null space. The resulting torques from Equation 5 affect only the change of the singular direction via internal joint motions while maintaining the position and orientation of the end-effector.

Since a small end-effector motion in the singular direction can cause a large finite internal joint motion in the null space, following a time-dependent trajectory can be difficult. A solution to this difficulty is to use a simple path planning to keep the singular direction orthogonal to the end-effector motion in the neighborhood of singularities. Then, the control can be done as in Case 2 where no motion is required in the singular direction. This path planning will avoid the time delay caused by finite internal joint motions of the manipulator. A practical approach to eliminate this path planning is to use time-independent controls such as goal position method [5].

Case 2: Control is done in the same way as the corresponding case of Type 1.

V. Experimental Result

The strategy presented in this paper has been implemented to control a PUMA 560 with the goal position method [5]. Figure 3 shows the three basic singularities in a PUMA 560: elbow lock, wrist lock, and head lock from left to right. Elbow lock is Type 1 and the other two are Type 2. Since these basic singularities can occur at the same time, the rank of the Jacobian matrix can vary from 3 to 6. The minimum rank of the Jacobian corresponds to the configuration at which the end-effector reaches the highest point directly above the base.

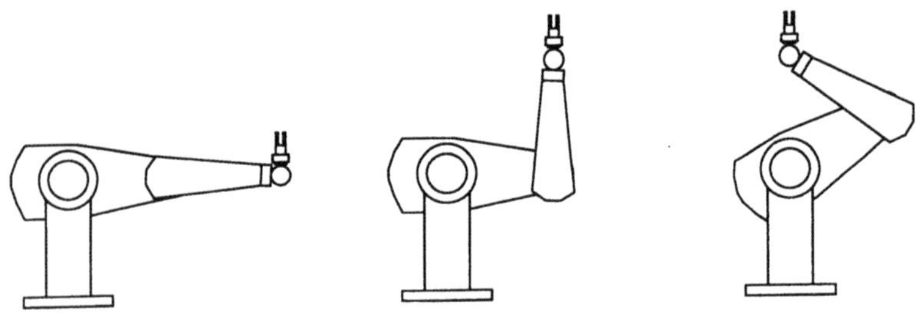

Figure 3: Three Basic Singular Configurations in PUMA 560

In Figure 4, the end-effector of a PUMA 560 is simultaneously moving out of two singularities, elbow lock (Type 1: Case 1) and wrist lock (Type 2: Case 2). The goal is to translate along the singular direction, x, while maintaining all other positions and orientations. Since internal joint motions are not needed, the end-effector motion is fully controlled as shown in the plot in Figure 4.

In Figure 5, the end-effector of a PUMA 560 is moving out of the singular configuration of wrist lock (Type 2: Case 1) along the singular direction, x, while maintaining all other positions and orientations. The initial configuration is shown in Figure 5 and the goal is to rotate 45° about x-axis. The motion of joint 4 and 6 is the finite internal joint motion and the motion of joint 5 accounts for the end-effector motion. The symmetric motion of joint 4 and 6 ensures the decoupled behavior. The end-effector motion remains smooth (solid line), even though there is a velocity change in joint 4 and 6 at the boundary of the singular configuration.

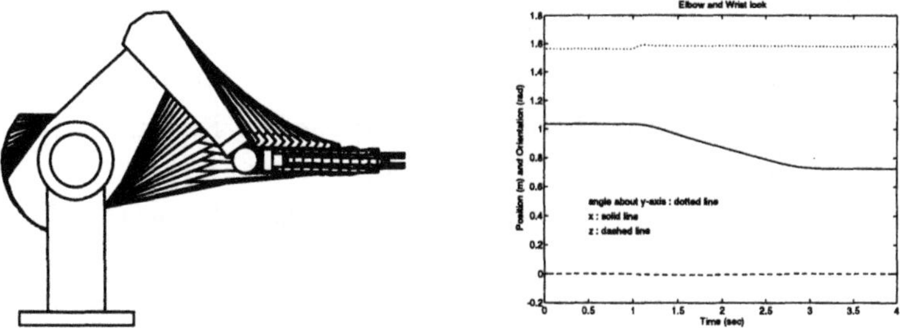

Figure 4: Compound Singular Configuration

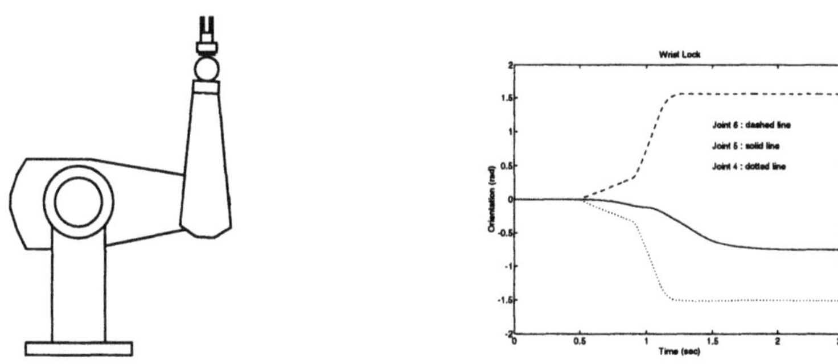

Figure 5: Wrist Lock

As the results show, this strategy effectively controls the manipulator in the neighborhood of singularities.

VI. Conclusions

We have presented a dynamically consistent strategy for manipulator control at singular configurations. This strategy is based on the characteristics of null spaces associated with two different types of singularities and the *dynamically consistent* force/torque relationship that guarantees decoupled behavior between end-effector control and null space control of redundant mechanisms. Conversion from non-redundant control to redundant control is achieved using singular directions and singular frames. The experimental result with a PUMA 560 shows the effectiveness of this strategy.

VII. Acknowledgments

Many thanks to Sean Quinlan, Diego Ruspini, and David Williams for their comments and help during the preparation of the manuscript.

VIII. References

[1] S. Chiaverini, B. Siciliano, O. Egeland "Experimental Results on Controlling a 6-DOF Robot Manipulator in the Neighborhood of Kinematic Singularities" Proc. 3rd International Symposium on Experimental Robotics Kyoto, Japan (1993)

[2] K. Doty, "A Theory of Generalized Inverses Applied to Robotics", International Journal of Robotics Research, Vol.12, No. 1 (1993)

[3] H. Hanafusa, T. Yoshikawa, Y. Nakamura, "Analysis and Control of Articulated Robot Arms with Redundancy" 8th IFAC, Vol. XIV (1981) p. 38-83.

[4] J.M. Hollerbach, K.C. Suh, "Redundancy Resolution of Manipulators Through Torque Optimization", 1985 (March 25-28). Proc. IEEE Int. Conf. Robot. & Autom. (1985), p. 1016-1021.

[5] O. Khatib, "A Unified Approach to Motion and Force Control of Robot Manipulators: The Operational Space Formulation", IEEE J. Robotics and Automation, Vol. 3, No. 1 (1987)

[6] O. Khatib, "Motion/Force Redundancy of Manipulators", Japan-U.S.A. symposium on Flexible Automation (1990)

[7] Z.C. Lai, D.C.H. Yang, "A New Method for the Singularity Analysis of Simple Six-link Manipulators", International Journal of Robotics Research (1986)

[8] Y. Nakamura, "Kinematical Studies on the Trajectory Control of Robot Manipulators", Ph.D. Thesis, Kyoto, Japan (1985)

Third-Order Control of a Planar System Tracking Constant Curvature Paths

S.J. Lorenc and M.M. Stanišić

Department of Aerospace and Mechanical Engineering
University of Notre Dame
Notre Dame, IN 46556, USA

Abstract - This paper develops the Cubic of Stationary Curvature for a planar two-revolute open chain. The Cubic is then used to develop a third-order of resolved motion rate control for constant curvature paths. The third-order of control increases the tracking accuracy of the two-revolute system and raises the question of optimal order of control.

I Introduction

Curvature Theory can be used in the motion synthesis of planar one degree of freedom systems, [1], [2], [3], [4], [5]. These are the English references wherein one can find the earlier references in other languages. The theory can also be extended to planar two degree of freedom systems, [6], [7], [8]. The two degree of freedom Curvature Theory has been developed only up to second-order path properties while the one degree of freedom theory has been developed to fifth-order path properties.

Curvature Theory has further been used to control two degree of freedom planar manipulators, essentially developing a *geometric based* second-order *resolved motion rate control* (RMRC). A *time-based* second-order RMRC has been developed by [9] and others. A second-order RMRC has been shown to significantly enhance the path tracking capabilities of a manipulator. In this paper, the Cubic of Stationary Curvature will be developed for a planar two-revolute open chain and will be used to develop a *third-order* RMRC for tracking paths of constant curvature. The path tracking accuracy is enhanced as compared to the second-order RMRC, though to a lesser degree than the second-order RMRC compared to the first-order RMRC. This comes at a cost of additional computation and raises the fundamental question, "*what is the computationally optimal order of resolved motion rate control?*".

II Brief Review of Second-Order Control of Planar Systems

In this section, a brief review of second-order control of a planar two degree of freedom system [7] will be presented. The general plane motion of an arbitrary point Q on a moving rigid body is given by the equations, (see Fig.(1))

$$X = a + x\cos\theta - y\sin\theta \tag{1}$$

$$Y = b + x\sin\theta + y\cos\theta \tag{2}$$

Σ and E represent the fixed and moving coordinate systems (bodies), respectively. The

A. J. Lenarčič and B. B. Ravani (eds.), Advances in Robot Kinematics and Computationed Geometry, 229–238.

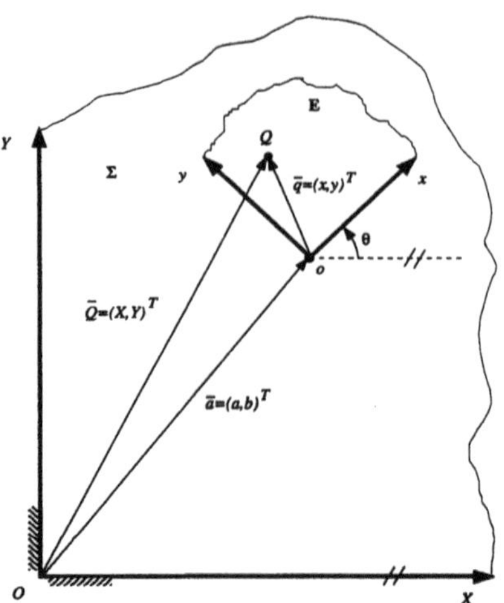

Figure 1: A General Planar Motion

coordinates of Q in Σ and E are denoted by $\overline{Q} = (X, Y)^T$ and $\bar{q} = (x, y)^T$, respectively. A rigid body assumption implies that $\bar{q}$ is a constant. The location of the origin of E in the fixed coordinate system Σ is denoted by $\bar{a} = (a, b)^T$.

In a two degree of freedom (d.o.f.) motion, the *motion variables* θ, a, and b are functions of two dimensionless *motion parameters*, λ and μ, which are the system's two degrees of freedom.

$$\theta = \theta(\lambda, \mu) \quad , \quad a = a(\lambda, \mu) \quad \text{and} \quad b = b(\lambda, \mu) \ .$$

A Notation

At the instant under consideration, corresponding to a setpoint on the work-space trajectory, θ, a and b are assumed to be instantaneously equal to zero, i.e. Σ and E are coincident. λ and μ are also assumed as instantaneously equal to zero. This is referred to as the *zero-position*.

$$\theta(0,0) = 0 \quad , \quad a(0,0) = 0 \quad \text{and} \quad b(0,0) = 0 \ .$$

In regards to notation, for any quantity such as X,

$$\left.\frac{\partial X}{\partial \lambda}\right|_{\substack{\lambda=0\\ \mu=0}} = X_\lambda \ , \ \left.\frac{\partial X}{\partial \mu}\right|_{\substack{\lambda=0\\ \mu=0}} = X_\mu ,$$

$$\left.\frac{\partial^2 X}{\partial \lambda^2}\right|_{\substack{\lambda=0\\ \mu=0}} = X_{\lambda\lambda} \ , \ \left.\frac{\partial^2 X}{\partial \mu^2}\right|_{\substack{\lambda=0\\ \mu=0}} = X_{\mu\mu} \ , \ \left.\frac{\partial^2 X}{\partial \lambda \partial \mu}\right|_{\substack{\lambda=0\\ \mu=0}} = X_{\lambda\mu} ,$$

$$\text{and} \qquad \left.\frac{dX}{dt}\right|_{\substack{\lambda=0\\ \mu=0}} = X_1 \ , \ \left.\frac{d^2 X}{dt^2}\right|_{\substack{\lambda=0\\ \mu=0}} = X_2 \ .$$

B Second-Order Instantaneous Kinematic Equations of Motion

From eqs.(1,2), in terms of the defined notation,

$$\begin{array}{ll} X_\lambda = a_\lambda - y\theta_\lambda & Y_\lambda = b_\lambda + x\theta_\lambda \\ X_\mu = a_\mu - y\theta_\mu & Y_\mu = b_\mu + x\theta_\mu \\ X_{\lambda\lambda} = a_{\lambda\lambda} - x{\theta_\lambda}^2 - y\theta_{\lambda\lambda} & Y_{\lambda\lambda} = b_{\lambda\lambda} - y{\theta_\lambda}^2 + x\theta_{\lambda\lambda} \\ X_{\mu\mu} = a_{\mu\mu} - x{\theta_\mu}^2 - y\theta_{\mu\mu} & Y_{\mu\mu} = b_{\mu\mu} - y{\theta_\mu}^2 + x\theta_{\mu\mu} \\ X_{\lambda\mu} = a_{\lambda\mu} - x\theta_\lambda\theta_\mu - y\theta_{\lambda\mu} & Y_{\lambda\mu} = b_{\lambda\mu} - y\theta_\lambda\theta_\mu + x\theta_{\lambda\mu} \end{array} \tag{3}$$

from which a second-order Taylor series expansion of the planar two degree of freedom motion is given by,

$$X = x + X_\lambda\lambda + X_\mu\mu + \frac{1}{2}(X_{\lambda\lambda}\lambda^2 + 2X_{\lambda\mu}\lambda\mu + X_{\mu\mu}\mu^2) \tag{4}$$

$$Y = y + Y_\lambda\lambda + Y_\mu\mu + \frac{1}{2}(Y_{\lambda\lambda}\lambda^2 + 2Y_{\lambda\mu}\lambda\mu + Y_{\mu\mu}\mu^2) \tag{5}$$

The truncated second-order Taylor series is used for a second-order of resolved motion rate control (RMRC). The six first-order constants, $(a_\lambda,\ a_\mu,\ b_\lambda,\ b_\mu,\ \theta_\lambda,\ \theta_\mu)$ and nine second-order constants, $(a_{\lambda\lambda},\ a_{\lambda\mu},\ a_{\mu\mu},\ b_{\lambda\lambda},\ b_{\lambda\mu},\ b_{\mu\mu},\ \theta_{\lambda\lambda},\ \theta_{\lambda\mu},\ \theta_{\mu\mu})$ which appear in the second-order equations of motion depend on the motion parameters of the particular system to be controlled, and the choice of Σ and E. The choice of Σ and E will be the canonical system which will yield the algebraically simplest results.

Differentiating eqs.(4,5) with respect to time, yields the instantaneous velocity distribution of E,

$$X_1 = X_\lambda\lambda_1 + X_\mu\mu_1 \tag{6}$$

$$Y_1 = Y_\lambda\lambda_1 + Y_\mu\mu_1 \tag{7}$$

Substituting eq.(3) into eqs.(6,7), and solving for the coordinates x_p and y_p of the Pole yields:

$$x_p = -\frac{b_\lambda + nb_\mu}{\theta_\lambda + n\theta_\mu} \tag{8}$$

$$y_p = \frac{a_\lambda + na_\mu}{\theta_\lambda + n\theta_\mu} \tag{9}$$

where n is the first-order speed ratio defined as,

$$n = \left.\frac{d\mu}{d\lambda}\right|_{\substack{\lambda=0\\ \mu=0}}$$

and is also the slope of the tangent to the joint-space trajectory. The canonical coordinate systems have their (Y, y) axes on the Polar Line, in which case x_p in eq.(8) is instantaneously equal to zero and therefore it must be true that

$$b_\lambda = b_\mu = 0$$

Using eq.(9), n can be found as a function of y_p.

$$n = -\frac{a_\lambda - \theta_\lambda y_p}{a_\mu - \theta_\mu y_p} \tag{10}$$

Eq.(10) is a mapping of y_p (found from the *work-space* trajectory tangent) to n, (the slope of the *joint-space* trajectory tangent).

A second-order speed ratio n' can be defined as,

$$n' = \left.\frac{dn}{d\lambda}\right|_{\substack{\lambda=0\\ \mu=0}} = \left.\frac{d^2\mu}{d\lambda^2}\right|_{\substack{\lambda=0\\ \mu=0}},$$

which, through calculus, determines the *curvature of the joint-space trajectory*. The relationship between n' and the *curvature of the work-space trajectory* is found through the pole velocity and the Inflection Circle, which are well-known second-order properties of constrained motion, [3].

The pole velocity is found by rederiving eqs.(6,7) for the position of the Pole, retaining first-order terms.

$$\frac{dX}{dt} = X_\lambda\frac{d\lambda}{dt} + X_\mu\frac{d\mu}{dt} + X_{\lambda\lambda}\lambda\frac{d\lambda}{dt} + X_{\lambda\mu}\left(\frac{d\lambda}{dt}\mu + \lambda\frac{d\mu}{dt}\right) + X_{\mu\mu}\mu\frac{d\mu}{dt} \tag{11}$$

$$\frac{dY}{dt} = Y_\lambda\frac{d\lambda}{dt} + Y_\mu\frac{d\mu}{dt} + Y_{\lambda\lambda}\lambda\frac{d\lambda}{dt} + Y_{\lambda\mu}\left(\frac{d\lambda}{dt}\mu + \lambda\frac{d\mu}{dt}\right) + Y_{\mu\mu}\mu\frac{d\mu}{dt} \tag{12}$$

Substituting eq.(3) into eqs.(11,12), and solving for the coordinates of the Pole yields,

$$x_p = \frac{\lambda\left[a_\lambda\theta_\lambda - b_{\lambda\lambda} + \frac{d\mu}{d\lambda}\left(a_\mu\theta_\lambda - b_{\lambda\mu}\right)\right] + \mu\left[\left(a_\lambda\theta_\mu - b_{\lambda\mu}\right) + \frac{d\mu}{d\lambda}\left(a_\mu\theta_\mu - b_{\mu\mu}\right)\right]}{\left(\theta_\lambda + \theta_\mu\frac{d\mu}{d\lambda}\right) + 2\lambda\left(\theta_{\lambda\mu}\frac{d\mu}{d\lambda} + \theta_{\lambda\lambda}\right) + 2\mu\left(\theta_{\mu\mu}\frac{d\mu}{d\lambda} + \theta_{\lambda\mu}\right)} \tag{13}$$

$$\begin{aligned} y_p = &\Big\{ \left(\theta_\lambda + \theta_\mu\tfrac{d\mu}{d\lambda}\right)\left[\lambda\left(a_{\lambda\lambda} + a_{\lambda\mu}\tfrac{d\mu}{d\lambda}\right) + \mu\left(a_{\lambda\mu} + a_{\mu\mu}\tfrac{d\mu}{d\lambda}\right) + \left(a_\lambda + a_\mu\tfrac{d\mu}{d\lambda}\right)\right] \\ &+ \left(a_\lambda + a_\mu\tfrac{d\mu}{d\lambda}\right)\left[\lambda\left(\theta_{\lambda\lambda} + \theta_{\lambda\mu}\tfrac{d\mu}{d\lambda}\right) + \mu\left(\theta_{\lambda\mu} + \theta_{\mu\mu}\tfrac{d\mu}{d\lambda}\right)\right]\Big\} \\ &\Big/\Big\{\left(\theta_\lambda + \theta_\mu\tfrac{d\mu}{d\lambda}\right)\left[\left(\theta_\lambda + \theta_\mu\tfrac{d\mu}{d\lambda}\right) + 2\lambda\left(\theta_{\lambda\lambda} + \theta_{\lambda\mu}\tfrac{d\mu}{d\lambda}\right) + 2\mu\left(\theta_{\lambda\mu} + \theta_{\mu\mu}\tfrac{d\mu}{d\lambda}\right)\right]\Big\} \end{aligned} \tag{14}$$

In the zero-position, x_p is zero as expected. Time differentiation of eqs.(13,14) yields the instantaneous pole velocity,

$$(x_p)_1 = \frac{\lambda_1[n^2(\theta_\mu a_\mu - b_{\mu\mu}) + n(\theta_\lambda a_\mu + \theta_\mu a_\lambda - 2b_{\lambda\mu}) + \theta_\lambda a_\lambda - b_{\lambda\lambda}]}{(\theta_\lambda + n\theta_\mu)} \tag{15}$$

$$(y_p)_1 = \frac{\lambda_1[n^3(\theta_\mu a_{\mu\mu} - \theta_{\mu\mu}a_\mu) - n^2\Gamma - n\Lambda + \theta_\lambda a_{\lambda\lambda} - \theta_{\lambda\lambda}a_\lambda + n'(\theta_\lambda a_\mu - \theta_\mu a_\lambda)]}{(\theta_\lambda a_\mu - \theta_\mu a_\lambda)^2} \tag{16}$$

where

$$\Gamma = -2\theta_\mu a_{\lambda\mu} + 3a_\mu\theta_{\lambda\mu} + \theta_{\mu\mu}a_\lambda - \theta_\lambda a_{\mu\mu}$$
$$\Lambda = -\theta_\mu a_{\lambda\lambda} + \theta_{\lambda\lambda}a_\mu + 2\theta_{\lambda\mu}a_\lambda - 2\theta_\lambda a_{\lambda\mu}$$

The only appearance of n' is in the (Y, y) component of the pole velocity, $(y_p)_1$. For a constrained motion,

$$(y_p)_1 = (PJ)_x\,\omega \tag{17}$$

where (PJ) is the directed line segment P to J, the diameter of the Inflection Circle, and $(PJ)_x$ is the (X, x) component of this directed line segment, [3]. In general,

$$\omega = \theta_\lambda\lambda_1 + \theta_\mu\mu_1 \tag{18}$$

Substituting eq.(18) into eq.(17) yields,

$$(y_p)_1 = (PJ)_x\,(\theta_\lambda\lambda_1 + \theta_\mu\mu_1) \tag{19}$$

Equating eq.(16) with eq.(19) and solving for n' in terms of (PJ) yields:

$$n' = \frac{(PJ)_x\,(\theta_\lambda + \theta_\mu n)^3 + n^2\Gamma + n^3(-\theta_\mu a_{\mu\mu} + \theta_{\mu\mu}a_\mu) + n\Lambda - \theta_\lambda a_{\lambda\lambda} + \theta_{\lambda\lambda}a_\lambda}{\theta_\lambda a_\mu - \theta_\mu a_\lambda} \tag{20}$$

Eq.(20) maps the *second-order work-space trajectory property*, (PJ), to the *second-order joint-space trajectory property*, n'.

III Cubic of Stationary Curvature for a Two-Revolute

In order to develop a third-order control for a planar two-revolute open chain, the Cubic of Stationary Curvature is derived for the two degree of freedom system. The curvature, κ, of the work-space path trajectory that a point (x, y) on the manipulator is to follow, is defined by:

$$\kappa = \frac{\frac{dX}{d\lambda}\frac{d^2Y}{d\lambda^2} - \frac{d^2X}{d\lambda^2}\frac{dY}{d\lambda}}{\left[\left(\frac{dX}{d\lambda}\right)^2 + \left(\frac{dY}{d\lambda}\right)^2\right]^{\frac{3}{2}}} \tag{21}$$

For a generalized two degree of freedom system, the derivatives with respect to λ, are defined, for example, as:

$$\frac{dX}{d\lambda} = \frac{\partial X}{\partial\lambda} + \frac{\partial X}{\partial\mu}\frac{d\mu}{d\lambda} \tag{22}$$

$$\frac{d^2X}{d\lambda^2} = \frac{\partial^2X}{\partial\lambda^2} + 2\frac{\partial^2X}{\partial\lambda\partial\mu}\frac{d\mu}{d\lambda} + \frac{\partial^2X}{\partial\mu^2}\left(\frac{d\mu}{d\lambda}\right)^2 + \frac{\partial X}{\partial\mu}\frac{d^2\mu}{d\lambda^2} \tag{23}$$

Rewriting eq.(21) in the *zero-position* using eqs.(22,23) and the defined notation yields the following:

$$\begin{aligned}\kappa \;=\; & [(X_\lambda + X_\mu n)(Y_{\lambda\lambda} + 2Y_{\lambda\mu}n + Y_{\mu\mu}n^2 + Y_\mu n') \\ & -(Y_\lambda + Y_\mu n)(X_{\lambda\lambda} + 2X_{\lambda\mu}n + X_{\mu\mu}n^2 + X_\mu n')] \\ & /\left[(X_\lambda + X_\mu n)^2 + (Y_\lambda + Y_\mu n)^2\right]^{\frac{3}{2}}\end{aligned} \tag{24}$$

Differentiating eq.(21) with respect to λ, yields the rate of change of the path curvature of the point (x, y) on the manipulator being controlled.

$$\frac{d\kappa}{d\lambda} = \frac{\left[\left(\frac{dX}{d\lambda}\right)^2 + \left(\frac{dY}{d\lambda}\right)^2\right]\left[\frac{dX}{d\lambda}\frac{d^3Y}{d\lambda^3} - \frac{d^3X}{d\lambda^3}\frac{dY}{d\lambda}\right] - 3\left[\frac{dX}{d\lambda}\frac{d^2Y}{d\lambda^2} - \frac{d^2X}{d\lambda^2}\frac{dY}{d\lambda}\right]\left[\frac{dX}{d\lambda}\frac{d^2X}{d\lambda^2} + \frac{dY}{d\lambda}\frac{d^2Y}{d\lambda^2}\right]}{\left[\left(\frac{dX}{d\lambda}\right)^2 + \left(\frac{dY}{d\lambda}\right)^2\right]^{\frac{5}{2}}} \tag{25}$$

From this equation, the points on the moving body which are instantaneously following a path with constant curvature must satisfy the condition $\frac{d\kappa}{d\lambda} = 0$. Setting $\frac{d\kappa}{d\lambda}$ equal to zero and rewriting the equation yields the Cubic of Stationary Curvature for a generalized two degree of freedom system.

$$\begin{aligned} &\left[\left(\frac{dX}{d\lambda}\right)^2 + \left(\frac{dY}{d\lambda}\right)^2\right]\left[\frac{dX}{d\lambda}\frac{d^3Y}{d\lambda^3} - \frac{d^3X}{d\lambda^3}\frac{dY}{d\lambda}\right] \\ &\quad = 3\left[\frac{dX}{d\lambda}\frac{d^2Y}{d\lambda^2} - \frac{d^2X}{d\lambda^2}\frac{dY}{d\lambda}\right]\left[\frac{dX}{d\lambda}\frac{d^2X}{d\lambda^2} + \frac{dY}{d\lambda}\frac{d^2Y}{d\lambda^2}\right] \end{aligned} \tag{26}$$

Substituting eqs.(22,23) into eq.(26) along with the appropriate notation yields:

$$\begin{aligned} & [(X_\lambda + X_\mu n)(Y_{\lambda\lambda\lambda} + 3Y_{\lambda\lambda\mu}n + 3Y_{\lambda\mu\mu}n^2 + Y_{\mu\mu\mu}n^3 + 3Y_{\mu\mu}nn' + 3Y_{\lambda\mu}n') \\ & -(Y_\lambda + Y_\mu n)(X_{\lambda\lambda\lambda} + 3X_{\lambda\lambda\mu}n + 3X_{\lambda\mu\mu}n^2 + X_{\mu\mu\mu}n^3 + 3X_{\mu\mu}nn' + 3X_{\lambda\mu}n') \\ & +n''(X_\lambda Y_\mu - X_\mu Y_\lambda)]\left[(X_\lambda + X_\mu n)^2 + (Y_\lambda + Y_\mu n)^2\right] \\ = \; & 3\big[(X_\lambda + X_\mu n)(Y_{\lambda\lambda} + 2Y_{\lambda\mu}n + Y_{\mu\mu}n^2) \\ & -(Y_\lambda + Y_\mu n)(X_{\lambda\lambda} + 2X_{\lambda\mu}n + X_{\mu\mu}n^2) + n'(X_\lambda Y_\mu - X_\mu Y_\lambda)\big] \\ & *\big[(X_\lambda + X_\mu n)(X_{\lambda\lambda} + 2X_{\lambda\mu}n + X_{\mu\mu}n^2 + X_\mu n') \\ & +(Y_\lambda + Y_\mu n)(Y_{\lambda\lambda} + 2Y_{\lambda\mu}n + Y_{\mu\mu}n^2 + Y_\mu n')\big] \end{aligned} \tag{27}$$

where n'' is the *third-order* instantaneous speed ratio defined as:

$$n'' = \left.\frac{dn'}{d\lambda}\right|_{\substack{\lambda=0\\ \mu=0}} = \left.\frac{d^2n}{d\lambda^2}\right|_{\substack{\lambda=0\\ \mu=0}} = \left.\frac{d^3\mu}{d\lambda^3}\right|_{\substack{\lambda=0\\ \mu=0}}$$

The third-order terms in eq.(27) need to be defined next.

$$\begin{aligned} X_{\lambda\lambda\lambda} &= a_{\lambda\lambda\lambda} - 3x\theta_\lambda\theta_{\lambda\lambda} - y\left(\theta_{\lambda\lambda\lambda} - {\theta_\lambda}^3\right) \\ Y_{\lambda\lambda\lambda} &= b_{\lambda\lambda\lambda} - 3y\theta_\lambda\theta_{\lambda\lambda} + x\left(\theta_{\lambda\lambda\lambda} - {\theta_\lambda}^3\right) \\ X_{\lambda\mu\mu} &= a_{\lambda\mu\mu} - x\left(2\theta_\mu\theta_{\lambda\mu} + \theta_{\mu\mu}\theta_\lambda\right) - y\left(\theta_{\lambda\mu\mu} - {\theta_\mu}^2\theta_\lambda\right) \\ Y_{\lambda\mu\mu} &= b_{\lambda\mu\mu} - y\left(2\theta_\mu\theta_{\lambda\mu} + \theta_{\mu\mu}\theta_\lambda\right) + x\left(\theta_{\lambda\mu\mu} - {\theta_\mu}^2\theta_\lambda\right) \\ X_{\lambda\lambda\mu} &= a_{\lambda\lambda\mu} - x\left(2\theta_\lambda\theta_{\lambda\mu} + \theta_\mu\theta_{\lambda\lambda}\right) - y\left(\theta_{\lambda\lambda\mu} - \theta_\mu{\theta_\lambda}^2\right) \\ Y_{\lambda\lambda\mu} &= b_{\lambda\lambda\mu} - y\left(2\theta_\lambda\theta_{\lambda\mu} + \theta_\mu\theta_{\lambda\lambda}\right) + x\left(\theta_{\lambda\lambda\mu} - \theta_\mu{\theta_\lambda}^2\right) \\ X_{\mu\mu\mu} &= a_{\mu\mu\mu} - 3x\theta_\mu\theta_{\mu\mu} - y\left(\theta_{\mu\mu\mu} - {\theta_\mu}^3\right) \\ Y_{\mu\mu\mu} &= b_{\mu\mu\mu} - 3y\theta_\mu\theta_{\mu\mu} + x\left(\theta_{\mu\mu\mu} - {\theta_\mu}^3\right) \end{aligned} \tag{28}$$

These terms, (X_λ, X_μ, etc.), define the particular mechanical system. In this paper, the system of interest is a planar two-revolute system. This system is defined by its first, second, and third-order instantaneous invariants. These invariants provide a geometric description of the planar two-revolute open chain. For this system, referenced to the canonical system, the only *non-zero* first, second, and third-order instantaneous invariants are:

$$a_\lambda = -l_1 \quad \theta_\lambda = 1 \quad \theta_\mu = 1 \quad b_{\lambda\lambda} = -l_1 \quad a_{\lambda\lambda\lambda} = l_1 \tag{29}$$

Substituting eqs.(3,28,29) into eq.(27) yields the Cubic of Stationary Curvature for a planar two-revolute open chain:

$$(Ax + By + C)(x^2 + y^2) + Dxy + Ex + Fy = 0 \tag{30}$$

where

$$\begin{aligned}
A &= -n''(n+1)^2 + 3n'^2(n+1) - n(2n+1)(n+1)^3 \\
B &= 3nn'(n+1)^2 \\
C &= 6l_1nn'(n+1) \\
D &= -l_1n^2(n^2+4n+5) + 3l_1n'^2 - 2l_1\left[n''(n+1) + n\right] \\
E &= l_1{}^2\left[n(n^2-1) - n''\right] \\
F &= 3l_1{}^2nn'
\end{aligned}$$

Alternatively, the Cubic of Stationary Curvature for a planar two-revolute open chain can be written in polar coordinates:

$$\frac{Ar^2 + E}{\sin\theta} + \frac{Br^2 + F}{\cos\theta} + \frac{Cr}{\sin\theta\cos\theta} + Dr = 0 \tag{31}$$

IV Third-Order of Resolved Motion Rate Control

In this section, a third-order of resolved motion rate control for a planar two-revolute system will be demonstrated. In order to control the system, a third-order Taylor series will be used. Up to the third-order instantaneous speed ratios are needed in order to write the coordinating Taylor series. Two of the speed ratios have been previously defined by [7,8] and presented in section II. The first-order instantaneous speed ratio, n, ensures that the system will instantaneously match the tangent of the work-space path trajectory. This is developed with knowledge of the location of the Pole. The second-order instantaneous speed ratio, n', will ensure that the system will instantaneously match the curvature of the work-space trajectory, and is found as a function of the diameter of the Inflection Circle. Finally, a third-order instantaneous speed ratio, n'', will provide the information necessary for the system to develop a work-space trajectory of *constant curvature*. To determine n'', the Cubic of Stationary Curvature for a planar two-revolute open chain, which was developed in the previous section, will be used. Since the only unknown in the Cubic of Stationary Curvature is n'', eq.(30) can be solved for n''. The third-order instantaneous speed ratio, n'', is found to be:

$$n'' = \frac{(Ix + Jy + K)(x^2 + y^2) + Lxy + Mx + Ny}{G^2x^3 + (l_1 + Gy)^2x} \tag{32}$$

where

$$\begin{array}{ll} G = n + 1 & H = 3nn' \\ I = -G[2n^4 + 5n^3 + 4n^2 + n - 3(n')^2] & J = HG^2 \\ K = 2l_1 HG & L = -l_1[n^4 + 4n^3 + 5n^2 + 2n - 3(n')^2] \\ M = -{l_1}^2 n(n-1)G & N = {l_1}^2 H \end{array}$$

Similarly, n'' can be solved from eq.(31) using polar coordinates. The third-order speed ratio, n'', can additionally be found geometrically as a function of the slope of the asymptote of the Cubic of Stationary Curvature.

When the three instantaneous speed ratios have been determined, a third-order Taylor series can be used to constrain the system to travel along the work-space path trajectory with a third-order approximation of the path. The third-order Taylor series can be written as:

$$\mu = n\lambda + \frac{1}{2!}n'\lambda^2 + \frac{1}{3!}n''\lambda^3 \tag{33}$$

V Numerical Example and Simulation Results

A numerical example will be used to demonstrate the increased path tracking accuracy due to the higher order of control. The system, along with the work-space path trajectory is shown in Fig.(2). The link dimensions of the system are taken as $l_1 =$

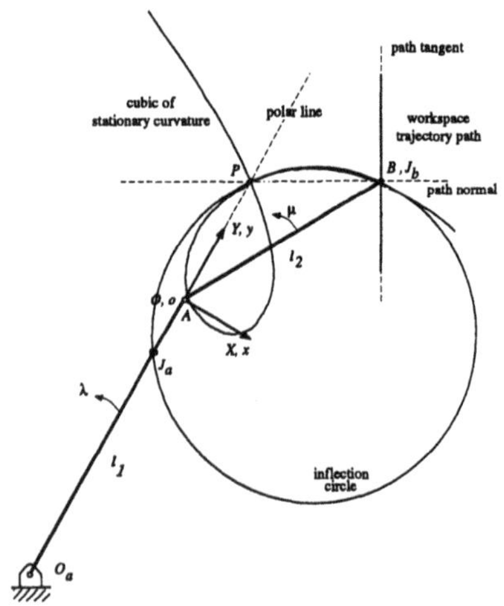

Figure 2: A Two-Revolute System

20.0 in. and $l_2 = 15.0$ in. The point on the system which is to be controlled is the point B. The work-space path trajectory is to be a straight line and is also shown in fig.(2). The pole, P, and the Inflection Circle shown are found, and this information is used to find the first and second-order instantaneous speed ratios, n, and n'. In this configuration with the desired path trajectory, the speed ratios are:

$$n = -3.31 \qquad n' = -10.57 \tag{34}$$

$$\begin{cases} \ddot{q}_r = \tau' \\ M_{ee}.\ddot{q}_e - T_e = -M_{er}.\tau' \end{cases} \tag{3}$$

The joint dynamics are now decoupled and linear so that we can perfectly control them with simple local P.I.D. controllers in tracking / regulation : $\tau' = \ddot{q}_{rd} + k_v.\delta\dot{q}_r + k_p.\delta q_r$, where $\ddot{q}_{rd}$ is the feedforward desired joints accelerations vector, k_p and k_v are the diagonal matrices of proportional and derivative gains and $\delta q_r = (q_{rd} - q_r), \delta\dot{q}_r = (\dot{q}_{rd} - \dot{q}_r)$.

V. Zero dynamics

We see from Eq. (3) that a dynamic sub-system remains non linear due to the non square nature of the input-ouput map. In fact this dynamic which is not observable from the controlled ouputs is interpreted as the zero-dynamics. This is easily explained by the fact that such non linear laws solve the input-output inversion problem [7]. Where the zeros at infinity are represented by the joint dynamics, the transmission zeros are the elastic dynamics [8]. Thus, it is normal, after we have applied this law, to see the zeros become dynamics. In regulation around a desired joint configuration, these dynamics reduce to :

$$M_{ee}.\delta\ddot{q}_e + K_{ee}.\delta q_e = -M_{er}.\tau'$$

Here, δq_e is the dynamic component of the global elastic deflection since the gravity forces are compensated by static deformations of links. Therefore, with our choice of the ouputs, the zero-dynamics are the dynamics of the structure with locked joints excited by the auxiliary control. Hence we conclude at this step of the design that if we do not do anything else the structure is going to be damped passively due to some internal links dissipations.

VI. Simulation results "without active damping"

We first report simulation results achieved with this first control layer represented by Eq.(2). The investigated kinematic is a two links planar manipulator. The links are one meter cylindrical beams with ten centimeters of diameter made of an usual steel. The elastic fields are decomposed on the embedded-free modes basis of each link considered as an Euler - Bernouilli beam. Each flexural field is approximated with one mode. Some passive dissipations are added as modal dampings on each link. The stiffness of the material is artificially decreased in order to have modal stiffenesses equal to $10.N/m$. We report a test in regulation around two configurations on fig.(1). Due to the identity of auxiliary control gains, the decoupled joint responses are merged. On the other hand we see that the elastic subsystem oscillates freely like the passive locked joints structure in these two configurations.

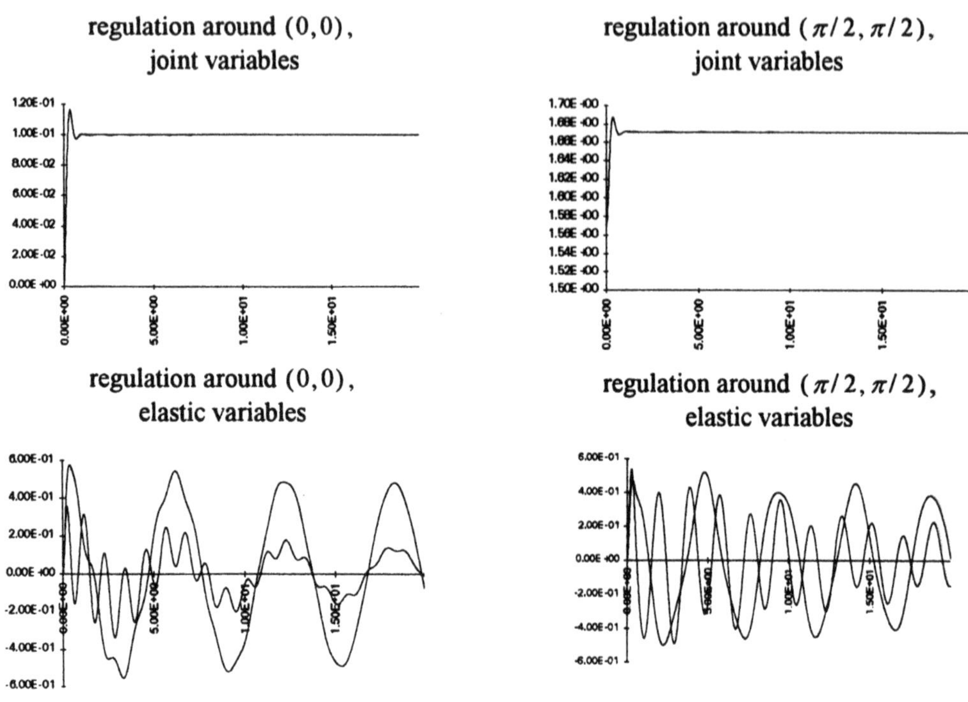

fig. 1 : Simulation results without active damping

VII. Design of the "elastic control component"

Next we note that the passive damping of the structure is not enough. Thus we decide to stabilize it by the control. This is achieved by adding to the previous auxiliary control a stabilizing elastic component : $\tau' = \tau'_r + \delta\tau'_e$. The $\delta\tau'_e$ elastic component is usually a complete regulation state feedback [1]: $\delta\tau'_e = -k_{pr}.\delta q_r - k_{vr}.\delta\dot{q}_r - k_{pe}.\delta q_e - k_{ve}.\delta\dot{q}_e$, where $\delta q_r = q_{rd} - q_r$, $\delta\dot{q}_r = \dot{q}_{rd} - \dot{q}_r$, $\delta q_e = q_{es} - q_e$ $\delta\dot{q}_e = \dot{q}_e$, and q_{es} is the static deflection which is compensated from the tool point of view, by the correction of the desired joint positions. In fact, with such an additional ("elastic") control we recouple the two subsystems via the control in order to pump out the energy of the modes of the stucture excited by the rigid control. Let us remark that the control problem now occurs as a compromise solution between the joint objectives that we have perfectly achieved before, and the damping of the modes that requires to give up the perfect joint decoupling. The feedback gains are usually computed [1] by a linear system deduced from an approximated linearization of Eq. (3) around nominal rigid trajectories. This process is justified by the nearly linear nature of the small perturbator elastic motions. We choose here to compute these gains from the regulation system in the target configuration of the trajectory. This system is deduced from Eq. (3) after we have applied the auxiliary control of Eq. (4):

$$\begin{pmatrix} \delta\dot{q}_r \\ \delta\ddot{q}_r \\ \delta\dot{q}_e \\ \delta\ddot{q}_e \end{pmatrix} = \begin{pmatrix} 0 & I & 0 & 0 \\ -k_p & -k_v & 0 & 0 \\ 0 & 0 & 0 & I \\ -M_{eed}^{-1}.M_{erd}.k_p & -M_{eed}^{-1}M_{erd}.k_v & -M_{eed}^{-1}K_{ee} & 0 \end{pmatrix} \cdot \begin{pmatrix} \delta q_r \\ \delta\dot{q}_r \\ \delta q_e \\ \delta\dot{q}_e \end{pmatrix} + \begin{pmatrix} 0 \\ -I \\ 0 \\ -M_{eed}^{-1}.M_{erd} \end{pmatrix} \tag{4}$$

where, due to the small motions around a steady state, we neglect coriofugal forces in Eq.(3), and "$_d$" indicates that the model is computed in the desired target configuration.

VIII. Gains computation

We have tested and compared the two usual approaches of designing a linear feedback. Our conclusions are the following.

A. Pole placement

The pole placement approach, if proper care is not taken, leads to state feedbacks which do not respect the natural dynamics of the structural modes and try to modify the stiffnesses of the modes. This results in very energetically expensive controls.

B. L.Q.R. approach

This computational method leads to real damping controllers since feedbacks naturally tend not to stiffen the modes but rather to damp them. For this reason this approach is prefered compared to the first one. We have looked for a criterion depending only on one scalar parameter which tunes the active damping. The following criterion, based on that used in [9] for a one link manipulator, has given the best results :

$$J = \int_0^{\infty} {}^T\delta y.I.\delta y + \tau^2.{}^T\delta\dot{y}.I.\delta\dot{y} + \rho.{}^T\delta\tau_e'.I.\delta\tau_e' \ .dt$$

where δy is the regulation error vector in the operational space and ρ is a weighting scalar of the control capacities. With such a criterion we first fix the control weight and then tune the damping weight τ. This scalar parameter is homogeneous with a response time and by increasing it one adds more damping to the closed loop poles of the controlled structure. The computations of the regulation state in the operationnal space use expressions :

$$\delta y = J_{rd}.\delta q_r + J_{ed}.\delta q_e \quad , \quad \delta\dot{y} = J_{rd}.\delta\dot{q}_r + J_{ed}.\delta\dot{q}_e$$

where, the second expression is an approximation justified by the fact that coriofugal forces are negligible in regulation.

IX. Simulation results with active damping

We now report simulation results achieved in the same conditions as those of fig. (1) except that we add now the damping component. With this planar kinematic , the operational vector has as components the absolute cartesian positions of the tip. We choose $\rho = 10$ and $\tau^2 \approx 500$. We see from fig. (2) that we have completely lost the ideal joint dynamics. However, the stucture is now well damped. In fact the global dynamic of the controlled polyarticulated structure is now as fast as the first structural mode that we damp rather than to stiffen. We are now going to investigate the case of a robot with flexible links and joints.

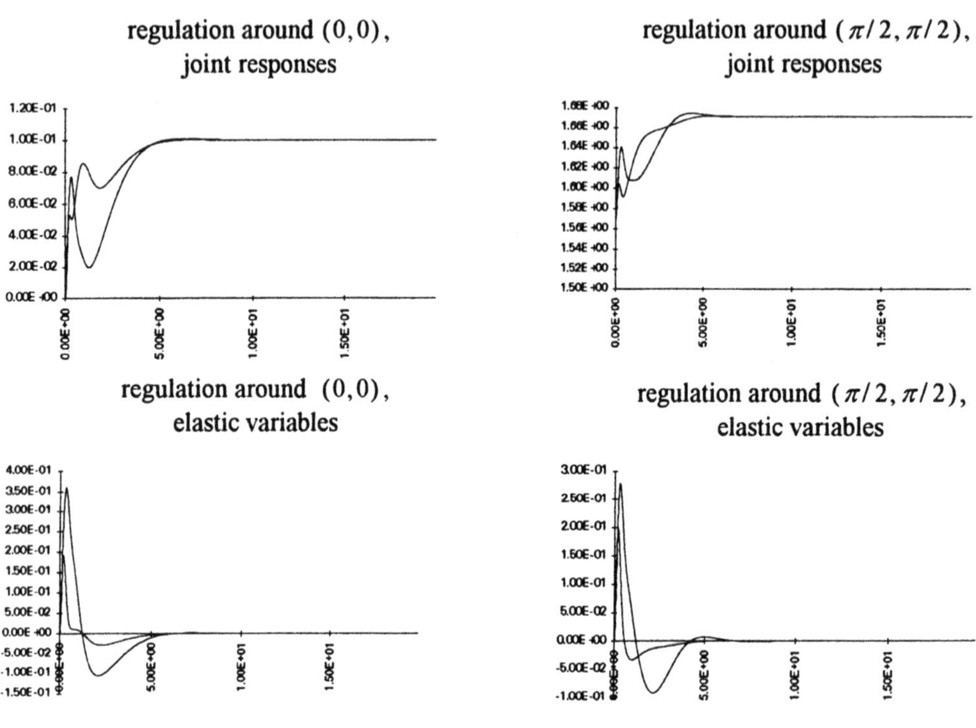

fig. 2 : Simulation results with active damping

X. Model of a manipulator with flexible links and joints

Here, we only consider transmissions with strong gear ratios as strings of gears or harmonic drives. From now on q_r denotes the vector of joint positions on the links side, q_e denotes the vector of links Rayleigh-Ritz positions , and q_m denotes the vector of rotor positions on the links side. The assumption of the strong gear ratios allows one, as Spong [10] done, to neglect all the transmission contributions to the whole kinetic energy except that of rotors motions around themselves. This leads us to the model :

$$\begin{pmatrix} M_{rr} & M_{re} \\ M_{er} & M_{ee} \end{pmatrix} \cdot \begin{pmatrix} \ddot{q}_r \\ \ddot{q}_e \end{pmatrix} + \begin{pmatrix} C_r(q,\dot{q}) \\ C_e(q,\dot{q}) \end{pmatrix} + \begin{pmatrix} 0 & 0 \\ 0 & K_{ee} \end{pmatrix} \cdot \begin{pmatrix} q_r \\ q_e \end{pmatrix} + \begin{pmatrix} Q_r(q) \\ Q_e(q) \end{pmatrix} = \begin{pmatrix} K.(q_m - q_r) \\ 0 \end{pmatrix} \tag{5}$$

$$J_m \cdot \ddot{q}_m + K.(q_m - q_r) = \tau_m$$

where J_m is the diagonal matrix of rotor inertias evaluated on the links side, K is the diagonal stiffness matrix of spring transmissions, τ_m is the vector of the motor torques on the link sides which becomes the new control inputs vector.

XI. Control of a manipulator with flexible links and joints

A first control solution consists of extending the non linear decoupling control to the map which relates the motor torques with the joints positions (on the links side). This leads to unrealistic control laws which require accelerations, jerks, and twists on the links side [10]. Deluca [11] showed that it is possible to decouple and linearize this input-output map via a dynamic non linear feedback. Other authors [2] put Eq. (5) under the same form as that of Eq. (1). This is achieved via a transformation from absolute coordinates to relative ones. This allows one to apply exactly the same control as in Eq. (2) with rigid joints and flexible links. This strategy leads to a non linear control of rotor motions coupled with an active damping of both links and joints flexibilities. The drawback of this approach is that it increases the dimension of the dynamic model used for the control design. Moreover it requires one to update the joints state feedback gains. On the other hand it is energetically economical since it damps all the flexibilities. We choose here to apply a method based on a frequency separation of the two sub-systems , (the transmission sub-system and the rigid joints manipulator sub-system), as two time scales controllers [10] do. This principle has the advantage not to compromise the control design of the manipulator with rigid joints and flexible links. Moreover the feedback added to control the transmissions does not depend on the current configuration, contrary to the previous approach.

XII. Torque control of transmissions

The principle is simple : The law of Eq. (2) is still valid, if we succeed in forcing the elastic torques transmited via the elastic springs to be instantaneously equal to the control torques based on the model with rigid joints. By "Instantaneously", we means sufficiently fast not to interact with the controller based on the model with rigid joints. Let us recall at this point that the bandwidth of this controller has been increased up to the first structural mode with rigid joints and flexible links, i.e. : the "first distributed mode". Thus we decide to control the elastic torques transmitted after the springs via some fast local loops. In order to design this law, we rewrite the transmission dynamics with respect to the elastic torques that we want to control, i.e. the torques $\tau = K.\Delta q$, where $\Delta q = q_m - q_r$. Thus, from Eq.(5):

$$(J_m.K^{-1}).K.\Delta\ddot{q} + K.\Delta q = \tau_m - J_m \cdot \ddot{q}_r \tag{6}$$

Let us isolate $\ddot{q}_r$ in the first equation of Eq. (5):

$$\ddot{q}_r = -W_{rr}.(C_r(q,\dot{q}) + Q_r(q) - K.\Delta q) - W_{re}.(C_e(q,\dot{q}) + Q_e(q) + K_{ee}.q_e)$$

, where:

$$\begin{pmatrix} M_{rr} & M_{re} \\ M_{er} & M_{ee} \end{pmatrix}^{-1} = \begin{pmatrix} W_{rr} & W_{re} \\ W_{er} & W_{ee} \end{pmatrix}$$

Inserting this expression into Eq. (6), allows us to obtain the expression for the torques:

$$(J_m.K^{-1}).K.\Delta\ddot{q} + (I + J_m.W_{rr})K.\Delta q = \tau_m + J_m.W_{rr}.(C_r + Q_r) + J_m.W_{re}.(C_e + Q_e + K_{ee}.q_e$$

We first compensate the non linear pertubations with the control. This is achieved by applying controls of the following form, where τ'_m is the new auxiliary control vector :

$$\tau_m = \tau'_m - J_m.W_{rr}.(C_r + Q_r) - J_m.W_{re}.(C_e + Q_e + K_{ee}.q_e) \qquad (7)$$

We obtain the new dynamics of the torques transmitted via the elastic joints :

$$(J_m.K^{-1}).K.\Delta\ddot{q} + (I + J_m.W_{rr})K.\Delta q = \tau'_m$$

The matricial term $(I + J_m.W_{rr})$ is bounded on the configurations set. It is then possible to make this dynamic as fast as we want with a feedback of the following form :

$$\tau'_m = K_\tau.(\tau_d - \tau) - K_{\dot{\tau}}.\dot{\tau} \qquad (8)$$

where τ_d is the desired torques vector based on the model with rigid joints, and K_τ and $K_{\dot{\tau}}$ are diagonal matrices of proportional and derivative gains. We finally find the controlled dynamic of the elastic torques:

$$J_m.K^{-1}.\ddot{\tau} + K_{\dot{\tau}}.\dot{\tau} + (I + J_m.W_{rr} + K_\tau)\tau = K_\tau.\tau_d$$

We take proportional gains large enough to make the dynamic invariant and fast enough not to interact with the first distributed mode. We form the global control from Eqs. (7, 8):

$$\tau_m = \left(K_\tau.\tau_d - J_m.W_{rr}.(C_r + Q_r) - J_m.W_{re}.(C_e + Q_e + K_{ee}.q_e)\right) - \left(K_\tau.\tau + K_{\dot{\tau}}.\dot{\tau}\right)$$

That is a superimposition of a slow and a fast component :

$$\tau_s = K_\tau.\tau_d - J_m.W_{rr}.(C_r + Q_r) - J_m.W_{re}.(C_e + Q_e + K_{ee}.q_e) \ , \ \tau_f = -(K_\tau.\tau + K_{\dot{\tau}}.\dot{\tau})$$

In practise, we first tune the local loops on torques such as they are more than ten times faster than the first distributed mode (3 octaves).

XIII. Simulation of the control with flexible links and joints

We now report results obtained in simulation with the previous two links planar kinematic. The links parameters are unchanged. The joint parameters are the following : $K = 10.I$, $J_m = I$. Note that these unusual weak inertia (compared to our initial modeling assumption) have to be related to the very strong flexibilities. Choosing $K_\tau = 1000.I$ and $K_\tau = 8.I$, the controlled transmissions are 100 times faster than the first distributed mode.

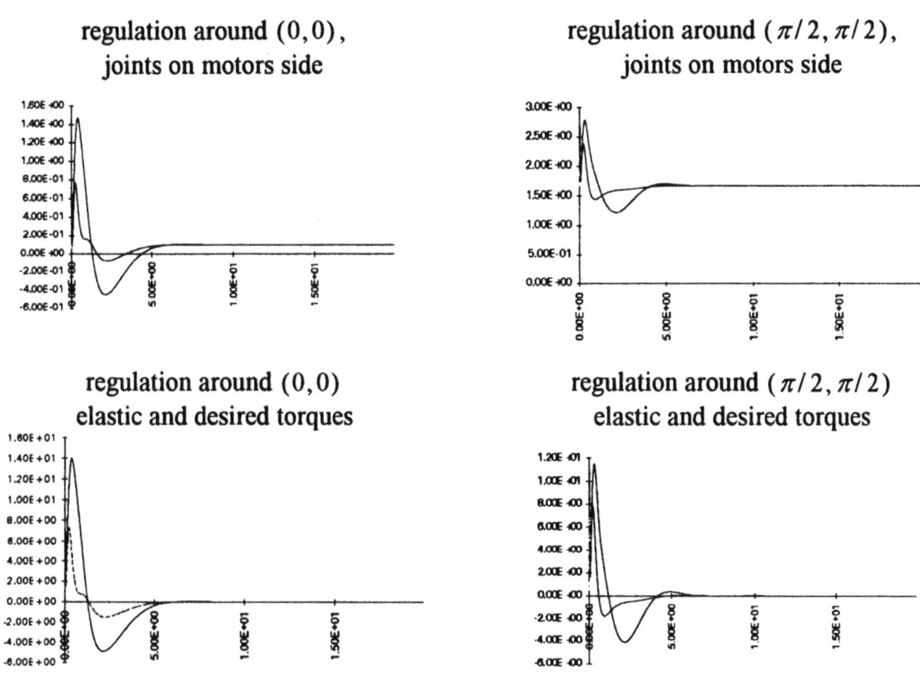

fig. 3 : Simulation of the control with flexible links and joints

From fig. (3) we see that the transmitted torques track the desired ones in such a way that the responses of the state variables after the joints are the same as in fig. (2) (this explains why we did not report them). The joints on the motors side oscillate in order to transmit the desired torques to the links through the springs. The high level oscillations have to be related to the weak joint flexibilities. Nevertheless such a controller is energetically more expensive than those based on a damping of the pertubator oscillations in the joints. Practically we have to deal with actuators bandwidths and saturations.

XIV. Conclusion

Let us sum up the transformations of the controller frequency band for the case of such a control strategy is applied to an industrial manipulator. We denote the first mode of the manipulator with flexible joints and rigid links as "first located mode", and we denote the first mode of the structure with rigid joints and flexible links as "first distributed mode". On some industrial robots these two modes are close enough to be coupled and create a first global mode at low frequencies. Let us recall that local controllers bandwidths have to be far enough below this mode in order not to excite it. With our strategy we first destroy the located mode by stiffening the joints with the control. We thus shift this mode to a higher frequency so that it becomes energetically decoupled with the first distributed mode which is now the first global mode. Then we apply the second control layer which increases the control bandwidth up to the first distributed mode that we choose to damp. Since the local loops are tuned once for all, this is achieved without too much complicating the controller as compared to the pure "distributed case" .

XV. References

[1] Singh S.N. , Schy A. A. : " Control of elastic robotic systems by nonlinear inversion and modal damping". Transct. of the ASME J. of Dynamic systems, Measurement and control Vol. 108 , no 5, pp 180-189, september 1986.

[2] Lin L.Y. and Yuan K. : "Motor based control of manipulators with flexible joints and links" . IEEE Proceedings on Robotics and Autom. 1990.

[3] Cetinkunt S. and Wayne J. Book : "Symbolic modelling of flexible manipulators" IEEE int. conf. Robotics Autom., pp. 2074-2080, 1987.

[4] Singh R.P.Vander Voort, R.J.Likins : "Dynamics of flexible bodies in tree topology - A computer oriented approach" Proceedings of the AIAA Dynamics specialist conference, Palm Springs, California May pp.327-337.1984.

[5] Kane T.R, Levinson D.A. " Formulation of equations of motion for complex spacecraft ", J.Guidance and control, vol. 3, No. 2, March-April 1980.

[6] De Luca A. Siciliano B. Joint based control of a nonlinear model of a flexible arm. American control conf. Atlanta, 1988.

[7] Hirschorn R. M. : Invertibility of multivariable nonlinear control systems, IEEE transactions on automatic control, vol. AC-24, pp 855-865 , December 1979.

[8] Desscusse J. Dion J.M. : On the structure at infinity of linear square decoupled system, IEEE transaction on Autom. control, vol. Ac-27, No. 4, pp. 971-974, August 1982.

[9] Schmitz J.E. : "Experiments on the end-point position control of a very flexible one-link manipulator" , these de P.H.D. de l'universite de Stanford, Guidance and control laboratory, Department of Aeronautics and astronautics. 1985.

[10] Spong M. W. : Modelling and control of elastic joint robots , transaction of the A.S.M.E.J. of dynamic systems, measurement and control, vol 109, pp 310-319 december 1987.

[11] De Luca A. Dynamic control of robots with joint elasticity, 5th. IEEE Int. Conf. Robotics and Automation, Philadelphia, 1988.

Inverse Kinematic Solution and Mixed Control Law for Redundant Robot in the Cartesian Space

A. AIT MOHAMED AND C. CHEVALLEREAU

Laboratoire d'Automatique de Nantes, U.A. CNRS 823
Ecole Centrale de Nantes, Universite de Nantes,
1 rue de la Noe, 44072 Nantes Cedex 03, France

Abstract - This paper yields a complete method to solve the inverse kinematic model for a redundant robot. An algorithm to the inverse kinematic problem is presented, based on criteria optimization, three types of criteria are studied. This algorithm gives a joint profile for the manipulator when it tracks a given path in the Cartesian space. The proposed method is analytic. In order to control on-line the robot by using the obtained joint profile, we suggest a new approach based on a mixed control law with a correction in the joint space and a correction in the Cartesian space. Simulation results using an RRR planar robot are presented.

1 Introduction

A lot of research have been carried out in the area of redundant robots, since these robots offer several advantages in dexterous motion tasks. Redundant manipulators have more degrees of freedom than necessary to locate the end-effector at the desired situation in the workspace. This increases the reachable workspace and provide more flexibility to accomplish complex tasks. Redundant manipulators can be used to optimize a performance index as a secondary criterion besides carring out the primarily desired task in the workspace. Two approaches can be applied to optimize the specified criterion. The first one is a global optimization using an integral cost function along the whole trajectory. The second approch uses a local optimization to determine an instantaneous optimal solution to the considered cost criterion. This method requires far less online calculation.
Many authors have used the extra degrees of freedom of the redundant robots, to optimize additional criteria when the given path in the workspace is tracked. Such criteria are the performance index that allows to satisfy : obstacles avoidance [1], keeping the joint coordinates within their joint limits [2,3], avoiding singularities [4,5], improving dexterity [6,7], optimizing joint torques [4], minimizing energy, minimizing the joint efforts....
The relation between the Cartesian space and the joint space, and then the optimization, can be studied at position, velocity or acceleration level. Like Wampler [8], we have chosen to use the position level. In our case the model can be used in all the workspace. This is well suited for criteria which depend on the joint position of the robot. In effect, the drift with periodic task in the Cartesian space is dramatically reduced. A complete analytic algorithm to get the joint position corresponding to one Cartesain position and minimizing a criterion is given in section 2. In section 3 the way to use the desired joint trajectory in a mixed control law is presented. This control law uses two feedbacks, one in the Cartesian space and one in the joint space. It gives very satisfactory results because the tracking of the Cartesian trajectory appears as a primary task and the optimization of the criterion as a secondary task with dynamic which can be chosed independently. Some perturbations on the Cartesian tracking, due to high joint velocity involved by the optimization, can be avoided. In section 4, simulation results are presented for a RRR planar robot avoiding the joint limits with a mixed control law at the joint velocity level.

A. J. Lenarčič and B. B. Ravani (eds.), Advances in Robot Kinematics and Computationed Geometry, 249–258.

Remark 2: An analytical method insures to use the same computation time for all positions of the robot, so the choice of a sampling period for on-line implementation is easier than with a numerical method.

2-2 Case of a quadratic criterion when W is a positive definite matrix

The criterion is quadratic and defined by Eq.(5) where **W** is positive definite. Such a criterion to be minimized can be the joint avoidance objective function where $\tilde{q}=(q_{max}+q_{min})/2$, $\tilde{q}$ is the mean value, q_{max} and q_{min} are the maximun and minimum joint limits. The matrix **W** is such that $w_{ii}=1/(q_{max}-q_{min})^2$ for i=1,..,n and $w_{ij}=0$ if $i\neq j$. The redundant inverse kinematic problem can be formulated as an optimization problem with constraints as follows:

$$\text{Minimize } \phi(q_L)=(q_L-\tilde{q})^T W(q_L-\tilde{q}) \tag{6}$$

$$\text{Subject to } J(q_c)\,(q_L-q_c)=0 \tag{7}$$

The solution q_L satisfaying Eqs.(6,7) is given by

$$q_L=q_c+[W^{-1}\,J^T\,(J\,W^{-1}\,J^T)^{-1}\,J-I_{nxn}]\,(q_c-\tilde{q}) \tag{8}$$

2-3 Case of a quadratic criterion when W is positive semi-definite

The quadratic criterion to be optimized is defined by (5) where **W** is a positive semi-definite matrix. For a planar robot with three revolute joints and three arms (RRR), in order to avoid singularities, Hsu [10] suggests to minimize $\phi(q)=(q_3-q_2)^2$, this criterion can be written as follows (**W** is positive semi-definite):

$$\phi(q) = q^T\,W\,q \quad \text{with} \quad W=\begin{bmatrix} 0 & 0 & 0 \\ 0 & 1 & -1 \\ 0 & -1 & 1 \end{bmatrix} \tag{9}$$

The variation Δq_k cannot be calculated by Eq.(8) since **W** is not invertible. Another way to determine the joint variation is to partition the jacobian matrix **J** into two submatrices J_p and J_s, where J_p is an (m,m) square non singular matrix and J_s is an (m,n-m) matrix. The jacobian matrix can be noted as $J = [\,J_p \quad J_s\,]$. The variation on position Δq can be partitioned into: primary variation Δq_p and secondary variation Δq_s, where Δq_p is an (m,1) vector, Δq_s is an (n-m,1) vector.

$$\Delta q = [\Delta q_p \quad \Delta q_s] \tag{10}$$

From the linearized Eq.(4), we deduce the variation Δq as a function of Δq_s by:

$$\Delta q= -\,E\Delta q_s \quad \text{where} \quad E=[(J_p^{-1}\,J_s)^T\,-I_{mxm}]^T \tag{11}$$

The criterion $\phi(q_L)$ is minimized at instant t_k if $\partial\phi(q_L)/\partial\Delta q_s=0$, the joint configuration which minimizes the objective function is given by

$$q_L = q_c - E\,(E^T\,W\,E)^{-1}\,E^T\,W\,(q_c - \tilde{q}) \tag{12}$$

Eq.(12) is identical to the solution given by (8). This solution is available for all weighted matrices **W** (positive definite or positive semi-definite) but we must choose at each sampling period the primary submatrix $J_p(q_c)$ which has the largest determinant in absolute value in order to avoid algorithmic singularities [11]. The algorithm proposed in section 2.1 above is applied to solve the inverse kinematic model, only step 3 is modified since before calculing Δq we must choose the candidate primary submatrix $J_p(q_c)$.

2-4 Case of a non quadratic criterion

An exemple of non quadratic criterion is:

$$\phi(q) = - \sqrt{\det(JJ^T)} \tag{13}$$

The criterion $\phi(q)$ is known as the manipulability measure [7]. Minimization of this non quadratic criterion permits to avoid kinematic singularities.

In order to obtain an analytical solution which minimizes the criterion $\phi(q)$,while keeping the desired path in the Cartesian space, we choose the variation Δq which minimizes the following criterion [11,12]:

$$C(q_L)=(\Delta q-\alpha \nabla\phi(q_c))^T(\Delta q-\alpha \nabla\phi(q_c)) \tag{14}$$

which is a quadratic criterion with an identity weighted matrix W and $\nabla\phi(q_c)$ is the gradient of $\phi(q)$ at joint position q_c. The solution optimizing the defined criterion $C(q_L)$ subject to Eq.(7) is given by:

$$q_L=q_c+ \alpha [I_{nxn} - J^T (J J^T)^{-1} J] \nabla\phi(q_c) \tag{15}$$

The five steps of the proposed algorithm are applied with Eq.(15) in step (3).

In the three cases described above, from a desired trajectory in the Cartesian space and a criterion on joint position $\phi(q)$, we obtain a desired trajectory in the joint space by solving the inverse kinematic model of the redundant manipulator.

3 Mixed Control Law at the Joint Velocity Level

The main task of the robot is to follow a given Cartesian trajectory. By solving the inverse kinematic problem (section 2), we have determined a proper joint profile q_d which enables the robot to describe the given Cartesian trajectory. The following section addresses the problem of using the joint profile in order to control on-line the robot. We have studied two kinds of control laws depending on the type of correction we used.

3-1 Joint velocity control

The first law is a classical control law at joint velocity level based solely on a correction in the joint space, so the velocity $\dot{q}$ of the manipulator is calculated by

$$\dot{q} = \alpha \dot{q}_d + K_n (q_d - q) \tag{16}$$

where K_n is a constant matrix, $\dot{q}_d$ is the derivative of the joint profile q_d and α is a scalar such that α=1 when $\dot{q}_d$ is computed and α=0 otherwise. Even if the tracking error $e_q=q-q_d$ is small the tracking errors $e_x=x-x_d$ in the Cartesian space can be large. The result is worst when α=0 since e_q is greater that with α=1.

3-2 Cartesian control at joint velocity level

We focus our interest on the control law defined in the Cartesian space since the task to be accomplished is specified in this space. Therefore it is necessary to define the input control $\dot{q}$ in the joint space as function of the Cartesian velocity $\dot{X}$. By differentiating Eq.(1), we obtain a linear relation between joint velocity $\dot{q}$ and Cartesian velocity $\dot{X}$.

$$\dot{X} = J\,\dot{q} \tag{17}$$

In case of a redundant robot, the rank of the jacobian matrix is at most equal to m (m<n), the general solution $\dot{q}$ of Eq.(17) is:

$$\dot{q} = J^{+}\ \dot{X} + (I_{nxn} - J^{+}J)\ Z \tag{18}$$

where Z is an arbitrary (n,1) vector and J^{+} is the pseudo inverse of the nonsquare jacobian matrix. If we want to regulate the tracking error in the Cartesian space we take the following control law defined at the joint velocity level

$$\dot{q} = J^{+}\ [\dot{X}_d + K_p\ (X_d - X)\ ] + (I_{nxn} - J^{+}J)\ Z \tag{19}$$

where K_p is a constant positive definite feedback gain matrix, $X_d(t)$ and $\dot{X}_d(t)$ are the desired position and velocity to be tracked. When J has full rank:

$$\dot{e}_x + K_p\ e_x = 0 \tag{20}$$

From Eq.(20), e_x goes exponentially towards zero, therefore the desired trajectory X_d will be correctly tracked.

3-3 Mixed control law

To obtain small tracking errors e_x and e_q, we suggest a new approach based on a mixed correction, on the one hand in the Cartesian space and on the other hand in the joint space, i.e., the joint velocity $\dot{q}$ is calculated as follows:

$$\dot{q} = J^{+}[\ \dot{X}_d + K_p(X_d - X)] + (I - J^{+}J)\ [\ \alpha\dot{q}_d + K_n(q_d - q)\ ] \tag{21}$$

Where α is defined as in Eq.(16). The first term in (21) allows to follow the specified Cartesian path with small tracking errors e_x and the second term does not cause any end-effector velocity, it is introduced in order to track the joint profile q_d. The above control laws at joint velocity Eq.(21) can be represented by Fig.1.
To compute $\dot{q}_d$, we use a numerical derivation:

$$\dot{q}_d = [q_d(k) - q_d(k-1)]/\tau \tag{22}$$

Where τ is the sampling period. This numerical derivation use merely the previous and the current joint profile, so it needs less calculation. Other efficient methods can be used to compute $\dot{q}_d$. In the following section we compare the tracking errors e_x and $e_q = q - q_d$ using the classical control law given by Eq.(16) or the mixed control law.
The mixed control law can be compared to a classical control law in the Cartesian space [3]. The Cartesian errors are of course of the same order but the mixed control law is more convenient for criterion depending merely on the joint position (joint limits, obstacles avoidance,...) and have the advantage to give naturally a solution with no drift for cyclic task. Such criteria are difficult to optimize with a classical method at the acceleration level

[11]. If the criterion to optimize leads to a joint profile with some discontinuity, taking $\alpha=0$ in Eq.(21) is very convenient since the discontinuity is filtered and the dynamic in the joint space can be chosen different from the dynamic in the Cartesian space, this is not the case with the classical Cartesian control law where high value of the joint velocity in the null space will perturb the tracking of Cartesian trajectory.

4 Simulation Results

We apply the suggested methods to a planar robot with three revolute joints and three arms of same lengths l=1m.

4-1 Results of the inverse kinematic model

The desired path is a circle given by $X_d=1+\sin(\pi t+\varphi)$ and $Y_d=1+\cos(\pi t+\varphi)$ where φ is a constant depending on the initial configuration.
For a joint availability objective criterion $\phi(q)$ defined in the section 2-1, the resolution of the inverse kinematic by using Eq.(8) and the proposed algorithm give good results. The joint position q is at vinicity of its mean value $\tilde{q}$ thus far from its joint limits (fig.2a), the criterion $\phi(q)$ is correctly optimized (fig. 2b). The joint limits are: for joint 1: $[-\pi/2;\pi]$, for joint 2: $[-\pi/2;\pi/2]$, for joint 3: $[0;2\pi]$. This example will be used to test the classical and mixed control law.

4-2 Mixed control law at the joint level

Using the classical control law Eq.(16) with $\alpha=0$ to follow the joint profile q_d leads to large tracking errors e_q fig.(3a) in the joint space which implies important tracking errors e_x in the Cartesian space fig.(3b). For the mixed control law Eq.(21) with $\alpha=0$, we obtained better result in the Cartesian space (see Fig.4b) for the same order of error in the joint space (see Fig.4a). Even with $\alpha=1$ the classical control law Eq.(16) does not allow to obtain the same performance in the Cartesian space as with the mixed control law and $\alpha=0$ (see fig.5a and Fig.5b). Best results are obtained by the mixed control law Eq.(21) with $\alpha=1$ for the optimization proposed since the joint position q_d and velocity $\dot{q}_d$ are known (see Fig.6a). In the Cartesian space the performances are identical (see Fig.6b). When the joint profile is discontinuous, the mixed control law with $\alpha=0$ gives better results than with $\alpha=1$ since velocities $\dot{q}_d$ are large and become infeasibile. In this case the results are much better than with a classical method even when $\alpha=0$ or $\alpha=1$. Using a classical control law, if $\alpha=0$ the Cartesian errors are high, if $\alpha=1$ the joint velocities are unaceptable.

5 Conclusion

For redundant manipulators, an efficient algorithm to solve the inverse kinematic problem has been presented. When the end-effector tracks the Cartesian path, the robot's position in the joint space is obtained by criteria optimization. The considered criteria depend only on joint position. In order to control online the manipulator, by exploiting the joint profile solution, we have proposed a mixed control law which involves small tracking errors in the joint and Cartesian spaces. When the Cartesian path $X(t)$ is cyclic, we can determine easily a cyclic joint profile $q_d(t)$, so by applying the mixed control law we obtain a response in the joint space with no drift. Mixed control can be also used at the joint acceleration level. The efficiency of these methods have been demonstrated with simulation results.

References

[1] Maciejewski, A.A., C.A. Klein (1985). Obstacle avoidance for kinematically redundant manipulators in dynamically varying environments. Int.J Rob. Research.Vol.4,No.3,109-117.
[2] Klein, C.A., C. Huang (1983). Review of pseudo inverse control for use with kinematically redundant manipulators. IEEE Trans. on Syst., Man, and Cybernetics. Vol. SMC-9, 245-250.
[3] Liégeois, A. (1977). Automatique supervisory control of the configuration and behavior of multi-body mechanisms. IEEE Trans. on Syst., Man, and Cybemitics. Vol. SMC-9, 245-250.
[4] Baillieul, J.,J. Hollerbach, and R. Brockett (1984). Programming and control of kinematically redundant manipulators. Proc. of 23rd Conf. on Decision and Control., 768-774.
[5] Baillieul, J. (1985). Kinematic programming alternatives for redundant manipulators. Proc. IEEE Conf. on Robotics and Automation, 722-728.
[6] Chang, P.H. (1986). A closed form solution for the control of manipulators with kinematic redundancy. Proc. IEEE Conf. on Robotics and Automation, 9-14.
[7] Yoshikawa, T., Analysis and control of robot manipulators with redundancy. (1983) Robotics research: The First Inter. Symp. MIT press, pp. 735-748.
[8] Wampler Ch W. (1987). Inverse kinematic function for redundant manipulators, Proc. IEEE Conf on Robotics and Automation, Raleigh, pp 610-617.
[9] Borrel, P., (1986). A study of multiple manipulator inverse kinematic solutions with applications to trajectory planning and workspaces determination. IEEE Conf. on Rob. and Aut., pp 1181-1185.
[10] Hsu, P., J. Hauser, S. Sastry (1988). Dynamic control of redundant manipulator. Proc. IEEE Conf on Robotics and Automation, 183-187.
[11] Chevallereau,C., A.Ait Mohamed (1992). Dynamic control of redundant robots in the cartesian space optimization by gradient method. 23rd Int. Sym on Indusriel Robots, 6th-9th October 1992, Brcelona (Spain), 81-86
[12] Ait Mohamed, A., C. Chevallereau (1993). Resolution of robot redundancy in the cartesian space by criteria optimization. Proc.IEEE Conf. on Robotics and Automation, Vol. 3, 646-651.

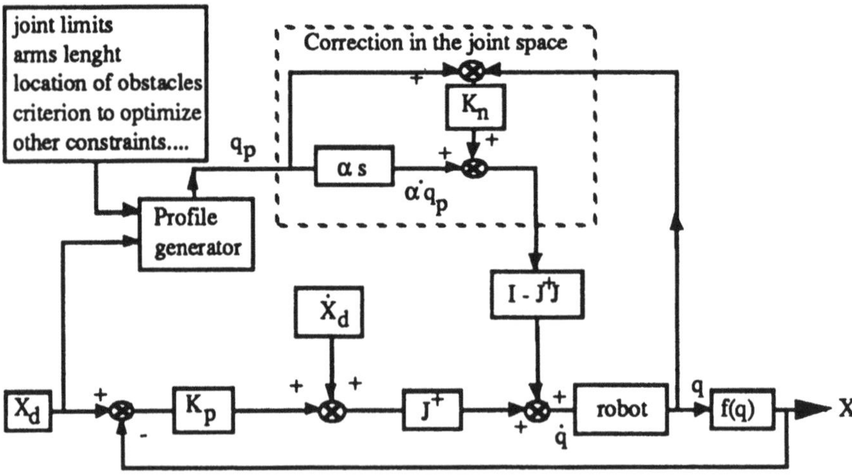

Fig.1: Mixed Control Law at Joint Velocity Level

Fig 2a: Joint position q (rad)

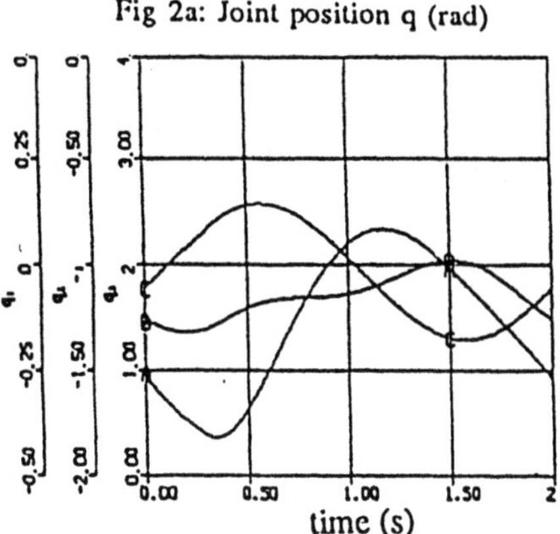

Fig 2b: Joint availibality function $\phi(q)$

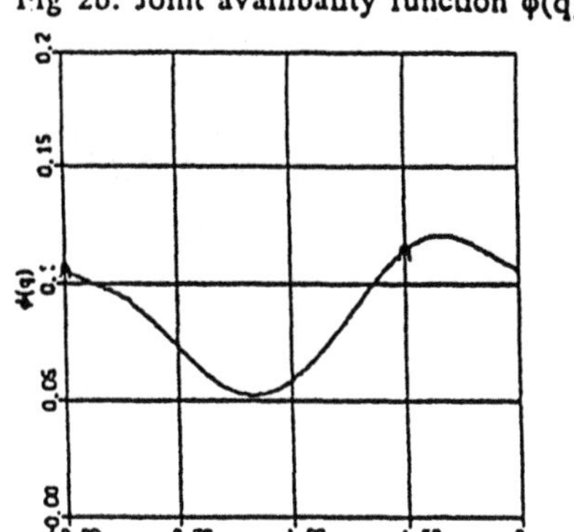

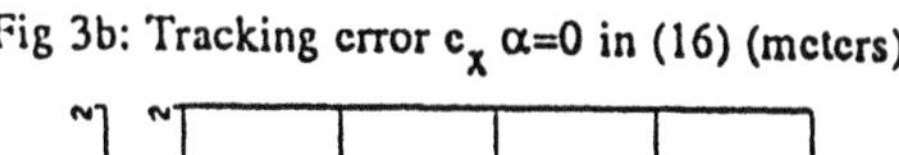

Fig 3a: Tracking error e_q $\alpha=0$ in (16) (rad)

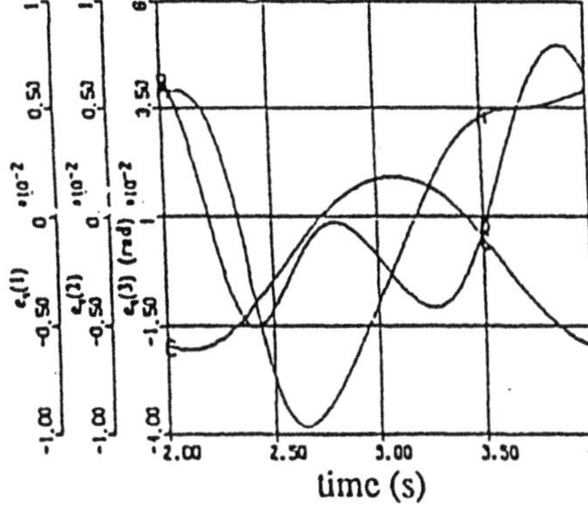

Fig 3b: Tracking error e_x $\alpha=0$ in (16) (meters)

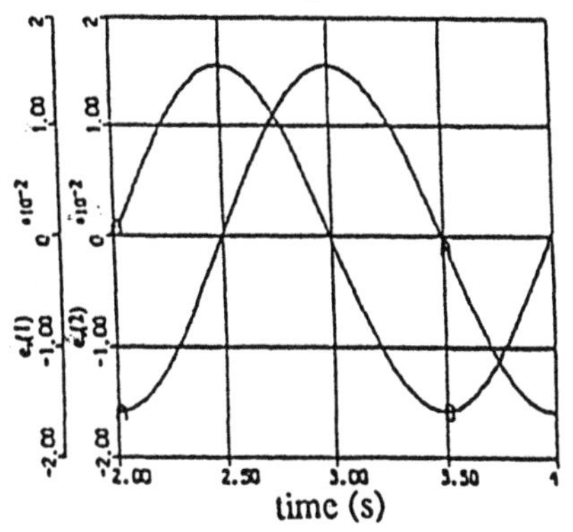

Fig 4a: Tracking error e_q $\alpha=0$ in (21) (rad)

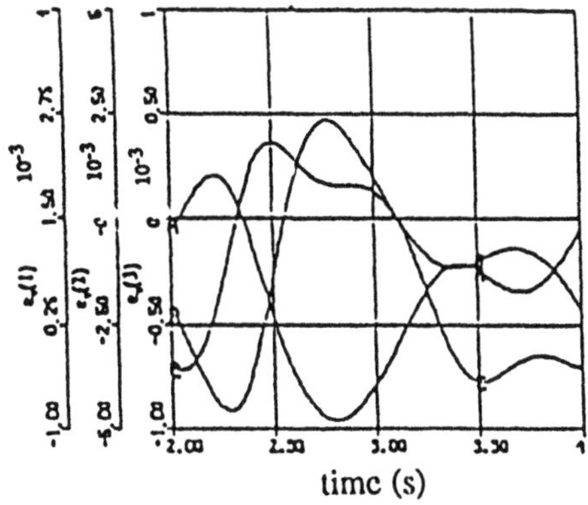

Fig 4b: Tracking error e_x $\alpha=0$ in (21) (meters)

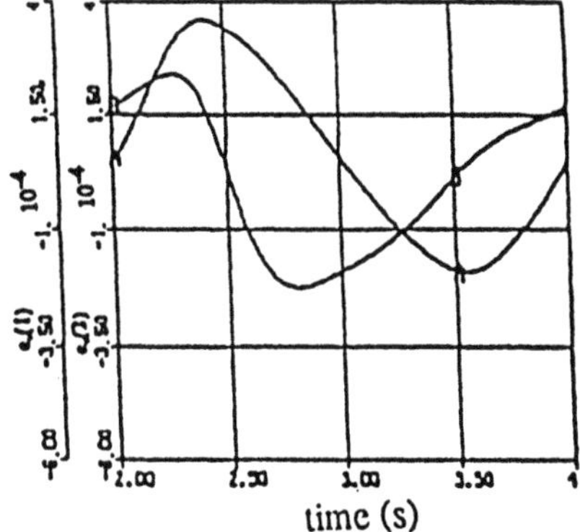

Fig 5a: Tracking error c_q α=1 in (16) (rad)

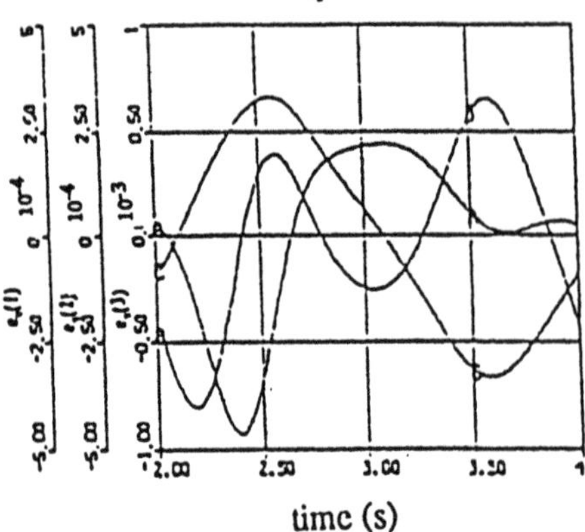

Fig 5b: Tracking error c_x α=1 in (16) (meters)

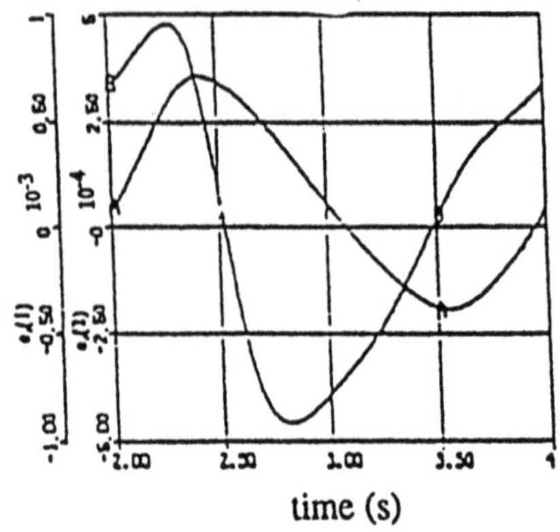

Fig 6a: Tracking error c_q α=1 in (21) (rad)

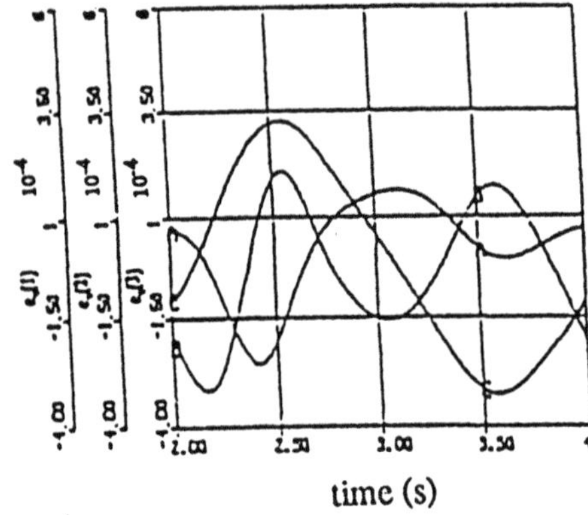

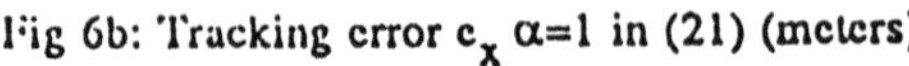
Fig 6b: Tracking error c_x α=1 in (21) (meters)

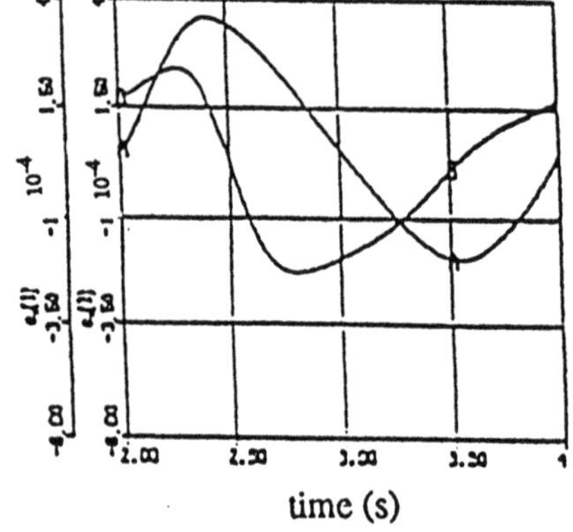

7. Force and Elasticity Analysis

An Inverse Force Analysis of a Spatial Three–Spring System

P. Dietmaier

Institut für Mechanik
Technische Universität Graz
Graz, Austria

e-mail: dietmaie@fmechds01.tu-graz.ac.at

Abstract — In this paper a closed–form solution for a spatial three–spring system under a single load has been deduced. One end of each spring was attached to a fixed pivot in space, while the other three ends are joined in a common pivot. The characteristics of the springs are assumed to be linearly elastic. The task was to compute all possible equilibrium configurations of the system when a given force is applied to the common pivot. First the easier symmetric and unloaded case was studied. For a free spring length greater than a specific value a maximum of 22 real equilibrium configurations was found. Then the general, loaded case was solved, revealing again up to 22 real equilibrium configurations. Finally, the stability criteria for this three–spring system were established and the analytical results were illustrated by a numerical example.

I. Introduction

The subject of our investigation is the spatial three–spring system exhibited in Fig. (1). There are three compliant coupling elements which can be viewed as springs or actuators connecting three fixed pivots attached to a common base with a fourth, common moving pivot. The springs or actuators are assumed to be linearly elastic. A pay load is applied to the moving pivot which determines the equilibrium position(s) of the system and the forces acting in the compliant coupling elements.

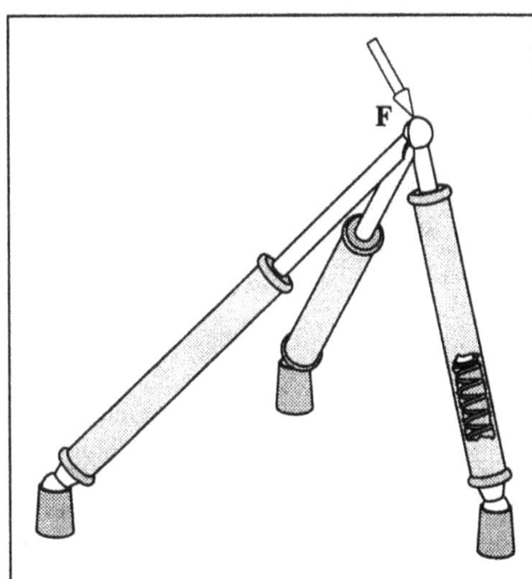

Fig. 1: The spatial three–spring system

This and similar compliant mechanical structures are widely used in all kinds of mounting or support structures as well as for parallel mechanisms or robot manipulators. While structural mechanics usually deals with more complex systems mechanism theory is more interested in systems with just a few parts. However, simple structures like the one we are investigating, have occasionally been used in structural mechanics to test numerical procedures concerning their ability to follow the path of the moving pivot. One such example is the system used by Bergan[1] which represents a similar three–spring system. But these

A. J. Lenarčič and B. B. Ravani (eds.), Advances in Robot Kinematics and Computationed Geometry, 261–270.

investigations usually focus primarily on how good a certain algorithm performs and not on how many equilibrium positions or paths exist or on how to obtain these results analytically. In the field of robotics, force control of compliant parallel robots is of a certain interest although the problem of the inverse force analysis does not seem to have been investigated for most of those systems. Recently, however, Pigosky and Duffy[2] solved the inverse force analysis problem of a planar two–spring system. A meeting with one of the authors became the starting point for this paper.

Before we start our analysis we want to clarify the meaning of a negative spring length which will appear in our calculations. If a spring is used as a representation of a solid body, negative spring lengths are not usually taken into consideration because a negative spring length would relate to a negative volume of this body, which is not realistic. However, if we think about a real spring or an actuator it might be possible to push them through the pivot or at least to operate them on the other side of the pivot as if the spring length were negative. Therefore a solution with negative spring lengths is possible and has to be taken into account. On the other hand some of the configurations may not be attainable because of self–intersection and other mechanical limitations as it always is the case for a mechanical system.

II. The symmetric, unloaded system

First we will analyze this much simpler system, where all springs have the same properties (linear spring characteristic with the same free length l_0 and the same spring elasticity constant k). The fixed pivots form an equilateral triangle, and no force is applied to the common pivot. Isolating the common pivot and applying the required equilibrium condition yields $\mathbf{F}_1 + \mathbf{F}_2 + \mathbf{F}_3 = \mathbf{0}$ with $\mathbf{F}_i$ denoting the individual spring forces. From this equation it is obvious to see that equilibrium is either established if all three forces are zero, or if the three lines of action lie in the same plane, the plane determined by the fixed pivots. For the first case, where all $\mathbf{F}_i$ are zero and the length of the springs are equal to the free spring length l_0, we obtain two equilibrium positions as the intersection of the three spheres centered at the fixed pivots with a radius l_0. These two positions lie above and below the center of the triangle formed by the fixed pivots and are at a distance of $\sqrt{l_0^2 - 1/3a^2}$ from the plane of the triangle.

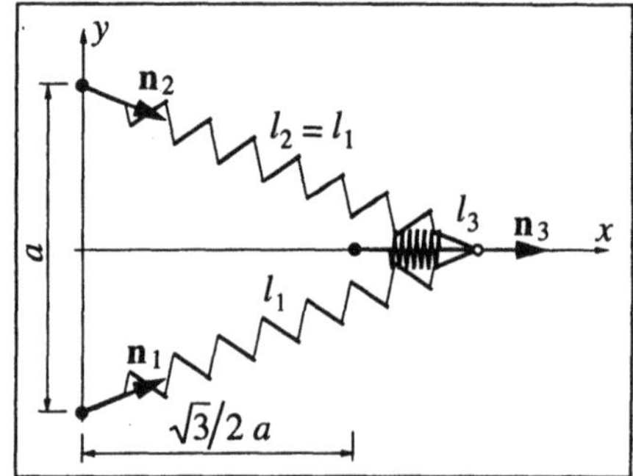

Fig. 2: Parameters of the symmetric and unloaded system configuration in the plane

To investigate the second case, where all three forces act in the plane of the fixed pivots, we will assume that two of the three spring lengths are equal and therefore the common pivot lies on one of the three lines of symmetry of the equilateral triangle (see Fig. 2). Although this assumption does not hold for a general nonlinear spring characteristic it can be shown that it is valid for this particular case. Furthermore, we will see that the number of equilibrium positions obtained by this approach is the same as the number for the general, loaded case, which justifies this procedure. From Fig. (2) we can read the conditions for all three springs connected at the common pivot as

$$\bar{l}_1 n_{1x} = \sqrt{3}/2 + \bar{l}_3 n_{3x} \tag{1}$$

and

$$\bar{l}_1 n_{1y} = 1/2 \tag{2}$$

with $\bar{l}_i = l_i/a$.
Equilibrium in the direction of the x–axis requires

$$-2k(\bar{l}_1 - \bar{l}_0)n_{1x} - k(\bar{l}_3 - \bar{l}_0)n_{3x} = 0\,, \tag{3}$$

and the coordinates of the unit vectors $\mathbf{n}_i$ pointing in the direction of the springs for positive spring length must comply with the restrictions

$$n_{1x}^2 + n_{1y}^2 = 1 \tag{4}$$

and

$$n_{3x}^2 = 1\,. \tag{5}$$

In order to solve this system of algebraic equations we can split the solution into two cases: $n_{3x} = 1$ and $n_{3x} = -1$. For the case $n_{3x} = 1$ we solve Eqs. (1–3) for $\bar{l}_3$, n_{1x}, and n_{1y} (a set of linear equations) and substitute the result for n_{1x} and n_{1y} into Eq. (4). After clearing the denominators we obtain for $\bar{l}_0 \neq 0$ a fourth order polynomial equation for $\bar{l}_1$, namely

$$9\bar{l}_1^4 - 12\bar{l}_0\bar{l}_1^3 + (3\bar{l}_0^2 - \sqrt{3}\,\bar{l}_0 - 3)\bar{l}_1^2 + 3\bar{l}_0\bar{l}_1 - \bar{l}_0^2 = 0 \tag{6}$$

and analogous for $n_{3x} = -1$

$$9\bar{l}_1^4 - 12\bar{l}_0\bar{l}_1^3 + (3\bar{l}_0^2 + \sqrt{3}\,\bar{l}_0 - 3)\bar{l}_1^2 + 3\bar{l}_0\bar{l}_1 - \bar{l}_0^2 = 0 \tag{7}$$

while for $\bar{l}_0 = 0$ both cases yield

$$3\bar{l}_1^2 - 1 = 0\,. \tag{8}$$

One of the solutions of Eq. (6) and (7) can be obtained quite easily. It is this the equilibrium position where the common pivot is located in the center of the triangle which is given by the solution $\bar{l}_1 = -1/\sqrt{3}$ for Eq. (6) and $\bar{l}_1 = 1/\sqrt{3}$ for Eq. (7). Dividing Eq. (6) and (7) by the respective solution yields two cubic equations, which read

$$9\bar{l}_1^3 - 3(\sqrt{3} + 4\bar{l}_0)\bar{l}_1^2 + 3\bar{l}_0(\sqrt{3} - \bar{l}_0)\bar{l}_1 - \sqrt{3}\,\bar{l}_0^2 = 0 \tag{9}$$

and

$$9\bar{l}_1^3 + 3(\sqrt{3} - 4\bar{l}_0)\bar{l}_1^2 - 3\bar{l}_0(\sqrt{3} - \bar{l}_0)\bar{l}_1 + \sqrt{3}\,\bar{l}_0^2 = 0 \tag{10}$$

respectively. It remains to be investigated how the number of real solutions of these cubic equations changes with respect to the parameter $\bar{l}_0$. Therefore we calculate the discriminant of Eq. (9) and (10), which is for a cubic equation $a_3 x^3 + a_2 x^2 + a_1 x + a_0 = 0$ determined by, $D = (q/2)^2 + (p/3)^3$, $p = (3a_1a_3 - a_2^2)/(3a_3^2)$ and $q = 2a_2^3/(27a_3^3) - a_1a_2/(3a_3^2) + a_0/a_3$ (see e.g. [3]). Applying this formulas to Eq. (9) and (10) we obtain the two discriminants

$$D_1 = \bar{l}_0^2(9 + 36\bar{l}_0^2 + 28\sqrt{3}\,\bar{l}_0^3 - 12\bar{l}_0^4) \tag{11}$$

and

$$D_2 = \bar{l}_0^2(9 + 36\bar{l}_0^2 - 28\sqrt{3}\,\bar{l}_0^3 - 12\bar{l}_0^4)\,. \tag{12}$$

Now, if $D_i \leq 0$ the cubic equation gives three real solutions while for $D_i > 0$ just one of the three solutions is real. Using this criteria for D_1 and D_2 we get three real solutions from Eq. (9) if either $\bar{l}_0 > 4.688581\ldots$ or $\bar{l}_0 < -0.835465\ldots$ and from Eq. (10) if either $\bar{l}_0 > 0.835465\ldots$ or $\bar{l}_0 < -4.688581\ldots$. (Note, that D_1 and D_2 are not valid for $\bar{l}_0 = 0$). Consequently the maximum of 6 real solutions is obtained for a value of $|\bar{l}_0| \geq 4.688581...$. Together with the 6 equilibrium positions on each of the other two lines of symmetry, the two solutions in the center of the triangle ($\bar{l}_1 = \pm 1/\sqrt{3}$), and the two out–of–plane solutions this yields a maximum of 22 real and different configurations.

Fig. (3) exhibits these 22 positions of the common pivot for $\bar{l}_0 = 5$. To emphasize the spatial impression the three fixed pivots (represented by small cones) and the three lines of symmetry together with the line connecting the two out of plane positions are shown as well. Positions denoted by black points represent stable positions while those marked by circles are not stable. Note also that the circle in the center of the triangle is half black half white denoting a stable and unstable configuration at this position.

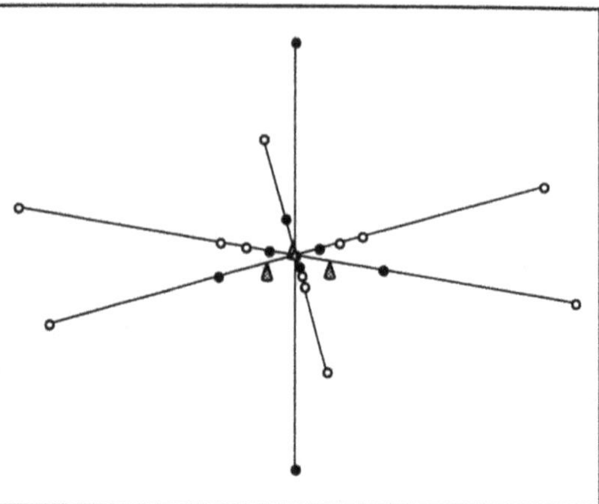

Fig. 3: Equilibrium positions of the symmetric, unloaded system for $\bar{l}_0 = 5$

Range of $\lvert\bar{l}_0\rvert$			Real Configurations
$1/\sqrt{3}$	$> \lvert\bar{l}_0\rvert \geq$	0	8
0.835465...		$1/\sqrt{3}$	10
4.688581...		0.835465...	16
∞		4.688581...	22

Tb. 1: Number of real equilibrium configurations with respect to $\bar{l}_0$

If $|\bar{l}_0|$ is less than 4.688581... the number of real configurations drops in three steps from 22 down to 8 according to Tb. 1. In the special case in which all free spring lengths are equal to zero, Eq. (8) gives $\bar{l}_1 = \pm 1/\sqrt{3}$ (and $\bar{l}_3 = \pm 1/\sqrt{3}$). Since this holds for the other two symmetric configurations as well, we obtain, by combining all possible combinations of the signs of the spring length, $2^3 = 8$ configurations with the common pivot at the center of the triangle. It may look as if all these 8 configurations were the same: However, in 7 of these configurations at least one spring length is negative, which means that the corresponding actuator is turned through 180°. Therefore the 8 configurations are actually different.

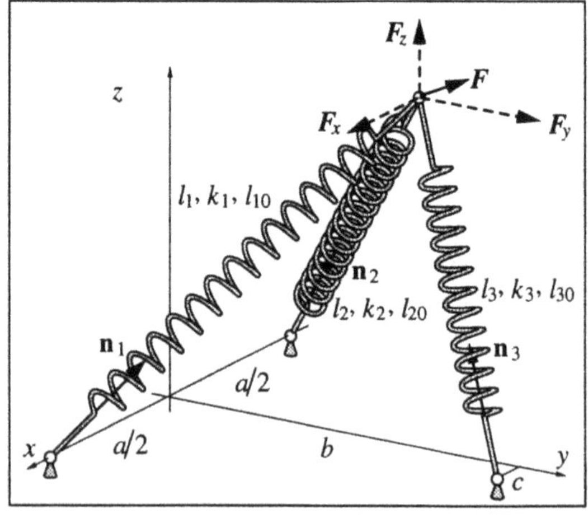

Fig. 4: Parameters of the general system configuration

III. The general, loaded system

In contrast to the relatively simple symmetric, unloaded case we now turn to the general and loaded case, which turns out to be not as easy to solve as we might expect. Fig. 4 illustrates the situation and all parameters relevant to computation. For the position analysis of

such a system it is very important that the system is described in a non–redundant way. Therefore we have chosen three direction vectors and the spring lengths, similar to the planar case, to identify a position unambiguously.
From Fig. 4 we can read the equilibrium conditions of the common pivot in vectorial form as

$$\mathbf{F} = \sum_{i=1}^{3} k_i (l_i - l_{i0}) \mathbf{n}_i \,. \tag{13}$$

In analogy to the symmetric case the k_i denote the spring elasticity constant, l_i the actual spring length, l_{i0} the free spring length and $\mathbf{n}_i$ the unit direction vectors of the springs. The geometric constraints, i.e. the fact that all three springs are joined at the common pivot, can be expressed by the two vector equations

$$a\mathbf{e}_x + l_1\mathbf{n}_1 = l_2\mathbf{n}_2 \tag{14}$$

and

$$(a/2 - c)\mathbf{e}_x + l_1\mathbf{n}_1 = b\mathbf{e}_y + l_3\mathbf{n}_3 \tag{15}$$

with $\mathbf{e}_x$ and $\mathbf{e}_y$ as the unit vectors in the x– and y–directions respectively. Finally we have to express the fact that each $\mathbf{n}_i$ is a unit vector:

$$\mathbf{n}_i \circ \mathbf{n}_i = 1, \quad \text{for} \quad i = 1,2,3\,. \tag{16}$$

Equations (13–16) represent a system of 12 algebraic equations for the 3 spring lengths l_i and the 9 coordinates of the direction vectors $\mathbf{n}_i$.This set of algebraic equations could basically be solved straight away by applying the Gröbner Bases Algorithm [4]. For our three–spring system we carried out such calculations for a few example data sets in modular arithmetic to obtain the number of positions to be expected. To determine the actual positions, arbitrary integer or floating point arithmetic would be required. However, it would take much longer and would not be as efficient as the analytic way proposed below.
In order to solve Eqs. (13–16) we first solve the three vector Eqs. (13–15) for the 9 coordinates of the direction vectors $\mathbf{n}_i$, which represent a system of 9 linear algebraic equations in these unknowns, where the remaining three unknown l_i's just show up in the coefficients. For the coordinates of the $\mathbf{n}_i$'s this yields fractions of multilinear polynomials in the l_i's. Notice, however, that the numerators of the fractions for the coordinates of the $\mathbf{n}_i$'s do not contain the corresponding spring length l_i since Eqs. (13–15) only contain products of l_i and the coordinates of the corresponding $\mathbf{n}_i$.
Substituting the 9 fractions for the coordinates of the unit vectors into the three scalar Eqs. (16) yields, after clearing the common denominators, three polynomial equations for the l_i's as linear combinations of the following power products:

$$\begin{aligned} f_1(&l_2^2, l_1^2 l_2^2, l_2 l_3, l_1^2 l_2 l_3, l_2^2 l_3, l_1 l_2^2 l_3, l_1^2 l_2^2 l_3, l_3^2,\\ &l_1^2 l_3^2, l_2 l_3^2, l_1 l_2 l_3^2, l_1^2 l_2 l_3^2, l_2^2 l_3^2, l_1 l_2^2 l_3^2, l_1^2 l_2^2 l_3^2) = 0\,, \end{aligned} \tag{17}$$

$$\begin{aligned} f_2(&l_1^2, l_1^2 l_2^2, l_1 l_3, l_1^2 l_3, l_1^2 l_2 l_3, l_1 l_2^2 l_3, l_1^2 l_2^2 l_3, l_3^2,\\ &l_1 l_3^2, l_1^2 l_3^2, l_1 l_2 l_3^2, l_1^2 l_2 l_3^2, l_2^2 l_3^2, l_1 l_2^2 l_3^2, l_1^2 l_2^2 l_3^2) = 0\,, \end{aligned} \tag{18}$$

and

$$\begin{aligned} f_3(&l_1^2, l_1 l_2, l_1^2 l_2, l_2^2, l_1 l_2^2, l_1^2 l_2^2, l_1^2 l_2 l_3, l_1 l_2^2 l_3,\\ &l_1^2 l_2^2 l_3, l_1^2 l_3^2, l_1 l_2 l_3^2, l_1^2 l_2 l_3^2, l_2^2 l_3^2, l_1 l_2^2 l_3^2, l_1^2 l_2^2 l_3^2) = 0\,. \end{aligned} \tag{19}$$

Eqs. (17–19) are homogeneous in the l_i's which means that a solution to these three equations is that all l_i's are simultaneously zero. It is therefore obvious that we are already faced with an extraneous root. In the next step we will eliminate l_3 from this set of three equations. The necessary and sufficient condition for the three Eqs. (17–19) to have the same solutions for l_3 is that the resultant of one of the equations and any linear combination of the other two has to vanish [5]. Choosing f_1 as the first equation and the linear combination $u f_2 + v f_3$ as the second, this condition reads

$$u^2 p_{12}(l_1,l_2) + v u p_{14}(l_1,l_2) + v^2 p_{16}(l_1,l_2) = 0 \tag{20}$$

with p_{12}, p_{14}, and p_{16} being multinomials of l_1 and l_2 with total degrees of 12, 14, and 16 respectively and l_1, l_2 appearing in monomials of degree 8 at the most. Since this condition has to hold for all values of u and v, the p_i's have to vanish, which yields three equations in l_1 and l_2.On closer examination of the p_i's we find a common factor which can be split off from the three multinomials. This factor turns out to be

$$\frac{\mathbf{n}_3 \circ \mathbf{n}_3}{[\mathbf{n}_1 \circ (\mathbf{n}_2 \times \mathbf{n}_3)]^2} = \frac{1}{V^2} \tag{21}$$

which is positive and therefore does not contribute any solutions. The main reason for the appearance of this factor is due to the fact that the coordinates of the direction vectors do not contain the corresponding l_i as already mentioned above. Dividing the multinomials of Eq. (20) by this factor results in the multinomial equations

$$p_8 = 0,\ p_{10} = 0, \text{and } p_{12} = 0 \tag{22}$$

with l_1 and l_2 appearing at most in monomials of degree 6.
In order to eliminate the next unknown, l_2, from these three equations we could proceed in the same way as we did with Eqs. (17–19). But this would be rather inefficient since the solutions (22 as we will see) would have to be found as the common roots of 7 polynomial equations of degree 72. Therefore we choose a different approach and eliminate first of all again l_3 from Eqs. (17–19) by writing these equations in matrix form as

$$\underline{A}(l_1,l_2) \cdot (l_3{}^2, l_3, 1)^{\mathrm{T}} = \underline{0}_{3\times1}. \tag{23}$$

A necessary, but not sufficient, condition of Eq. (23) to be solvable for l_3 is that

$$\det(\underline{A}) = 0 \tag{24}$$

which yields, again after splitting off the factor of Eq. (21), a multinomial of a total degree of 6 and a maximum degree of 4 in l_1 and l_2 respectively. This equation of degree 4 in l_2 together with the polynomials p_8 and p_{10} from (22) of degree 6 in l_2 are now used to set up a necessary condition for l_1 so that this system becomes solvable for l_2. Therefore we multiply Eq. (24) by l_2, $l_2{}^2$, and $l_2{}^3$ and p_8 as well as p_{10} by l_2. These additional 5 equations together with the three original equations can be written with the help of an 8×8 matrix $\underline{B}$ as a matrix equation

$$\underline{B}(l_1) \cdot (l_2{}^7, l_2{}^6, l_2{}^5, l_2{}^4, l_2{}^3, l_2{}^2, l_2, 1)^{\mathrm{T}} = \underline{0}_{8\times1}\,. \tag{25}$$

Notice that the matrix elements of those rows of $\underline{B}$ which were formed by using Eq. (24) are polynomials of degree 4 while the other matrix elements are polynomials of degree 6. Again, for Eq. (25) to be solvable a necessary condition is that $\det(\underline{B}) = 0$. According to the special structure of matrix $\underline{B}$ this condition yields a polynomial equation of degree 36 in l_1 which can be written as a product in the following way:

$$l_1{}^4 (a + l_1)^2 (a - l_1)^2 (a/2 + l_1{}^2)\, p_4(l_1)\, p_{22}(l_1) = 0\,. \tag{26}$$

The first three factors of Eq. (26) yield solutions with at least one of the l_i's equal to zero, independently of some of the parameters, especially the load. These are no valid solutions to the initial equations but were introduced by the first elimination step. In a similar way the fourth factor entered Eq. (26) after the last step because $\det(\underline{B}) = 0$ is not a sufficient condition. The factor $p_4(l_1)$ a polynomial of degree 4, again produces unwanted roots which turn out to be common solutions to p_8 and p_{10} of (22) but not of Eq. (24). So all the solutions to our problem are contained in the last factor $p_{22}(l_1)$, which, in agreement with the symmetric and unloaded case, may yield up to 22 real and different configurations of our three–spring system.

IV. Stability criteria

For the sake of completeness we shall briefly discuss the stability conditions of the specific three–spring system with respect to small dislocations of the common pivot. Therefore we linearize the three components of the equilibrium Eq. (13) and the three Eqs. (16) with respect to the coordinates of the direction vectors $\mathbf{n}_i$, the spring length l_i, and the coordinates of the applied force $\mathbf{F}$. Furthermore we can write the position vector pointing to the common pivot in three different ways as

$$\mathbf{x} = a/2\,\mathbf{e}_x + l_1\mathbf{n}_1,\ \mathbf{x} = -a/2\ \mathbf{e}_x + l_2\mathbf{n}_2,\ \mathbf{x} = c\,\mathbf{e}_x + b\,\mathbf{e}_y + l_3\mathbf{n}_3 \tag{27}$$

which, by linearizing the coordinate equations, gives 9 additional equations:

$$\begin{aligned}\Delta x &= \Delta l_i\, n_{ix} + l_i\, \Delta n_{ix},\\ \Delta y &= \Delta l_i\, n_{iy} + l_i\, \Delta n_{iy},\\ \Delta z &= \Delta l_i\, n_{iz} + l_i\, \Delta n_{iz} \quad \text{with } (i = 1,2,3).\end{aligned} \tag{28}$$

Together with the linearized Eqs. (13) and (16) we now have 15 linear equations for the three increments of the components of the applied force, the three increments of the spring lengths and the nine increments of the components of the three direction vectors. Solving these equations, we can express the increments of the force by homogeneous linear functions of the increments of the dislocation in matrix form by

$$\underline{\Delta F} = \underline{m}\ \underline{\Delta x} \tag{29}$$

with the elements of the symmetric 3×3 stiffness matrix being

$$m_{ij} = \left[\sum_{r=1}^{3} k_r \left(1 - \frac{l_{r0}}{l_r}\right)\right] \delta_{ij} + \sum_{r=1}^{3}\left(k_r \frac{l_{r0}}{l_r} n_{ri}\, n_{rj}\right) \tag{30}$$

and

$$\delta_{ij} = \begin{cases} 1, \text{ for } i = j \\ 0, \text{ for } i \neq j \end{cases},\ n_{i1} = n_{ix},\ n_{i2} = n_{iy},\ n_{i3} = n_{iz}. \tag{31}$$

For a position to be called stable we demand that the additional force required to dislocate the common pivot has a positive component in the direction of the dislocation. This condition can be expressed by the matrix inequality

$$\underline{\Delta x}^{\mathrm{T}}\ \underline{\Delta F} = \underline{\Delta x}^{\mathrm{T}}\ \underline{m}\ \underline{\Delta x} > 0\,. \tag{32}$$

Hence the matrix $\underline{m}$ has to be positive definite which results in the three conditions (see e.g. [6])

$$m_{11} > 0,\ \det\begin{bmatrix} m_{11} & m_{12} \\ m_{21} & m_{22}\end{bmatrix} > 0,\ \text{and } \det(\underline{m}) > 0 \tag{33}$$

for a configuration to be called stable.

V. A numerical example

For the general case a final numerical example shows that there may exist up to 22 real configurations of our three–spring system. Although we tried to find a parameter set as general as possible it turned out that only parameter sets very close to the symmetric case yield the desired maximum number of configurations. Therefore we have chosen the system parameters according to Tb. 2 with λ being a scaling factor representing different pay loads. For a chosen value of λ we first solve $p_{22}(l_1)=0$ from Eq. (26) for l_1, we then substitute each of the solutions l_1 into Eq. (25) and solve these sets of linear equations for l_2. Each of the pairs (l_1, l_2) are then substituted into Eq. (23) the solution of which yields the corresponding values for l_3. Substituting each triplet (l_1, l_2, l_3) into Eqs. (13–15) and solving them for the 9 coordinates of the direction vectors yields the remaining parameters which unambiguously determine the configuration.
This procedure allows us to compute and plot the l_i's with respect to λ, which is shown in Fig. 5 (note the nonlinear scaling of λ). From this dia-

$a=1$	$b=0.9$	$c=0.02$
$l_{10}=5$	$l_{20}=4.95$	$l_{30}=5$
$k_1=10$	$k_2=10$	$k_3=10.2$
$F_x=0.02\lambda$	$F_y=-0.04\lambda$	$F_z=-0.03\lambda$

Tb. 2: Parameters of the numerical example

Path	Pos.	l_1	l_2	l_3
1	1	4.97745	4.94736	5.02193
	2	4.71562	4.66209	5.56161
	3	5.25609	4.37730	5.31330
	4	5.51273	4.55115	4.88342
	5	4.97962	4.95127	5.02189
2	6	5.23443	5.35200	4.37493
	7	4.49023	5.48636	4.97506
	8	4.41856	5.40083	5.13255
	9	0.56503	0.59798	0.60327
	10	1.84668	1.86427	-0.88665
3	11	0.99550	0.95931	-1.74016
	12	0.91929	0.87823	-1.64719
	13	1.03941	-0.93129	-0.13834
4	14	-0.56483	-0.50754	1.09449
	15	-1.46784	0.91218	0.68125
	16	-1.89501	1.19818	1.10788
5	17	1.27547	-1.99318	1.16748
	18	0.90483	-1.39691	0.54633
6	19	1.92801	-0.99819	1.97039
7	20	-0.57009	-0.59528	-0.60092
8	21	-0.90308	1.05750	-0.14744
9	22	-0.96810	1.89374	1.91895

Tb. 3: The 22 configurations for

Fig. 5: Spring lengths as functions of the load factor

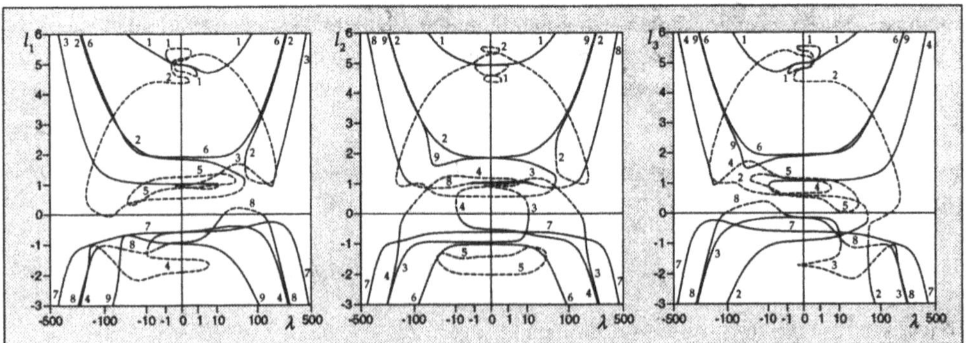

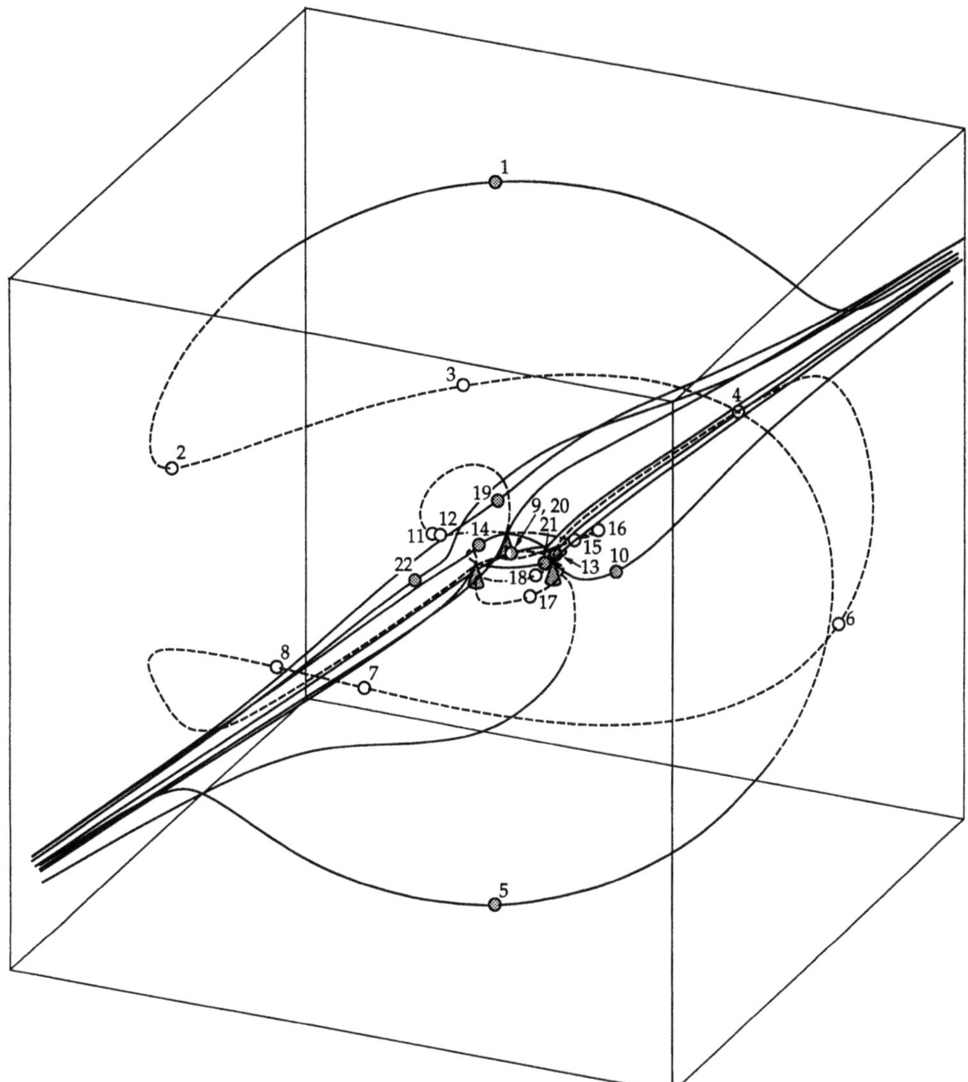

Fig. 6: Paths of the common pivot and equilibrium positions for $\lambda = 0.1$

gram we can read that there are 9 paths for our three–spring system. Eight of them start and end at infinity (resulting from the $2^3 = 8$ different combinations of the signs of the spring lengths) while path 5 forms a closed loop. Dashed parts of the paths denote unstable configurations according to the conditions derived above. In a small interval around $\lambda = 0$ these 9 paths yield 22 real and different configurations. For $\lambda = 0.1$. the spring lengths characterizing the configurations are given in Tb. 3 while the coordinates of the direction vectors are suppressed for the sake of brevity. To complete this example, Fig. 6 shows the spatial paths traced by the common pivot when the pay load changes with respect to λ as well as the 22 positions for $\lambda = 0.1$. A comparison with Fig. 3 shows that some of the positions change quite a lot although the parameter set is very close to the symmetric case. This is also a hint that with greater changes in the parameters some of the positions no longer stay real.

VI. Conclusions

For a particular three–spring system an inverse force analysis to determine the location of the moving pivot and the forces in the actuators, was performed. First the symmetric and unloaded system was studied which revealed the relation between the free spring length and the number of real equilibrium positions of such a system. A maximum of 22 real configuration was found. The analysis of the general, loaded case disclosed again the existence of up to 22 real configurations. The stability conditions of such a system were established and a numerical example clarified the analytic results. The most significant result of this work lies in the analytical solution to the problem, which also allows a very efficient numerical solution. Furthermore the exact number of configurations is obtained in this way and the motion of the moving pivot with respect to a changing force can be computed. The appearance of negative spring lengths was discussed and shown to be relevant.

Acknowledgment

I wish to express my gratitude to my colleague Dipl.–Ing. S. Lösch for carrying out various numerical examples with his modular Gröbner Bases Algorithm.

VII. References

[1] P.G. Bergan, "Solution by Iteration in Displacements and Load Spaces", *Nonlinear Finite Element Analysis in Structural Mechanics*, p. 551–571, (Wunderlich, Stein, Bathe, Eds.), Proceedings of the Europe-US Workshop Ruhr-Universität Bochum, Germany, July 28-31 1981, Springer, (1981).

[2] T. Pigoski and J. Duffy, "An Inverse Force Analysis of a Planar Two–Spring System", presented at the first Austrian IFToMM Symposium, Seggauberg, Austria, 4–9 July 1993, (no proceedings).

[3] Bronstein–Semendjajew, *Taschenbuch der Mathematik*, p. 117, Teubner Verlagsgesellschaft, Leipzig, (1960).

[4] B. Buchberger, "Gröbner Bases: An Algorithmic Method in Polynomial Ideal Theory", (N. K. Bose, Ed.), *Multidimensional System Theory*, p. 184–232, D. Reidel, Dordrecht, Boston, Lancaster, (1985).

[5] O. Perron, *Algebra I (Die Grundlagen)*, pp. 266, W. de Gruyter&Co., Berlin, 1951.

[6] F. R. Gantmacher, *Matrizentheorie*, p. 315, Springer–Verlag, Berlin, Heidelberg, New York, Tokyo, (1986).

A two-arm robot can be stronger than two arms
Experiments on optimal force distribution in two-arm robot

Yann Bouffard-Vercelli*[1], Pierre Dauchez** and François Pierrot**

*Universität-GH Paderborn, FB 10, Automatisierungstechnik, Prof. Lückel
Pohlweg 55,
D-33098 Paderborn - Germany -

**LIRMM - Université Montpellier II / CNRS N° 9928
161, rue Ada
34392 Montpellier Cedex 5 - France -

Abstract - The work presented in this paper is mainly based on the concept of Force Manipulability of Two-arm Robots. We show that theoretically the maximum external force that a two-arm robot can provide depends on the internal force in the carried object. One consequence of such an assertion is that it is sometimes possible to exert an external force greater than the sum of forces produced by the two robots. The aim of this paper is to experimentally validate such an assertion.

I. Introduction

Nowadays it becomes common to consider two cooperative robots as a convenient way to perform complex tasks in order to shorten cycle-times or to improve reachability: the two arms are only coordinated in a "loose" manner [1]. We have been working for many years on the case where two arms are coordinated in a "tight" manner: in our approach the manipulators are controlled by a single computing system and become a two-arm robot. In the past years we used such a system in order to carry cumbersome rigid objects [2], to deform flexible objects [3] or to perform assembly "in space" [4]. One of the first industrial application of "tightly-coordinated" robots is now working in an aeronautics company [3]. However all these applications have at least one common point: the fact that we mainly used force sensing to improve accuracy or efficiency. But recently, Bouffard [5] and Tabarah [6] stressed another aspect: the fact that we can use two-arm robots in order to provide higher forces than the sum of forces produced by the two robots. The aim of this paper is to experimentally verify this idea that could at first-sight seem strange. However we first recall some of the basic theoretical ideas. Then some experiments are described regarding a single robot-arm in order to precisely evaluate its force capabilities. Finally experiments with two arms are conducted that prove the previous assertion.

[1] Yann Bouffard-Vercelli is currently a *graduierten* post doctorate fellow at HNI

A. J. Lenarčič and B. B. Ravani (eds.), Advances in Robot Kinematics and Computationed Geometry, 271–280.

II. Force manipulability of single arm robots

The *manipulability* of a robot is a way of measuring its manipulating ability. Following this concept, a force manipulability ellipsoid and a velocity manipulability ellipsoid can be defined for a single arm. In the rest of this paper, we will only be concerned with the force manipulability concept. With this assumption, let us consider a n-dof arm and a m-dimensional taskspace. The well-known static relationship between the task force vector **h** and the joint torque vector τ implies the use of the transpose of the robot jacobian matrix **J** as follows:

$$\tau = \mathbf{J}^t \mathbf{h} \tag{1}$$

Then, the unit sphere in the joint torque space (normalized torques):

$$\tau^t \tau = 1 \tag{2}$$

maps into the task force space ellipsoid (for a given robot configuration):

$$\mathbf{h}^t [\mathbf{J}\ \mathbf{J}^t]\ \mathbf{h} = 1 \tag{3}$$

which is called *force manipulability ellipsoid*. It of course depends on the robot configuration, via $(\mathbf{J}\ \mathbf{J}^t)$. It is usually said that this ellipsoid represents the maximum force the robot can exert in all the task directions, for a given configuration. Actually, if we consider a subspace of the joint torque space limited by the actual maximum amplitudes of the robot torques (i.e. actuators capabilities), we obtain a so-called *torque polytope*, approximated by a unit sphere in equation 2. This polytope maps into a *force polytope* by transformation (1). The force ellipsoid is again only an approximation of this force polytope. It can be numerically shown that this approximation is less and less accurate when n increases. However, it is still interesting to use it, because of the analytical form (equation 3) that the polytope does not have.

The approximation is quite obvious, and is even worse if the actuators do not have the same maximum torques, as will be seen now. Let us consider a 2-dof arm with revolute joints and unit length links evolving in a (x,y) plane; consider also the robot configuration given by the joint angles values $q_1 = +127$ degrees and $q_2 = -127$ degrees; finally assume that $\tau_{2max} = 1.5\ \tau_{1max}$ (τ_1 and τ_2 are the two components of the torque τ). In this case, Figure 1 presents the torque polytope (dotted surface), of which the unit sphere (here a circle) becomes a terrible approximation, and Figure 2 presents the corresponding force polytope. A better solution is then to define a *weighted ellipsoid* instead of a unit sphere by:

$$\tau^t [\mathbf{W}\ \mathbf{W}^t]\ \tau = 1 \tag{4}$$

where **W** is a weighting diagonal matrix, the ith component of which is $1/\tau_{imax}$. This torque ellipsoid (Figure 1) maps into a weighted force manipulability ellipsoid (Figure 2) defined by:

$$\mathbf{h}^t [\mathbf{J}\ \mathbf{W}\ \mathbf{W}^t\ \mathbf{J}^t]\ \mathbf{h} = 1 \tag{5}$$

The principal axes of this weighted force manipulability ellipsoid are different from that of the non-weighted one. This means that the non-weighted force ellipsoid generally gives wrong results on the possible force magnitude in every task space direction.

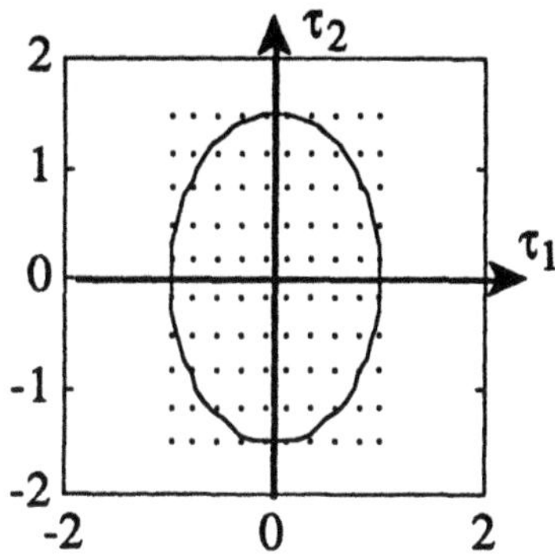

Figure 1: Weighted torque ellipsoid

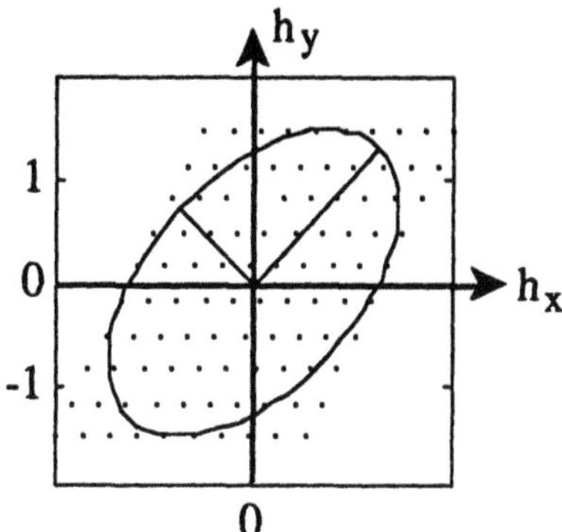

Figure 2: Weighted force ellipsoid

III. Statics of two-arm robots

In order to extend the concept of force manipulability to two-arm robots, one must find a definition of a task force vector **h**. In order to be consistent with the approach proposed in [7] we choose **h** such as it includes a so-called external force/moment vector $\mathbf{f}_a$ (producing the motion of the object) and a so-called internal force/moment vector $\mathbf{f}_r$ (creating constraints in the object). The definition of these vectors is as follows:

$$\left.\begin{aligned} \mathbf{f}_a &= \mathbf{f}_{b1}+\mathbf{f}_{b2} \\ \mathbf{f}_r &= 1/2\ (\mathbf{f}_{b1}-\mathbf{f}_{b2}) \end{aligned}\right| \qquad (6)$$

where $\mathbf{f}_{bi}$ represents the force/moment vector exerted by arm i on the object at the tip of *the virtual stick i.* The virtual stick i is defined as an imaginary rigid body joining the end effector i to a user-defined point on the object. Therefore, the tips of the virtual sticks coincide (and more strictly two reference frames attached to these tips coincide), which yields the extremely simple expressions reported in equation 6. For the arm i, there exists a jacobian matrix $\mathbf{J}_i$ such that:

$$\tau_i = \mathbf{J}_i^t\, \mathbf{f}_{bi} \qquad (7)$$

where τ_i represents the joint torque vector of arm i. Therefore, from the definitions of equation 6, we obtain the following equation:

$$\tau = \begin{bmatrix} \tau_1 \\ \tau_2 \end{bmatrix} = \begin{bmatrix} 1/2 J_1^t & J_1^t \\ 1/2 J_2^t & -J_2^t \end{bmatrix} \begin{bmatrix} \mathbf{f}_a \\ \mathbf{f}_r \end{bmatrix} = \mathbf{J}_{2A}^t \, \mathbf{h} \tag{8}$$

which represents the mapping between the torque space and the force space for the two-arm robot, through the two-arm robot jacobian matrix $\mathbf{J}_{2A}$. Thus, there is conceptually no difference between a single-arm robot and a two-arm robot, as far as force transfer is concerned. This feature will be used in the next section.

IV. Force ellipsoids for two-arm robots

The simplest way to study this problem was presented in [8]: the author proposed to compute the two-arm-ellipsoid as the intersection of each arm ellipsoid. In fact this solution cannot take into account interactions between the two robots and it leads to a non-realistic representation of the two-arm robot capabilities. On the other hand the approach used in [9] defined two manipulabilities: one in the absolute space, one in the relative space; however the obtained ellipsoids were computed respectively with a null relative component, and a null absolute component. We will propose here another definition. Assume, for the sake of simplicity, that all the actuators (of the two arms) have the same torque capabilities. From equation 8, we can define a force ellipsoid whose equation is given by equation 3, with the definitions of $\mathbf{h}$ and $\mathbf{J}_{2A}$ of equation 8. If the actuators capabilities are different, it is not difficult to define a weighted force ellipsoid for the two-arm robot, as presented before for the single arm case. We call it a *generalized force ellipsoid*, because $\mathbf{h}$ is a generalized force/moment vector which includes external and internal forces and moments. One must notice that this ellipsoid is defined in the $\mathbf{h}$ task space, which is the cartesian product of the "classical" force task space (the one appearing in equation 3 for example) by itself.

For example considering a 12-degree-of-freedom two-arm robot leads to a generalized force ellipsoid of dimension 12; then one can define an infinity of external-force ellipsoids (dimension 6) corresponding to various cuts in the generalized force ellipsoid, i.e. for various given values of the internal force $\mathbf{f}_{r0}$:

$$\left| \begin{array}{l} \mathbf{h}^t \, [\mathbf{J}_{2A} \, \mathbf{J}_{2A}^t] \, \mathbf{h} < 1 \\ \mathbf{f}_r = \mathbf{f}_{r0} \end{array} \right. \tag{9}$$

$$\text{with } \mathbf{h} = \begin{bmatrix} \mathbf{f}_a \\ \mathbf{f}_{r0} \end{bmatrix}$$

One important consequence of equation 9 is that, depending on the configuration and <u>depending on the internal force</u>, the external force capability of the two-arm robot can dramatically change, and could be greater or smaller than the sum of each robot external force capability. The experiments reported in the next section are aimed at validating this theoretical result.

V. Experimental validation

The basic idea of the experiments we did was to produce an internal force in order to modify the greatest external force available on the two-arm robot. We decided to begin with experiments on a single PUMA arm in order to precisely evaluate the maximum capabilities of the robot. In fact it is not that easy to know *a priori* what is the maximum force capability. As far as the official documentation is concerned the PUMA is supposed to be able to only carry a 2.5 kg load; but this is the nominal load, supposed to be moved with nominal speed and acceleration.

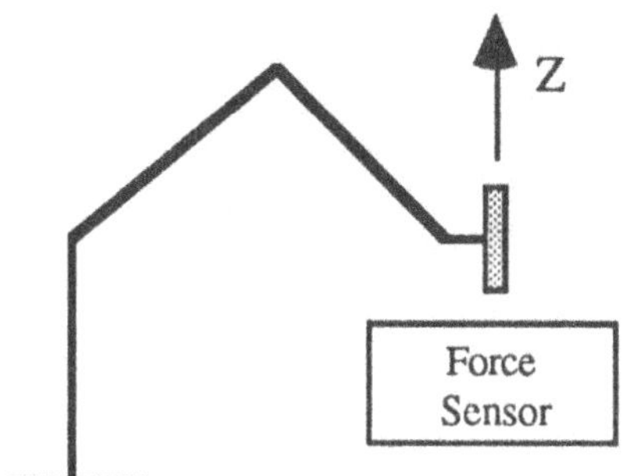

Figure 3: A PUMA pushing on a force sensor.

Since we are here concerned with statics, we decided to measure the PUMA maximum force capability by pushing on a force sensor (Figure 3) until the robot reaches its physical limits, namely a current saturation on (at least) one of the actuators.

Moreover, we wanted to reach the robot physical limits: then we did not need any force servoing. Consequently the force sensor was only used to measure forces, not to produce any force feedback. We simply controlled the robot in position and requested the robot to do small displacements along Z axis: when the robot stopped moving, the physical limits were reached. We then computed the joint torques with equation 1. We just present here the results of two experiments corresponding to two different robot configurations. For the first one we measured a maximum force of 310 N along Z and the computation (equation 1) shows that, according to Unimation documentation, both actuators #2 and #5 reach their physical limits (Table 1); for the second experiment, we measured a maximum force of 320 N but only acuator #5 reaches its physical limit (Table 2). These two experiments confirm that the maximum cartesian force capabilities of a robot depend on its configuration, via the jacobian matrix.

Joint #	1	2	3	4	5	6
Position (deg.)	10	35	-185	12	60	41
Max. Torque (Unim.)	98	186	90	24	20	21
Torque (with Eq. (1))	21	190	81	0	21	0

Table 1: Maximum capabilities of a PUMA 560 - Configuration 1

Joint #	1	2	3	4	5	6
Position (deg.)	15	41	-201	16	72	42
Max. Torque (Unim.)	98	186	90	24	20	21
Torque (with Eq. (1))	15	163	57	0	21	0

Table 2: Maximum capabilities of a PUMA 560 - Configuration 2

Of course, these experiments have to be done for each arm because they cannot be regarded as exactly identical: the maximal force obtained with each arm (for a given configuration of the object) can then be added to obtain the maximal external force of the two-arm robot (for the same configuration of the object) when the internal force is set to zero. This is a kind of "reference value" for our experiments: for three different configurations, we exerted different internal forces in the object (along X axis) in order to study if the maximal external force can differ from this reference value (Figure 4).

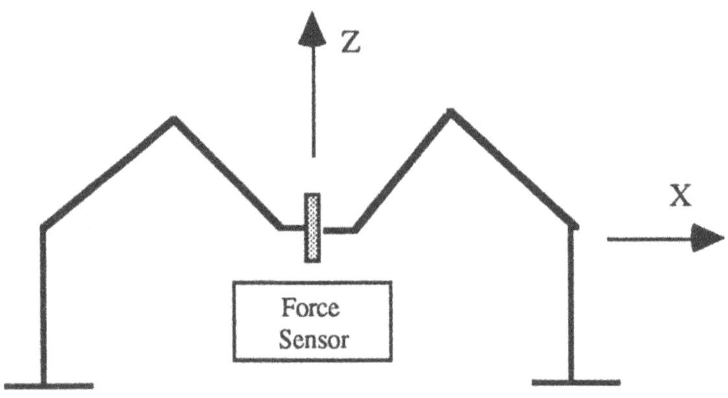

Figure 4: Two PUMA's pushing on a single force sensor.

However we were not able to evaluate this internal force in terms of "Newtons" because it is too dangerous to use very expensive robotics 6-axis force sensors in such experiments: we decided to evaluate this internal force by taking into account the relative position of the two arms. The object was rigidly fixed to one of the arms and the tip of the other could fit into a hole inside the object (Figure 5): this allowed us to exert external forces along Z axis while setting various internal forces (including a null internal force) by requesting various relative positions.
In order to change the two-arm robot configuration we changed the force sensor position along Z axis; we placed it at 0.41m, 0.66m and 0.91m above the robots bases; the corresponding robots configurations are given respectively in Table 3, Table 4 and Table 5.

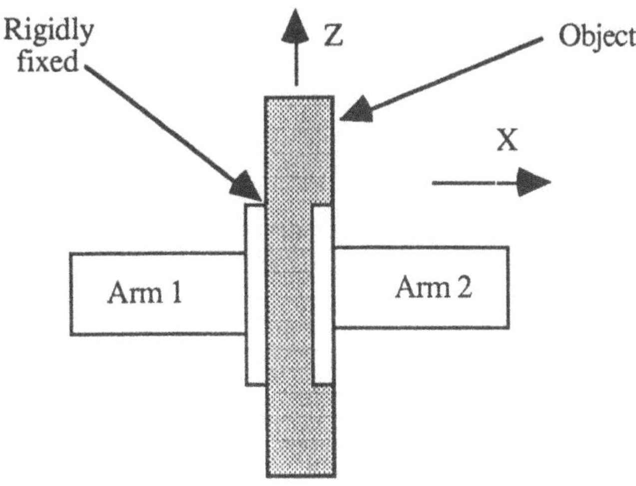

Figure 5: Relative location of the two end-effectors

Z_{sens}=0.41m	Robot 1						Robot 2					
Joint #	1	2	3	4	5	6	1	2	3	4	5	6
Position (deg)	7	5	-132	12	38	35	-6	175	-55	-11	-37	-36

Table 3: Coordination of two PUMA's - Configuration 1

Z_{sens}=0.66m	Robot 1						Robot 2					
Joint #	1	2	3	4	5	6	1	2	3	4	5	6
Position (deg)	7	25	-143	16	28	31	-6	160	-40	-13	-30	28

Table 4: Coordination of two PUMA's - Configuration 2

Z_{sens}=0.91m	Robot 1						Robot 2					
Joint #	1	2	3	4	5	6	1	2	3	4	5	6
Position (deg)	7	40	-132	73	8	-28	-6	146	-51	-54	-8	9

Table 5: Coordination of two PUMA's - Configuration 3

In the first case, the sum of maximum forces that can be exerted by each robot is 600 N (measured with internal force = 0). With the two robots coordinated in order to be a two-arm robot we were able to provide up to 850 N by increasing the internal force (Figure 6).

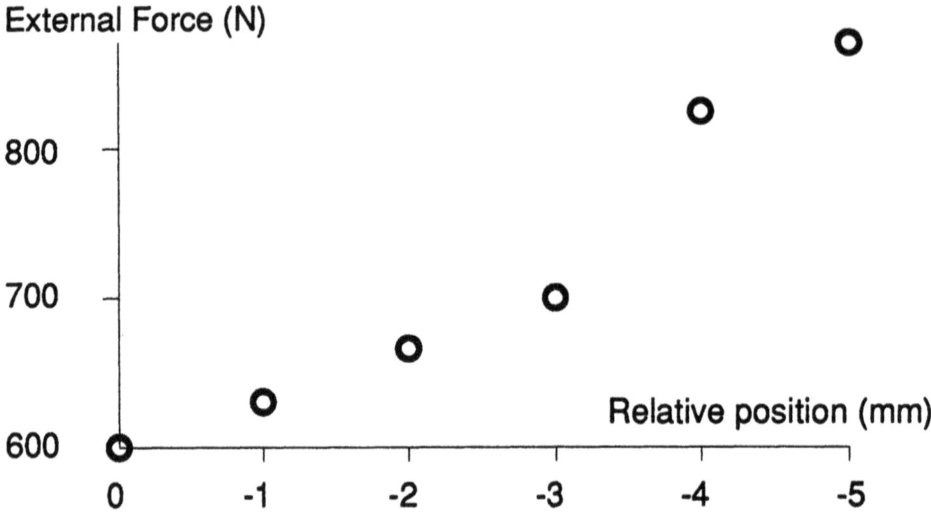

Figure 6: External force vs. relative position. Configuration 1

In the second experiment, the sum of maximum forces that can be exerted by each robot is 330 N. With the two robots coordinated in order to be a two-arm robot we did not observe any sensitive change while increasing the internal force (Figure 7).

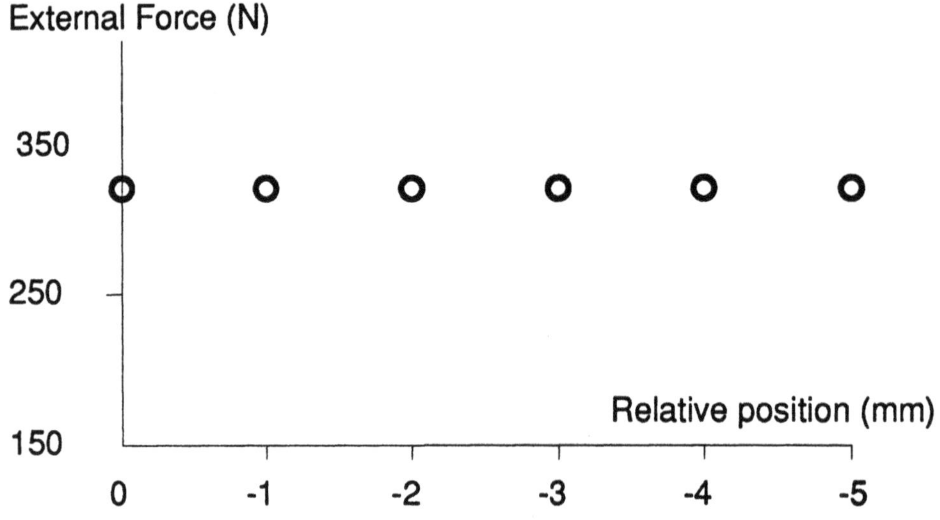

Figure 7: External force vs. relative position. Configuration 2

In the last experiment, the sum of maximum forces that can be exerted by each robot is 350 N. With the two robots coordinated in order to be a two-arm robot we observed that the external force decreases down to 220 N while increasing the internal force (Figure 8).

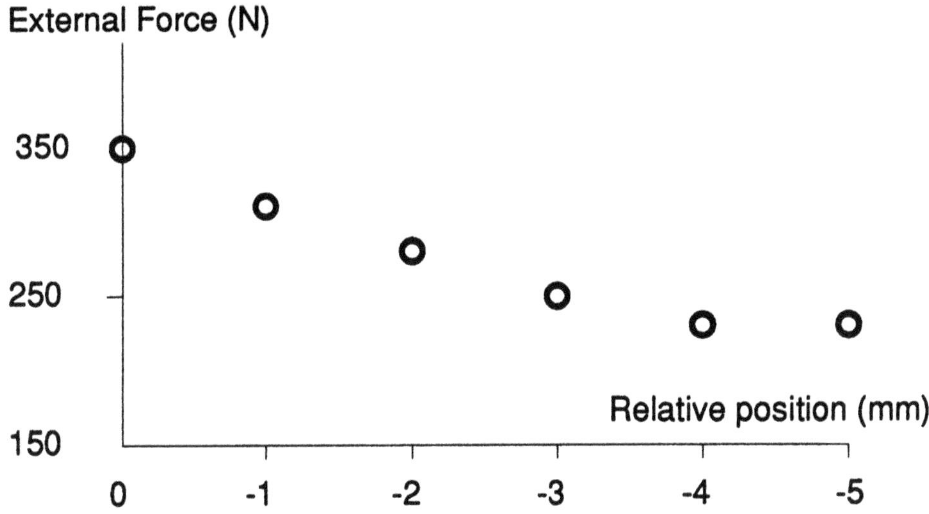

Figure 8: External force vs. relative position. Configuration 3

VI. Conclusion

Starting with the definition of the force manipulability ellipsoid for single-arm robots, we have proposed its extension to the two-arm robot case. This extension has been made possible by considering the two arms as a single robot, and by using a very simple static model of such a two-arm robot, based upon an *ad hoc* formulation of the task. We have thus shown the existence of an infinite number of external force ellipsoids, one per value of internal forces and moments. The real experiments reported in this paper have validated this theoretical approach in showing that the maximum external force that a two-arm robot can exert on its environment depends on the internal force it exerts simultaneously on a rigid object it manipulates. More precisely, we have also demonstrated that, for some robot configuration in space, this maximum external force is greater than the sum of the forces that each robot could exert in the same configuration. This is an interesting result which makes the use of coordinated arms even more attractive for manipulating a single object. But one must be careful about choosing the absolute location where to perform this task, because some of our experiments also showed that the maximum external force can be smaller than the sum of the forces of each robot . Therefore, we can conclude that in a two-arm robot, we cannot really treat internal and external forces and moments independently.

VII. References

[1] J.Y.S. Luh, "Coordinated motion of Two-arm Robots", Report of NSF Workshop on Coordinated Multiple Robot Manipulators: Planning, Control and Application, San Diego, USA (1987).

[2] X. Delebarre, "Commandes position et force de deux bras manipulateurs pour l'exploration planétaire", Thèse de Doctorat, Université Montpellier II, 1992 (in French).

[3] E. Dégoulange, "Commande en effort d'un robot manipulateur à deux bras : application au contrôle de la déformation d'une chaîne cinématique fermée", Thèse de Doctorat, Université Montpellier II, 1993 (in French).

[4] Y.E. Bouffard-Vercelli, "Commande hybride et coordination de deux bras manipulateurs en environnement non structuré : manipulabilité et assemblage", Thèse de Doctorat, Université Montpellier II, 1992 (in French).

[5] Y.E. Bouffard-Vercelli and Pierre Dauchez, "On the manipulability of two-arm robots", Proc. of 4th ISRAM, 313-318, Santa-Fe, USA (1992).

[6] E. Tabarah, "Trajectory planning for the coordinated continuous-path motion of two-robot systems", PhD Thesis, University of Toronto, 1993.

[7] M. Uchiyama and P. Dauchez, "A symmetric hybrid position/force control scheme for the coordination of two robots", Proc. of the IEEE Conf. on Robotics and Automation, 350-356, Philadelphia, USA (1988).

[8] S. Lee, "Dual redundant arm configuration optimization with task-oriented dual arm manipulability", IEEE Transactions on Robotics and Automation, 5(1), 78-97 (1989).

[9] P. Chiacchio, S. Chiaverini, L. Sciavico and B. Siciliano, "Reformulation of dynamic manipulability ellipsoid for robotic manipulators", Proc. of IEEE Conf. on Robotics and Automation, 2192-2197, Sacramento, USA (1991).

Minimum Joint Torque Configurations of Planar Multiple-Link Manipulator

J. Lenarčič

The Jožef Stefan Institute
University of Ljubljana
61111 Ljubljana, Slovenia

Abstract - Optimum configurations of a hyper-redundant *n-R* multiple-link planar robot manipulator are studied in this work. The problem consists of optimising the joint torques caused by an external force and the gravity of links in a given position of the end effector. In order to minimise the joint torques the manipulator produces configurations of typical forms which can easily be recognised and mathematically described. This leads to extremely effective algorithms that can be used for real-time kinematic control of robot manipulators which possess several degrees of freedom.

I. Introduction

A diversity of tasks of manipulation require sophisticated mechanical motion in an unstructured, dynamically varying environment. Since there is an infinity of possible motions that a redundant manipulator can make, all of which correspond to a given task, redundant manipulators possess several advantages over the non redundant ones. In addition to positioning and orienting the end effector, redundant mechanisms are capable of optimising different performance criteria, such as minimisation of joint velocities and accelerations, joint torques and energy. Simultaneously, they can also be used in avoiding obstacles and ill-conditioned or other undesired configurations. Redundancy is a common feature in nature. From the viewpoint of positioning and orienting the wrist, also the human arm can be regarded as redundant.

The hyper-redundant multiple-link manipulators are referred to as the mechanisms which contain an elevated number of degrees of freedom [1]. These manipulators are analogous in morphology to snakes or tentacles and are therefore more versatile that usual redundant manipulators. Unfortunately, a practical realisation of multiple-link manipulators is still far away. In the first place, this is because of the enormous technical and technological problems that arise from their mechanical design and control. The numerical treatment of these mechanisms is also very difficult and complex. As reported in [2], for example, the most elementary inverse kinematics problem includes several thousands of arithmetical operations. Hence, efficient real-time algorithms and control schemes are still to be investigated.

The objective of the present article is to study the optimum configurations of a hyper-redundant *n-R* multiple-link robot manipulator with respect to a set of optimisation criteria related to the joint torques caused by static external forces and gravity of the links. We show that the resulting optimum configurations achieve typical forms which can easily be recognised. The main practical advantage of knowing these forms, as it was anticipated in

A. J. Lenarčič and B. B. Ravani (eds.), Advances in Robot Kinematics and Computationed Geometry, 281–288.

[3], is that we can simplify the procedure of calculating the optimum solutions useful for control and motion planning of multiple-link robot manipulators. In the first part of this article we establish very simple optimum configurations which are related to an external force applied to the end effector. In the second part other forms are computed by adding the gravity of the links, and at last, a combination of both is considered.

II. End effector position

As shown in Fig. 1, the *n-R* mechanism is assumed to move in the *x-y* plane. In the reference configuration all the links are parallel in the direction of *x* and the joint angles q_1, q_2, ... q_n are zero. They are measured about the axes perpendicular to the *x-y* plane. If the manipulator possesses identical links, the position of the end effector can be expressed by the following backward recursions

$$x_i = x_{i+1} + d\cos(\alpha_i), \qquad i = n, n-1, \ldots, 1, \tag{1}$$
$$y_i = y_{i+1} + d\sin(\alpha_i), \qquad i = n, n-1, \ldots, 1,$$

where d is the length of links, x_{n+1}, $y_{n+1} = 0$, and $x = x_1$, $y = y_1$. The orientation of the ith link, given in absolute angles, is

$$\alpha_i = \alpha_{i-1} + q_i, \qquad i = 1, 2, \ldots, n, \ \alpha_0 = 0. \tag{2}$$

It is also assumed that the mechanism possesses unlimited joint angles and that there are no collisions between the links or other obstacles inside the workspace.

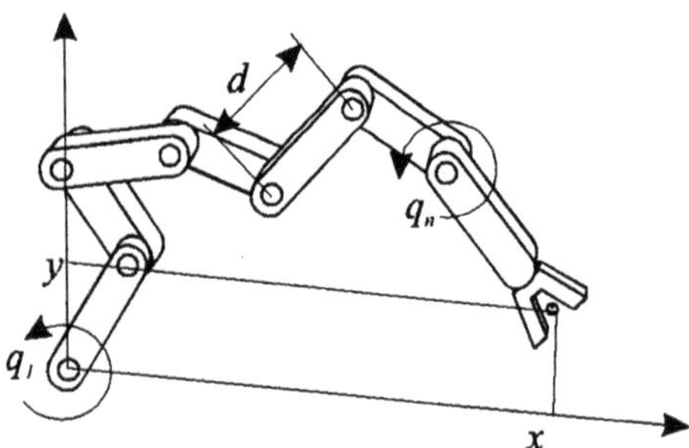

Fig. 1: *n-R* planar manipulator

The derivatives of x and y can be obtained with no additional computations by

$$\frac{\partial x_i}{\partial \alpha_k} = -y_j, \ \frac{\partial y_i}{\partial \alpha_k} = x_j, \qquad j = \max(i,k), \tag{3}$$

and the corresponding Jacobian matrix is

$$J = \begin{bmatrix} -y_1 & -y_2 & \cdots & -y_n \\ x_1 & x_2 & \cdots & x_n \end{bmatrix}. \tag{4}$$

III. Joint torques caused by external force

In this section, we presuppose that the force applied to the end effector is much larger than the gravity of the links. If the weight of the links is neglected, the following well known equation can be used to obtain the corresponding joint torques

$$\boldsymbol{\tau} = \boldsymbol{J}^T \boldsymbol{f}, \tag{5}$$

where f is the force applied to the end effector. Since the work area of the mechanism in Fig. 1 is symmetric about the first joint angle, we can rotate the coordinate frame x-y without any lost of generality and we can imply that $f = F$ is vertical and parallel to y. It then follows from (5)

$$\tau_i = Fx_i \,, \qquad i = 1,2,\ldots,n \,. \tag{6}$$

The aim of this work is to minimise the sum of the joint torques specified by the following cost function

$$\mathscr{F} = \sum_{i=1}^{n} |\, \tau_i \,| \tag{7}$$

which can be rewritten as

$$\mathscr{F} = Fd \sum_{i=1}^{n} \left| \sum_{j=i}^{n} c_j \right| \tag{8}$$

and simplified when all x_i in (6) are positive (or negative)

$$\mathscr{F} = Fd \sum_{i=1}^{n} i c_i \,. \tag{9}$$

By joining Eqs. (1,6-9), we can impose a linear programming problem in the form

$$\underset{c_i,\; i=1,n}{Minimise} \{\mathscr{F}\} \tag{10}$$

subject to

$$x = d\sum_{i=1}^{n} c_i, \quad -1 \le c_i \le 1 \,, \quad i = 1,2,\ldots,n \,. \tag{11}$$

Here, c_i are the cosines of the absolute joint angles α_i. To solve this problem, a standard scheme of linear programming can be practised. It is clear, however, that to minimise (10) we must minimise the number of non zero values c_i that begin with the lowest index i. A first guess for the optimum solution, where $x = kd$, $k < n$, is as follows

$$c_1 = c_2 = \cdots = c_k = 1, \quad c_{k+1} = c_{k+2} = \cdots = c_n = 0 \,, \tag{12}$$

and the corresponding value of the cost function

$$\mathscr{F} = Fd \frac{k(k+1)}{2} \,. \tag{13}$$

It is possible to show that other combinations of c_i is more costly. For instance, suppose that $c_k = \xi$ and $0 < \xi < 1$. In order to satisfy (11), at least one additional c_i, $i = k+1,k+2,...,n$, must assume a non zero value. Let this be $c_{k+1} = 1 - \xi$. The corresponding cost function is evidently greater than (13) for the given domain of ξ

$$\tilde{\mathcal{F}} = Fd\left\{\frac{k(k-1)}{2} + k\xi + (k+1)(1-\xi)\right\} = \mathcal{F} + Fd(1-\xi) \Rightarrow \tilde{\mathcal{F}} > \mathcal{F}. \tag{14}$$

Suppose now that $c_k = -\xi$ and $0 < \xi < 1$. In this case, two additional c_i, $i = k+1,k+2,...,n$, must assume non zero values to solve (11). Let these be $c_{k+1}, c_{k+2} = (1+\xi)/2$. The related cost function is

$$\tilde{\mathcal{F}} = Fd\left\{\frac{k(k-1)}{2} - k\xi + \frac{1+\xi}{2}(2k+3)\right\} = \mathcal{F} + \frac{3Fd}{2}(1+\xi) \Rightarrow \tilde{\mathcal{F}} > \mathcal{F}. \tag{15}$$

If (12) is the optimum solution in terms of cosines, we can get the optimum absolute joint angles in the following form

$$\alpha_1 = \alpha_2 = \cdots = \alpha_k = 0, \quad \alpha_{k+1} = \alpha_{k+2} = \cdots = \alpha_n = \pm 90° . \tag{16}$$

Let the y coordinate of the end effector position be specified as jd, $j \leq n-k$. Then, the difference between the number of positive and the number of negative joint angles $\alpha_{k+1}, \alpha_{k+2}, ..., \alpha_n$ in the optimum solution (16) must be equal to j. Besides, if n-k is even, j must be even, and if n-k is odd, j must be odd.

Unfortunately, the optimum solution (16) is valid only when $j+k \leq n$. An extension to the cases when $j+k > n$ (note that $j^2 + k^2 \leq n^2$ to stay inside the work area) is as follows

$$\alpha_1 = \alpha_2 = \cdots = \alpha_{\hat{k}} = \alpha, \quad \alpha_{\hat{k}+1} = \alpha_{\hat{k}+2} = \cdots = \alpha_n = \pm 90° , \tag{17}$$

where

$$\alpha = \cos^{-1}\left\{\frac{2k(n-j)}{k^2 + (n-j)^2}\right\}, \qquad \frac{k}{\cos(\alpha)} + 1 \geq \hat{k} \geq \frac{k}{\cos(\alpha)} . \tag{18}$$

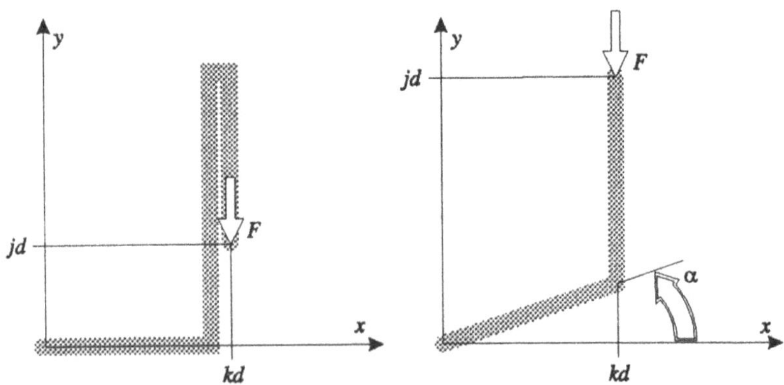

Fig. 2: Optimum solutions with respect to external force

Evidently, in utilising the above equations one should be aware of the error that originates from discretising the values of the end effector positions $x = kd$ and $y = jd$. This error decreases by increasing n and vanishes for all positions of the end effector inside the work area of the manipulator when n is infinite. Fig. 2 shows two situations, one corresponds to the solution (16) and the other to the solution (17). In practice, these are the best static configurations when the manipulator is grasping a heavy load.

IV. Joint torques caused by gravity of links

In this section, the external force is zero and the joint torques are caused only by the gravity forces concentrated in the centres of the links. It is presupposed that all the links possess the same masses denoted by m and the related gravity forces mg are in the direction of the axis y. Considering the x components of the positions of the links established in (1) we can develop the following equation for the joint torques that are relative to the gravity forces of the links

$$\tau_i = mg\left\{(n-i+\tfrac{1}{2})x_i - \sum_{j=i+1}^{n} x_j\right\} . \tag{19}$$

If the cost function is specified as a series of absolute joint torques analogously to (7), we have in this case

$$\mathcal{G} = mgd\sum_{i=1}^{n}\left|\sum_{j=i}^{n}(n-j+\tfrac{1}{2})c_j\right| . \tag{20}$$

Similarly to the previous section, we can obtain the following simplified form of the cost function when all x_i are positive (or negative)

$$\mathcal{G} = mgd\sum_{i=1}^{n} i(n-i+\tfrac{1}{2})c_i . \tag{21}$$

The objective is, therefore, to find a combination of the cosines c_i, $i = 1,2,\ldots,n$ in order to minimise the cost function $\mathcal{G}$ with respect to the given end effector position and the imposed limits of the cosine functions (11). Apparently, in contrast to the optimisation (10,11), this solution can be found by minimising the number of non zero values c_i which end with the last index i

$$c_1 = c_2 = \cdots = c_{n-k} = 0,\quad c_{n-k+1} = c_{n-k+2} = \cdots = c_n = 1 , \tag{22}$$

where $x = kd$, and $k < n$. The corresponding minimum value of the cost function $\mathcal{G}$ can be obtained by

$$\mathcal{G} = mgd\sum_{i=n-k+1}^{n} i(n-i+\tfrac{1}{2}) = mgd\sum_{i=n-k+1}^{n} i(n-i) + mgd\sum_{i=n-k+1}^{n} i(\tfrac{1}{2}) , \tag{23}$$

where

$$\sum_{i=n-k+1}^{n} i(n-i) = \sum_{i=1}^{k-1} i(n-i) = \sum_{i=1}^{k-1} i(k-i) + (n-k)\sum_{i=1}^{k-1} i \quad \text{and} \quad \sum_{i=n-k+1}^{n} i(\tfrac{1}{2}) = \tfrac{1}{2}\sum_{i=1}^{n} i - \tfrac{1}{2}\sum_{i=1}^{n-k} i . \tag{24}$$

Since

$$\sum_{i=1}^{r} i(r-i) = \sum_{i=1}^{r-1} i(r-i) = \frac{r}{6}(r-1)(r+1) \text{ and } \sum_{i=1}^{r} i = \frac{r}{2}(r+1) , \tag{25}$$

we finally get

$$\mathcal{G} = mgd\left\{\frac{k}{6}(k-1)(3n-2k+1)+\frac{k}{4}(2n-k+1)\right\} . \tag{26}$$

It can be demonstrated that other combinations of c_i are more expensive. The set of joint angles relative to (22) is then

$$\alpha_1 = \alpha_2 = \cdots = \alpha_{n-k} = \pm 90^\circ, \quad \alpha_{n-k+1} = \alpha_{n-k+2} = \cdots = \alpha_n = 0 \tag{27}$$

which is valid if the position of the end effector $x = kd$, $y = jd$ satisfies $j+k \leq n$. An extension to the cases inside the work area of the manipulator where $j+k > n$ is as follows

$$\alpha_1 = \alpha_2 = \cdots = \alpha_{n-\hat{k}} = \pm 90^\circ, \; \alpha_{n-\hat{k}+1} = \alpha_{n-\hat{k}+2} = \cdots = \alpha_n = \alpha , \tag{28}$$

and where α and $\hat{k}$ are defined as in (18). Examples of the optimum solutions (27) and (28) are presented in Fig. 3.

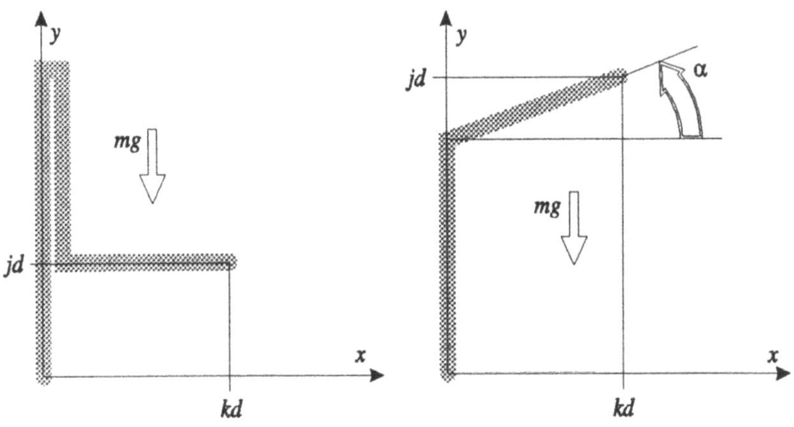

Fig. 3: Optimum solutions with respect to gravity forces of links

V. Joint torques caused by gravity of links and external force

In a more general situation, the gravity of the links is combined with an external force applied to the end effector of the manipulator. Here, it is additionally assumed that this external force is parallel to the gravity forces of the links. In practice, this is the case when the multiple-link manipulator grippes a load in a given position of the work area, and the gravity of the load is considered as the external force F. Thus, the joint torques are the sum of (6) and (19)

$$\tau_i = mg\left\{(n-i+\tfrac{1}{2})x_i - \sum_{j=i+1}^{n} x_j\right\} + Fx_i \,. \tag{29}$$

In correspondence with (7) and by considering (8) and (20) we can develop the total cost function in the following form

$$\mathcal{T} = d\sum_{i=1}^{n}\left|\sum_{j=i}^{n}\left\{mg(n-j+\tfrac{1}{2})+F\right\}c_j\right| , \tag{30}$$

where mg is the gravity force of each link and F the gravity force of the load attached to the end effector. When all cosines in (30) are positive (or negative) the cost function $\mathcal{T}$ can be expressed as the sum of the cost functions $\mathcal{F}$ and $\mathcal{G}$ that are developed in the previous sections and specified in (9) and (21)

$$\mathcal{T} = \mathcal{F} + \mathcal{G} = d\sum_{i=1}^{n} i\left\{mg(n-i+\tfrac{1}{2})+F\right\}c_i \,. \tag{31}$$

The goal is to minimise the cost function $\mathcal{T}$ subject to the desired end effector position inside the work area of the manipulator and subject to the inequality conditions imposed by the limits of the cosine functions in (11).

In the same way as in (9) and (21), the minimum of the cost function (31) is achieved by minimising the number of non zero values c_i. Consequently, the solution can be found either in the forms (16,17) illustrated in Fig. 2 or in the forms (27,28) illustrated in Fig. 3. A numerical analysis of this linear programming problem shows that the optimum solution depends on the ratio between the gravity force of the links mg and the gravity force of the load F. The following value of the total cost function (31) is obtained if we introduce the solution (16)

$$\mathcal{T}_k = mgd\left\{\sum_{i=1}^{k} i(n-i) + \tfrac{1}{2}\sum_{i=1}^{k} i\right\} + Fd\sum_{i=1}^{k} i \,, \tag{32}$$

which can be rewritten according to (24) as follows

$$\mathcal{T}_k = mgd\left\{\sum_{i=1}^{k} i(k-i) + (n-k)\sum_{i=1}^{k} i + \tfrac{1}{2}\sum_{i=1}^{n} i - \tfrac{1}{2}\sum_{i=1}^{n-k} i - \tfrac{1}{2}k(n-k)\right\} + Fd\left\{\sum_{i=1}^{n} i - \sum_{i=1}^{n-k} i - k(n-k)\right\}. \tag{33}$$

The same can be done for the solution (27)

$$\mathcal{T}_{n-k} = mgd\left\{\sum_{i=n-k+1}^{n} i(n-i+\tfrac{1}{2})\right\} + Fd\left\{\sum_{i=n-k+1}^{n} i\right\} , \tag{34}$$

and after introducing (24) we get

$$\mathcal{T}_{n-k} = mgd\left\{\sum_{i=1}^{k} i(k-i) + (n-k)\sum_{i=1}^{k-1} i + \tfrac{1}{2}\sum_{i=1}^{n} i - \tfrac{1}{2}\sum_{i=1}^{n-k} i\right\} + Fd\left\{\sum_{i=1}^{n} i - \sum_{i=1}^{n-k} i\right\} . \tag{35}$$

By taking into account

$$(n-k)\sum_{i=1}^{k} i-(n-k)\sum_{i=1}^{k-1} i=k(n-k) \tag{36}$$

we obtain the difference

$$\mathcal{T}_k-\mathcal{T}_{n-k}=k(n-k)(\tfrac{1}{2}mg-F)d \ . \tag{37}$$

Since n is greater than k inside the work area of the manipulator, the following inequalities are assembled

$$\begin{aligned}&\mathcal{T}_k>\mathcal{T}_{n-k}, \text{ if } mg>2F \ ,\\ &\mathcal{T}_k<\mathcal{T}_{n-k}, \text{ if } mg<2F \ ,\\ &\mathcal{T}_k=\mathcal{T}_{n-k}, \text{ if } mg=2F \ .\end{aligned} \tag{38}$$

This means that we will choose (16,17) as the optimum solutions if the weight of the load is greater than one half of the weight of the link. When the weight of the load is less, (27,28) are the optimum solutions. The solutions (16,17) and (27,28) are equivalent when the gravity of the load is equal to one half of the weight of the link.

VI. Conclusions

Redundant manipulators can change their motion or static properties in a desired position (and orientation) of the end effector by modifying their configurations. In the present work, we studied the optimum configurations of a multiple-link hyper-redundant planar *n-R* robot manipulator with respect to the sum of the absolute torques in the joints. In general, these can be found by solving an adequately defined linear programming problem. Unfortunately, such an approach is time consuming and cannot be efficiently used in control.

The fundamental idea in this work is to analyse the geometrical characteristics of the optimum configurations. We demonstrated that, if the manipulator possesses identical links, the optimum configurations achieve typical and recognisable forms which can be described and reconstructed by very simple geometrical relations. For a desired position of the end effector and the given length of the manipulator we can establish the optimum configurations, expressed in terms of the values of joint angles, with very few arithmetical operations. This approach is extremely efficient and stable and is, therefore, adequate for real-time control. Yet, its utilisation is restricted to static situations or to a very slow motion of the manipulator.

VII. References

[1] G.S. Chirikjian and J.W. Burdick, "Kinematics of Hyper-Redundant Manipulators", *Advances in Robot Kinematics* (S. Stifter and J. Lenarčič, Eds.), Springer-Verlag, Wien (1991).

[2] J. Lenarčič, "Computational Considerations on Kinematics Inversion of Multi-Link Redundant Robot Manipulators", *Computational Kinematics* (J. Angeles, G. Hommel and P. Kovacs, Eds.), Kluwer Academic Publishers, Dordrecht (1993).

[3] J. Lenarčič, "Optimum Configurations of Planar n-R Hyper-Redundant Mechanism", *Proc. ICAR'93 - International Conference on Advanced Robotics*, Tokyo, Japan (1993).

8. Inverse Kinematics

Positional Inverse Kinematic Problems in $T^n \times \Re^m$ Solved in $T^{2(n+m)}$

Federico Thomas and Carme Torras

Institut de Cibernètica (UPC - CSIC)
Diagonal 647, 08028-Barcelona, SPAIN

Abstract - Solving the inverse kinematic problem for a closed spatial mechanism with n translational and m rotational links is here reduced to the problem of navigating in the configuration space of the spherical orthogonal mechanism with, at most, $2(n+m)$ degrees of freedom.

A recursive algorithm to find the analytic solution of spherical orthogonal mechanisms is provided. This solution is thus amenable to differentiation, leading to a characterization of the tangent space of the self-motion manifold of such mechanisms. It is precisely this tangent space that provides the solution of the translational part of arbitrary spatial mechanisms.

The approach taken for navigating in the configuration space of the spherical orthogonal mechanism is numerical in nature, with the advantage of working in a space with a well-defined norm.

The problem of finding points on the self-motion manifold satisfying a set of extra constraints, such as joint limits, might be addressed through reasonable extensions of the algorithm presented.

I. Introduction

It is generally accepted that no satisfactory solution has been found for the general positional inverse kinematic problem. This is why the redundant manipulator literature has focused on the linearized first-order instantaneous kinematic relation between joint velocities (for recent advances in this area see [1] and [2]). Given the position and velocity states, the set of joint coordinates can be obtained either by directly solving positional equations (for a classical reference see [3]) or by solving the first-order differential equations derived from the linearization (see, for example, [4]). The latter alternative is relatively easier than the former. Nevertheless, it exhibits important difficulties to provide a description, at least a local one, of the self-motion manifold of the mechanism to be analyzed.

This paper deepens on the former alternative providing a way around these difficulties. To this end, we exploit the following two facts: (a) any kinematic loop equation can be modeled as the loop equation derived from the so-called n-bar mechanism by taking as many bars as needed and constraining some of the resulting degrees of freedom; and (b) the solution of the translational component of the loop equation of the n-bar mechanism is provided by the tangent bundle of the self-motion manifold of its spherical indicatrix. These two facts lead to a unified approach for the analysis of any closed loop containing independent revolute, prismatic and cylindrical pairs, which has been already published elsewhere [5] [6]. Herein, we concentrate ourselves on the description of a numerical algorithm derived from this analysis that permits converging,

A. J. Lenarčič and B. B. Ravani (eds.), Advances in Robot Kinematics and Computationed Geometry, 291–300.

from an unfeasible point of the configuration space of the spherical indicatrix, to the *nearest* solution point. The relevance of this algorithm derives from the fact that it only requires knowledge of the self-motion manifold of the indicatrix, that is, it looks for points with certain characteristics in T^n, where n is not greater than twice the number of degrees of freedom (d.o.f.) of the mechanism.

The paper is structured as follows. Section II briefly describes the theory and notation used throughout the paper, showing the great relevance of the study of the n-bar mechanism and, in particular, of its spherical indicatrix: the orthogonal spherical mechanism. Section III is devoted to the analysis of this latter mechanism. In particular, a recursive analytic procedure for obtaining a local description of its self-motion manifold and its tangent space, at the lowest computational cost, is derived in this section. Section IV shows the application of the previous two sections to solving inverse kinematic problems of any closed loop containing independent revolute, prismatic and cylindrical pairs, through the definition of two error functions in T^n and, finally, Section V provides a summary of the main points in the paper, as well as the conclusions and prospects for future research.

II. Basics

A closed kinematic chain is determined by a sequence $X_1, \ldots, X_p$ of screws of the corresponding links through their Plücker coordinates. Its *geometry* is determined by the dual quantities [7] $\hat{\alpha}_1, \hat{\alpha}_3, \ldots, \hat{\alpha}_{2p-1}$, where $\hat{\alpha}_i = \alpha_i + \varepsilon a_i$, α_i being the angle from $X_{(i+1)/2}$ to $X_{(i+3)/2}$ and a_i, the distance from $X_{(i+1)/2}$ to $X_{(i+3)/2}$. Its *configuration* is determined by the dual quantities $\hat{\theta}_2, \hat{\theta}_4, \ldots, \hat{\theta}_{2p}$, where $\hat{\theta}_i = \theta_i + \varepsilon t_i$, θ_i being the angle around $X_{i/2}$ and t_i, the offset along $X_{i/2}$. Note that α_i, a_i, θ_i and t_i are the Denavit-Hantenberg parameters of the mechanism. Then, by assigning $\hat{\phi}_i = \hat{\alpha}_i$ when i is odd and $\hat{\phi}_i = \hat{\theta}_i$ when i is even, the loop equation of a closed kinematic chain can be expressed as:

$$\mathbf{F}(\hat{\Phi}) = \prod_{i=1}^{2p} \mathbf{Rx}(\hat{\phi}_i)\mathbf{Rz}(\pi/2) = \prod_{i=1}^{n} \mathbf{B}(\hat{\phi}_i) = \mathbf{I}, \tag{1}$$

where $\hat{\Phi} = (\hat{\phi}_1, \hat{\phi}_2, \ldots, \hat{\phi}_n) = (\phi_1 + \varepsilon t_1, \phi_2 + \varepsilon t_2, \ldots, \phi_n + \varepsilon t_n)$ is called the *vector of displacements*; $\Phi = (\phi_1, \phi_2, \ldots \phi_n)$, the *vector of rotations*; and $D = (d_1, d_2, \ldots d_n)$, the *vector of translations*. This equation corresponds to the loop equation of what in [5] and [6] is called the n-bar mechanism.

A. Fundamental Theorems

Theorem I. The solution of the non-dual part of (1)

$$\prod_{i=1}^{n} \mathbf{B}(\phi_i) = \mathbf{I}, \tag{2}$$

is a connected $(n-3)$-dimensional pseudomanifold that can be characterized, outside of its singular points and at least locally, using $r = n-3$ parameters, $\Psi = (\psi_1, \psi_2, \ldots, \psi_r)$.

Proof. The fact that the solution is a connected pseudomanifold is proved in subsection B and local parameterizations are discussed in subsection C. □

Theorem II. *Spatial to spherical transference.* The solution of the dual part of (1), outside of the singular points of the non-dual part, can be expressed as:

$$D = \mathbf{K}\Lambda, \quad \forall \Lambda = (\lambda_1, \ldots, \lambda_r)^T \in \Re^r, \tag{3}$$

where

$$\mathbf{K} = \begin{bmatrix} \frac{\partial \phi_1}{\partial \psi_1} & \cdots & \frac{\partial \phi_1}{\partial \psi_r} \\ \vdots & & \vdots \\ \frac{\partial \phi_n}{\partial \psi_1} & \cdots & \frac{\partial \phi_n}{\partial \psi_r} \end{bmatrix}. \tag{4}$$

Proof. For a full proof of this theorem see [5]. □

Equation (2) corresponds to the loop equation of an orthogonal spherical mechanism. This theorem shows the great relevance of deepening on the structure of the self-motion manifold of the orthogonal spherical mechanisms, and how a thorough understanding of them is very helpful in the study of spatial mechanisms.

B. The Self-Motion Set of the Orthogonal Spherical Mechanism as a Punched Manifold

The configuration space, C, of a spherical mechanism is a product space formed by the n-fold product of the individual variables of rotation, that is, $C = S^1 \times S^1 \times \ldots \times S^1 = T^n$, where T^n is an n-torus, which is a compact n-dimensional manifold.

An orthogonal spherical mechanism becomes redundant for $n > 3$. Then, let the redundant inverse kinematic solution of equation (2) be expressed as a $(n-3)$-dimensional algebraic set or *self-motion set*, M, embedded in T^n. This self-motion set, however, is not a $(n-3)$-manifold but rather a pseudomanifold or *punched manifold* because of the presence of singular points. Singular points correspond to those situations in which the mechanism becomes planar, that is when all axes of rotation lie on the same plane (see [5]). Then, from a topological point of view, splitting all these singular points yields a $(n-3)$-manifold M'. Thus, M can be obtained from M' by pinching M' at certain pairs of points.

In order to prove these facts, observe that equation (2) has a straightforward geometric interpretation as an n-sided spherical polygon. Consider a unit sphere centered at the coordinate origin. As a result of applying successive rotations, the z−axis will describe on the surface of the sphere a spherical polygon with sides of length ϕ_i and exterior angles equal to $\pi/2$. Alternatively, the y-axis will describe a spherical polygon with sides of length $\pi/2$ and exterior angles equal to ϕ_i.

Lemma I. The self-motion set of an orthogonal spherical mechanism with $n > 3$ is connected.

Proof. Consider the spherical polygon described by the y-axis, i.e. a polygon with all its sides of length $\pi/2$. Note that, by varying the angle between two sides of length

$\pi/2$, one can form a triangle whose third side may attain every value between 0 and π. Therefore, one can fix arbitrarily $n-3$ consecutive variables of an orthogonal spherical mechanism, this leading to a spherical chain with $n-2$ sides that can always be closed with the remaining two sides.

Moreover, if the resulting angle between these two sides is different from 0 and π, then there are two alternative solutions corresponding to the two possible symmetric placements of the two sides on the sphere. As the angle approaches 0 or π, the two solution branches fuse into one precisely at these two values.

To prove that the solution set is connected, it suffices to show that a reference configuration can be reached from every other configuration. Let us choose the reference configuration as $\phi_i = \pi, \forall i = 1, \ldots n$, when n is even, and as $\phi_1 = \phi_2 = \phi_3 = \pi/2, \phi_i = \pi, \forall i = 4, \ldots n$, when n is odd.

Now, from any initial configuration, one can make the exterior angles approach sequentially their values at the reference configuration. Thanks to the last two sides, the chain will remain closed throughout the process. □

Lemma II. $\Phi = (\phi_1, \ldots, \phi_n)$ is a singular point iff (i) $\phi_i (\text{mod } \pi) = 0$, $i = 1, \ldots, n$; and (ii) $(\sum_{i=1}^{n} \phi_i) (\text{mod } 2\pi) = 0$.

Proof. The former condition ensures that the mechanism lies on the plane defined by the first and the last axis, and the latter is a simplified version of the closure equation (2) when the former holds. Both provide a necessary and sufficient condition for the mechanism to be in a planar configuration and, hence, for the configuration point to be singular. □

Corollary I. When n is odd, the orthogonal spherical mechanism has no singularities.

Corollary II. When n is even, the number of singularities is 2^{n-2}.

Corollary III. When $n > 4$ the self-motion set remains connected after removing its singular points. (See [5] for a complete analysis of the 4-bar mechanism.)

C. Parameterizations and Symmetries of the Self-Motion Manifold

After removing the singular points, the self-motion set becomes an r-dimensional smooth manifold, M_c, of class C^∞, which will be called *self-motion manifold.* Then, r coordinates of the surrounding space T^n can be taken as local coordinates in the neighborhood of each point $\Phi_0 \in M_c$. This is, in fact, the implicit function theorem formulated in convenient terms, whose proof can be found in any textbook on differential geometry. In what follows we will study this simple parameterization.

Let us take r consecutive variables in the chain as parameters. Without loss of generality, let $\{\phi_1, \phi_2, \ldots, \phi_r\}$ be the set of parameters. Hence, the equation of rotations can be expressed as:

$$\mathbf{Rx}(\phi_{r+1})\ \mathbf{Rz}(\pi/2)\ \mathbf{Rx}(\phi_{r+2})\ \mathbf{Rz}(\pi/2)\ \mathbf{Rx}(\phi_{r+3}) = \mathbf{A}, \tag{5}$$

which has always solution for any proper orthogonal matrix $\mathbf{A}$ encompassing all the parameters. In general, this equation has the following two discrete solutions:

$$
\begin{aligned}
\phi_{r+1} &= \mathrm{atan2}(\pm a_{21}, \mp a_{31}) \\
\phi_{r+2} &= \mp \mathrm{acos}(-a_{11}) \\
\phi_{r+3} &= \mathrm{atan2}(\mp a_{12}, \mp a_{13})
\end{aligned} \tag{6}
$$

where a_{ij} denotes the element (i, j) of **A**. One solution is obtained by taking the upper row of signs, and the other, by taking the lower one.

When $a_{11} = \pm 1$, there appear infinite solutions. The points of the self-motion manifold where this happens are called singularities of the parameterization, and it can be easily shown that they correspond to those situations in which the last three rotation axes are coplanar.

As it is shown in the next section, the above formulation, although correct, can be greatly improved using geometric arguments to reduce computational overhead during the computation of variables in terms of parameters, and partial derivatives of variables with respect to parameters.

Lemma III. *Symmetries.* Given a point $\Phi_0 = (\phi_1, \ldots, \phi_n)$ on the self-motion manifold, points $\Phi_1 = (\phi_1 + \pi, -\phi_2, \phi_3 + \pi, \phi_4, \ldots, \phi_n)$ and $\Phi_2 = (\phi_n, \phi_1, \ldots, \phi_{n-1})$ are also on it.

Proof. The first symmetry can be derived by analyzing (6). The second one is obvious. □

Corollary IV. The iterative computation of the symmetries in *Lemma III* leads at most to $n \cdot 2^n$ symmetric points for any point on the self-motion manifold.

III. Inverse Kinematics of Orthogonal Spherical Mechanisms

We have already seen that equation (2) has a geometric interpretation as an n-sided spherical polygon. In this section, after introducing some basic relations from Spherical Trigonometry, we will derive a recursive algorithm to find the inverse kinematics of any spherical orthogonal mechanism. The recursive nature of the solution allows us to compute the derivatives needed to find the solution for the translational component of the n-bar mechanism from the inverse kinematics of its spherical indicatrix, by applying equation (3).

A. Spherical Trigonometry Preliminaries

Let us denote ϕ_1, ϕ_2 and ϕ_3 the sides of an spherical triangle and α_{12}, α_{23} and α_{31} its exterior angles, the *cosine*, *sine-cosine* and *sine* laws are [8]:

$$
\begin{aligned}
&\cos\alpha_{12} = \cos\alpha_{31}\cos\alpha_{23} - \sin\alpha_{31}\sin\alpha_{23}\cos\phi_3 \\
&-\sin\alpha_{12}\cos\phi_2 = \cos\alpha_{31}\sin\alpha_{23} + \sin\alpha_{31}\cos\alpha_{23}\cos\phi_3 \\
&\sin\alpha_{12}\sin\phi_2 = \sin\phi_3\sin\alpha_{31}.
\end{aligned} \tag{7}
$$

From this, one can derive the following relation for a triangle having two exterior angles equal to $\pi/2$. Taking $\alpha_{31} = \alpha_{23} = \pi/2$, we find that $\alpha_{12} = \pi - \phi_3$. Moreover, $\phi_1 = \phi_2 = \pi/2$.

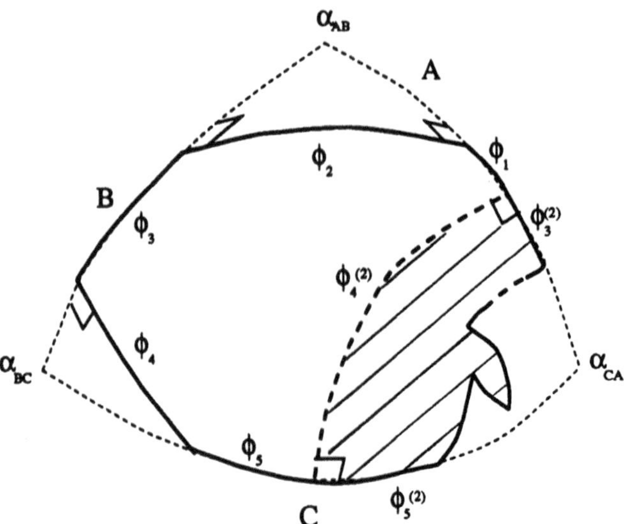

Fig. 1: Triangle ABC constructed by prolonging the sides ϕ_1, ϕ_3 and ϕ_5 of the original n-gon. Using the fact that the three small triangles have 2 exterior angles equal to $\pi/2$, the solution of the n-gon reduces to that of the $(n-2)$-gon, which is shown shaded.

Note that a sequence of $n > 2$ points on a unit sphere defines in general a unique spherical n-gon with sides $\leq \pi$, but a total of 2^n n-gons if we take into account sides of length $> \pi$. In the triangle relation above, we have taken the determination with sides $\leq \pi$.

B. A Recursive Algorithm for the Solution of Spherical Orthogonal Mechanisms

The spherical polygon corresponding to a spherical orthogonal loop has all its exterior angles equal to $\pi/2$. The crucial idea of the algorithm is to reduce the solution of one such n-gon to that of an $(n-2)$-gon. To this end, we construct a triangle by prolonging three alternate sides of the n-gon, as shown in *fig. 1*.

Note that two exterior angles in each of the two small triangles formed in this way have value $\pi/2$, and we can apply the relation derived above for spherical triangles to express two exterior angles and one side of the big spherical triangle as functions of ϕ_2, ϕ_3 and ϕ_4. Moreover, we can construct the third small triangle by drawing a side perpendicular to both ϕ_1 and ϕ_5. In this way, the three new sides $\phi_3^{(2)}$, $\phi_4^{(2)}$ and $\phi_5^{(2)}$ are originated. In sum, we have the following expressions for the sides of the big triangle:

$$\begin{aligned} A &= \phi_1 - \phi_3^{(2)} + \pi \\ B &= \phi_3 + \pi \\ C &= \phi_5 - \phi_5^{(2)} + \pi, \end{aligned} \tag{8}$$

and the following expressions for its exterior angles:

$$\begin{aligned} \alpha_{AB} &= \pi - \phi_2 \\ \alpha_{BC} &= \pi - \phi_4 \\ \alpha_{CA} &= \pi - \phi_4^{(2)}. \end{aligned} \tag{9}$$

Now, we can use the three laws of Spherical Trigonometry included in the preceding subsection to relate the new variables $\phi_3^{(2)}$, $\phi_4^{(2)}$ and $\phi_5^{(2)}$ to the old ones ϕ_1, ϕ_2, ϕ_3, ϕ_4 and ϕ_5 as follows:

$$\begin{aligned} \phi_3^{(2)} &= \phi_1 - \mathrm{atan2}(\sin\phi_4\sin\phi_3, -\cos\phi_4\sin\phi_2 + \sin\phi_4\cos\phi_3\cos\phi_2) \\ \phi_4^{(2)} &= \mathrm{acos}(-\cos\phi_2\cos\phi_4 - \sin\phi_2\sin\phi_4\cos\phi_3) \\ \phi_5^{(2)} &= \phi_5 - \mathrm{atan2}(\sin\phi_2\sin\phi_3, -\cos\phi_2\sin\phi_4 + \sin\phi_2\cos\phi_3\cos\phi_4) \end{aligned} \tag{10}$$

As a result of this process, we have reduced the solution of the original n-gon to the solution of the $(n-2)$-gon with sides $\phi_3^{(2)}$, $\phi_4^{(2)}$, $\phi_5^{(2)}$, $\phi_6, \ldots \phi_n$. Note that the resulting $(n-2)$-gon has all its exterior angles also equal to $\pi/2$.

The following recurrence equations can thus be easily derived:

$$\begin{aligned} \phi_{2i-1}^{(i)} &= \phi_{2i-3}^{(i-1)} - \mathrm{atan2}(f(\phi_{2i-1}^{(i-1)}, \phi_{2i}^{(1)}), g(\phi_{2i-2}^{(i-1)}, \phi_{2i-1}^{(i-1)}, \phi_{2i}^{(1)})), \\ \phi_{2i}^{(i)} &= \mathrm{acos}(h(\phi_{2i-2}^{(i-1)}, \phi_{2i-1}^{(i-1)}, \phi_{2i}^{(1)})), \\ \phi_{2i+1}^{(i)} &= \phi_{2i+1}^{(1)} - \mathrm{atan2}(f(\phi_{2i-2}^{(i-1)}, \phi_{2i-1}^{(i-1)}), g(\phi_{2i}^{(1)}, \phi_{2i-1}^{(i-1)}, \phi_{2i-2}^{(i-1)})), \end{aligned} \tag{11}$$

where, to ease the notation, we have introduced the following functions:

$$\begin{aligned} f(\alpha,\beta) &= \sin\alpha \ \sin\beta, \\ g(\alpha,\beta,\gamma) &= -\sin\alpha \ \cos\beta + \cos\alpha \ \sin\gamma \ \cos\beta, \\ h(\alpha,\beta,\gamma) &= -\cos\alpha \ \cos\beta - \sin\alpha \ \sin\gamma \ \cos\beta \end{aligned} \tag{12}$$

and this is valid $\forall i = 2, \ldots \lceil (n-4)/2 \rceil$.

The initial conditions can be derived in the same way depending on whether n is even or odd. In the former case, one has to consider the equations for an hexagon, which are essentially the same as the recurrence equations. Taking $n = 2i+4$, these equations are:

$$\begin{aligned} \phi_n &= \mathrm{atan2}(f(\phi_{2i}^{(i)}, \phi_{2i+1}^{(i)}), g(\phi_{2i-1}^{(i)}, \phi_{2i}^{(i)}, \phi_{2i+1}^{(i)})), \\ \phi_{n-1} &= \mathrm{acos}(h(\phi_{2i-1}^{(i)}, \phi_{2i+1}^{(i)}, \phi_{2i}^{(1)})), \\ \phi_{n-2} &= \mathrm{atan2}(f(\phi_{2i-1}^{(i)}, \phi_{2i}^{(i)}), g(\phi_{2i+1}^{(i)}, \phi_{2i}^{(i)}, \phi_{2i-1}^{(i)})). \end{aligned} \tag{13}$$

While, when n is odd, one has to consider the equations for a pentagon, which follow from those of the hexagon by forming a small triangle (with two exterior angles equal to $\pi/2$) limited by two consecutive sides of the pentagon. This amounts to making the following substitutions in the equations for the hexagon: $\phi_{2i-1}^{(i)}, \phi_{2i}^{(i)}$ and $\phi_{2i+1}^{(i)}$ have to be replaced by $\phi_{2i-1}^{(i)} - \pi/2$, $\pi/2$ and $\phi_{2i}^{(i)} - \pi/2$, respectively. Moreover, $n = 2i + 3$ in this case.

The above recurrence equations, together with the initial conditions, provide a general analytic expression of the inverse kinematics of spherical orthogonal mechanisms. The functions involved can be easily differentiated, leading to the general solution of spatial orthogonal mechanisms, as explained below.

C. Tangent Space

We begin by noting that most partial derivatives of the functions introduced in the preceding subsection can be written in terms of the functions themselves. The complete listing of these derivatives follows:

$$\begin{aligned} \frac{\partial f}{\partial \alpha} &= \cos\alpha \ \sin\beta, & \frac{\partial g}{\partial \alpha} &= h(\alpha,\beta,\gamma), & \frac{\partial h}{\partial \alpha} &= -g(\alpha,\beta,\gamma), \\ \frac{\partial f}{\partial \beta} &= \sin\alpha \ \cos\beta, & \frac{\partial g}{\partial \beta} &= -\cos\alpha \ f(\alpha,\beta), & \frac{\partial h}{\partial \beta} &= \sin\gamma \ f(\alpha,\beta), \\ & & \frac{\partial g}{\partial \gamma} &= f(\alpha,\beta) + \cos\alpha \ \cos\gamma \ \cos\beta, & \frac{\partial h}{\partial \gamma} &= -g(\alpha,\beta,\gamma). \end{aligned} \tag{14}$$

With these derivatives and those of the functions atan2 and acos, the partial derivatives of the variables $\phi_{n-2}, \phi_{n-1}, \phi_n$ with respect to the parameters $\phi_1, \ldots \phi_{n-3}$ can be found using the same recursive structure of the algorithm described above. Space limitations prevent us from giving an exhaustive listing of all these derivatives, but a sample of three:

$$\begin{aligned} \frac{\partial \phi_{2i-1}^{(i)}}{\partial \phi_{2i-2}^{(i-1)}} &= \frac{f(\phi_{2i-1}^{(i-1)}, \phi_{2i}^{(1)})\ h(\phi_{2i-2}^{(i-1)}, \phi_{2i-1}^{(i-1)}, \phi_{2i}^{(1)})}{f^2(\phi_{2i-1}^{(i-1)}, \phi_{2i}^{(1)}) + g^2(\phi_{2i-2}^{(i)}, \phi_{2i-1}^{(i)}, \phi_{2i}^{(1)})}, \\ \frac{\partial \phi_{2i}^{(i)}}{\partial \phi_{2i-2}^{(i-1)}} &= \frac{-g(\phi_{2i-2}^{(i-1)}, \phi_{2i-1}^{(i-1)}, \phi_{2i}^{(1)})}{\sin\phi_{2i}^{(i)}}, \\ \frac{\partial \phi_{2i+1}^{(i)}}{\partial \phi_{2i+1}^{(1)}} &= 1. \end{aligned} \tag{15}$$

Now note that it is not necessary to derive the full analytic expression of these derivatives, since only the values of the derivatives at particular points are required and these can be recursively evaluated using the formulas above.

These derivatives conform the matrix $\mathbf{K}$ in equation (3), leading to the solution for the translational component of the n-bar mechanism.

In the next section we will show how the solution of the n-bar mechanism can be used to find the inverse kinematics of arbitrary spatial mechanisms.

IV. Inverse kinematics of Arbitrary Single Closed-Loop Mechanisms

By constraining some of the variables in (1), one can model any closed kinematic loop containing independent translational and rotational pairs. Thus, we define T

and $\mathcal{R}$ as the set of indices of the constrained translational and rotational d.o.f. and $|\mathcal{T}|$ and $|\mathcal{R}|$ their cardinalities, respectively. Note that, if the Denavit-Hantenberg parameters are properly taken, $|\mathcal{T}|+|\mathcal{R}|\leq n$.

Let $\tilde{\Phi} = (\ldots, \phi_i, \ldots)$, $i \in \mathcal{R}$, and $\tilde{D} = (\ldots, d_j, \ldots)$, $j \in \mathcal{T}$, where the elements of $\tilde{D}$ are scaled so that $||\tilde{D}|| = 1$.

Once we choose a stating point, $\Phi^0 = (\phi_1^0, \ldots, \phi_n^0)$, on the self-motion manifold of the spherical indicatrix of the corresponding n-bar mechanism and a set of values for the constrained d.o.f., we introduce two errors, called translational and rotational, that will allow us to direct the search from Φ^0 towards a solution of the mechanism under analysis.

In general, we have to find Λ so that

$$\tilde{D} = \tilde{\mathbf{K}}\Lambda, \tag{16}$$

where

$$\tilde{\mathbf{K}} = \begin{bmatrix} \vdots & & \vdots \\ \frac{\partial \phi_i}{\partial \psi_1} & \cdots & \frac{\partial \phi_i}{\partial \psi_r} \\ \vdots & & \vdots \end{bmatrix}, \quad i \in \mathcal{T}. \tag{17}$$

Nevertheless, this is not always possible and the value of Λ that provides the closest value of $\tilde{D}$ to the desired one, in the least squares sense, which will be called Λ^0 is that which minimizes the residual $r = ||\tilde{\mathbf{K}}\Lambda - \tilde{D}||$. Then, the translational error $\mathcal{E}_t$ is defined as:

$$\mathcal{E}_t(\Phi) = ||\tilde{\mathbf{K}}\Lambda^0 - \tilde{D}||. \tag{18}$$

There are several numerical approaches for obtaining Λ^0 (see [9]).

On the other hand, the rotational error, $\mathcal{E}_r$, is simply defined as:

$$\mathcal{E}_r(\Phi) = \frac{\sum_{i\in\mathcal{R}}(\phi_i - \phi_i^0)(\mathrm{mod}\ 2\pi)}{2\pi\,|\mathcal{R}|}. \tag{19}$$

It is clear that $0 \leq \mathcal{E}_t(\Phi) \leq 1$ and $0 \leq \mathcal{E}_r(\Phi) \leq 1$, and a solution of the analyzed mechanism is found iff $\mathcal{E}_t = \mathcal{E}_r = 0$

The implementation is now at the level that permits finding inverse kinematic solutions for the n-bar mechanism when all translations are constrained and all rotations ramain free. Then, only the translational error is considered. The starting point Φ_0 is randomly generated. But, in order to improve speed, the translational error function is evaluated in all its symmetric points. Then, the symmetric point with minimum translational error is effectively used as starting point.

Since, for the moment, partial derivatives of errors are not available, we are bound to using minimization methods requiring only function evaluations, not derivatives. Obviously, this is not very efficient. The *downhill simplex method* [9] has been chosen to get something working quickly.

V. Conclusions

This paper presents a generalized solution to the inverse kinematics of single closed chains with arbitrary number of degrees of freedom using the concept of *spatial to spherical transference.*

The procedure requires points on the self-motion manifold of the spherical indicatrix of the n-bar mechanism and their tangent planes, which can be either analytically or numerically computed. An analytic solution is provided.

A foreseen extension of this work is finding a trajectory between two predetermined configurations of the mechanism. This can become more difficult if joint limits are to be taken into account.

Acknowledgements: This work has been partially supported by the ESPRIT III Basic Research Action Program of the EC under contract No. 6546 (PROMotion).

References

[1] Chen, Y-Ch., and Walker, D., 1993. A Consistent Null-Space Based Approach to Inverse Kinematics of Redundant Robots, *1993 IEEE Int. Conf.on Robotics and Automation*, pp. 374-382.

[2] Deo, A.S., and Walker, I.D. 1993. Adaptive Non-linear Least Squares for Inverse Kinematics, *1993 IEEE Int. Conf. on Robotics and Automation*, pp. 187-194.

[3] Uicker, J.J., Denavit, J., and Hartenberg, R.S. 1964. An Iterative Method for the Displacement Analysis of Spatial Mechanisms, *Transactions of the ASME, Journal of Applied Mechanics*, June, pp. 309-314.

[4] Goldenberg, A.A., Benhabib, B., and Fenton, R.G. 1985. A Complete Generalized Solution to the Inverse Kinematics of Robots, *IEEE Journal of Robotics and Automation*, Vol. RA-1, No.1, pp. 14-21, March.

[5] Thomas, F. 1992. On the N-bar Mechanism, or How to Find Global Solutions to Single Loop Spatial Kinematic Chains, *1992 International Conference on Robotics and Automation*, Nice, France, May.

[6] Thomas, F., 1993. The Self-Motion Manifolds of the N-bar Mechanism, in *Computational Kinematics*, J. Angeles et al. (eds.), Kluwer Academic Publishers, October.

[7] Veldkamp, G.R. 1976. On the Use of Dual Numbers, Vectors and Matrices in Instantaneous, Spatial Kinematics, *Mechanisms and Machine Theory*, vol. 11, pp. 141-156.

[8] Duffy, J., and Rooney, J. 1975. A Foundation for a Unified Theory of Analysis of Spatial Mechanisms, *Transactions of the ASME, Journal of Engineering for Industry*, November, pp. 1159-1165.

[9] Press, W.H., Flannery, B.P., Teukolsky, S.A., and Vetterling, W.T., 1990, Numerical Recipes in C, Cambridge University Press, Cambridge, U.K.

Explicit Modelling of General Task Spaces for Inverse Kinematics

R. Featherstone*
Dept. of Engineering Science
Oxford University
Parks Road, Oxford, OX1 3PJ, England

Abstract - This paper proposes a new architecture for model-based inverse kinematics, in which the task space is described by an explicit model; and it describes a specific implementation of this architecture in which the task-space model is a Lie group representing an n-tuple of displacements to be realized by the mechanism. The objective of this work is to increase the generality of model-based kinematics software.

Introduction

A typical model-based inverse kinematics system has an architecture like that shown in Figure 1. The input is a command describing the desired position or motion of the mechanism, and the output is a vector of joint variables that satisfy the command. The calculation is performed by a general-purpose inverse kinematics algorithm, using a kinematic model of a specific mechanism. Some examples are described in [1, 5, 7, 8].

The advantage of a model-based approach is that it is possible to write a single program embodying the kinematics algorithm, test it, debug it, and then apply it to a wide variety of different mechanisms; thus avoiding the need to write and test separate software for each individual mechanism.

However, the purpose of an inverse-kinematics system is to perform a transformation from a task space to the configuration space of a particular mechanism; so it follows that a complete description of an inverse-kinematics problem must include a description of the task space as well as the kinematics of the mechanism.

The lack of an explicit task-space model means that assumptions about the task space have to be built into the software. For example, a program written to control the position and orientation of a robot's end effector will assume that the task space is six-dimensional, containing three translational and three rotational components specifying the location of a particular link in the mechanism. In contrast, a program written to control the fingertip positions of a dexterous hand will assume that the task space is $3n$-dimensional, specifying the position of a single point in each of n different links.

*This work was supported by SERC Advanced Fellowship number B92/AF/1466 and starter grant number GR/J20272. My thanks to Bill Triggs for his comments on this paper.

A. J. Lenarčič and B. B. Ravani (eds.), Advances in Robot Kinematics and Computationed Geometry, 301–308.

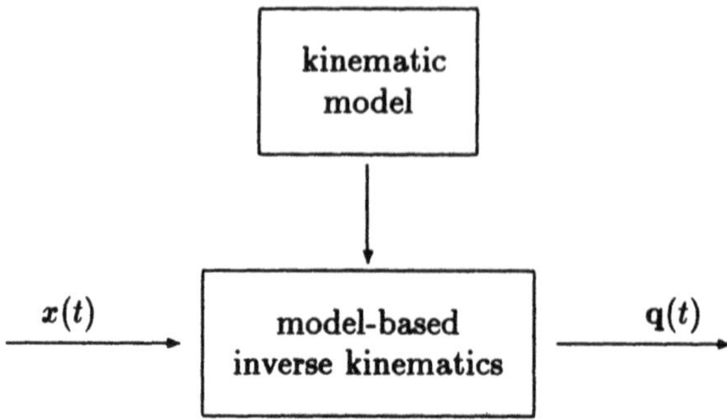

Figure 1: Architecture of a typical model-based inverse kinematics system.

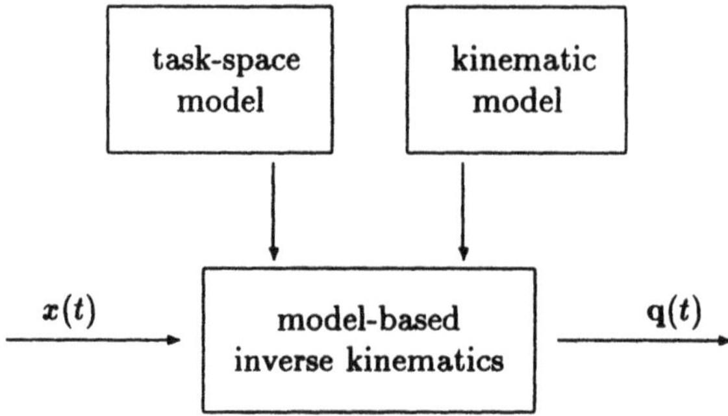

Figure 2: Model-based inverse kinematics system with task-space model.

Both programs are performing inverse kinematics, but neither can perform the task of the other. If the two task spaces could be described with explicit models then it would be possible to write a single program to do both tasks.

This paper proposes that an explicit task-space model be included in the architecture, as shown in Figure 2; and describes a specific implementation of this concept, in which the task space is a Lie group, and its elements specify n-tuples of relative displacements between pairs of coordinate frames located in the mechanism. The architecture of the model is described, a method of constructing trajectories, and an algorithm to perform trajectory tracking. Execution timing tests show that the modelling overhead is significant, but not excessive.

The Task-Space Model

Any computer model of a physical or mathematical system must have a structure, or architecture, which is imposed on it by the modelling system. For example, one popular architecture for a kinematic model consists of a table of Denavit-Hartenberg parameters and an array of joint type codes. A more elaborate model might include connectivity data, so that kinematic trees and loops can be described. The architecture defines what data can be included in a model, and hence what variety of systems can be modelled.

In general, there are two aspects of a task space that must be modelled: its mathematical properties and its kinematic interpretation. The latter, in conjunction with a kinematic model, defines the relationship between the task space and the configuration space of a particular mechanism. A task-space model can only be used with a kinematic model that is compatible with its kinematic interpretation. (For example, you can't control the fingertips of a robot with no fingers.)

The model architecture used in this paper comprises a mathematical structure that is a Lie group, and a kinematic interpretation based on pairs of coordinate frames placed in the mechanism. Each pair is associated with a displacement group, so that elements of the group specify the displacement between the two frames; and the task space itself is defined as the Cartesian product of these groups, so that an element of the task space specifies all of the displacements simultaneously.

This architecture allows a fairly general specification of positional data for use in motion commands, but it does not allow for non-positional data, such as contact forces, compliances, behavioural responses to sensor data, or cost functions (for avoiding obstacles, singularities, etc.). Also, the use of Lie groups places some restrictions on the types of spaces that can be described (e.g., a pointing task can not be modelled because the set of pointing directions in 3D does not constitute a Lie group).

A task-space model is implemented as a data structure specifying the number of pairs and, for each pair, the location of each frame and a group-identifier code. A frame location is specified by a body-identification number, indicating which body the frame is attached to, and a transformation matrix giving the location of the frame in a body-fixed coordinate system defined by the kinematic model. A group-identifier code identifies an entry in the library of displacement-group models. In the author's implementation, this library contains descriptions of the groups T(2), T(3), E(2) and E(3). These descriptions define the internal representation of a group element, functions to perform various operations on group elements, and functions to convert between the internal representation and a standard, common representation for spatial displacements. (Spatial algebra [3] was chosen as the common representation.)

The purpose of the group-model library is to allow the user to select the nature and degree of constraint imposed on each pair. For example, selecting E(3) allows the user to specify both position and orientation, thereby imposing six constraints on the pair, whereas selecting T(3) allows the user to specify only the relative positions of the two origins, thus imposing only three constraints. The 2-D groups are intended for 2-D displacements, and are defined in the x–y plane.

As the author is not familiar with group theory, only a few simple definitions and properties will be used here in connection with task spaces. Let X be the task space and $\mathsf{TX} = T(e \in \mathsf{X})$ be the associated Lie algebra (the space of tangent vectors at the identity). If $a_1, a_2 \in \mathsf{X}$ then $a_1 \circ a_2$ is defined as the result of performing displacements a_1 and a_2 in the order written. Let $\exp : \mathsf{TX} \mapsto \mathsf{X}$ be the exponential mapping of tangent vectors onto the task space. If $\mathbf{v} \in \mathsf{TX}$ is interpreted as a velocity, then $\exp(\mathbf{v})$ is the displacement resulting from the action of $\mathbf{v}$ over one unit of time. If $a \in \mathsf{X}$ does not contain any rotational component with a magnitude of π then $\log(a)$ is defined as the unique smallest vector $\mathbf{v}$ satisfying $\exp(\mathbf{v}) = a$, and a^p is defined as $\exp(p \log(a))$. The special cases a^1, a^0 and a^{-1} (the inverse of a) are defined for all a.

Trajectories

A trajectory is a continuous curve in task space, parameterized by a single variable, and usually required to be differentiable with respect to that variable. This section describes a method of constructing trajectories in a task space that is a Lie group.

The traditional approach to trajectory synthesis treats rotation and translation separately [2, chapter 4]. Since we don't know in advance which components of a task-space element will be rotations and which translations, we must use a method that treats them both in the same way. This can be done by constructing curves directly in the task space. There are a variety of formulas that can be used, such as

$$x(s) = a_1^{p_1(s)} \circ a_2^{p_2(s)} \circ \cdots \circ a_n^{p_n(s)} \tag{1}$$

and

$$x(s) = \exp(p_1(s)\,\mathbf{v}_1 + p_2(s)\,\mathbf{v}_2 + \cdots + p_n(s)\,\mathbf{v}_n), \tag{2}$$

where $a_i \in \mathsf{X}$, $\mathbf{v}_i \in \mathsf{TX}$ and $p_i(s)$ are scalar polynomials in s. These curves can be designed to pass through a sequence of locations in much the same way as an interpolating polynomial; but it is generally better to use them as splines. The author's trajectory generator implements a cubic spline satisfying $x(0) = a_1$, $\frac{dx}{ds}(0) = \mathbf{v}_1$, $x(1) = a_2$ and $\frac{dx}{ds}(1) = \mathbf{v}_2$ with the formula

$$x(s) = a_1 \circ \exp(s(s-1)^2\,\mathbf{v}_1) \circ (a_1^{-1} \circ a_2)^{s^2(3-2s)} \circ \exp(s^2(s-1)\,\mathbf{v}_2).$$

Some care is required when using these formulas. With Eq. (1) none of the a_i may have a rotational component equal to π, while Eq. (2) breaks down if the desired trajectory contains a rotational component that passes through π. The safest approach is to design curves that are relative displacements from one set point to the next, and ensure beforehand that every component of relative rotation between two adjacent set points is less than π.

Kinematics Calculations

As there is no symbolic solution to the general inverse kinematics problem, a numerical method must be used. This section explains how to implement numerical inverse kinematics in a Lie-group task space. In particular, Newton's method is implemented for the task of following a trajectory.

The first step is to formulate a suitable forward (or direct) kinematics function. What is needed is a function of the form $k : \mathsf{Q} \mapsto \mathsf{X}$, where Q is the configuration space of the mechanism. Recall that an element of the task space is an n-tuple of displacements (implemented as an array), with each displacement referring to the relative locations of a particular pair of bodies; so the computation carried out by k is to calculate the values of these displacements as a function of the joint variables. This can be done in three stages:

1. Calculate the position and orientation of each body in the mechanism, using the data in the kinematic model and a conventional, model-based kinematics algorithm.

2. Calculate the displacements between the pairs of coordinate frames identified in the task-space model.

3. Convert each displacement into the internal representation for the group associated with that displacement, and assemble the results into an array to form the task-space element.

Notice that both the kinematic and task-space models are used in this calculation.

Next we need the Jacobian of k. There is more than one way to define this quantity, but the most convenient definition in the current context is as follows: $\mathbf{J}(\mathbf{q}) : T(\mathbf{q} \in \mathsf{Q}) \mapsto T(k(\mathbf{q}) \in \mathsf{X})$, which means that the Jacobian of k, evaluated at $\mathbf{q}$, is a mapping from the tangent space of Q at $\mathbf{q}$ (the space of infinitesimal joint displacements around the value $\mathbf{q}$) into the tangent space of X at $k(\mathbf{q})$. It follows from this definition that

$$k(\mathbf{q}+\boldsymbol{\delta}\mathbf{q}) \approx k(\mathbf{q}) \circ \exp(\mathbf{J}(\mathbf{q})\, \boldsymbol{\delta}\mathbf{q}). \tag{3}$$

We are now in a position to describe the implementation of Newton's method in task space. The equation we wish to solve is

$$k(\mathbf{q}) = x, \tag{4}$$

given an initial guess, $\mathbf{q}_0$, somewhere near the desired solution. Substituting $\mathbf{q} = \mathbf{q}_0 + \boldsymbol{\delta}\mathbf{q}$ in Eq. (4) and applying Eq. (3) gives

$$k(\mathbf{q}_0) \circ \exp(\mathbf{J}(\mathbf{q}_0)\, \boldsymbol{\delta}\mathbf{q}) \approx x.$$

Premultiplying both sides by $k(\mathbf{q}_0)^{-1}$ and linearizing gives

$$\mathbf{J}(\mathbf{q}_0)\, \boldsymbol{\delta}\mathbf{q} \approx \log(k(\mathbf{q}_0)^{-1} \circ x);$$

hence

$$\mathbf{q} \approx \mathbf{q}_0 + \mathbf{J}(\mathbf{q}_0)^{-1} \log(k(\mathbf{q}_0)^{-1} \circ x). \tag{5}$$

In these equations, $k(\mathbf{q}_0)^{-1}$ is the inverse of the group element $k(\mathbf{q}_0)$ and $\mathbf{J}(\mathbf{q}_0)^{-1}$ is the matrix inverse of $\mathbf{J}(\mathbf{q}_0)$. Eq. (5) breaks down at singularities; but if the Jacobian is not invertible because of redundancy then the pseudoinverse (or a weighted pseudoinverse) may be used instead. The algorithm for Newton's method is therefore

$\mathbf{q} \leftarrow \mathbf{q}_0$; (A)
repeat
  $\mathbf{q} \leftarrow \mathbf{q} + \mathbf{J}(\mathbf{q})^{-1} \log(k(\mathbf{q})^{-1} \circ x)$
until $|k(\mathbf{q})^{-1} \circ x| < \epsilon$;

where the left-arrow indicates assignment. Note the need for a magnitude operator to determine whether $k(\mathbf{q})$ is sufficiently close to x.

Let us now consider the task of following a trajectory. The problem is defined as follows: given a continuous curve $x(s) \in \mathsf{X}$, $s \in [s_0, s_1]$, and a configuration $\mathbf{q}_0$ satisfying $k(\mathbf{q}_0) = x(s_0)$, find a corresponding continuous curve $\mathbf{q}(s) \in \mathsf{Q}$ such that $k(\mathbf{q}(s)) = x(s)\ \forall s \in [s_0, s_1]$ and $\mathbf{q}(s_0) = \mathbf{q}_0$.

To solve this problem, let $x(s)$ be sampled to produce an ordered sequence of points $\{x_i\}$, $i = 0 \ldots n$, where $x_i = x(s_0 + i\,h)$ and h is the sampling interval. The objective is then to find a corresponding sequence $\{\mathbf{q}_i\}$ (given $\mathbf{q}_0$) such that $k(\mathbf{q}_i) = x_i$, and such that a continuous curve interpolated through these points solves the original problem with sufficient accuracy for a sufficiently small value of h. This last condition implies that $\mathbf{q}_i$ and $\mathbf{q}_{i+1}$ must be close together, in the sense that the distance between them (given a suitable metric) should be $O(h)$.

The following simple modification of algorithm A will solve this problem. It performs one iteration per sample point, and achieves the necessary closeness between $\mathbf{q}_i$ and $\mathbf{q}_{i+1}$ by using the former as the starting point for calculating the latter.

$$\begin{aligned} &\textbf{for } i = 1 \textbf{ to } n \textbf{ do} \qquad\qquad (\mathrm{B}) \\ &\quad \mathbf{q}_i \leftarrow \mathbf{q}_{i-1} + \mathbf{J}(\mathbf{q}_{i-1})^{-1} \log(k(\mathbf{q}_{i-1})^{-1} \circ x_i); \end{aligned}$$

This is not a very sophisticated algorithm. Better algorithms have been published that work at singularities, have adaptive step-size control, etc. (e.g. see [5, 6, 9]). The important point here is not the quality of the algorithm, but to demonstrate how it can be adapted to work in a generalized task space.

Implementation

In order to test the feasibility of using a task-space model, the author has written a program containing the following components:

- a kinematics modeller for spatial kinematic trees with revolute and prismatic joints;
- a general forward-kinematics function, implemented using spatial arithmetic;
- a task-space modeller as described above, but limited to specifying a single displacement;
- a function to calculate k and $\mathbf{J}$ from the forward-kinematics data and the task-space model;
- a task-space trajectory generator producing sampled position and velocity data; and
- a collection of numerical inverse-kinematics algorithms, including algorithm B. (See [4] for a description of the others).

test case	fwd kin		ts	solve	total
	pos	ts			
2R, T(2)	83	71	2	40	196
3R, T(2)	127	82	2	87	299
3R, E(2)	127	101	34	73	349
5RP, T(3)	251	119	3	337	710
5RP, E(3)	256	160	166	319	909

Table 1: Execution timing results for inverse kinematics via Newton's method.

The program allows the user to select the kinematic model, the task-space model and the kinematics algorithm at will; and it has turned out to be a good testbed for developing kinematics algorithms. The main limitations of this program are that the task-space model allows only a single displacement, that the mechanism can not contain kinematic loops, and that acceleration calculations are not implemented. The limitation in the task-space model was the result of lack of time rather than any inherent difficulty in implementation.

Some execution timing tests were performed on algorithm B, and the results are shown in Table 1. Each test case is identified by a robot model and a displacement group. 2R and 3R are planar robots with two and three revolute joints, respectively, and 5RP is a 6-DoF polar robot. The remaining columns give execution times for various parts of the calculation. The first two columns give the time to calculate the forward kinematics (k and $\mathbf{J}$), divided into a conventional position calculation (bodies and joint axes), and the calculation of task-space data from the position data. The next column refers to the task-space calculation done by the inverse-kinematics algorithm ($\log(k(\mathbf{q}_{i-1})^{-1} \circ x_i)$), the next to solving the linear equation, and the final column gives the total time for one iteration of the algorithm. All times are in microseconds, and the program was executed on a MIPS processor. Owing to measurement errors, the total times are not always the sums of the constituent parts.

A program that uses an explicit task-space model is likely to be slower than one designed to operate in a single task space. The model overhead is the difference between the time it takes the model-based program to do its task-related calculations and the time it takes a dedicated program to do the same. From the figures in Table 1, the model-based program spends between 17% and 38% of its time doing task-related calculations. These figures set an upper limit on the size of the model overhead, from which we can conclude that the use of an explicit task-space model is efficient enough to be a practical proposition, at least for problems of a similar nature to the test cases in Table 1.

Conclusions

This paper proposes that a general-purpose software system for performing inverse kinematics calculations should include an explicit model of the task space, on the grounds that a complete description of the inverse kinematics problem requires information on the relationship between task-space elements and configurations of the mechanism, and such information is better placed in an explicit model data structure rather than hard-wired into the code.

To illustrate this concept, a particular example of a task-space model architecture was described. In this architecture, the task space itself is a Lie group, and each element specifies an n-tuple of relative displacements between pairs of coordinate frames attached to the mechanism. Methods were described for representing trajectories and for performing inverse kinematics calculations in this task space; and execution timing tests showed that it is efficient enough for practical use. However, there are some defects in this architecture that need to be fixed. For example, there is no provision for cost functions and metrics. This kind of data is needed to establish the correct weighting of joint motions for pseudoinverse control, and the relative weighting of linear and angular displacements in task space. Another problem is that the use of Lie groups restricts the types of task space that can be modelled.

References

[1] Angeles, J., On the Numerical Solution of the Inverse Kinematics Problem, *Int. Jnl. Robotics Research*, 4(2):21–37, 1985.

[2] Brady, J. M. *et al.* (eds.), *Robot Motion: Planning and Control*, MIT Press, Cambridge, Mass., 1982.

[3] Featherstone, R., *Robot Dynamics Algorithms*, Kluwer Academic Publishers, Boston/-Dordrecht/Lancaster, 1987.

[4] Featherstone, R., Accurate Trajectory Transformations for Redundant and Nonredundant Robots, *to appear in* IEEE Int. Conf. Robotics and Automation, San Diego, 1994.

[5] Goldenberg, A. A., Benhabib, B., and Fenton, R. G., A Complete Generalized Solution to the Inverse Kinematics of Robots, *IEEE Jnl. Robotics & Automation*, RA-1(1):14–20, 1985.

[6] Gupta, K. C. and Kazerounian, K., Improved Numerical Solutions of Inverse Kinematics of Robots, *Proc. IEEE Int. Conf. Robotics and Automation*, pp. 743–748, St. Louis, 1985.

[7] Kazerounian, K., On the Numerical Inverse Kinematics of Robotic Manipulators, *Trans. ASME Jnl. Mechanisms, Transmissions & Automation in Design*, 109:8–13, 1987.

[8] Oh, S. Y., Orin, D., and Bach, M., An Inverse Kinematic Solution for Kinematically Redundant Robot Manipulators, *Jnl. Robotic Systems*, 1(3):235–249, 1984.

[9] Tsai, Y. T. and Orin, D. E., A Strictly Convergent Real-Time Solution for Inverse Kinematics of Robot Manipulators, *Jnl. Robotic Systems*, vol. 4, no. 4, pp. 477–501, 1987.

On the General Solution of the Inverse Kinematics of Six-Degrees-of-Freedom Manipulators

W. KHALIL, D. MURARECI

Laboratoire d'Automatique de Nantes URA C.N.R.S 823
1 Rue de la Noë, 44072, Nantes cedex, FRANCE
Khalil@lan.ec-nantes.fr

Abstract - This paper presents a method for the calculation of the inverse kinematics of six-degree-of-freedom robots. The method is derived from the general method presented by Raghavan and Roth [1], [2]. The new contribution resides on the possibility of taking into account the special values of the geometric parameters in order to derive the characteristic polynomial in the simplest form and using the minimum number of equations. In such a special geometry the characteristic polynomial may be obtained symbolically for general desired position and orientation of the end effector. The given method has been programmed using Mathematica.

I. Introduction

The inverse kinematic model is the problem of calculating the joint variables corresponding to a given situation (position and orientation) of the end effector.The inverse kinematics of particular manipulators has been studied in a lot of works such as Pieper [1] and Paul [2]. Raghavan and Roth [3] [4] have presented a general method to calculate the inverse kinematics of all general geometry manipulators having 6 d.o.f (**6R, 5R1P, 4R2P, 3R3P**). It has been shown that the maximum number of solutions is equal to 16 in the case of **6R** manipulators, the solution using the general method is given in the form of a univariant polynomial in one of the joint variables that gives all the solutions to the inverse kinematics problem for this variable.This polynomial is called the characteristic polynomial. For some special values of the geometric parameters the method of Raghavan and Roth may not work (Mavroidis [5]). The aim of this paper is to modify the general method to take into account the special geometry and to use the most convenient starting equation in order to facilitate the solution and to reduce the degree of the polynomial of the first joint solved.

II. Description of the manipulator

The manipulator is described using the modified Denavit and Hartenberg notation [6],[7]. The coordinate frame j is assigned fixed with respect to link j. The z_j axis is along the axis of joint j, the x_j axis is along the common perpendicular of z_j and z_{j+1}. The geometry of the robot is defined by the following geometric parameters (for j=1,..., 6):

- α_j, d_j, θ_j, r_j which define frame j with respect to frame j-1, the joint variable q_j is equal to θ_j if j is rotational, or r_j if j is prismatic.
- σ_j defines the type of joint j, with $\sigma_j = 0$ for j rotational, $\sigma_j = 1$ for j prismatic.

A. J. Lenarčič and B. B. Ravani (eds.), Advances in Robot Kinematics and Computationed Geometry, 309–318.

The matrix ${}^{j-1}T_j$, defining frame j with respect to frame j-1, is given as:

$${}^{j-1}T_j = \text{Rot}(x, \alpha_j)\ \text{Trans}(x, d_j)\ \text{Rot}(z, \theta_j)\ \text{Trans}(z, r_j)$$

$${}^{j-1}T_j = \begin{bmatrix} Cj & -Sj & 0 & d_j \\ C\alpha_j Sj & C\alpha_j Cj & -S\alpha_j & -r_j S\alpha_j \\ S\alpha_j Sj & S\alpha_j Cj & C\alpha_j & r_j C\alpha_j \\ 0 & 0 & 0 & 1 \end{bmatrix} \quad (1)$$

with $Cj = \text{Cos}(\theta_j)$, $Sj = \text{Sin}(\theta_j)$, $C\alpha_j = \text{Cos}(\alpha_j)$, $S\alpha_j = \text{Sin}(\alpha_j)$

The direct kinematic model defining the end effector frame in the base frame is given as :

$$U_0 = {}^0T_1(q_1)\,{}^1T_2(q_2)\,{}^2T_3(q_3)\,{}^3T_4(q_4)\,{}^4T_5(q_5)\,{}^5T_6(q_6) \quad (2)$$

III. Raghavan and Roth general method

A complete description of the general method of Raghavan and Roth for calculating the inverse kinematic model using the classical Denavit and Hartenberg notation can be found in Raghavan and Roth [1]. Here we describe the outline of this method for **6R** robots using our notation of robot description. Therefore some modifications are introduced. The main steps of the method are :

1. The transformation equation (2) of a **6R** manipulator can be written under the form:

$${}^0T_1\ {}^1T_2\ {}^2T_3\ {}^3T_4 = U_0\ {}^6T_5\ {}^5T_4 \quad (3)$$

Since the forth and third columns of ${}^{j-1}T_j$ are not functions of θj, the forth and third columns of the previous equation are not functions of θ_4.

2. By equating the following elements on both sides of Eq.(3) we get 14 scalar equations devoided of θ_4 and s_j^2, sc_j, c_j^2:

$$\mathbf{a},\ \mathbf{P},\ \mathbf{a.P},\ \mathbf{P},\ \mathbf{axP},\ \mathbf{a(P.P)-2P(a.P)} \quad (4)$$

where "." stands for the scalar product and "x" for the vector product, **a** and **P** are the vectors formed of the first three components of columns 3 and 4 respectively of Eq. (3).

3. Eqs. (4) form the following system :

$$\mathbf{A\ X1 = B\ Y} \quad (5)$$

where: **A** is a (14x9) matrix, whose elements are functions of S1 and C1,

B is a (14x8) matrix whose elements are constant,
X1=[S2S3, S2C3, C2S3, C2C3, S2, C2, S3, C3, 1 $]^T$
Y=[S5S6, S5C6, C5S6, C5C6, S5, C5, S6, C6 $]^T$

4. Elimination of $\theta 5$ and $\theta 6$: using 8 equations out of the 14 equations of Eq.(5) to eliminate **Y**, the resulting system of 6 equations takes the form :

$$\mathbf{D\ X1} = \mathbf{0} \tag{6}$$

where the elements of the (6x9) matrix **D** are functions of S1, C1, the constant geometric parameters and the elements of $\mathbf{U_0}$.

5. Elimination of $\theta 2$ and $\theta 3$: using the following substitutions in Eq.(6):
$Si = 2x_i \ / \ (1 + x_i^2)$ and $Ci = (1 - x_i^2) \ / \ (1 + x_i^2)$
with $x_i = \tan(q_i/2)$, we obtain the following system:

$$\mathbf{E\ X2} = \mathbf{0} \tag{7}$$

where: **E** is a (6 x 9) matrix, whose elements are quadratic functions of x_1, and

$$\mathbf{X2} = [\, x_2^2x_3^2, x_2^2x_3, x_2^2, x_2x_3^2, x_2x_3, x_2, x_3^2, x_3, 1 \,]^T \tag{8}$$

6. Multiplying Eq.(7) by x_2, we obtain 6 new equations with 3 additional variables:

$$\mathbf{E\ X3} = \mathbf{0} \tag{9}$$

where: $\mathbf{X3} = [\, x_2^3x_3^2, x_2^3x_3, x_2^3, x_2^2x_3^2, x_2^2x_3, x_2^2, x_2x_3^2, x_2x_3, x_2 \,]^T \quad (10)$

Regrouping Eqs.(8,10) we obtain the following system of equations:

$$\mathbf{\Sigma\ X} = \mathbf{0} \tag{11}$$

where

$$\mathbf{X} = [\, x_2^3x_3^2, x_2^3x_3, x_2^3, x_2^2x_3^2, x_2^2x_3, x_2^2, x_2x_3^2, x_2x_3, x_2, x_3^2, x_3, 1]^T \tag{12}$$

Σ is (12 x 12) matrix, whose elements are quadratic functions of x_1, such that

$$\Sigma = \begin{bmatrix} \mathbf{E} & \mathbf{0}_{6x3} \\ \mathbf{0}_{6x3} & \mathbf{E} \end{bmatrix} \tag{13}$$

where $\mathbf{0}_{6x3}$ is the null 6x3 matrix.

7. The condition ($\det\Sigma=0$) gives the characteristic polynomial of the manipulator in x_1. For a **6R** general manipulator its degree is equal to 16.

8. Substituting back each real solution of θ_1 in Eq.(11) gives a unique value for θ_2 and θ_3 from the solution of a linear system. Further substitution of $\theta_1, \theta_2, \theta_3$ in Eq.(5) gives a unique value for θ_5, θ_6. Finally, θ_4 follows from two elements of the first two columns of the equation:

$$^{3}T_{4} = {}^{3}T_{2}\, {}^{2}T_{1}\, {}^{1}T_{0}\, {}^{0}T_{6}\, {}^{6}T_{5}\, {}^{5}T_{4}$$

It has been shown that the same method can be applied to solve the inverse kinematics problem of general **5R1P**, **4R2P**, and **3R3P** robots [4], [5]. It is to be noted that in the elements of vectors **X1** and **Y** of Eq.(5), Cj is replaced by r_j and Sj is replaced by r_j^2 for j prismatic, and that Eq.(3) cannot be used if joint 4 is prismatic. One of the following six different ways of the loop closure equation can be used :

$$^{4}T_{5}\, {}^{5}T_{6}\, {}^{6}T_{7}\, {}^{0}T_{1} = {}^{4}T_{3}\, {}^{3}T_{2}\, {}^{2}T_{1} \quad (14)$$
$$^{5}T_{6}\, {}^{6}T_{7}\, {}^{0}T_{1}\, {}^{1}T_{2} = {}^{5}T_{4}\, {}^{4}T_{3}\, {}^{3}T_{2} \quad (15)$$
$$^{6}T_{7}\, {}^{0}T_{1}\, {}^{1}T_{2}\, {}^{2}T_{3} = {}^{6}T_{5}\, {}^{5}T_{4}\, {}^{4}T_{3} \quad (16)$$
$$^{0}T_{1}\, {}^{1}T_{2}\, {}^{2}T_{3}\, {}^{3}T_{4} = {}^{7}T_{6}\, {}^{6}T_{5}\, {}^{5}T_{4} \quad (17)$$
$$^{1}T_{2}\, {}^{2}T_{3}\, {}^{3}T_{4}\, {}^{4}T_{5} = {}^{1}T_{0}\, {}^{7}T_{6}\, {}^{6}T_{5} \quad (18)$$
$$^{2}T_{3}\, {}^{3}T_{4}\, {}^{4}T_{5}\, {}^{5}T_{6} = {}^{2}T_{1}\, {}^{1}T_{0}\, {}^{7}T_{6} \quad (19)$$

with $^{7}T_{6} = U_0$ and $^{6}T_{7} = U_0^{-1}$

These equations gives the matrix Σ and the characteristic polynomial as function of xi, with i=1,...,6, respectively

In general, for **6R** manipulators any of the equations (14) to (19) can be used as starting equation, but owing to some conditions between the geometric parameters, or if the robot has prismatic joints, some of these equations cannot be used. Mavroidis [5] has given some conditions between the geometric parameters to avoid some of these equations, because the resulting Σ matrix will contain dependent columns. Instead of eliminating such equations we propose to take them and to reduce the number of variables in **X1** and the dimension of **A**; this choice leads to reducing the dimension of Σ and, consequently, the characteristic polynomial which may be obtained completely in symbolic and reduced form.

IV. General form of starting equation

Eqs.(17) to (20) can be put on the the following general form:

$$^{i-1}T_{i}\, {}^{j-1}T_{j}\, {}^{k-1}T_{k}\, {}^{m-1}T_{m} = {}^{r-1}T_{r}\, {}^{s-1}T_{s}\, {}^{t}T_{t-1} \quad (20)$$

where j=i+1, k=j+1, m=k+1, thus i can be equal to 1, 2 or 3. The characteristic polynomial thus obtained is a function of the joint variable i. To consider the characteristic polynomial using the variables 4, 5, and 6 the inverse manipulator will be used (Appendix 1).

It is to be noted that joint m cannot be prismatic because r_m will appear in the forth column of Eq.(20) .

V. Reducing the dimension of matrix Σ

The following cases are considered to reduce the dimension of the Σ matrix.

A. The matrix A contains dependent columns

If some columns of the matrix **A** are linearly dependent, then the matrix Σ will contain also dependent columns and all the characteristic polynomial coefficients turn to 0. In this case new reduced **A** and Σ can be obtained by eliminating the dependent columns and regrouping the corresponding elements of the vector **X1**, such that Eq.(5) becomes:

$$\mathbf{A_R}(14xa)\ \mathbf{X1_R} = \mathbf{B}(14x8)\ \mathbf{Y} \tag{21}$$

where: a is the number of independent columns in the matrix **A**.

$\mathbf{A_R}$ is obtained from **A** by taking only the independent columns.

$\mathbf{X1_R}$ is the (ax1) vector, obtained from **X1** after regrouping the elements corresponding to dependent columns on those corresponding to the independent columns.

For instance if $A(;i) = p\ A(;j)$, and $A(;m) = 0$, where $A(;j)$ is the jth column of the matrix **A**. Then the reduced $\mathbf{A_R}$ is formed from A after eliminating the columns j and m, while $\mathbf{X1_R}$ is formed from **X1** after eliminating the jth and m^{th} elements and replacing the ith element by the sum $\mathbf{X1}(i)+p\ \mathbf{X1}(j)$.

For the solution of the regrouped system two cases are considered:

1) If a is less than or equal to 6

This means that the effective number of unknowns equal to a, then the calculation of **E** and Σ is not needed, the following system of equations can be used directly:

$$\mathbf{C}\ \mathbf{X1_R} = \mathbf{0} \tag{22}$$

where the matrix **C** of dimension (axa), is obtained from **D** (step 4) after :

a- eliminating arbitrarily (6-a) rows

b- replacing $C(\theta_i)$ and $S(\theta_i)$ as function of x_i.

As the last element of the system (22) is unity, thus (det **C** = 0) gives the solution of the variable i. Substituting it in (22) permits to get the regrouped vector $\mathbf{X1_R}$ by solving a system of linear equations with a-1 unknowns. The solution of the joint variables j and k will be consequently obtained. It is to be noted that since the elements of $\mathbf{X1_R}$ are linear combinations of the elements of **X1**, multiple solutions may be obtained in this step. The variables of the right hand side (R.H.S.) of Eq.(21) will be obtained by solving a system of linear equations.

2) If a is greater than 6

The number of independent columns of A is greater than 6. In this case the calculation of Σ (step 6) must be carried out. But its dimension will be reduced.

B. The robot has more than one prismatic joint:

The following two cases has been given:

- It has been pointed out in Raghavan [2] that in the case of the **4R2P** manipulator, the characteristic polynomial can be obtained as the determinant of Σ either of 12x12 matrix if i is prismatic or 6x6 matrix if i is rotational. Thus the second choice must be taken if possible.

- if the manipulator is of **3P3R** the characteristic polynomial is obtained from the determinant of Σ either of 6x6 matrix if i is prismatic or 3x3 matrix if i is rotational. Thus this simplification must be considered.

VI. Simplified solution of special robots

The special robots considered in this paper are those leading to dependent columns of **A**. They will be defined as function of the type of the joints i,j,k,m of Eq.(20) and of the values of the geometric parameters of links j and k. The following 8 combinations for the types of the joints (i,j,k) may exist **RRR, PRR, RRP, PRP, RPR, PPR, RPP** and **PPP**.

A. The joints i, j and k are ***XRR***

X stands for rotational or prismatic joint. The following cases are considered:

1) if $d_m = d_k = r_k = 0$ with $S\alpha k \neq 0$, $S\alpha m \neq 0$,

This condition means that the joints j, k and m form a spherical joint. The following relations are verified:

$A(;4) = -C\alpha k\, A(;1)$

$A(;6) = -(C\alpha_m S\alpha_k / S\alpha_m)\, A(;1)$

$A(;2) = C\alpha_k\, A(;3)$

$A(;5) = (C\alpha_m S\alpha_k / S\alpha_m)\, A(;3)$

$A(;7) = 0$

Thus, the column 7 of the matrix **A** will be eliminated . While the columns 4 and 6 are eliminated because proportional to the column 1, the corresponding unknowns will be regrouped to the first one, similarly the columns 2 and 5 are eliminated, the corresponding variables are regrouped to the third unknown. Thus, the vector $\mathbf{X1_R}$ is given by:

$\mathbf{X1_R} = [\, S\theta j\, S\theta k - C\alpha k\, C\theta j\, C\theta k - \beta\, C\theta j,\ C\theta j\, S\theta k + C\alpha k\, S\theta j\, C\theta k + \beta\, S\theta j,\ C\theta k,\ 1\,]^T$

with $\beta = C\alpha_m S\alpha k / S\alpha_m$

The polynomial of the variable i is determined by det **C** = 0, with **C** is of dimension (4 x 4). If the solution of the vector $\mathbf{X1_R}$ is supposed equal to:

$\mathbf{X1s} = [xs1, xs2, xs3, 1]^T$,

then we get:

$\theta k = ArcCos(xs_3)$

$S\theta j = (H\, xs_1 + Z\, xs_2)/(H^2+Z^2)$

$C\theta j = (H\, xs_2 - Z\, xs_1)/(H^2+Z^2)$

where $H = S\theta k$, $Z = C\alpha_k C\theta k + \beta$

Two solutions are given for θj and θk .

2) If $\alpha_k = p\,\pi$ and $\alpha_m = p'\pi$ (p and p' are integer)
This mean that the axes j, k and m are rotational and parallel. We get

$A(;4) = -C\alpha_k\,A(;1)$, $A(;6) = -(C\alpha_k\,d_k/d_m)\,A(;1)$
$A(;2) = C\alpha_k\,A(;3)$, $A(;5) = -(C\alpha_k\,d_k/d_m)\,A(;3)$
$A(;7) = 0$

Repeating the same procedure as before, we obtain that the columns 5,6,7,8 of **A** will be eliminated. The elements of the vector $\mathbf{X1_R}$ are given by:

$\mathbf{X1_R} = [\,S\theta j\,S\theta k - C\alpha k\,C\theta j\,C\theta k - \beta\,C\theta j,\ C\theta j\,S\theta k + C\alpha j\,S\theta j\,C\theta k + \beta\,S\theta j,\ C\theta k,\ 1\,]^T$
with $\beta = C\alpha_k\,d_k\,/\,d_m$
The matrix **C** is of dimension (4 x 4).
If the solution of the vector $\mathbf{X1_R}$ is supposed equal to:
$\mathbf{X1s} = [xs1,\ xs2,\ xs3,1]^T$,
then we get:
$\theta k = \mathrm{ArcCos}(xs_3)$
$S\theta j = (H\,xs_1 + Z\,xs_2)/(H^2+Z^2)$
$C\theta j = (H\,xs_2 - Z\,xs_1)/(H^2+Z^2)$
where $H = S\theta k$, $Z = C\alpha_k\,C\theta k + \beta$
Two solutions are given for θj and θk.

3) if $\alpha_k = \alpha_m$ and $d_k = d_m$ and $r_k = 0$
In this case the axes j, k and m verify the conditions of Bennett, then:

$A(;5) = A(;2)$, and $A(;6) = A(;4)$

This turns out the column 1 is equal to 7 and column 4 is equal to column 10 in Σ. Thus the columns 7 and 10 will be eliminated. The Σ matrix is reduced to dimension (10x10) by eliminating two arbitrary rows.
The vector $\mathbf{X_R}$ is given by:

$\mathbf{X_R} = [\,x_j^3\,x_k^2 + x_j\,x_k^2,\ x_j^3\,x_k,\ x_j^3,\ x_j^2 x_k^2 + x_k^2,\ x_j^2\,x_k,\ x_j^2,\ x_j\,x_k,\ x_j,\ x_k,\ 1]^T$
The variables x_j and x_k are obtained directly in the solution of $\mathbf{X_R}$.

B. The joints i, j and k are RRP
In this case if $\alpha_k = p\,\pi$ and $\alpha_m = p'\pi$ (the axes j, k and m are parallel) then:

$A(;1) = 0$, $A(;3) = 0$

The columns 7 and 10 of the matrix Σ are equal to the columns 1 and 4 respectively. By eliminating two rows of Σ, it becomes of dimension (10 x 10). The vector $\mathbf{X_R}$ is given as:

$\mathbf{X_R} = [x_j^3 r_k^2 + x_j r_k^2, x_j^3 r_k, x_j^3, x_j^2 r_k^2 + r_k^2, x_j^2 r_k, x_j^2, x_j r_k, x_j, r_k, 1]^T$
The variables x_j and r_k are obtained directly in the solution of the vector $\mathbf{X_R}$.

C. If the joints i, j, k are RPR
The following cases are considered:

1). If $\alpha_k = (2p+1)\pi/2$ and $\alpha_m = p'\pi$
In this case the axes j and k are perpendicular and the axes k and m are parallel, then we get :
$A(;1) = 0$, $A(;3) = 2 S\alpha_k d_m A(;5)$
$A(;2) = 0$, $A(;4) = 0$, $A(;7) = S\alpha_k d_m A(;6)$

- thus the columns 1,2 and 4 of **A** are eliminated. The columns 3 and 7 are depending on 5 and 6 respectively, thus they will be eliminated and the vector $\mathbf{X1_R}$ is given by :

$$\mathbf{X1_R} = [r_j^2 + 2 S\alpha_k d_m r_j S\theta_k , r_j + S\alpha_k d_m S\theta_k, C\theta_k, 1]^T$$

The solution of x_i is given by det C = 0, with C is a 4x4 matrix.
Assuming that the solution of $X1_R$ is given as:
$\mathbf{X1s} = [xs_1, xs_2, xs_3, 1]^T$, thus we get:
$\theta_k = \mathrm{ACos}(xs1)$, and $r_j = xs_2/(S\alpha_k d_m S\theta_k + xs_3)$
thus for each solution of x_i, we get two values for q_k and r_j .

2). If $\alpha_m = p\pi$ (the axes k and m are parallel and j is prismatic) then:
$A(;1) = 0$ $A(;2) = 0$
These relations turn the columns (1,3) of Σ to be equal and the column 2 to be equal to zero. By eliminating the columns 2 and 3, and eliminating two arbitrary rows the reduced Σ will be of dimension (10 x 10), with:
$\mathbf{X_R} = [r_j^3 x_k^2 + r_j^3, r_j^2 x_k^2, r_j^2 x_k, r_j^2, r_j x_k^2, r_j x_k, r_j, x_k^2, x_k, 1]^T$
The variables r_j and x_k are obtained directly in the solution of $\mathbf{X_R}$.

D. The joints i, j and k are RPP or PRP
In this case, the variables of the vector X will be composed of those of the two prismatic joints: (r_j, r_k) for RPP and (r_i,r_k) for PRP, the variable of the final polynomial will be x_i in the first case and x_j in the second one.

$$\mathbf{X1} = [r_j^2 r_k^2, r_j^2 r_k, r_j r_k^2, r_j r_k, r_j^2, r_j, r_k^2, r_k, 1]^T \text{ for RPP}$$
$$\mathbf{X1} = [r_i^2 r_k^2, r_i^2 r_k, r_i r_k^2, r_i r_k, r_i^2, r_i, r_k^2, r_k, 1]^T \text{ for PRP}$$

We get: $A(;1) = 0$, $A(;2) = 0$, $A(;3) = 0$, thus the vector $\mathbf{X1_R}$ is given by :

$$\mathbf{X1_R} = [r_j r_k, r_j^2, r_j, r_k^2, r_k, 1]^T \text{ for RPP}$$
$$\mathbf{X1_R} = [r_i r_k, r_i^2, r_i, r_k^2, r_k, 1]^T \text{ for PRP}$$

Thus the polynomial of the variable x_i or x_j is obtained using the determinant of a 6x6 matrix. The variables r_j and r_k or r_j and r_k are obtained directly in the solution of $\mathbf{X1_R}$.

E. Two prismatic joints separated by 3 or 4 rotational joints
Since the optimum choice is to put the two variables together with vector **X1**, Eq.(21) cannot be used.
We use as starting equation :

$$^1T_2\,^2T_3\,^3T_4 = {}^1T_0\,^7T_6\,^6T_5\,^5T_4 \tag{23}$$

where the prismatic variables are on the R.H.S.
The following system of 14 equations is obtained:

$$\mathbf{B\ Y} = \mathbf{A\ X} \tag{24}$$

The general method is applied to this system where the left hand side becomes the R.H.S. and vice-versa:

$\mathbf{Y}$=[S2S3, S2C3, C2S3, C2C3, S2, C2, S3, C3 $]^T$
$\mathbf{X1_R} = [\, r_1\, r_5, r_1^2, r_1, r_5^2, r_5, 1\,]^T$ for **PRRRPR**
$\mathbf{X1_R} = [\, r_1\, r_6, r_1^2, r_1, r_6^2, r_6, 1\,]^T$ for **PRRRRP**

The characteristic polynomial is given by the determinant after eliminating **Y** from Eq.(24).

F. The robot has three prismatic joints.
The method used here is analogous to Pieper's [1] and Raghavan's [4] methods: the position and the orientation are decoupled so the rotational joint angles can be solved independently of the prismatic variables.
Once the rotational joints are taken in the left hand of the equation, we have to arrange it so that the (3;3) element is only dependent on one of the angles. The (3,3) element of the R.H.S. is constant.

VII. Automatic programming of the method

The presented method has been programmed using Mathematica system in order to get the inverse kinematic model automatically and if possible under symbolic form; this program is integrated to the software package SYMORO+ [8]. Depending on the order of complexity the symbolic calculation can stop after calculating Σ, **D**, or even the calculation of **A** and **B**, then for given numerical values for the desired location the corresponding joint variables can be obtained by completing the algorithm as given in section 3. For manipulators with reduced equations such that the matrix Σ is not needed the determinant of **C** can be symbolically obtained as function of general desired situation $\mathbf{U_0}$.

VIII. Conclusion

This paper presents a method for the calculation of the inverse kinematics of general six degrees of freedom robots and take into account the special values of the geometric parameters in order to get the polynomial characteristic in the most simplest form and using the minimum number of equations. In some cases it can be seen that the polynomial of the first resolved joint is obtained as the determinant of a 4x4 matrix. In such special geometry the polynomial characteristic may be obtained symbolically for general desired position and orientation of the end effector. The given method has been programmed using Mathematica to get automatically the inverse kinematic model of general or special robots. The special robots resolved in this paper concern the relation between the types and the geometric parameters of 3 consecutive joints or if the number of prismatic joints are equal to 2 or 3.

IX. References

[1] Pieper D.L., "The kinematics of manipulators under computer control", Ph. D. Thesis, Stanford University, Stanford, 1968.

[2] Paul R.C.P., *Robot manipulators : mathematics, programming and control*, MIT Press, 1981.

[3] Raghavan M., Roth B., "Kinematic Analysis of the 6R Manipulator of General Geometry", *Proc. of the 5th ISRR*, MIT press, Cambridge, 1990, p.236-270 .

[4] Raghavan M., Roth B., "A General Solution for the Inverse Kinematics of all chains", Romansy 90, Cracow, Poland 1990 .

[5] Mavroidis C., "Inverse Kinematics of six-degree of freedom General and Special Manipulators using symbolic computation", *thesis*, Université Pierre et Marie Curie, 1993 .

[6] Khalil W., Kleinfinger J.-F., "A new geometric notation for open and closed-loop robots", *Proc. IEEE Conf. on Robotics and Automation*, San Francisco, 1986,p. 1174-1180.

[7] Dombre E., Khalil W., "Modélisation et commande des robots", *Edition Hermes*, Paris, 1988.

[8] Khalil W., Creusot D., Chevallereau C., Bennis F., "SYMORO+, User Guide", L.A.N., April 1994.

Appendix 1: Inverse manipulator

The robot (σ_j', α_j', d_j', θ_j', r_j') is said to be the inverse of the robot (σ_j, α_j, d_j, θ_j, r_j) if the transformation matrix ${}^0T_6$ (σ_j', α_j', d_j', θ_j', r_j') is equal to ${}^0T_6^{-1}$ (σ_j, α_j, d_j, θ_j, r_j).

Since the parameters α_1, d_1 can be taken always to zero in the modified notations then the parameters of the inverse robot can be obtained as given in Table 1 as functions of (σ_j, α_j, d_j, θ_j, r_j).

j'	σ_j'	α_j'	j'	θ_j'	j'
1	σ_6	0	0	$-\theta_6$	$-r_6$
2	σ_5	$-\alpha_6$	$-d_6$	$-\theta_5$	$-r_5$
3	σ_4	$-\alpha_5$	$-d_5$	$-\theta_4$	$-r_4$
4	σ_3	$-\alpha_4$	$-d_4$	$-\theta_3$	$-r_3$
5	σ_2	$-\alpha_3$	$-d_3$	$-\theta_2$	$-r_2$
6	σ_1	$-\alpha_2$	$-d_2$	$-\theta_1$	$-r_1$

Table 1: Geometric parameters of the inverse manipulator

Coping with Joint Velocity Limits in First-Order Inverse Kinematics Algorithms

Pasquale Chiacchio Stefano Chiaverini

Dipartimento di Informatica e Sistemistica
Università degli Studi di Napoli "Federico II"
via Claudio 21, 80125 Napoli, Italy
E-mail: {chiacchio,chiaverini}@disna.dis.unina.it

Abstract — A major problem of inverse kinematics algorithms is that the generated joint velocities may violate the speed limits of the joint actuators; this occurs when the assigned task-space trajectory is too fast or the manipulator is required to move close to a kinematic singularity. In this paper, it is shown how to properly cope with joint velocity limits in first-order inverse kinematics algorithms, while keeping path-tracking capabilities. This goal is achieved by suitably slowing down the task-space trajectory when joint velocity limits are encountered. The time law is modified through a time warp such that the introduced virtual time allows fulfillment of the velocity constraints. A number of case studies are developed to show application of the proposed technique.

I. Introduction

The desired motion of a manipulator is typically specified in the task space by assigning a geometric path to be followed by the end effector with a given time law. On the other hand, the robot arm is actuated at the joints thus requiring control actions to be performed in the joint space. For this reason, a considerable research effort has been devoted to solve the inverse kinematics problem.

A closed-form analytical solution can be found only for non-redundant manipulators having simple geometric structures [1]. For all those structures that cannot be solved in closed-form, a number of algorithmic techniques have been devised mainly based on the use of the manipulator's Jacobian matrix. In this work, first-order inverse kinematics algorithms are investigated, which are aimed to invert the first-order direct differential kinematics.

The resolved motion rate control technique [2] uses the pseudoinverse of the Jacobian matrix to obtain the joint velocities corresponding to a given end-effector velocity; the joint angles are then obtained by numerical integration. To overcome the problem of kinematic singularities a damped least-squares inverse of the Jacobian matrix has been proposed in lieu of the pseudoinverse [3,4]. These and other related techniques suffer from errors due to both long-term numerical integration drift and incorrect initial joint angles.

Algorithmic solutions that circumvent these problems are based on considering a closed-loop dynamic system whose input is the task-space trajectory and whose output are the corresponding joint angles and velocities. A closed-loop inverse kinematics

A. J. Lenarčič and B. B. Ravani (eds.), Advances in Robot Kinematics and Computationed Geometry, 319–328.

scheme based on the Jacobian transpose was first independently proposed in [5,6]. Similar schemes can be devised that use either the Jacobian pseudoinverse, e.g. [7], or a damped least-squares inverse, e.g. [8,9].

A common problem of the above inverse kinematics algorithms is that the generated joint velocities may violate the speed limits of the joint actuators; this occurs when the assigned task-space trajectory is too fast or the manipulator is required to move close to a kinematic singularity. A possible solution would be to include saturations on the joint velocity variables in the algorithms; however, path tracking would no longer be guaranteed.

In this paper, it is shown how to properly cope with joint velocity limits in first-order inverse kinematics algorithms, while keeping path-tracking capabilities. This goal is achieved by suitably slowing down the task-space trajectory when joint velocity limits are encountered. The time law is modified through a time warp such that the introduced virtual time allows fulfillment of the velocity constraints. In addition, the task-space trajectory may be speeded up, if possible, to recover the delay caused by the previous slow down of the trajectory. A number of case studies are developed to show application of the proposed technique.

II. Inverse Kinematics Algorithms

The direct kinematics of a general serial-link manipulator can be written in the well-known form

$$\boldsymbol{x} = \boldsymbol{f}(\boldsymbol{q}) \tag{1}$$

$$\dot{\boldsymbol{x}} = \boldsymbol{J}(\boldsymbol{q})\dot{\boldsymbol{q}} \tag{2}$$

where $\boldsymbol{q}$ is the vector of the joint-angle variables, $\boldsymbol{x}$ is the vector of the end-effector variables, and $\boldsymbol{J} = d\boldsymbol{f}/d\boldsymbol{q}$ is the manipulator's Jacobian matrix.

The motion of the manipulator is specified by assigning a desired end-effector trajectory; for the first-order inverse kinematics problem this means that the functions

$$\boldsymbol{x}_d(t) \tag{3}$$

$$\dot{\boldsymbol{x}}_d(t) = \frac{d\boldsymbol{x}_d}{dt}(t), \tag{4}$$

are known.

Algorithmic solutions to the inverse kinematics problem are based on the inversion of the first-order direct differential kinematics (2) and are usually implemented in discrete-time form. In this case, the joint angles are obtained by numerical integration of the joint velocities; this causes drift of the solution that can be overcome by resorting to closed-loop versions of the algorithms. In the remainder of this section the discrete-time version of first-order inverse kinematics algorithms is reported.

First-order algorithms output the joint angles and velocities corresponding to a given end-effector trajectory. It can be recognized that they can be written in the general form

$$\dot{q}_k = H_k(q_{k-1})w_k \tag{5}$$

$$q_k = q_{k-1} + \dot{q}_k \Delta t \tag{6}$$

where the subscript k denotes quantities evaluated at the k-th time instant t_k, and Δt is the time interval. The expressions for H_k and w_k depend on the particular scheme and are listed in Table I in which the matrix K is a suitable constant positive-definite gain matrix, and λ is the damping factor. The vector $e_k = x_{d,k} - x_{k-1}$, is the closed-loop algorithms' error due to numerical integration of the joint velocity solution.

Algorithm	$H_k(q_{k-1})$	w_k
O.l. pseudoinverse	$J^{\mathrm{T}}(JJ^{\mathrm{T}})^{-1}$	$\dot{x}_{d,k}$
O.l. DLS inverse	$J^{\mathrm{T}}(JJ^{\mathrm{T}} + \lambda^2 I)^{-1}$	$\dot{x}_{d,k}$
C.l. transpose	KJ^{T}	e_k
C.l. pseudoinverse	$J^{\mathrm{T}}(JJ^{\mathrm{T}})^{-1}$	$\dot{x}_{d,k} + Ke_k$
C.l. DLS inverse	$J^{\mathrm{T}}(JJ^{\mathrm{T}} + \lambda^2 I)^{-1}$	$\dot{x}_{d,k} + Ke_k$

Table 1

III. Coping with Joint Velocity Limits

A common problem of the inverse kinematics algorithms is that the generated joint velocities may violate the speed limits of the joint actuators; this occurs when the assigned task-space trajectory is too fast or the manipulator is required to move close to a kinematic singularity. A possible solution would be to include saturations on the joint velocity variables in the algorithms; however, path tracking would no longer be guaranteed.

To keep path-tracking capabilities, the only possibility is to slow down the time law of the desired trajectory when speed limits are encountered. This can be done by introducing a virtual time τ via a time warp in the trajectory generation, i.e.

$$\tilde{x}_d(t) = x_d(\tau(t)), \tag{7}$$

where $\tau(t)$ is an increasing function such that $\tau(0) = 0$ and $\tau(t) \leq t$. The latter condition implies that the virtual time is not allowed to overtake the real time; in other words, the execution time of the trajectory can never be shorter than its planned duration.

To embed the virtual time into the above first-order inverse kinematics, an algorithm to generate the time warp must be devised in the form

$$\tau_k = \tau_{k-1} + \Delta\tau_k, \tag{8}$$

where at each step the increment $\Delta\tau_k$ solves the problem

$$\begin{aligned} &\max \Delta\tau_k \\ &\textit{subject to} \qquad \Delta\tau_k \geq 0 \\ &\qquad\qquad\qquad t_k \geq \tau_{k-1} + \Delta\tau_k \\ &\qquad\qquad -\dot{q}_{i,lim} \leq \dot{q}_{i,k} \leq \dot{q}_{i,lim} \qquad i = 1, \ldots, n \end{aligned} \tag{9}$$

in which $\dot{q}_{i,lim}$ is the i-th joint speed limit and n is the number of joints. Note that the second constraint would allow the virtual time to recover the real time in a single step if the speed limits are not violated.

The general solution to the problem (9) is complex due to the relationship between the joint velocities $\dot{q}_{i,k}$ and the increment $\Delta\tau_k$ established through (5,7), which is nonlinear because of (7) in all practical cases. However, since all the above algorithms are based on reduced-order approximations of the inverse kinematics, a reduced-order approximation of (7) can be consistently adopted to simplify the problem's solution.

For first-order algorithms, it is possible to assume that at each step the increment of the desired trajectory $\tilde{\boldsymbol{x}}_d$ is a linear function of the virtual time increment as follows

$$\Delta\tilde{\boldsymbol{x}}_{d,k} = \frac{d\boldsymbol{x}_d}{d\tau}(\tau_{k-1})\Delta\tau_k \tag{10}$$

$$\tilde{\boldsymbol{x}}_{d,k} = \tilde{\boldsymbol{x}}_{d,k-1} + \Delta\tilde{\boldsymbol{x}}_{d,k}. \tag{11}$$

Based on (7), it can be recognized that eqs. (10,11) can be rewritten in the form

$$\Delta\tilde{\boldsymbol{x}}_{d,k} = \dot{\boldsymbol{x}}_d(\tau_{k-1})\Delta\tau_k \tag{12}$$

$$\tilde{\boldsymbol{x}}_{d,k} = \boldsymbol{x}_d(\tau_{k-1}) + \dot{\boldsymbol{x}}_d(\tau_{k-1})\Delta\tau_k, \tag{13}$$

which shows that at each step the reference trajectory can be computed by using the known functions (3,4) evaluated at the current virtual time instant.

The actual desired velocity, seen with respect to the real time as needed for the velocity feedforward terms, is given by

$$\dot{\tilde{\boldsymbol{x}}}_{d,k} = \frac{\Delta\tilde{\boldsymbol{x}}_{d,k}}{\Delta t}. \tag{14}$$

Notice that the actual velocity is increased or decreased with respect to the planned motion depending on $\Delta\tau_k$ being greater or lower than Δt.

By resorting to (12,13), it is possible to write the expression of the joint velocities for all the above first-order algorithms in the form

$$\dot{\boldsymbol{q}}_k = \boldsymbol{a}_k\Delta\tau_k + \boldsymbol{b}_k \tag{15}$$

which is linear in $\Delta\tau_k$. The expressions for $\boldsymbol{a}_k$ and $\boldsymbol{b}_k$ relative to each algorithm are given in Table II.

Algorithm	$\boldsymbol{a}_k$	$\boldsymbol{b}_k$
O.l. pseudoinverse	$\boldsymbol{H}_k \frac{d\mathbf{x}_d}{d\tau}(\tau_{k-1})\frac{1}{\Delta t}$	$\mathbf{0}$
O.l. DLS inverse	$\boldsymbol{H}_k \frac{d\mathbf{x}_d}{d\tau}(\tau_{k-1})\frac{1}{\Delta t}$	$\mathbf{0}$
C.l. transpose	$\boldsymbol{H}_k \frac{d\mathbf{x}_d}{d\tau}(\tau_{k-1})$	$\boldsymbol{H}_k(\boldsymbol{x}_d(\tau_{k-1}) - \boldsymbol{x}_{k-1})$
C.l. pseudoinverse	$\boldsymbol{H}_k(\frac{d\mathbf{x}_d}{d\tau}(\tau_{k-1})\frac{1}{\Delta t} + \boldsymbol{K}\frac{d\mathbf{x}_d}{d\tau}(\tau_{k-1}))$	$\boldsymbol{H}_k\boldsymbol{K}(\boldsymbol{x}_d(\tau_{k-1}) - \boldsymbol{x}_{k-1})$
C.l. DLS inverse	$\boldsymbol{H}_k(\frac{d\mathbf{x}_d}{d\tau}(\tau_{k-1})\frac{1}{\Delta t} + \boldsymbol{K}\frac{d\mathbf{x}_d}{d\tau}(\tau_{k-1}))$	$\boldsymbol{H}_k\boldsymbol{K}(\boldsymbol{x}_d(\tau_{k-1}) - \boldsymbol{x}_{k-1})$

Table 2

At this point, the problem (9) can be reformulated in a linear form as

$$
\begin{aligned}
&\max \Delta\tau_k \\
&\textit{subject to} \qquad \Delta\tau_k \geq 0 \\
&\qquad\qquad\qquad t_k \geq \tau_{k-1} + \Delta\tau_k \\
&\qquad\qquad -\dot{q}_{i,lim} \leq a_{i,k}\Delta\tau_k + b_{i,k} \leq \dot{q}_{i,lim} \\
&\qquad\qquad\qquad\qquad i = 1, \ldots, n \\
&\qquad\qquad\qquad \Delta\tau_k \leq m\Delta t
\end{aligned}
\tag{16}
$$

The last constraint has been added to the original problem to guarantee consistence of the desired trajectory increment with the linear approximation used, by upper-bounding the virtual time increment to a suitable integer multiple m of the real time increment. In the case $m = 1$, the virtual time is never allowed to recover the real time; if $m > 1$, the task-space trajectory is speeded up, if possible, to recover the delay caused by the previous slow down of the trajectory.

The solution to (16) can be easily found since the problem is one-dimensional. In detail, the set of constraints implies a set of upper and lower bounds on $\Delta\tau_k$. The solution $\Delta\tau_{sol,k}$ exists if the smallest upper bound is greater than the largest lower bound; in this case, the sought $\Delta\tau_{sol,k}$ is the smallest upper bound.

It can be recognized that if a solution does not exist, it is $|b_{i,k}| > q_{i,lim}$ for some i. This happens only in the case of closed-loop algorithms when the norm of the deviation $\boldsymbol{x}_d(\tau_{k-1}) - \boldsymbol{x}_{k-1}$ is too large. Since the closed-loop algorithms guarantee low tracking errors, in all practical cases large deviations may be experienced at the beginning of the motion due to the initial joint configuration $\boldsymbol{q}_0$ being inconsistent with the starting point of the desired trajectory, i.e. $\boldsymbol{x}_{d,0} = \boldsymbol{f}(\boldsymbol{q}_0)$; in this case, however, the path-tracking problem is ill-posed and the algorithm should be re-run with consistent inital conditions.

Finally, the first-order inverse kinematics algorithms become

$$\dot{q}_k = a_k \Delta\tau_{sol,k} + b_k \tag{17}$$

$$q_k = q_{k-1} + \dot{q}_k \Delta t \tag{18}$$

which use all quantities already computed to solve (16) and require little extra computation with respect to the corresponding original inverse kinematics algorithms (5,6).

IV. Case Studies

Without loss of generality, a simple two-degree-of-freedom RR planar manipulator is considered to test the proposed technique.

The manipulator has two equal links of 0.3 m; the initial configuration is $q(0) = [pi/3 \quad -2\pi/3]^{\mathrm{T}}$ rad corresponding to the end-effector position $p(0) = [0.3 \quad 0]^{\mathrm{T}}$ m. The joint speed limits are both equal to 3 rad/s.

The desired path is a circle in the plane with center at $[0.2 \quad 0]^{\mathrm{T}}$ m and radius 0.1 m; the path does not require the arm to cross kinematic singularities. The assigned time law is a quintic polynomial ensuring null initial and final velocity and acceleration. The planned duration of the motion is 1 s.

In the following, the analisys will be limited to the open-loop pseudoinverse, closed-loop transpose, closed-loop pseudoinverse algorithms. The damped least-squares inverse algorithms are not considered since they substantially differ from pseudoinverse algorithms only if kinematic singularities are involved. For each considered algorithm, the new proposed solution (17,18) is compared to the joint velocity solution (5), saturated and then integrated to obtain the joint angles. The time interval is $\Delta t = 0.001$ s.

First, the open-loop pseudoinverse algorithm is considered. Figure 1 reports the results when the joint velocities are saturated; notice that the manipulator cannot track the desired path with the imposed time law. The results obtained with the proposed technique in the case $m = 1$ are shown in Figure 2; now path tracking has been achieved while the joint velocities remain into their limits. As can be seen, the virtual time τ has realized the sought time warp; the actual duration of the motion is 1.26 s. In Figure 3, the results obtained with $m = 5$ are plotted; as anticipated the virtual time is now allowed to recover the real time leading to a shorter duration of motion (1.05 s).

Then, the case of closed-loop transpose algorithm is examined. Figure 4 presents the results of the algorithm with saturation of the joint velocities while Figure 5 reports those obtained with the proposed technique in the case $m = 1$. An analougus behavior can be observed, confirming the improved performance.

Finally, the same considerations can be made in the case of closed-loop pseudoinverse algorithm (see Figures 6 and 7).

V. Conclusions

The path-tracking problem for first-order inverse kinematics algorithms has been investigated when joint velocity limits are present. It has been shown that good tracking performance can be kept by suitably slowing down the the task-space trajectory when joint velocity limits are encountered. The time law is modified through a time warp such that the introduced virtual time allows fulfillment of the velocity constraints. This has been obtained by solving a one-dimensional constrained problem for the virtual time increment; remarkably, the problem has been keenly reformulated to obtain a computationally inexpensive solution. In addition, the task-space trajectory may be speeded up, if possible, to recover the delay caused by the previous slow down of the trajectory. Case studies have been developed which have demonstrated the effectiveness of the proposed technique.

VI. Acknowledgements

The work reported in this paper has been supported by *Consiglio Nazionale delle Ricerche* under contracts n. 93.00904.PF67 and n. 93.00946.PF67.

VII. References

[1] D.L. Pieper, *The kinematics of manipulators under computer control*, Ph.D. Dissertation, Stanford University, Stanford, CA, 1969.

[2] D.E. Whitney, "Resolved motion rate control of manipulators and human prostheses," *IEEE Transactions on Man-Machine Systems*, Vol. 10, pp. 47–53, 1969.

[3] Y. Nakamura and H. Hanafusa, "Inverse kinematic solutions with singularity robustness for robot manipulator control," *Transactions of the ASME Journal of Dynamic Systems, Measurements, and Control*, vol. 108, pp. 163–171, 1986.

[4] C.W. Wampler II, "Manipulator inverse kinematic solutions based on vector formulations and damped least-squares methods," *IEEE Transactions on Systems, Man, and Cybernetics*, vol. 16, pp. 93-101, 1986.

[5] A. Balestrino, G. De Maria and L. Sciavicco, "Robust control of robotic manipulators," *Preprints of the 9th IFAC World Congress*, Budapest, H, vol. 6, pp. 80–85, July 1984.

[6] W.A. Wolovich and H. Elliott, "A computational technique for inverse kinematics," *Proceedings of the 23rd IEEE Conference on Decision and Control*, Las Vegas, NV, pp. 1359–1363, Dec. 1984.

[7] P. Chiacchio, S. Chiaverini, L. Sciavicco and B. Siciliano, "Closed-loop inverse kinematics schemes for constrained redundant manipulators with task-space augmentation and task-priority strategy," *The International Journal of Robotics Research*, vol. 10, pp. 410–425, 1991.

[8] C.W. Wampler and L.J. Leifer, "Applications of damped least-squares methods to resolved-rate and resolved-acceleration control of manipulators," *Transactions of the ASME Journal of Dynamic Systems, Measurements, and Control*, vol. 110, pp. 31–38, 1988.

[9] S. Chiaverini, B. Siciliano, and O. Egeland, "Experimental results on controlling a 6-dof robot manipulator in the neighborhood of kinematic singularities," *Preprints of the 3rd International Symposium on Experimental Robotics*, Kyoto, J, Oct. 1993.

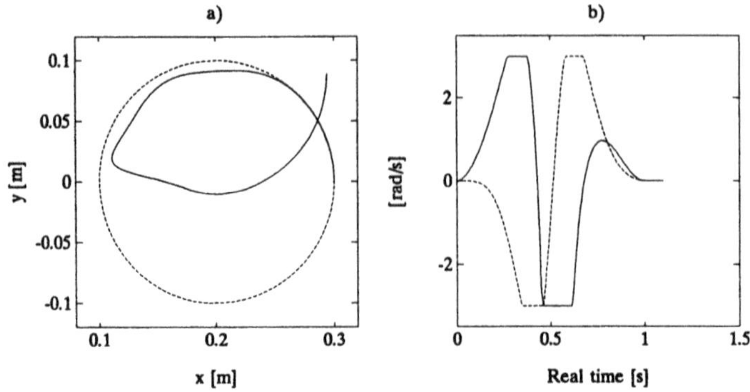

Figure 1: Open-loop pseudoinverse algorithm with saturation of joint velocities: a) desired (dashed) and actual (solid) path; b) joint 1 (solid) and 2 (dashed) velocities.

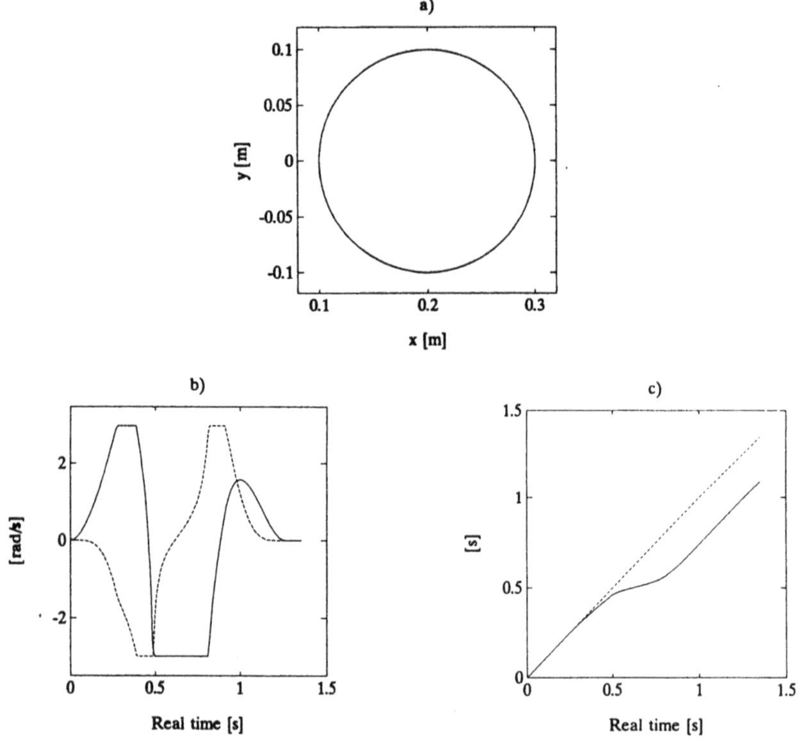

Figure 2: Open-loop pseudoinverse algorithm with time warp ($m = 1$): a) desired (dashed) and actual (solid) path; b) joint 1 (solid) and 2 (dashed) velocities; c) real time (dashed) and virtual time (solid).

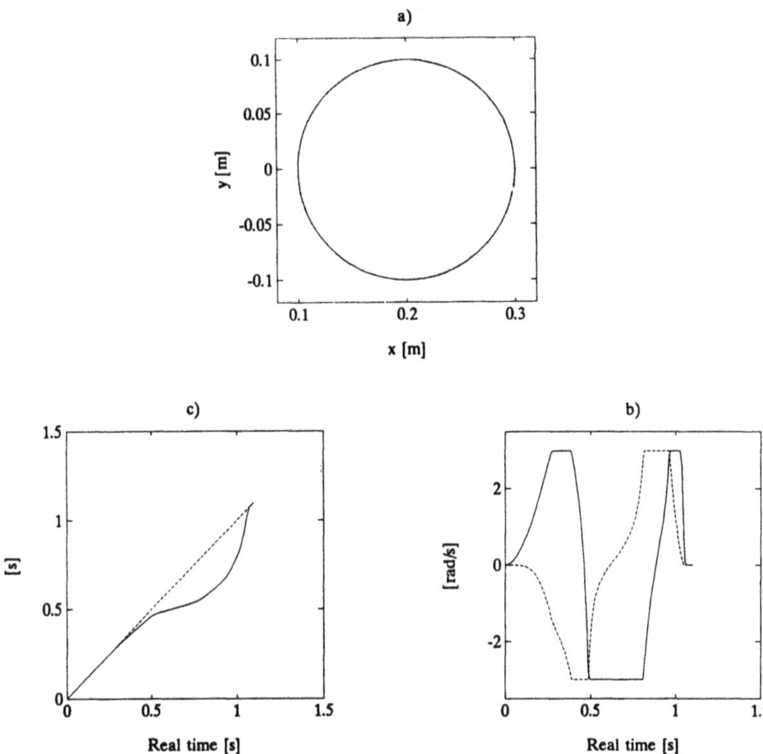

Figure 3: Open-loop pseudoinverse algorithm with time warp ($m = 5$): a) desired (dashed) and actual (solid) path; b) joint 1 (solid) and 2 (dashed) velocities; c) real time (dashed) and virtual time (solid).

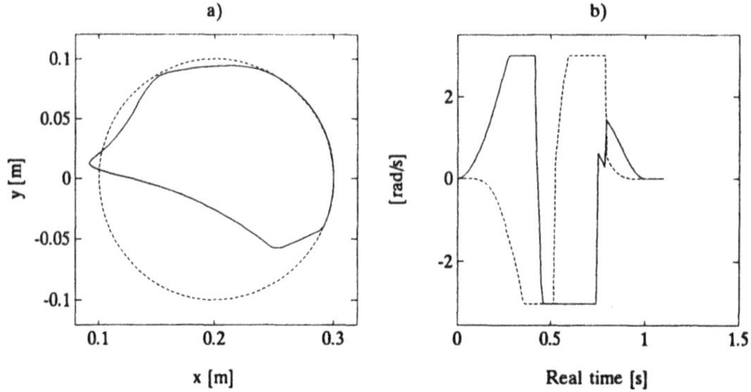

Figure 4: Closed-loop transpose algorithm with saturation of joint velocities: a) desired (dashed) and actual (solid) path; b) joint 1 (solid) and 2 (dashed) velocities.

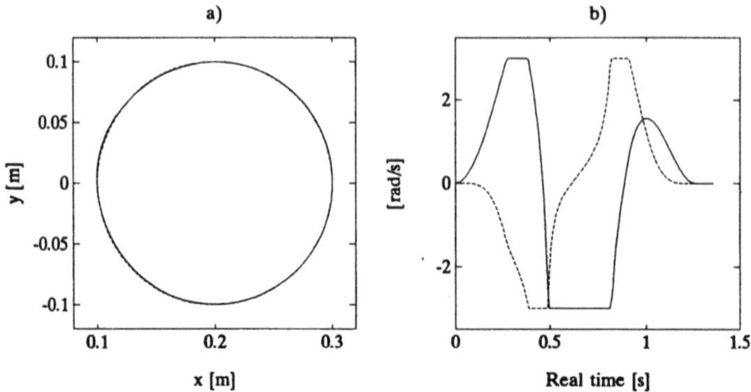

Figure 5: Closed-loop transpose algorithm with time warp ($m = 1$): a) desired (dashed) and actual (solid) path; b) joint 1 (solid) and 2 (dashed) velocities.

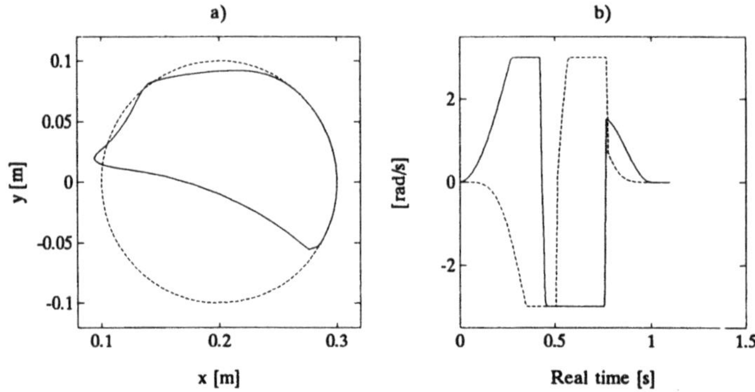

Figure 6: Closed-loop pseudoinverse algorithm with saturation of joint velocities: a) desired (dashed) and actual (solid) path; b) joint 1 (solid) and 2 (dashed) velocities.

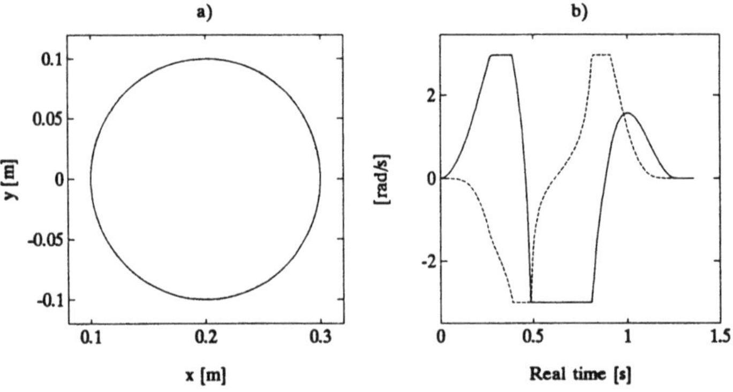

Figure 7: Closed-loop pseudoinverse algorithm with time warp ($m = 1$): a) desired (dashed) and actual (solid) path; b) joint 1 (solid) and 2 (dashed) velocities.

Hybrid Position-Orientation Decoupling in Robot Manipulators

Marcelo H. Ang Jr.
Department of Mechanical and Production Engineering
National University of Singapore
Singapore 0511, Republic of Singapore
E-mail: mpeangh@leonis.nus.sg

Abstract – Kinematic decoupling in robotic manipulators implies that there is a subset of joints primarily responsible for the completion of a subset of the manipulator task. In this paper, we take a general view of kinematic decoupling in which we identify link subsystems primarily responsible for completion of a hybrid combination of *both* position and orientation. Our analysis leads to the discovery of decoupled manipulator geometries, for which closed-form inverse kinematic solutions are guaranteed. Kinematic decoupling also implies singularity decoupling wherein singularities of decoupled subsystems are equivalent to the manipulator singularities. The analysis leads to another method for identifying the singularities and solving the inverse kinematics problem of six-axes, serial-chain manipulators with decoupled geometries. The practicality of the concepts introduced is demonstrated through an industrial robot example involving a hybrid position and orientation decoupling.

I. Introduction

It is known that the general six-revolute (6R) series-chain manipulator has at most sixteen solutions to the inverse kinematics problem [1]. By systematic component selection and/or manipulation of the kinematic equations, the problem can be reduced to root finding of a 16th degree polynomial (i.e., *characteristic polynomial*) in one unknown [2]. *Decoupling* can be used to reduce the degree of this polynomial for certain manipulator geometries. Decoupling means finding a subset of joints primarily responsible for the completion of a subset of the manipulator task. Decoupling therefore involves the identification of the *decoupled task* (i.e., subset of Cartesian degrees-of-freedom) *and* the *decoupled robot subsystem* (i.e., subset of robot joints) primarily responsible for the completion of the decoupled task.

Besides spherical wrists and Cartesian arms, it is known that the following geometries guarantee analytic inverse kinematic solutions [4,2,5,3]: 6R with 3 parallel axes; 6R with 3 pairs of intersecting axes and all axes are normal to their neighbors; 4R2T (2 translational joints) with 1 pair of parallel revolutes; 5R1T with 2 pairs of parallel revolutes; 6R with 3 pairs of parallel revolutes; and 4R2P with 2 pairs of revolute-prismatic parallel axes with the pairs separated by a revolute joint. Furthermore, certain structural parameters further reduce the order of the characteristic polynomial (hence, the number of possible solutions) to less than four. A comprehensive enumeration of these is presented in [4].

In this paper, we analyze task decoupling in robot manipulators from a robot motion point-of-view. That is, we relate the end-effector motion to the joint motion and find robot geometries for which the end-effector motion components are unaffected by

A. J. Lenarčič and B. B. Ravani (eds.), Advances in Robot Kinematics and Computationed Geometry, 329–338.

a certain subset of joint motion. We consider only six-axes, serial-chain, and non-redundant manipulators. Our analysis leads to the discovery of two other decoupled manipulator geometries, wherein the decoupled system is responsible for a subset of degrees-of-freedom involving a hybrid combination of *both* position and orientation. These geometries are manipulators with 2 parallel translational joints normal to a rotational joint, and those with a translational joint normal to a pair of parallel rotational joints. Furthermore, we use this framework to also prove that 6R manipulators with 3 parallel revolute axes represent hybrid position-orientation decoupling. Closed-form solutions are therefore guaranteed. Our analysis also leads to another method for identifying the singularities of six-axes manipulators with decoupled geometries.

II. Kinematic Model

The links are numbered consecutively from the base (link 0) to the final end (link N). The joints are numbered so that joint i connects links $(i-1)$ and i. We assign a Cartesian coordinate frame $(O_i; \mathbf{x}_i, \mathbf{y}_i, \mathbf{z}_i)$ to each link $(i = 0, 1, 2, \ldots, N)$ according to the Denavit–Hartenberg convention in [6, pp. 36–41]. Each link translates along or rotates around the $\mathbf{z}$-axis of the coordinate frame assigned to the previous link. The position and orientation of frame i with respect to frame $i-1$ is characterized by the four kinematic parameters θ_i, d_i, a_i, and α_i; d_i and θ_i are the *distance* and *angle* between links i and i-1, a_i and α_i are the *length* (offset distance) and *twist* (offset angle) of link i. The joint coordinate q_i is the *angular* displacement (θ_i) around $\mathbf{z}_{i-1}$ if joint i is *rotational*, or the *linear* displacement (d_i) along $\mathbf{z}_{i-1}$ if joint i is *translational*. The end-effector translational velocity $\mathbf{u}_N$ and angular velocity $\boldsymbol{\omega}_N$ are related to the joint velocities $\dot{\mathbf{q}}$ through the configuration-dependent Jacobian matrix $\mathbf{J}(\mathbf{q})$. Each column $\mathbf{J}_i$ of the Jacobian matrix is computed depending upon the type of joint [6]:

$$[\mathbf{J}]_i = \sigma_i \begin{pmatrix} \mathbf{z}_{i-1} \times (\mathbf{p}_N - \mathbf{p}_{i-1}) \\ \mathbf{z}_{i-1} \end{pmatrix} + (1 - \sigma_i) \begin{pmatrix} \mathbf{z}_{i-1} \\ 0 \end{pmatrix}, \tag{1}$$

where $\sigma_i = 1$ if joint i is a rotational joint and $\sigma_i = 0$ if joint i is a translational joint.

III. Decoupled Robot Geometries

The Jacobian matrix forms our basis for identifying decoupled robot geometries. We perform coordinate transformations on the Jacobian to reveal necessary and sufficient conditions for decoupling. Specifically, we seek special robot geometries for which the Jacobian matrix, transformed in an appropriate way, would have a (3×3) zero block matrix. Using this approach, we are able to identify five decoupled robot geometries. Each decoupled system has three degrees-of-freedom involving a subset of the components of the manipulator task (i.e., position and orientation). Closed-form inverse kinematic solutions are therefore guaranteed. Two geometries are well-known decoupled robot geometries for which the decoupled subsystem involves position and orientation only. These two geometries involve manipulator configurations in which any three rotational axes co-intersect (manipulators with spherical wrists) and geometries in which there are three translational joints (Cartesian arms). The other three decoupled geometries have their decoupled subsystems consist of a hybrid combination of position and orientation components of the task. In the following, we analyze these geometries from hybrid decoupling point of view.

A. Two Translational Joints Normal to a Rotational Joint

We define the "arm" to consist of the two translational joints and the rotational joint normal to the translational joints. The remaining joints belong to the "wrist". Without loss of generality, we formulate the Jacobian matrix with its first three columns representing the motion contribution of the arm joints. The arm therefore consists of joints 1 to 3 while the wrist consists of joints 4 to 6. The arm and wrist joints need not be consecutive. The columns of the Jacobian can be interchanged to achieve these arm and wrist definitions (and joint numberings). We express the Jacobian in the arm coordinate frame whose $\mathbf{z}$ axis is the rotational joint axis of interest (i.e., whose direction is normal to two translational joint axis directions). Let $\mathbf{z}_0$ be this rotational joint axis, and $\mathbf{z}_1$ and $\mathbf{z}_2$ be the two translational joint axes normal to $\mathbf{z}_0$. We then formulate the Jacobian in frame 0. This means that the end-effector velocity vector and all the vectors used in computing the Jacobian are expressed in frame 0 and $\mathbf{z}_0$ is equal to $(0,0,1)^T$. Since the two translational joint axes and $\mathbf{z}_0 \times (\mathbf{p}_N - \mathbf{p}_0)$ are normal to $\mathbf{z}_0$, they will all lie in a plane parallel to the xy-plane of frame 0 and will therefore have a zero z-component in frame 0. The first 6×3 block column of the Jacobian matrix thus becomes:

$$[\mathbf{J}(\mathbf{q})]_{123} = \begin{pmatrix} q & z_{1x} & z_{2x} \\ r & z_{1y} & z_{2y} \\ 0 & 0 & 0 \\ 0 & 0 & 0 \\ 0 & 0 & 0 \\ 1 & 0 & 0 \end{pmatrix}, \tag{2}$$

where q, z_{1x}, and z_{2x} are the x-components and r, z_{1y}, z_{2y} are the y-components of $\mathbf{z}_0 \times (\mathbf{p}_N - \mathbf{p}_0)$, $\mathbf{z}_1$ and $\mathbf{z}_2$, respectively. We note the presence of a (3×3) zero matrix in (2) which means that decoupling is guaranteed. The position of the zero matrix in $\mathbf{J}(\mathbf{q})$ corresponds to the contribution of the arm joints to the z component of the end-effector translational velocity and to the x and y components of the end-effector angular velocity. The decoupled task space therefore consists of the z component of the end-effector translational velocity and the x and y components of the end-effector angular velocity. Motion of any of the arm joints would therefore not effect motion in any of the directions in the decoupled task space. All vectors in this decoupled task space, when expressed in frame 0 (the frame whose $\mathbf{z}$ axis is the rotational joint axis of interest), are therefore functions of the wrist joint coordinates only.

We can now obtain the manipulator singularities in terms of the decoupled subsystem singularities. We interchange rows 3 and 6 of the Jacobian matrix (2) to achieve the following decoupled Jacobian:

$$\begin{pmatrix} u_{N_x} \\ u_{N_y} \\ \omega_{N_z} \\ \cdots \\ \omega_{N_x} \\ \omega_{N_y} \\ u_{N_z} \end{pmatrix} = \begin{pmatrix} \mathbf{J}_A & \vdots & \mathbf{B}^* \\ \cdots & \cdots & \cdots \\ \mathbf{0} & \vdots & \mathbf{J}_W \end{pmatrix} \dot{\mathbf{q}}. \tag{3}$$

where $\mathbf{J}_A(\mathbf{q})$ and $\mathbf{J}_W(\mathbf{q})$ in (3) are the *arm* and *wrist* Jacobians respectively:

$$\mathbf{J}_A(\mathbf{q}) = \begin{pmatrix} q & z_{1x} & z_{2x} \\ r & z_{1y} & z_{2y} \\ 1 & 0 & 0 \end{pmatrix} \text{ and } \mathbf{J}_W(\mathbf{q}) = \begin{pmatrix} z_{3x} & z_{4x} & z_{5x} \\ z_{3y} & z_{4y} & z_{5y} \\ t_3 & t_4 & t_5 \end{pmatrix}, \tag{4}$$

where t_i is the z-component of $\mathbf{z}_i \times (\mathbf{p}_N - \mathbf{p}_i)$. (Although the arm and wrist Jacobians represent sole motion contributions of the arm and wrist joints respectively, the Jacobians are functions of all the manipulator joints, in general.) The symbolic expression for $\mathbf{B}^*$ is unimportant since we are interested in obtaining the manipulator singularities only. We note that with our definitions of the arm and the wrist, the arm is primarily responsible for two positional DOF, i.e., positioning the end-effector in a plane parallel to the xy plane of frame 0, *and* is also primarily responsible for one orientational DOF, i.e., rotating the end-effector around an axis parallel to the z axis of frame 0. The wrist is primarily responsible for two orientational and one positional DOF, i.e., rotating the end-effector about any axis lying on a plane parallel to the xy-plane of frame 0 and positioning the end effector along an axis parallel to the z-direction of frame 0, respectively. Singularity decoupling, therefore, is guaranteed and the manipulator singularities consist of the arm singularities plus the wrist singularities.

B. One Translational Joint Normal to Two Parallel Rotational Joints

As in the previous section, we interchange columns of the Jacobian to define the "arm" to consist of the translational joint and the two parallel rotational joints normal to the translational joint. Without loss of generality, we let $\mathbf{z}_0 = \mathbf{z}_1 = (0,0,1)^T$ be the two parallel rotational joint axes, and $\mathbf{z}_2 = (z_{2x}, z_{2y}, 0)^T$ be the translational axes normal to $\mathbf{z}_0$. The vectors $\mathbf{z}_0 \times (\mathbf{p}_N - \mathbf{p}_0)$, $\mathbf{z}_1 \times (\mathbf{p}_N - \mathbf{p}_1)$, and $\mathbf{z}_2$, therefore, all lie in a plane parallel to the xy-plane of frame 0 and will therefore have a zero z-component in frame 0. We interchange rows 3 and 6 of the Jacobian to achieve the decoupled Jacobian as in (3), where

$$\mathbf{J}_A(\mathbf{q}) = \begin{pmatrix} q & s & z_{2x} \\ r & t & z_{2y} \\ 1 & 1 & 0 \end{pmatrix}, \tag{5}$$

where q, s, and z_{2x} are the x-components and r, t, z_{2y} are the y-components of $\mathbf{z}_0 \times (\mathbf{p}_N - \mathbf{p}_0)$, $\mathbf{z}_1 \times (\mathbf{p}_N - \mathbf{p}_1)$ and $\mathbf{z}_2$ respectively. $\mathbf{J}_W(\mathbf{q})$ is as given in (4). The primary responsibilities of the arm and the wrist are the same as in Section III.A and described in Equation (3). The decoupled system is achieved when the end-effector velocity is expressed in frame 0. The z component of the translational velocity and the x and y components of the angular velocity are functions of the wrist joints coordinates only. Furthermore, singularity decoupling is guaranteed.

C. Any Three Rotational Joints Parallel

We define the "arm" to consist of the three parallel rotational joints. The remaining joints belong to the "wrist". We let $\mathbf{z}_0 = \mathbf{z}_1 = \mathbf{z}_2 = (0,0,1)^T$ be the parallel rotational joint axes. The vectors $\mathbf{z}_{i-1} \times (\mathbf{p}_N - \mathbf{p}_{i-1})$, for $i = 1,2,3$, therefore, all lie in a plane parallel to the xy-plane of frame 0 (or 1 and 2) and will therefore have a zero z-component in frame 0. We interchange rows 3 and 6 of the Jacobian to achieve the decoupled Jacobian as in (3), where

$$\mathbf{J}_A(\mathbf{q}) = \begin{pmatrix} v_1 & v_2 & v_3 \\ w_1 & w_2 & w_3 \\ 1 & 1 & 1 \end{pmatrix} \tag{6}$$

v_i, and w_i are the x- and y- components, respectively, of $\mathbf{z}_{i-1} \times (\mathbf{p}_N - \mathbf{p}_{i-1})$. $\mathbf{J}_W(\mathbf{q})$ is as given in (4). The primary responsibilities of the arm and the wrist are the same as in Section III.C and described in Equation (3). As in Sections III.A and III.B, the decoupled system is achieved when the end-effector velocity is expressed in frame 0. The z component of the translational velocity and the x and y components of the angular velocity are functions of the wrist joints coordinates only. Singularity decoupling, therefore, is guaranteed.

IV. Hybrid Position and Orientation Decoupling

We consider the hybrid decoupling case wherein one subsystem is responsible for two positional and one orientational degrees-of-freedom, and the other subsystem is responsible for two orientational and one positional degrees-of-freedom. For expository convenience, we refer to the "arm" as the subsystem responsible for two positional and one orientational degrees-of-freedom, while the "wrist" is responsible for the remainder of the degrees-of-freedom. For these decoupled geometries, the joint definitions for the arm and wrist subsystems vary while the task definitions for each subsystem is constant. The arm is defined to consist of the joint axes with the special geometry indicated in the previous section. The joint definitions for the arm and the wrist are consistent with their task specifications, as shown by the decoupled Jacobian in Equation (3). We note that the joints of the arm need not be consecutive; the same applies to the wrist joints. Decoupling, however, is guaranteed, and closed-form inverse kinematics solutions exist for any of these geometries.

Let frames A_0, A_1 and A_2 be defined such that their $\mathbf{z}$ axes coincide with the arm joint axes $(\mathbf{z}_{A_0}, \mathbf{z}_{A_1}, \mathbf{z}_{A_2})$, and according to the appropriate joint definitions for the arm corresponding to the decoupled geometry of interest. Let $\mathbf{z}_{A_0}$ be the arm joint axis of interest in the decoupled geometries (Sections III.A to III.C). Let frame A_3 be attached to the terminal link of the arm furthest from the arm base (frame A_0). Frame A_3 may be coincident to the wrist coordinate frame immediately *following* the arm. Let frame B be the inertial frame of reference where the manipulator task is described in.

We express the manipulator task in frame A_0 to achieve the decoupled system by the following coordinate transformation:

$$ {}^{A_0}\mathbf{T}_N \triangleq \begin{pmatrix} n_x & o_x & a_x & p_x \\ n_y & o_y & a_y & p_y \\ n_z & o_z & a_z & p_z \\ 0 & 0 & 0 & 1 \end{pmatrix} = {}^{B}\mathbf{T}_{A_0}^{-1}\, {}^{B}\mathbf{T}_N = {}^{A_0}\mathbf{T}_{A_3}\, {}^{A_3}\mathbf{T}_N. \tag{7} $$

The decoupled task space consists of the z-component of the end-effector translational velocity and the x and y components of the end-effector angular velocity. All these vectors, when expressed in frame A_0, are functions of the wrist joint coordinates only. In addition, according to the arm definition for any of the three "hybrid" (position and orientation) decoupled geometries (in Sections III.A to III.C), the arm can rotate only about $\mathbf{z}_{A_0}$. The orientation of frame A_3 is, therefore, constrained such that the z-component of each unit vector describing the direction of each coordinate axis of frame A_3 is *independent* of the arm joint coordinates. Regardless of where frame A_3 is, its orientation in frame A_0 can be represented by a rotation matrix ${}^{A_0}\mathbf{R}_{A_3}$ whose last row is *independent* of the arm joint coordinates. Similarly, the z component of ${}^{A_0}\mathbf{p}_{A_3}$ is not affected by the arm joint coordinates, since the arm can only position its terminal link in a plane parallel to the xy plane of frame A_0. Since the elements of the third row of ${}^{A_0}\mathbf{T}_{A_3}$ are independent of the arm joint coordinates (constant for consecutive arm joints, while functions of the wrist joint coordinates for non-consecutive arm joints), then n_z, o_z, a_z, and p_z are functions of the wrist joint coordinates only. This then is the three-degree-of-freedom decoupled system, where n_z, o_z, a_z, and p_z are the subset of the desired manipulator task expressed in frame A_0. Indeed, this implies that the z component of the translational velocity of the end-effector in frame A_0, and the x and y components of the end-effector angular velocity in frame A_0 are not affected by the arm joint motions, as shown in the decoupled Jacobians in Sections III.A to III.C. This is the decoupled low-order system for which a closed-form solution for the wrist joint coordinates is attainable. Once the wrist joint coordinates are solved for, then closed-form solutions are used to solve for the arm joint coordinates to satisfy the remainder subset of the task.

For the special case of the arm joints being consecutive, the last row of ${}^{A_0}\mathbf{R}_{A_3}$ and the z component of ${}^{A_0}\mathbf{p}_{A_3}$ are always constants rather than being functions of the wrist joint coordinates. In either case, the decoupled system consists of the last row of ${}^{A_0}\mathbf{R}_{A_3}$ and the z component of ${}^{A_0}\mathbf{p}_{A_3}$ which are functions of the wrist joint coordinates only. This important property guarantees the realization of a closed-form inverse kinematics solution.

V. Application Example

In this section, we demonstrate our approach using a six-joint industrial robot that

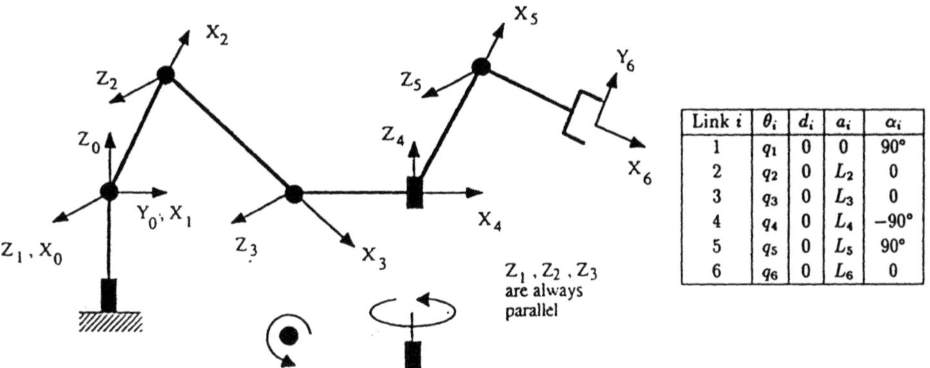

Link i	θ_i	d_i	a_i	α_i
1	q_1	0	0	90°
2	q_2	0	L_2	0
3	q_3	0	L_3	0
4	q_4	0	L_4	−90°
5	q_5	0	L_5	90°
6	q_6	0	L_6	0

Figure 1: *An Articulated Robot with Three Intermediate Rotational Axes Parallel.*

has an articulated robot geometry with three intermediate axes parallel and a non-spherical wrist. We derive a closed-form inverse kinematics solution and we identify the manipulator singularities. The schematic diagram of the articulated robot and the coordinate frames attached to each link are shown in Fig. 1. The three rotational joint axes $\mathbf{z}_1$, $\mathbf{z}_2$ and $\mathbf{z}_3$ are always parallel. Decoupling is therefore guaranteed (Section III.C) and the decoupled system involves a hybrid combination of position and orientation.

Let the manipulator task be defined as in Eq. (7). According to Section III.C, we define the "arm" to consist of the second, third and fourth joints, and the "wrist" to consist of the first, fifth and sixth joints. The configuration space of the arm is $\mathbf{q}_A = (\theta_2, \theta_3, \theta_4)^T$, while the configuration space of the wrist to be $\mathbf{q}_W = (\theta_1, \theta_5, \theta_6)$.

For decoupling, we express the manipulator task in the arm coordinate frame A_0 (whose z-axis coincides with the joint axis direction of interest indicated in Sections III.A to III.C). For this example, frame 1 coincides with A_0. We express the manipulator task in frame 1 using: $\mathbf{A}_1^{-1}\ {}^0\mathbf{T}_6 = \mathbf{A}_2\mathbf{A}_3\mathbf{A}_4\mathbf{A}_5\mathbf{A}_6$. Equating each element in the third row of this matrix equation, we have the following decoupling system involving the wrist joint coordinates only:

$$n_x s_1 - n_y c_1 = -s_5 c_6 \tag{8}$$

$$o_x s_1 - o_y c_1 = s_5 s_6 \tag{9}$$

$$a_x s_1 - a_y c_1 = c_5 \tag{10}$$

$$p_x s_1 - p_y c_1 = -L_6 s_5 c_6 - L_5 s_5 \tag{11}$$

where $s_i \triangleq \sin\theta_i$, and $c_i \triangleq \cos\theta_i$. Equations (8)-(11) represents a three-degree-of-freedom system involving only the wrist joint coordinates $\mathbf{q}_W = (\theta_1, \theta_5, \theta_6)$. A closed-form solution is therefore guaranteed. Algebraic manipulations and the half-angle transformation $t_i \stackrel{\Delta}{=} \tan\frac{\theta_i}{2}$ turn Equations (8) to (11) into a fourth order polynomial with constant coefficients:

$$at_1^4 + bt_1^3 + ct_1^2 - bt_1 + a = 0 \tag{12}$$

$$a = a_y^2 + \left(\frac{p_y - L_6 n_y}{L_5}\right)^2 - 1 \tag{13}$$

$$b = 4\left[a_x a_y - \left(\frac{-p_x + L_6 n_x}{L_5}\right)\left(\frac{p_y - L_6 n_y}{L_5}\right)\right] \tag{14}$$

$$c = 4\left[a_x^2 + \left(\frac{-p_x + L_6 n_x}{L_5}\right)^2\right] - 2\left[a_y^2 + \left(\frac{p_y - L_6 n_y}{L_5}\right)^2\right] - 2 \tag{15}$$

In general (i.e., for $a \neq 0$), there are four solutions for t_1 in (12). (Closed-form solutions for the roots of a fourth degree polynomial can be found in [8].) Once the four solutions for t_1 are computed, the joint coordinate is then computed according to

$$\theta_1 = \text{ATAN2}\left(\frac{s_1}{c_1}\right) = \text{ATAN2}\left(\frac{2t_1}{1-t_1^2}\right).$$

If $a = 0$ and $b \neq 0$, we have a singularity and the number of solutions for θ_1 reduces to three, namely:

$$t_1 = 0, \qquad t_1 = \frac{-c \pm \sqrt{c^2 + 4b^2}}{2b}.$$

If $a = b = 0$ and $c \neq 0$, we have a singularity and the number of solutions for θ_1 reduces to one, i.e., $t_1 = 0$. If $a = b = c = 0$, we have a singularity and there can be an infinite number of solutions for θ_1, i.e., t_1 can take any value. The fifth and sixth joint coordinates are computed from θ_1:

$$\theta_5 = \text{ATAN2}\left(\frac{s_1 \frac{-p_x + L_6 n_x}{L_5} + c_1 \frac{p_y - L_6 n_y}{L_5}}{a_x s_1 - a_y c_1}\right), \text{ and } \theta_6 = \text{ATAN2}\left(\frac{\frac{n_x s_1 - n_y c_1}{-s_5}}{\frac{o_x s_1 - o_y c_1}{s_5}}\right),$$

if $s_5 \neq 0$. If $s_5 = 0$, θ_6 can assume any value since one degree-of-freedom is lost.

Now that once the wrist joint coordinates $(\theta_1, \theta_5, \theta_6)$ are known, we can now solve for the arm joint coordinates $(\theta_2, \theta_3, \theta_4)$ to satisfy the remainder of the task (n_x, o_x, a_x, p_x, n_y, o_y, a_y, p_y). Again this is a low order (three-degree-of-freedom) system for which a closed-form solution is guaranteed. We first separate the contribution of the wrist subsystem (which is known) by:

$$\mathbf{A}_1^{-1}\,{}^0\mathbf{T}_N \mathbf{A}_6^{-1} \mathbf{A}_5^{-1} \stackrel{\Delta}{=} \begin{pmatrix} N_x & O_x & A_x & P_x \\ N_y & O_y & A_y & P_y \\ 0 & -1 & 0 & 0 \\ 0 & 0 & 0 & 1 \end{pmatrix} = \mathbf{A}_2 \mathbf{A}_3 \mathbf{A}_4. \tag{16}$$

Note that the third row in the above matrix is constant, because the arm joints are consecutive. In the general case of non-consecutive arm joints, this row would be a function of the wrist joint coordinates, which have been solved for in the previous steps.

We solve the arm joint coordinates using the N_y, A_y, P_x and P_y components of the matrix equation (16). There are two solutions for the arm joint coordinates:

$$\theta_3 = \mathrm{ATAN2}\left(\frac{\pm\sqrt{1-c_3^2}}{c_3}\right), \qquad c_3 = \frac{\sqrt{(P_x - L_4A_y)^2 + (P_y - L_4N_y)^2 - L_3^2 - L_2^2}}{2L_2L_3}.$$

$$\theta_2 = \mathrm{ATAN2}\left(\frac{(P_y - L_4N_y)(L_3c_3 + L_2) - (P_x - L_4A_y)L_3s_3}{(P_y - L_4N_y)L_3s_3 + (P_x - L_4A_y)(L_3c_3 + L_2)}\right).$$

The fourth joint coordinate θ_4 is obtained from N_y and A_y elements of (16) yielding:

$$\theta_4 = \mathrm{ATAN2}\left(\frac{N_y}{A_y}\right) - \theta_2 - \theta_3.$$

We now identify the manipulator singularities by exploiting the singularity decoupling for this manipulator as derived in Section III.C. In the following derivation of the arm and wrist Jacobians, we express all vectors in frame 3 and specify the origin of frame 3 as the velocity reference point for the end-effector to simplify the symbolic expressions for the Jacobian matrix [9,10].

We formulate the arm and wrist Jacobians according to (6) and (4) where v_i and w_i are the x- and y- components of $\mathbf{z}_i \times (\mathbf{p}_3 - \mathbf{p}_i)$, respectively; and t_i is the z-component of $\mathbf{z}_i \times (\mathbf{p}_3 - \mathbf{p}_i)$ yielding:

$$\mathbf{J}_A(\mathbf{q}) = \begin{pmatrix} L_2s_3 & 0 & 0 \\ L_3 + L_2c_3 & L_3 & 0 \\ 1 & 1 & 1 \end{pmatrix}, \quad \mathbf{J}_W(\mathbf{q}) = \begin{pmatrix} -L_2c_2 - L_3s_3 & L_4 & 0 \\ \sin(\theta_2+\theta_3) & -s_4 & c_4s_5 \\ \cos(\theta_2+\theta_3) & c_4 & s_4s_5 \end{pmatrix},$$

The manipulator singularities are then identified through singularity decoupling, i.e., the manipulator singularities consist of the arm singularities plus the wrist singularities:

$$\det[\mathbf{J}] = \underbrace{\{L_2L_3s_3\}}_{\det[\mathbf{J}_A]}\underbrace{\{s_5[L_2c_2 + L_3\cos(\theta_2+\theta_3) + L_4\cos(\theta_2+\theta_3+\theta_4)]\}}_{\det[\mathbf{J}_W]}.$$

VI. Conclusions

Although the concept of decoupling has been known since Pieper's pioneering work two decades ago, this is the first time it has been used in a *general sense* wherein the decoupled system involves a hybrid combination of position *and* orientation components. In this paper, we presented a complete analysis of task decoupling in robot manipulators. Our framework led to the discovery of other robot geometries for which decoupling is guaranteed. We highlighted the fact that task decoupling guarantees closed-form inverse kinematic solutions and derived an algorithm for the closed-form solutions.

We analyzed task decoupling from the novel point-of-view of robot motion. Our analysis led to the concept of ***singularity decoupling*** which provides another method for the identification of (simple six-axes) manipulator singularities from the decoupled subsystem singularities. Finally, we demonstrated the application of our framework to both inverse robot kinematics and singularity identification through an example involving a hybrid decoupling of position and orientation.

Although task decoupling is not guaranteed for general manipulator geometries, the concept of modular robotics wherein link subsystems are primarily responsible for task subsystems is still applicable. Indeed we exploited this modular concept of the "arm" and the "wrist" to derive physically-intuitive inverse kinematics algorithm and method for identifying manipulator singularities for non-simple manipulators in [11] and [10], respectively.

References

[1] M Raghavan and B. Roth. "Kinematic Analysis of the 6R Manipulator of General Geometry". In H. Miura and S. Arimoto, editors, *Proceedings of the Fifth International Symposium on Robotics Research*, MIT Press, Cambridge, MA, 1990.

[2] M Raghavan and B. Roth. "A General Solution for the Inverse Kinematics Feedback for Automatic Assembly". In *IFAC Symposium of Information and Control Problems in Manufacturing Technology*, Tokyo, 1977.

[3] D.L. Pieper. *The Kinematics of Manipulators Under Computer Control.* PhD thesis, Computer Science Department, Stanford University, Stanford, CA, October 1968.

[4] Mavroidis and Roth B. "Structural Parameters Which Reduce the Number of Manipulator Configurations". In *Robotics, Spatial Mechanisms, and Mechanical Systems (ASME 22nd Biennial Mechanisms Conference, DE-45)*, pages 359–366, ASME Design Engineering Division, Scottsdale, AZ, September 13-16 1992.

[5] J.Duffy. *Analysis of Mechanisms and Robot Manipulators.* Edward Arnold, London, UK, 1980.

[6] K.S. Fu, R.C. Gonzalez, and C.S.G. Lee. *Robotics: Control, Sensing, Vision, and Intelligence.* McGraw-Hill, New York, NY, 1987.

[7] R.P. Paul. *Robot Manipulators - Mathematics, Programmming and Control.* MIT Press, Cambridge, MA, 1981.

[8] *CRC Handbook of Mathematical Sciences.* CRC Press, West Palm Beach, FL, 1978.

[9] M. Renaud. "Geometric and Kinematic Models of a Robot Manipulator: Calculation of the Jacobian Matrix and its Inverse". In *Proceedings of the 11th International Symposium on Industrial Robots*, pages 757–763, Tokyo, Japan, October 7-9, 1981.

[10] V.D. Tourassis and M.H. Ang Jr. "Identification and Analysis of Robot Manipulator Singularities". *International Journal of Robotics Research*, 11(3):248–259, June 1992.

[11] V.D. Tourassis and M.H. Ang Jr. "A Modular Architecture for Inverse Robot Kinematics". *IEEE Transactions on Robotics and Automation*, 5(5):555–568, October 1989.

Dual Basis of Screw-Vectors for Inverse Kinestatic Problems in Robotics

Catherine Bidard
Service de Téléopération et Robotique du CEA/DTA,
Bat. 38, Route du Panorama, BP 6,
92 265 Fontenay-aux-roses Cedex FRANCE

Abstract - This paper introduces dual bases of screw-vectors, which are associated with the dual spaces of wrenches and twists. Dual bases are required in order to express general coordinates functions and transformations of screw-vectors. Analytical and geometrical methods for determining dual bases are proposed and applied to inverse kinestatic problems.

I. Introduction

Plücker screw coordinates are widely used for formulating the first-order kinematics and the statics of robots' mechanisms. Wrenches and twists are homogeneous with geometric screws [1,2]. They have in addition a vectorial nature and are hence called screw-vectors [3]. While the mathematical nature of screw-vectors is well established [4,5,6,7,8], while physical relations are known to be invariant by coordinates transformations, practical calculus are still done on non-intrinsic representations : Plücker coordinates or motor representations [8,9]. The practical and conceptual advantages of intrinsic vector and tensor calculus are although appreciated in engineering sciences as mechanics and kinematics [10]. But the vectors with which engineers are familiar are cartesian vectors, especially three-dimensional Euclidean vectors. Screw-vectors are different from cartesian vectors because they are not endowed with an Euclidean metric [3].

By using only Plücker coordinates or motors, one is limited to a class of particular bases of the screw-vector spaces, which we call cartesian bases. However the formulation of numerous kinestatic models involve mostly non-cartesian bases. For instance, the kinestatic models of a 6 degrees-of-freedom robot is defined by 6 screw-vectors which compose a basis in the regular configurations of the robot. Furthermore, the amount of computations required for inverting kinestatic models, may be significantly reduced by a convenient choice of the projection basis; non-cartesian bases may be convenient for some problems. However, the practical use of non-cartesian screw-vector bases relies on the definition of dual bases. The aim of this paper is to introduce dual bases of screw-vectors, and to illustrate their use in formulating and solving inverse kinestatic problems.

In the first part, dual bases of screw-vectors are naturally introduced for the dual spaces of wrenches and twists. The algebraic structure of the screw-vector spaces is briefly recalled. Then coordinate functions and transformations are expressed in intrinsic form, using the reciprocal product of screw-vectors. Finally, the inverse kinematic problem of serial six dof is shown to consist in determining the dual of a screw-vector basis. In the second part of the paper, the general problem of determining the dual of a screw-vector basis is considered. Analytical methods, based on the projection of the basis to be dualized, are presented. The geometrical determination of the dual basis is also considered, because it derives benefit from geometric reasonning about each particular problems. Two examples of applications are finally treated, the inverse kinematic problem of a three-cylindric serial robot arm, and the inverse static problem of a triangular parallel six-dof robot.

A. J. Lenarčič and B. B. Ravani (eds.), Advances in Robot Kinematics and Computationed Geometry, 339–348.

II. Coordinates and Dual Bases of Screw–vectors

A. *Dual Spaces of Wrenches and Twists*

Twists and wrenches are members of two six–dimensional dual vector spaces [3], which we shall denote by $\mathcal{T}$ and $\mathcal{W}$ respectively. A twist **T** is indeed a tangent vector of the space of rigid body displacements, while a wrench **W** is a co–tangent vector of the same space. A wrench is formally defined as a one–form over the space of twists which associates a real number, the power imparted to the rigid body, with a twist. As the space of twists $\mathcal{T}$ and the space of wrenches $\mathcal{W}$ are isomorphic, they are practically identified with a one vector space called the *screw–vector space* (1). Such an identification provides a geometrical analogy between wrenches and twists. The power imparted to the rigid body is then expressed as an inner scalar product of the twist **T** and the wrench **W** :

$$\text{Power} = \mathbf{W} \bigcirc \mathbf{T} = \mathbf{T} \bigcirc \mathbf{W} \tag{1}$$

The inner product denoted by $\bigcirc$ is the *reciprocal product of screw–vectors* (2) [7,11].

Moreover, screw–vector spaces have the structure of a Lie algebra. The space of twists is indeed the Lie algebra se(3) of the group of rigid body displacements SE(3) [4,5,6,7,8]. In addition to the vector space structure, the Lie algebra endows screw–vector spaces with 3 operators. The *screw–vector product*, denoted by $\times$, is the Lie bracket of the Lie algebra [4,5,6]. It is bilinear, anti–symmetric, and satisfies the Jacobi identity. The *reciprocal product* $\bigcirc$ of screw–vectors is the Klein form of the Lie algebra se(3) [7], while the *perpendicular product* $\diamond$ is the Killing form of the Lie algebra [6,7]. Both products, $\bigcirc$ and $\diamond$, are bilinear and symmetric. Notably, the reciprocal product is non definite. Self–reciprocal screw–vectors are *line–vectors* [3]. The perpendicular product is degenerate. *Free–line–vectors* are the self–perpendicular screw–vectors; they form a sub–class of line–vectors [3]. As any screw–vectors may be expressed as a linear combination of unit line–vectors, operations on screw–vectors may be reduced to operations on unit line–vectors, which may be interpreted in terms of geometry of lines. For a presentation of screw–vector calculus, readers are referred to classical presentations [9,12] which are developped in terms of "dual vectors"; concise presentations are given in [8,13,14].

B. *Dual Bases of Wrenches and Twists*

A basis of the screw–vector space may be composed of any six linearly independant screw–vectors $\mathbf{S}\mathfrak{B}_i$ (i = 1,6). A screw–vector **S**, may be represented by the column–vector $\underline{S}$ of its six coordinates $S_{[i]}$ in any basis $(\mathfrak{B}\mathcal{S}) = (\ \mathbf{S}\mathfrak{B}_1\ \mathbf{S}\mathfrak{B}_2\ \mathbf{S}\mathfrak{B}_3\ \mathbf{S}\mathfrak{B}_4\ \mathbf{S}\mathfrak{B}_5\ \mathbf{S}\mathfrak{B}_6\ )$:

$$\mathbf{S} = \sum_{i=1}^{6} S_{[i]}\ \mathbf{S}\mathfrak{B}_i = (\mathfrak{B}\mathcal{S})\ \underline{S} \tag{2}$$

(1) Remarks about the terminology - We choose the term **screw–vector** introduced by Lipkin [3] in order to put the emphasize on the vectorial nature of twists and wrenches. Screw–vectors are often called **screws** [1,11,15], but the term screw refers also to the geometrical object composed of a line with a pitch, that is an element of the projectivization of the screw–vector space [7,8]. Screw–vectors are also called **motors** [7,9,13] when represented by a pair of vectors.

(2) The **reciprocal product** is called the **mutual moment of screws** by numerous authors [1,2,15].

Let $(B\mathcal{W})$ and $(B\mathcal{T})$ be two screw-vector bases chosen for representing respectively wrenches and twists. Let $\underline{W}$ be the coordinates of a wrench **W** in the basis $(B\mathcal{W})$, and $\underline{T}$, the coordinates of a twist **T** in the basis $(B\mathcal{T})$. The two bases $(B\mathcal{W})$ and $(B\mathcal{T})$ are said to be **dual** if they ensure that :

$$\mathbf{W} \bigcirc \mathbf{T} = \underline{W}^{\prime}\ \underline{T} \tag{3}$$

Let the 6 screw-vectors of the basis $(B\mathcal{W})$ be denoted $S\mathcal{W}_i$ (i = 1,6) and the 6 screw-vectors of $(B\mathcal{T})$ be denoted $S\mathcal{T}_i$ (i = 1,6). In order to be dual, the basis $(B\mathcal{W})$ and $(B\mathcal{T})$ must satisfy :

$$S\mathcal{W}_i \bigcirc S\mathcal{T}_j = \delta_{ij} \qquad \text{for } i,j = 1,6 \tag{4}$$

where δ_{ij} is the Kronecker symbol. The condition of duality given by Eq. (4) may be expressed in matrix form :

$$(\mathfrak{B}\mathcal{W})^{\bigcirc}(\mathfrak{B}\mathcal{T}) = \begin{bmatrix} S\mathcal{W}_1 \bigcirc S\mathcal{T}_1 & \cdots & S\mathcal{W}_1 \bigcirc S\mathcal{T}_6 \\ \vdots & & \vdots \\ S\mathcal{W}_6 \bigcirc S\mathcal{T}_1 & \cdots & S\mathcal{W}_6 \bigcirc S\mathcal{T}_6 \end{bmatrix} = Id_6 \tag{4bis}$$

Screw-vector dual bases $(B\mathcal{W})$ and $(B\mathcal{T})$ are in a one-to-one correspondence. From now on, $(B\mathcal{W})$ and $(B\mathcal{T})$ will always denote dual bases.

C. Coordinate Functions of Screw-vectors

The dual of a screw-vector basis allows to express the coordinate functions of any screw-vector in this basis. The coordinates $T_{[i]}$ of a twist **T** in the basis $(B\mathcal{T})$ are given by :

$$T_{[i]} = \mathbf{T} \bigcirc S\mathcal{W}_i \qquad \text{for } i = 1,6 \tag{5}$$

and reciprocally, the coordinates $W_{[i]}$ of a twist **W** in the basis $(B\mathcal{W})$ are given by :

$$W_{[i]} = \mathbf{W} \bigcirc S\mathcal{T}_i \qquad \text{for } i = 1,6 \tag{6}$$

For a concise notation, we express the coordinates' vectors as follows :

$$\underline{T} = (B\mathcal{W})^{\bigcirc}\ \mathbf{T} \tag{5bis}$$

$$\underline{W} = (B\mathcal{T})^{\bigcirc}\ \mathbf{W} \tag{6bis}$$

Plücker coordinates of screw-vectors are the coordinates of screw-vectors in a basis composed of 3 unit line-vectors $(\pounds x_O, \pounds y_O, \pounds z_O)$ whose axes are orthogonal lines intersecting at a point O, and their 3 Euclidean polars [(3)] $(\pounds x', \pounds y', \pounds z')$ which are free-line vectors. Such a basis of screw-vectors is thus specified by a reference tetrahedron $(O;\ \vec{e}^{\,x}, \vec{e}^{\,y}, \vec{e}^{\,z})$ defining a coordinate system in $\mathcal{E}^3$. We call it a *cartesian basis* of screw-vectors.
The coordinates of a twist **T** in a cartesian basis $(B\mathcal{T}) = (\pounds x_O, \pounds y_O, \pounds z_O, \pounds x', \pounds y', \pounds z')$ are the Plücker ray coordinates [3]. The coordinates of a wrench **W** in the dual cartesian basis

(3) Lipkin [3] defines the Euclidean polarity of screws as an extension of the Euclidean polarity of lines in projective three-dimensional Euclidean space. The Euclidean polarity of a screw vector **S** is a free-line-vector **S'** = **L'** whose magnitude is the magnitude of the screw-vector **S** and whose direction is the direction of the screw-vector 's axis. The composition of the Euclidean polar operator with the reciprocal product is equivalent to the perpendicular product [3]. This equivalence is of practical importance for carrying out screw-vector calculus.

$(B\mathcal{W}) = (\pounds x', \pounds y', \pounds z', \pounds x_O, \pounds y_O, \pounds z_O)$ are the Plücker axis coordinates [3]. According to the general form of coordinate functions (Eq. 5 and 6), the Plücker ray and axis coordinates of a screw-vector may be expresssed as :

$$\underline{T}^{(ray)} = \begin{bmatrix} \mathbf{T} \bigcirc \pounds x' \\ \mathbf{T} \bigcirc \pounds y' \\ \mathbf{T} \bigcirc \pounds z' \\ \mathbf{T} \bigcirc \pounds x_O \\ \mathbf{T} \bigcirc \pounds y_O \\ \mathbf{T} \bigcirc \pounds z_O \end{bmatrix} = \begin{bmatrix} \mathbf{T} \diamond \pounds x_O \\ \mathbf{T} \diamond \pounds y_O \\ \mathbf{T} \diamond \pounds z_O \\ \mathbf{T} \bigcirc \pounds x_O \\ \mathbf{T} \bigcirc \pounds y_O \\ \mathbf{T} \bigcirc \pounds z_O \end{bmatrix} \qquad \underline{W}^{(axis)} = \begin{bmatrix} \mathbf{W} \bigcirc \pounds x_O \\ \mathbf{W} \bigcirc \pounds y_O \\ \mathbf{W} \bigcirc \pounds z_O \\ \mathbf{W} \bigcirc \pounds x' \\ \mathbf{W} \bigcirc \pounds y' \\ \mathbf{W} \bigcirc \pounds z' \end{bmatrix} = \begin{bmatrix} \mathbf{W} \bigcirc \pounds x_O \\ \mathbf{W} \bigcirc \pounds y_O \\ \mathbf{W} \bigcirc \pounds z_O \\ \mathbf{W} \diamond \pounds x_O \\ \mathbf{W} \diamond \pounds y_O \\ \mathbf{W} \diamond \pounds z_O \end{bmatrix} \quad (7)$$

This expression of Plücker coordinates was established by Parkin [15].

D. *Coordinate Transformations of Screws*

Let a twist **T** be represented by coordinates $\underline{T}^{(\alpha)}$ in a first basis $(B\mathcal{T}\alpha)$, and coordinates $\underline{T}^{(\beta)}$ in a second basis $(B\mathcal{T}\beta)$. The transformation from coordinates $\underline{T}^{(\alpha)}$ to coordinates $\underline{T}^{(\beta)}$ is written as follows :

$$\underline{T}^{(\beta)} = (B\mathcal{W}\beta)^{\circ}(B\mathcal{T}a)\ \underline{T}^{(\alpha)} \quad (8)$$

where $(B\mathcal{W}\beta)$ is the dual basis of $(B\mathcal{T}\beta)$. For instance, the well-known transformation matrix of ray Plücker coordinates [16] may be expressed according to Eq. (8). The 6 columns of the transformation matrix $(B\mathcal{W}\beta)^{\circ}(B\mathcal{T}\alpha)$ are the coordinates of the screw-vectors of $(B\mathcal{T}\alpha)$ in the basis $(B\mathcal{T}\beta)$.

The transformation from coordinates $\underline{W}^{(\alpha)}$ of a wrench **W** in the basis $(B\mathcal{W}\alpha)$, dual of $(B\mathcal{T}\alpha)$, to its coordinates $\underline{W}^{(\beta)}$ in the basis $(B\mathcal{W}\beta)$ is written as :

$$\underline{W}^{(\beta)} = (B\mathcal{T}\beta)^{\circ}(B\mathcal{W}a)\ \underline{W}^{(\alpha)} \quad (9)$$

The 6 columns of the transformation matrix $(B\mathcal{T}b)^{\circ}(B\mathcal{W}a)$ are the coordinates of the screw-vectors of $(B\mathcal{W}\alpha)$ in the basis $(B\mathcal{W}\beta)$. The two transformation matrices of Eq. (8) and (9) are related by :

$$(B\mathcal{T}\beta)^{\circ}(B\mathcal{W}a) = [(B\mathcal{W}\beta)^{\circ}(B\mathcal{T}a)]^{-t} = [(B\mathcal{T}a)^{\circ}(B\mathcal{W}\beta)]^{-1} \quad (10)$$

E. *The Dual Bases of Screw-vectors Associated with a 6 dof Serial Arm*

The twist **T** of the end-effector of serial 6 dof robot robot arm is a linear combination of the unit screw-vectors $\$_i$ (i = 1,6) defined by the joints. In a regular configuration of the robot arm, the 6 screw-vectors $\$_i$ (i = 1,6) constitute a basis $(B\mathcal{A}\mathcal{T})$ of the space of twists. $(B\mathcal{A}\mathcal{T})$ defines the direct (first-order) kinematic model of the robot (Eq. 11) and also its inverse static model (Eq. 12) :

$$\mathbf{T} = \sum_{i=1}^{6} \eta_i\ \$_i = (B\mathcal{A}\mathcal{T})\ \underline{\eta} \quad (11)$$

$$\underline{\lambda} = (B\mathcal{A}\mathcal{T})^{\circ}\ \mathbf{W} \quad (12)$$

where $\underline{\eta}$ is the column vector of the joints' velocities, $\underline{\lambda}$ is the column vector of the torques (or forces) applied by the joints, and **W** is the wrench applied at the end-effector which is equilibrated by the joints torques (or forces). The usual form of the direct kinematic model is the projection of Eq. (11) on a cartesian basis $(\mathrm{B}\mathcal{T})$ whose center P is a point on the end-effector. The jacobian matrix may thus be written as :

$$J = (\mathrm{B}\mathcal{W})^{\circ}(\mathrm{B}\mathcal{A}\mathcal{T}) \tag{13}$$

where $(\mathrm{B}\mathcal{W})$ is the dual basis of $(\mathrm{B}\mathcal{T})$. The orientation of the tetrahedron associated with $(\mathrm{B}\mathcal{T})$ is usually fixed with respect to the base of the arm; while a reference frame attached to the end-effector is needed for some applications. The columns of J are the Plücker ray coordinates of the joints' axis screw-vectors $\$_i$ when $(\mathrm{B}\mathcal{T}) = (\pounds x_P, \pounds y_P, \pounds z_P, \pounds x', \pounds y', \pounds z')$. This fact is now well-recognized [17,18].

The dual basis $(\mathrm{B}\mathcal{A}\mathcal{W})$ of $(\mathrm{B}\mathcal{A}\mathcal{T})$ define the inverse kinematic model of the robot arm, and also its direct static model :

$$\underline{\eta} = (\mathrm{B}\mathcal{A}\mathcal{W})^{\circ}\mathbf{T} \tag{14}$$

$$\mathbf{W} = (\mathrm{B}\mathcal{A}\mathcal{W})\,\underline{\lambda} \tag{15}$$

The transpose of the inverse jacobian matrix J may be written as :

$$J^{-t} = (\mathrm{B}\mathcal{T})^{\circ}(\mathrm{B}\mathcal{A}\mathcal{W}) \tag{16}$$

The columns of the matrix J^{-t} are the axis coordinates of the screw-vectors of $(\mathrm{B}\mathcal{A}\mathcal{W})$. This was observed by Hunt [18,19].

The determination of the dual basis $(\mathrm{B}\mathcal{A}\mathcal{W})$ allows to define in a coordinate-free form the inverse kinematic and direct static models of a 6-dof serial arm. The screw-vectors of $(\mathrm{B}\mathcal{A}\mathcal{W})$ have an important physical meaning. They are the "wrench supports" [1] along which the joints' actuators apply a wrench on the end-effector.

III. Derivation of the Dual of a Screw-vector Basis

A. *Different Methods for Deriving the Dual Basis of a Given Basis*

Let's suppose we have defined a basis $(\mathrm{B}\mathcal{T})$ of the space of twists, and we are searching for the dual basis $(\mathrm{B}\mathcal{W})$ of the space of wrenches. The following results may also be applied to the inverse problem, i.e. deriving $(\mathrm{B}\mathcal{T})$ from $(\mathrm{B}\mathcal{W})$, by reversing the symbols $\mathcal{T}$ and $\mathcal{W}$.

The coordinates of the screw-vectors of the basis $(\mathrm{B}\mathcal{W})$ in a basis of wrench $(\mathrm{B}\mathcal{W}_0)$ are, according to Eq. (9) and (10), given by the transformation matrix :

$$(\mathrm{B}\mathcal{T}_0)^{\circ}(\mathrm{B}\mathcal{W}) = [(\mathrm{B}\mathcal{W}_0)^{\circ}(\mathrm{B}\mathcal{T})]^{-t}$$

where $(\mathrm{B}\mathcal{T}_0)$ is the dual basis of $(\mathrm{B}\mathcal{W}_0)$. $(\mathrm{B}\mathcal{W})$ may thus be calculated as :

$$(\mathrm{B}\mathcal{W}) = (\mathrm{B}\mathcal{W}_0)\,[(\mathrm{B}\mathcal{W}_0)^{\circ}(\mathrm{B}\mathcal{T})]^{-t} \tag{17}$$

This method is efficient when the basis $(\mathrm{B}\mathcal{W}_0)$ is chosen so that the matrix $(\mathrm{B}\mathcal{W}_0)^{\circ}(\mathrm{B}\mathcal{T})$ is sparse, i.e. many coordinates of the screw-vectors of $(\mathrm{B}\mathcal{T})$ expressed in $(\mathrm{B}\mathcal{T}_0)$ are zero. It is not required that $(\mathrm{B}\mathcal{T}_0)$, the dual basis of $(\mathrm{B}\mathcal{W}_0)$ be known. We shall see §III.B that this method comprises the method of derivation of the inverse jacobian matrix of a 6 dof robot using a well-chosen reference frame [17,20].

The screw-vectors of the basis $(\mathrm{B}\mathcal{W})$ may also be represented in the basis $(\mathrm{B}\mathcal{T})$, to be more correct in the basis of wrenches that is associated with the basis of twists $(\mathrm{B}\mathcal{T})$

by the isomorphism of wrench and twist spaces. The calculus of such coordinates involves the inversion of the symmetric matrix $(B\mathcal{T})^{\circ}(B\mathcal{T})$:

$$(B\mathcal{W}) = (B\mathcal{T})\ [(B\mathcal{T})^{\circ}(B\mathcal{T})]^{-1} \tag{18}$$

For a general basis $(B\mathcal{T})$, the determination of $(B\mathcal{W})$ in the basis $(B\mathcal{T})$ is efficient when the many pairs of screw-vectors of $(B\mathcal{T})$ are reciprocal.

The regularity of the matrix $(B\mathcal{T})^{\circ}(B\mathcal{T})$ is a necessary and sufficient condition of the linear independance of the six screw-vectors of $(B\mathcal{T})$. In contrast with the condition given in [21] which uses the non intrinsic "orthogonal product" instead of the reciprocal product, this condition does not apply to a set of $n < 6$ screw-vectors. This limitation is due to the fact that a screw-vector space contains 3 dimensional sub-spaces that are isotropic with respect to the reciprocal product [3].

Each screw-vector of $(B\mathcal{W})$ may be expressed in a determinant form. For instance the first screw-vector is :

$$S\mathcal{W}_1 = \frac{1}{|(B\mathcal{T})^{\circ}(B\mathcal{T})|} \begin{vmatrix} S\mathcal{T}_1 & \cdots & S\mathcal{T}_6 \\ S\mathcal{T}_2 \circ S\mathcal{T}_1 & \cdots & S\mathcal{T}_2 \circ S\mathcal{T}_6 \\ \vdots & & \vdots \\ S\mathcal{T}_6 \circ S\mathcal{T}_1 & \cdots & S\mathcal{T}_6 \circ S\mathcal{T}_6 \end{vmatrix} \tag{18bis}$$

This expression is similar to the expression given in [21] for deriving a screw-vector reciprocal to a set of n independent screws. Because it uses the reciprocal product instead of the "orthogonal product", it doesn't apply to sets of $n < 6$ screw-vectors.

An alternative way of determining the dual basis is to consider independently the sub-problems of determining a screw-vector $S\mathcal{W}_i$ of the dual basis $(B\mathcal{W})$. The geometric screw $S\mathcal{W}_i$ of the screw-vector $S\mathcal{W}_i$ is defined from the 5 conditions of reciprocity: $S\mathcal{W}_i \circ S\mathcal{T}_j = 0$ $(i \neq j)$. Its magnitude and the orientation of its axis are defined by the normalization condition : $S\mathcal{W}_i \circ S\mathcal{T}_i = 1$. For very particular geometric configurations of the geometric screws of the five screw-vectors $S\mathcal{T}_j$, $S\mathcal{W}_i$ may be defined directly from geometrical reasoning. In the following example, we will show that it often efficient in practical robotics problems to combine geometric reasoning and analytical calculus.

B. *Application to the Inverse Kinematic Problem of a 6-dof Serial Robot Arm*

We consider the 6-dof robot composed of 3 cyclindric joints presented in [22]. The three axis of the cylindric joints are considered to be placed in general relative positions. They are represented by three unit line-vectors $\pounds_i$ $(i = 1,3)$ whose direction are given by the vectors $\vec{z}_i$ (Fig. 1). The screw-vector basis defined by the elementary joints'axis is :

$$(B\mathcal{A}\mathcal{T})\ (\ \pounds_1\ \ \pounds_1'\ \ \pounds_2\ \ \pounds_2'\ \ \pounds_3\ \ \pounds_3'\ ) \tag{19}$$

We denote the 6 screw-vectors of the dual basis $(B\mathcal{A}\mathcal{W})$ as follows :

$$(B\mathcal{A}\mathcal{W}) = (\ S\mathcal{W}_{1r},\ S\mathcal{W}_{1t}\ \ S\mathcal{W}_{2r},\ S\mathcal{W}_{2t}\ \ S\mathcal{W}_{3r},\ S\mathcal{W}_{3t}\ ) \tag{20}$$

We determine below the screw-vectors of $(B\mathcal{A}\mathcal{W})$ directly from the geometrical feature of the problem.

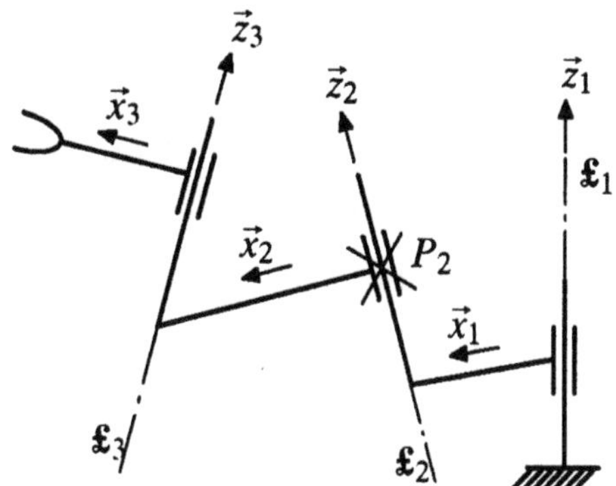

Fig. 1 : Kinematic skeleton of the 3–cylindric robot (example from [22])

Let i,j,k be a circular permutation of 1,2,3. $S\mathcal{W}_{ir}$ must be reciprocal with the 3 independent free–line vectors $\pounds_1'$, $\pounds_2'$ and $\pounds_3'$. Accordingly, $S\mathcal{W}_{ir}$ is a free–line vector. It must also be reciprocal with the 2 line–vectors $\pounds_j$ and $\pounds_k$. $S\mathcal{W}_{ir}$ is thus a free–line vector whose direction is the common perpendicular of the jth and k^th^ lines. The normalization condition, $S\mathcal{W}_{ir} \bigcirc \pounds_i = 1$, allows finally to determine the magnitude of $S\mathcal{W}_{ir}$. $S\mathcal{W}_{ir}$ is thus expressed as :

$$S\mathcal{W}_{ir} = \frac{\pounds_{jk}'}{\pounds_i \diamond \pounds_{jk}} \tag{21}$$

where $\pounds_{jk}'$ is the unit free–line–vector whose direction is given by the vector $\vec{z}_j \times \vec{z}_k$.

$S\mathcal{W}_{it}$ must be reciprocal with the 2 free–line–vectors $\pounds_j'$ and $\pounds_k'$. This implies that $S\mathcal{W}_{it}$ is a screw–vector whose axis is perpendicular with the jth and k^{th} lines. $S\mathcal{W}_{ir}$ must also be reciprocal with the 2 line–vectors $\pounds_j$ and $\pounds_k$. Accordingly, the axis of $S\mathcal{W}_{it}$ is the line intersecting perpendicularly the jth and k^{th} lines. The magnitude and the pitch of the screw–vector $S\mathcal{W}_{it}$. $S\mathcal{W}_{it}$ are defined by the two last conditions of duality : $S\mathcal{W}_{it} \bigcirc \pounds_i = 0$ and $S\mathcal{W}_{it} \diamond \pounds_i = 1$.

$$S\mathcal{W}_{it} = \frac{1}{\pounds_i \diamond \pounds_{jk}} \left(\pounds_{jk} - \frac{\pounds_i \bigcirc \pounds_{jk}}{\pounds_i \diamond \pounds_{jk}} \pounds_{jk}' \right) \tag{22}$$

where $\pounds_{jk}$ is the unit line–vector which represents the line intersecting perpendicularly the jth and k^{th} axis, and whose orientation is given by the vector $\vec{z}_j \times \vec{z}_k$.

In the previous development, the axes of the screw–vectors of the dual basis $(B\mathcal{A}\mathcal{W})$ have been derived directly from geometrical considerations. Then, the magnitude and the pitch (when not zero) have been derived from algebraic manipulations. The dual basis $(B\mathcal{A}\mathcal{W})$ could be obtained as efficiently in a pure analytical way, following the method based on the projection of $(B\mathcal{A}\mathcal{T})$ (§3.1, Eq. 17). Indeed, a basis $(B\mathcal{W}_0)$ may be composed of the three screw–vectors $\pounds_j \times \pounds_k$ and their Euclidean polars $(\pounds_j \times \pounds_k)'$. The screw–vectors of $(B\mathcal{W}_0)$ are linearly independent when the robot is in a regular configuration. The matrix to be inverted, $(B\mathcal{W}_0)^{\bigcirc}(B\mathcal{A}\mathcal{T})$, is then very sparse, because the screw–vector product of two screw-vectors is reciprocal and perpendicular with these two screw-vectors. It should be noticed that $(B\mathcal{W}_0)$ is not a cartesian basis. The projection of $(B\mathcal{AT})$ in a well-chosen cartesian basis would lead to significantly more calculus. Let for instance, the cartesian basis $(B\mathcal{T}_{cart.})$ be defined by $(P_2;\ \vec{x}_2, \vec{y}_2, \vec{z}_2)$, the matrix $(B\mathcal{W}_{cart.})^{\circ}(B\mathcal{AT})$ has 17 while $(B\mathcal{W}_0)^{\circ}(B\mathcal{AT})$ has 9 non zero components.

C. *Application to the Inverse Static Problem of a Parallel 6–dof Robot*

We consider a 6-dof fully parallel manipulator classified as Triangular Symmetric Simplified Manipulators [23,24]. In the following development, we do not require however that the center points of the Hooke joints form a symmetric plane hexagon and that the triangle composed by the centers of the spherical joints be an isoscele. The six linear actuators of the robot apply a wrench on the end-effector along six unit lines-vectors $\pounds_i$:

$$\mathbf{W} = (\ \pounds_1\ \ \pounds_2\ \ \pounds_3\ \ \pounds_4\ \ \pounds_5\ \ \pounds_6\ )\ (f_1\ f_2\ f_3\ f_4\ f_5\ f_6)^t \tag{23}$$

These six unit line-vector $\pounds_i$ constitute a basis $(B\mathcal{AW})$ of the space of wrenches when the robot is in a regular configuration. Each line represented by $\pounds_i$ joins a hooke joint center A_i with a spherical joint center B_i ; it intersect one of the other lines at point $B_{ij} = B_i = B_j$ (Fig. 2).

The dual basis $(B\mathcal{AT})$ of $(B\mathcal{AW})$ defines the direct kinematic model of the robot :

$$\mathbf{T} = (B\mathcal{AT})\ (v_1\ v_2\ v_3\ v_4\ v_5\ v_6)^t \tag{24}$$

where $\mathbf{T}$ is the twist of the end-effector and v_i are the linear actuators' velocities. We denote $S\mathcal{T}_i$ the ith screw-vector of the dual basis $(B\mathcal{AT})$.

Let's consider the two first screw-vectors of $(B\mathcal{AT})$: $S\mathcal{T}_1$ and $S\mathcal{T}_2$. $S\mathcal{T}_1$ and $S\mathcal{T}_2$ are both reciprocal with the 4 line-vector $\pounds_i$ (i = 3,4). The screw-vector sub-space $\mathcal{W}_{12}$ reciprocal with $\pounds_i$ (i = 3,4) is spanned by two unit-line vectors :

- $\pounds_{B_{34}B_{56}}$ which represents the line that joins the points B_{34} and B_{56},
- $\pounds_{\Pi_{34}\Pi_{56}}$ which represents the line that is the intersection between the two planes Π_{34} and Π_{56} ; the plane Π_{ij} being the plane that contains the two intersecting line $\pounds_i$ and $\pounds_j$.

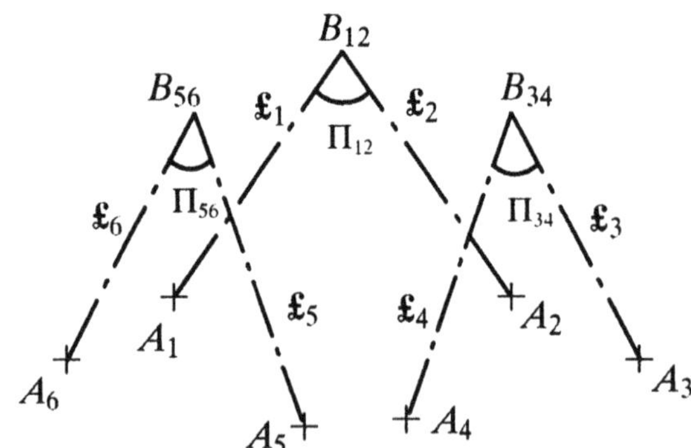

Fig. 2 : Geometric configuration of the six wrench-support lines of a triangular parallel robot [Merlet 89]

$S\mathcal{T}_1$ and $S\mathcal{T}_2$ may be expressed by their coordinates in the basis ($\pounds_{B_{34}B_{56}}$ $\pounds_{\Pi_{34}\Pi_{56}}$) of $\mathcal{W}_{12}$:

$$(S\mathcal{T}_1 \;\; S\mathcal{T}_2) = (\pounds_{B_{34}B_{56}} \;\; \pounds_{\Pi_{34}\Pi_{56}}) \begin{pmatrix} a_1 & a_2 \\ \beta_1 & \beta_2 \end{pmatrix} \tag{25a}$$

The remaining conditions of duality may be expressed as :

$$(S\mathcal{T}_1 \;\; S\mathcal{T}_2)^{\circ} (\pounds_1 \;\; \pounds_2) = Id_2$$

The coordinates α_i and β_i are then obtained by the inversion of a 2.2 matrix :

$$\begin{pmatrix} a_1 & a_2 \\ \beta_1 & \beta_2 \end{pmatrix} = \left((\pounds_{B_{34}B_{56}} \;\; \pounds_{\Pi_{34}\Pi_{56}})^{\circ} (\pounds_1 \;\; \pounds_2) \right)^{-1} \tag{25b}$$

A similar calculus may be done for deriving the remaining screw–vectors of the dual basis. The screw–vectors of the dual basis ($B\mathcal{T}$) are thus explicitely determined by combining geometric reasoning and formal screw–vector calculus. They define intrinsically the direct kinematic model and inverse static model of the parallel robot. These models are known to be dificult to be calculated explicitely [24]. This difficluty has been overcome here for triangular parallel robots by taking into account the special geometric arrangement of the wrench supports.

IV. Conclusions

Dual bases of screw–vectors have been introduced. They allows to express coordinates functions and transformations of screw–vectors in any kind of screw–vector bases. The coordinate of a screw–vector are expressed as the reciprocal products of the screw–vector with the screw–vector of the dual basis. This expression is a genralization of the expression of Plücker coordinates given by Parkin [15]. Matrix of screw–vector coordinates transformations are also composed by reciprocal products of screw–vectors.

Dual bases provide a concise analytical framework for formulating kinestatic models in intrinsic form. In this paper we show that the problems of inverting kinestatic models are equivalent to find a dual basis. Although different results published on serial robot–arms [18,19] and on serial chains of parallel robots [25,26] are related to this fact, it seems that the role of the dual bases of screw–vectors has not been previously outlined. This framework may be applied to other kinestatic problems, using partitioned dual bases [27,28] and successive reductions of the models of serial and parallel chains [28].

Two examples illustrate the practical resolution of inverse kinestatic problems. They show how intrinsic screw–vector algebraic notation may support and complete the geometric reasoning about screws and lines. The link between the screw–vectors and the geometric objects as lines and points could be made more explicit in the notation by using bi–vectors [11].

References

[1] K. H. Hunt, *Kinematic Geometry of Mechanisms*, Oxford University Press, Oxford (1978).

[2] J. Phillips, *Freedom in Machinery – II: Screw Theory Exemplified*, Cambridge University Press, Cambridge (1990).

[3] H. Lipkin, *Geometry and mappings of screws with applications to the hybrid control of robotic manipulators*, PhD dissertation, University of Florida (1985).

[4] J.-M. Hervé, "Intrinsic formulation of problems of geometry and kinematics", *Mechanism and Machine Theory*, *17*(3), 179–184 (1982).

[5] J.-M. Hervé, "The mathematical group structure of the set of displacements", *Mechanisms and Machine Theory*, *29*(1), 73–81 (1994).

[6] A. Karger and J. Novák, *Space Kinematics and Lie Groups*, translated by M. Basch, from Czech publication: *Prostorová Kinematika a Lieovy Grupy*, (SNTL, Prague, Czechoslovakia, (1978)) Gordon and Breach Pub., New York (1985).

[7] J. M. Selig and J. Rooney, "Reuleaux pairs and surfaces that cannot be gripped", *The Int. J. of Robotics Research*, *8*(5), 79–87 (1989).

[8] A. E. Samuel, P. R. MacAree and K. H. Hunt, "Unifying screw geometry and matrix transformations", Int. J. Robotics Research, *10*(5), 454–472 (1991).

[9] L. Brand, *Vector and Tensor Analysis*, John Wiley & Sons Publishing, New York, (1947).

[10] J. Angeles, Rational Kinematics, Springer Tracts in Natural Philosophy, vol. 34, (C. Truesdell Eds.), Springer Verlag, New York (1988).

[11] J. M. MacCarthy, *Introduction to theoretical kinematics*, J. Wiley & Sons, Chichester, (1990)

[12] F. M. Dimentberg, *The screw Calculus and its Application in Mechanics*, Translated and reproduced by the Clearinghouse for Federal Scientific and Technical Information, Va. 22151, Document N°. FTD-HT-23-1632, Springfield (1965).

[13] K.J. Waldron, "The use of motors in spatial kinematics", Proc. of IFToMM Int. Symp. on Linkages and Computer Design Methods, vol. B, 535–545, Bucharest (1973).

[14] A. T. Yang, "Calculus of screws", *Basic Questions of Design Theory* (W. R. Spiller Ed.), 265–281, North-Holland, Amsterdam (1974).

[15] J. A. Parkin, "Co-ordinate transformations of screws with applications to screw systems and finite twists", *Mechanism and Machine Theory*, *25*(6), 489–478 (1990).

[16] L. Woo and F. Freudenstein, "Application of line geometry to theoretical kinematic analysis of mechanical", *J. of Mechanisms*, *5*(3), 417–460 (1970).

[17] K. J. Waldron, S.-L. Wang and S. J. Bolin, "A study of the jacobian matrix of serial manipulators", *ASME J. of Mechanisms, Transmissions, and Automation in Design*, *111*, 230–237, (1985).

[18] K. H. Hunt, "Special configurations of robot arms via screw theory – part I : The jacobian and matrix of cofactors", *Robotica*, *4*, 171–179 (1986).

[19] K. J. Waldron and K. H. Hunt, "Series-parallel dualities in actively coordinated mechanism", *Robotics Research*, *10*(5), 473–480 (1991).

[20] M. Renaud, *Contribution à la Modélisation et à la Commande Dynamique des Robots Manipulateurs*, Thèse d'état, Université Paul Sabatier de Toulouse (1980)

[21] K. Sugimoto and J. Duffy, "Application of linear algebra to screw systems", *Mechanism and Machine Theory*, *17*(1), 73–83 (1982).

[22] G. R. Pennock and B. C. Viestra, "The inverse kinematics of a three-cylindric robot", *Int. J. of Robotics Research*, *9*(4), 75–85 (1990).

[23] J.-P. Merlet, "Singular configurations of parallel manipulators and Grassmann geometry", *The Int. J. of Robotics Research*, *8*(5), 45–56 (1989).

[24] J.-P. Merlet, *Les robots parallèles*, éditions Hermes, Paris (1990).

[25] Agrawal S. K. and Roth B., "Statics of in-parallel manipulators systems", *ASME J. of Mechanical Design*, *114*(4), 564–568, (1992).

[26] V. Kumar, "Instantaneous kinematics of parallel-chain robotic mechanisms", *ASME J. of Mechanical Design*, *114*(3), 349–358, (1992).

[27] C. Bidard, "Screw-vector bond graphs for multibody systems modelling", Int. Conf. on Bond Graph Modeling, Proc. of the Western Simulation Multiconference, 195–200, SCS, San Diego (1993).

[28] C. Bidard, *Graphes de liaisons torsoriels pour la modélisation et l'analyse ciné-statique des mécanismes de robots*, Thèse de Doctorat, Université de Lyon, to be presented in (1994).

9. Kinematic Design

A Comparison of Two Minimally-Singular Articulated Arm-Subassemblies

S.J. Remis* and M.M. Stanišić†

*Yoder Software, Inc.
430 Parkovash Ave.
South Bend, IN 46617 USA
email: 74670,3412@compuserve.com

†Aerospace and Mechanical Engineering Department
University of Notre Dame
Notre Dame, IN 46556 USA
email: stanisic@sparky.ame.nd.edu

Abstract - Recent developments in manipulator kinematics have led to the designs of singularity-free pointing systems, wrist-subassemblies, and both spherical and articulated arm-subassemblies. Determining a proper kinematic constraint for the joints of these redundant systems differs from simply ensuring the pointing subsystem is singularity-free. In this paper two different constraints of a redundant articulated arm-subassembly are compared: one constraint removes the singularities of the pointing subsystem and the second involves the three degree-of-freedom motion of the entire subassembly. It is shown that the constraint which eliminates the singularities of the pointing subsystem fails to produce a singularity-free articulated arm-subassembly, and this constraint must be modified to include the effects of the elbow rotation.

I. Introduction

In recent years researchers have coordinated the joint variables of redundant manipulators to develop wrist-subassemblies [1] [2] [3] [4], a spherical manipulator [5], and an articulated arm-subassembly (AAS) [6], all of which are without workspace singularities. These singularity-free systems are designed by enforcing a kinematic constraint on the redundant actuators that ensures the singularities become unreachable due to an unavoidable mechanical interference condition (it was shown in [7] that these singularities must theoretically exist). For the wrist-subassemblies it is possible to constrain the manipulator to make the singularities coincide with interference between the tool axis and the forearm of the arm-subassembly. For arm-subassemblies the singularities can be made to coincide with interference between the wrist-subassembly and the base of the arm-subassembly. When all singularities are made to coincide with these naturally-occurring, unreachable, interference conditions, then the entire reachable workspace is singularity-free.

Although the pointing subsystem of a manipulator subassembly causes the workspace singularities [7] [12] [13], it is shown in this paper that enforcing a kinematic constraint which eliminates the singularities of the pointing subsystem does not necessarily eliminate the singularities of the entire subassembly. Such a technique has been presented

A. J. Lenarčič and B. B. Ravani (eds.), Advances in Robot Kinematics and Computationed Geometry, 351–358.

[8], and this method is compared with a method that considers fully the required motion of the entire subassembly (not just the pointing subsystem) [6]. The analysis in this paper is for a redundant AAS when constrained by two different kinematic functions. It is shown that the constraint resulting in a singularity-free pointing subsystem does not achieve singularity-free positioning by the AAS, and the constraint must be modified to include the effect of the elbow rotation as part of the kinematic constraint function.

II. Singularity Elimination

For a given arm-subassembly, denote by $\overline{r}_c$ the position vector from the center of the shoulder (or the origin of the base coordinate system if the arm-subassembly does not contain a shoulder) to the center of the wrist-subassembly, and by $\overline{\theta}$ the vector of joint coordinates. The differential mapping

$$\dot{\overline{r}}_c = J \dot{\overline{\theta}} \tag{1}$$

determines the singularities of the subassembly [10]. These singularities occur when the rank of J is less than the dimension of the workspace vector $\overline{r}_c$. For redundant manipulators J is non-square, but can be reduced to a square matrix by constraining one joint variable to be a function of the remaining variables (this assumes a singly-redundant system), *e. g.*,

$$\theta_1 = g(\theta_2, \theta_3, \theta_4) . \tag{2}$$

This constraint corresponds to allowing only a subset of the possible jointspace parameters $\overline{\theta}$ to be achievable by the subassembly.

A column is eliminated from J by substituting the derivative of eqn. (2),

$$\dot{\theta}_1 = g_2 \dot{\theta}_2 + g_3 \dot{\theta}_3 + g_4 \dot{\theta}_4 , \tag{3}$$

into eqn. (1). In eqn. (3) g_i represents the partial derivative of θ_1 with respect to θ_i. This substitution yields the (square) reduced Jacobian J_r. This matrix reduction technique is a special case of Baillieul's extended Jacobian method [11], where the first column and the last row are eliminated from the extended Jacobian matrix. For a given kinematic constraint g the redundant subassembly is singular when $|J_r| = 0$. Some choices for g will prevent the manipulator from spanning the maximum workspace; these constraints are not *proper* [14]. Other choices for g may result in a maximum workspace, but will not eliminate the workspace singularities. The goal is to select a proper kinematic constraint g that: (1) does not reduce the workspace from its maximum size, and (2) ensures all singularities of the corresponding $|J_r|$ coincide with the unremovable interference conditions. When these two conditions are achieved the maximum workspace is without singularities.

III. Redundant Articulated Arm-Subassembly

The redundant AAS shown in fig. 1 is the same actuator geometry described in [6]. This geometry is also that of the arm-subassembly within the Omnidirectional Arm[1],

[1] In this paper we assume the shoulder of the Omnidirectional Arm contains a spherical pointing system, by reducing the small shoulder offset to zero.

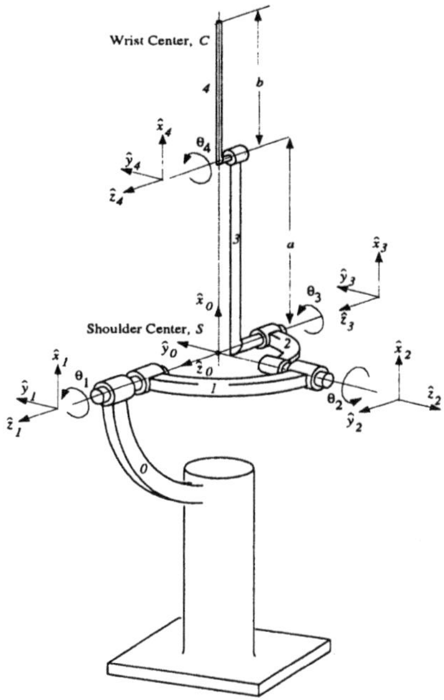

Fig. 1: A redundant articulated arm-subassembly (AAS). In this figure all joints are shown in their zero positions.

presented in [8]. For this arm-subassembly,

$$\overline{r}_c = \begin{pmatrix} c\theta_1 \; c\theta_2 \, (a \; c\theta_3 + b \; c\theta_{34}) - s\theta_1 \, (a \; s\theta_3 + b \; s\theta_{34}) \\ s\theta_1 \; c\theta_2 \, (a \; c\theta_3 + b \; c\theta_{34}) + c\theta_1 \, (a \; s\theta_3 + b \; s\theta_{34}) \\ s\theta_2 \, (a \; c\theta_3 + b \; c\theta_{34}) \end{pmatrix}, \tag{4}$$

where $s\theta$ and $c\theta$ represent the sine and the cosine, respectively, of θ, and θ_{34} is written in place of $\theta_3 + \theta_4$.

The Jacobian matrix for the unconstrained redundant AAS is:

$$J = \begin{bmatrix} -s\theta_1 c\theta_2 C_1 - c\theta_1 S_1 & -c\theta_1 s\theta_2 C_1 & -c\theta_1 c\theta_2 S_1 - s\theta_1 C_1 & -b\,(s\theta_1 c\theta_{34} + c\theta_1 c\theta_2 s\theta_{34}) \\ c\theta_1 c\theta_2 C_1 - s\theta_1 S_1 & -s\theta_1 s\theta_2 C_1 & -s\theta_1 c\theta_2 S_1 + c\theta_1 C_1 & b\,(c\theta_1 c\theta_{34} - s\theta_1 c\theta_2 s\theta_{34}) \\ 0 & c\theta_2 C_1 & -s\theta_2 S_1 & -b s\theta_2 s\theta_{34} \end{bmatrix}, \tag{5}$$

where

$$C_1 = a \; c\theta_3 + b \; c\theta_{34}$$

and

$$S_1 = a \; s\theta_3 + b \; s\theta_{34}.$$

Selecting θ_1 as the dependent actuator, as shown in eqns. (2) and (3), J is reduced to the following square matrix:

$$J_r = \begin{bmatrix} -(c\theta_1\, s\theta_2 \;+\; g_2\, s\theta_1\, c\theta_2)\, C_1 \;+ & -c\theta_1\,(c\theta_2 \;+\; g_3)\, S_1 \;- & -c\theta_1\, S_2 \;- \\ -g_2\, c\theta_1\, S_1 & s\theta_1\,(1 \;+\; g_3\, c\theta_2)\, C_1 & s\theta_1\, C_2 \\ & & \\ -(s\theta_1\, s\theta_2 \;-\; g_2\, c\theta_1\, c\theta_2)\, C_1 \;- & -s\theta_1\,(c\theta_2 \;+\; g_3)\, S_1 \;+ & -s\theta_1\, S_2 \;+ \\ g_2\, s\theta_1\, S_1 & c\theta_1\,(1 \;+\; g_3\, c\theta_2)\, C_1 & c\theta_1\, C_2 \\ & & \\ c\theta_2\, C_1 & -s\theta_2\, S_1 & -\, b\, s\theta_2\, s\theta_{34} \end{bmatrix} \tag{6}$$

where

$$C_2 \;=\; b\, c\theta_{34}\,(1 \;+\; g_4 c\theta_2) \;+\; a\, g_4\, c\theta_2 c\theta_3$$

and

$$S_2 \;=\; b\, s\theta_{34}\,(c\theta_2 \;+\; g_4) \;+\; a\, g_4\, s\theta_3 .$$

This reduced Jacobian matrix satisfies

$$\dot{\vec{r}}_c \;=\; J_r \left(\dot\theta_2 \;\; \dot\theta_3 \;\; \dot\theta_4\right)^{\mathrm{T}} . \tag{7}$$

The redundant AAS, constrained by an arbitrary function g, is singular when $|J_r| = 0$, which is computed to be:

$$a\, b\, s\theta_4\, \left[g_2\, s\theta_2\, S_1 \;+\; C_1\, (1 \;+\; g_3\, c\theta_2)\right] \;=\; 0. \tag{8}$$

IV. Two Kinematic Constraint Functions

To demonstrate the importance of selecting the proper kinematic constraint function g, consider two such functions:

$$\theta_1 \;=\; \theta_3, \tag{9a}$$

or

$$\theta_1 \;=\; \theta_3 \;+\; \frac{\pi}{2} \;-\; \operatorname{atan}\left(\frac{1 + Rc\theta_4}{Rs\theta_4}\right), \tag{9b}$$

where $R = \frac{b}{a}$, the ratio of the manipulator's link lengths. Eqn. (9a) will be referred to as the symmetry constraint. This is the kinematic constraint that eliminates the singularities of the pointing subsystem by ensuring these singularities occur only when the pointed axis must be coincident with the base of the subassembly [4]. The symmetry constraint is also the constraint used to coordinate the joints of the Omnidirectional Arm [8]. Although the symmetry constraint ensures that the upper arm of the redundant AAS can point anywhere in the reachable workspace without encountering a pointing singularity, this paper will show that there is a distinct difference between these pointing singularities and the singularities of positioning the wrist center. Eqn. (9b) is the kinematic constraint derived in [6] by considering the 3-dof motion of the subassembly instead of only the 2-dof motion of the pointing subsystem.

Both constraints (9a) and (9b) have the partial derivatives

$$g_2 \;=\; 0, \tag{10a}$$

$$g_3 = 1. \tag{10b}$$

However, the symmetry constraint of eqn. (9a) has the partial derivative

$$g_4|_{(9a)} = 0 \tag{11}$$

while the constraint given by eqn. (9b) has the partial derivative

$$g_4|_{(9b)} = \frac{R(R + c\theta_4)}{R^2 + 2Rc\theta_4 + 1}. \tag{12}$$

Note that in eqn. (8) $|J_r|$ does not include any terms containing g_4. Because of this fact, the expression for singularities of the subassembly is the same when constrained by either eqn. (9a) or (9b), which have identical partial derivatives g_2 and g_3. This expression for singularities reduces to:

$$a\,b\,s\theta_4\ (a\,c\theta_3 + b\,c\theta_{34})\ (1 + c\theta_2) = 0. \tag{13}$$

Factors a and b from eqn. (13) are the lengths of the upper arm and forearm, respectively, and are nonzero. Factor $s\theta_4$ equals zero only when the AAS is at a workspace boundary (either the minimum- or maximum-reach surface). These singularities are unavoidable, and they do not affect the performance of the subassembly inside its workspace. The remaining factors from eqn. (13) will be used to demonstrate the importance of selecting a proper kinematic constraint function.

V. Inverse Kinematics

A. The Symmetry Constraint $\theta_1 = \theta_3$

The geometry of the manipulator in the $a - b$ plane (the $\hat{x}_3 - \hat{y}_3$ plane) allows direct solution of θ_4 given the position of the wrist center $\overline{r}_c = (x_c, y_c, z_c)^T$ [9]:

$$\theta_4 = \pi - 2\ \mathrm{atan}(\frac{r}{s - h}), \tag{14}$$

where

$$h = \sqrt{x_c^2 + y_c^2 + z_c^2},$$

$$s = \frac{1}{2}(a + b + h),$$

$$r = \sqrt{\frac{(s-a)\,(s-b)\,(s-h)}{s}}.$$

θ_1, θ_2, and θ_3 can be computed once θ_4 is known:

$$\theta_1 = \theta_3 = \mathrm{atan}\left(\frac{y_c - bs\theta_4}{x_c + a + bc\theta_4}\right), \tag{15}$$

$$\theta_2 = \mathrm{atan}\left(\frac{z_c}{x_c\, c\theta_3 + y_c\, s\theta_3}\right). \tag{16}$$

B. *The Constraint* $\theta_1 = \theta_3 + \frac{\pi}{2} - \text{atan}\left(\frac{1+Rc\theta_4}{Rs\theta_4}\right)$

The inverse kinematics under this constraint function were presented in [6]. They are repeated here for completeness.

As with the symmetry constraint, θ_4 is computed using the geometry of the manipulator in the $a - b$ plane

$$\theta_4 = \pi - 2\,\text{atan}(\frac{r}{s-h}). \qquad (17)$$

We can determine θ_1 given θ_4 by first computing the intermediate function $f\,(\theta_4)$:

$$f\,(\theta_4) = \frac{\pi}{2} = \text{atan}\left(\frac{1+R\,c\theta_4}{R\,s\theta_4}\right) \qquad (18).$$

Next the terms S_3 and C_3 are determined:

$$S_3 = \sin\left[f\,(\theta_4)\right]$$

and

$$C_3 = \cos\left[f\,(\theta_4)\right].$$

The third component of eqn. (4) is then used to solve for the term $a\;c\theta_3 + b\;c\theta_{34}$. This is substituted into the first and second components of eqn. (4), with the result

$$\theta_1 = \text{atan}\left[\frac{y_c + a\;S_3 + b\,(S_3\;c\theta_4 - C_3\;s\theta_4)}{x_c + a\;C_3 + b\,(C_3\;c\theta_4 + S_3\;s\theta_4)}\right]. \qquad (19)$$

θ_3 is found from eqn. (9b) to be the following:

$$\theta_3 = \theta_1 - \frac{\pi}{2} + \text{atan}\left(\frac{1+R\,c\theta_4}{R\,s\theta_4}\right). \qquad (20)$$

Finally, eqn. (4) is used to compute θ_2 directly:

$$\theta_2 = \text{atan}\left(\frac{z_c}{x_c\;c\theta_1 + y_c\;s\theta_1}\right). \qquad (21)$$

VI. Workspace Comparison

Animations of the redundant AAS shown in fig. 1 were developed on a Silicon Graphics IRIS^{TM} workstation, for both the symmetry constraint of eqn. (9a) and the constraint given by eqn. (9b). These animations use the inverse kinematics developed above to examine the dexterity of points in the workspaces of the respective arm-subassemblies, and display the corresponding value of abs$\,(|J_r|)$ with a color index (grey-scale as shown in the figures). Figs. 2a and 2b present vertical slices through the center of the workspaces of these animations: fig. 2a for the symmetry constraint of eqn. (9a), and fig. 2b for the constraint given by eqn. (9b).

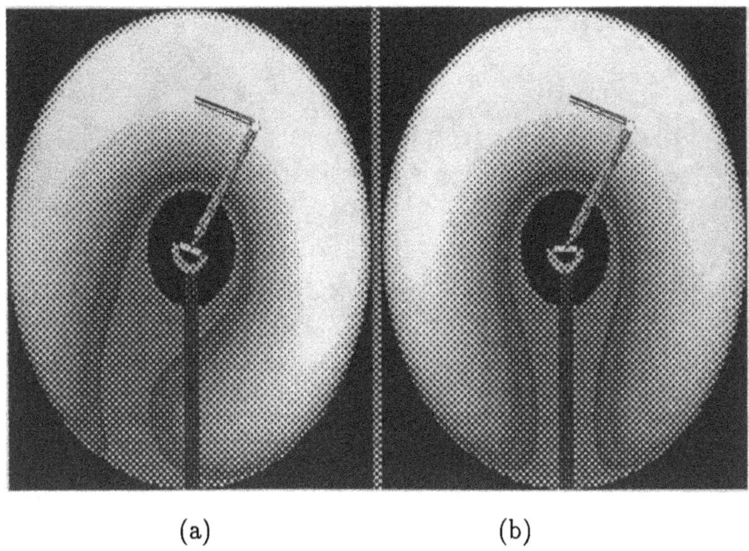

(a) (b)

Fig. 2: Images from animations of the redundant AAS under kinematic constraints: (a) eqn. (9a) and (b) eqn. (9b).

In these figures the regions in which the colors fade from white to grey represent those regions in which abs$(|J_r|) \geq 0.15$, with brighter shades of white corresponding to larger values. The solid grey regions represent workspace coordinates for which abs$(|J_r|) < 0.15$ (the regions of low dexterity), and the black regions are unreachable. In these figures it can be seen that the region of low dexterity under the symmetry constraint is inside the reachable workspace of the device. With the constraint given by eqn. (9b), however, this volume, which continues to theoretically exist, coincides with interference between the wrist center and the base of the subassembly. Since the base of the subassembly is not penetrable by any arm-subassembly, the singular configurations have been removed from the reachable workspace of the subassembly. Additional workspace slices show that the entire reachable workspace contains no singularities when this arm-subassembly is constrained through eqn. (9b).

VII. Conclusions

This paper analyzed a redundant AAS under two different kinematic constraints. It was shown that the kinematic constraint which eliminates the pointing singularities of the shoulder pointing subsystem fails to eliminate the workspace singularities of the arm-subassembly. Any constraint that eliminates the workspace singularities of the redundant AAS must include the effects of the elbow rotation. An animation of this subassembly was developed, and results from this animation were shown. These illustrations document: (1) the workspace singularities that exist when only the pointing subsystem is considered in developing a kinematic constraint function, and (2) the complete, singularity-free workspace that is achieved when the effects of the elbow rotation are included when determining the kinematic constraint function.

References

[1] J. P. Trevelyan, P. D. Kovesi, M. Ong, and D. Elford, "ET: A Wrist Mechanism Without Singular Positions," *Int. J. Robotics Res.*, 4(4):71-85, Winter 1986.

[2] V. Milenkovic, "New Nonsingular Robot Wrist Design," *Proc. 17th ISIR and Robots 11, RI of SME*, Chicago, Il., pp. 13.29-13.42, April 1987.

[3] M. E. Rosheim, "Four New Robot Wrist Actuators," *Robots 10, RI of SME*, Chicago, Il., pp. 8.1-8.45, April 1986.

[4] M. M. Stanišić and O. Duta, "Symmetrically Actuated Double Pointing Systems: The Basis of Singularity-Free Robot Wrists," *IEEE Trans. Robotics Automat.*, 6(5):562-569, Oct. 1990.

[5] F. Di Caprio and M. M. Stanišić, "The Kinematic Control Equations of a Singularity Free Six Degree of Freedom Manipulator," accepted for publication, ASME *J. Mechanical Design.*

[6] S. J. Remis and M. M. Stanišić, "Design of a Singularity-Free Articulated Arm-Subassembly," *IEEE Trans. Robotics Automat.*, 9(6):816-824, Dec. 1993.

[7] C. W. Wampler II, "Inverse Kinematic Functions for Redundant Spherical Wrists," *IEEE Trans. Robotics Automat.*, 5(1):106-111, Feb. 1989.

[8] M. E. Rosheim, "Design of an Omnidirectional Arm," *Proc. 1990 IEEE Int. Conf. Robotics Automat.*, Cincinnati, May 1990.

[9] W. H. Beyer, ed., *CRC Standard Mathematical Tables*, 25th ed., West Palm Beach, FL., CRC Press, Inc., 1978, chpt. V, p. 173.

[10] D. E. Whitney, "The Mathematics of Coordinated Control of Prosthetic Arms and Manipulators," *Trans. ASME, J. Dyn. Sys. Measurement Contr.*, 94G(4):303-309, Dec. 1972.

[11] J. Baillieul, "Avoiding Obstacles and Resolving Kinematic Redundancy," *Proc. 1986 IEEE Int. Conf. Robotics Automat.*, San Fransisco, Apr. 1986.

[12] M. M. Stanišić and J. W. Engelberth, "A Cartesian Description of Arm-Subassembly Singularities in Terms of Singular Surfaces," Proceedings of the 20th *Biennial ASME Mechanisms Conference*, DE-Vol. 43, Sept. 1988.

[13] M. M. Stanišić and J. W. Engelberth, "A Geometric Description of Manipulator Singularities in Terms of Singular Surfaces," Proceedings of the Intl. Meeting *Advances in Robot Kinematics*, Ljubljana, Yugoslavia, Sep. 1988.

[14] S. J. Remis, "Determining Proper Kinematic Constraints to Eliminate the Singularities of Redundant Manipulators," PhD dissertation, Aerospace and Mechanical Engineering Dept., The University of Notre Dame, 1994.

Design and Multi-Objective Optimization of a Linkage for a Haptic Interface

Vincent Hayward, Jehangir Choksi, Gonzalo Lanvin, and Christophe Ramstein*

McGill University
Electrical Engineering Department and
Center for Intelligent Machines
3480 University Street, Montréal, Québec, Canada, H3A 2A7

* Centre for Information Technologies Innovation
575, Chomedey Boulevard Laval, Québec, Canada, H7V 2X2

Abstract – A method to carry out the design of linkage for a haptic interface is described. Factors such as size, workspace, intrusion, inertia, response and structural properties are considered in this process. The dependencies of the various criteria are examined and a hierarchical method is applied. The result is a compact device which is easy to manufacture and which fulfills the requirements demanded by its application. Several quantitative measures designed to capture its principal properties are at the heart the process.

I. Introduction and Problem Statement

The design of a haptic interface is driven by many requirements. Because a haptic device is essentially a human-machine interface, it must have the general features of an ergonomic design. In particular, it should be compact and the operating workspace should be large in relation to the size of the device itself. It should also cause minimal spatial intrusion in the work area of the user. Thus, the size relations are the first general indicators of performance.

The frequency response must be wide since humans are known to perceive force stimuli well above 300 Hz. The device must also be accurate since the amplitude of force signals are sensed by most operators over many orders of magnitude [16]. This, ideally, requires the complete absence of backlash, friction, and other disturbing dynamics; in particular structural dynamics. Note however that precision is of no particular importance due to reasons including postural persistence and various phenomena in the human perception and control of the position and motion of limbs.

The response of the device must also be uniform throughout the usable workspace so the perceived signals will not be distorted as humans are sensitive to minute differences in the amplitude and nature of mechanical signals. Ideally, the inertia of the device should be much less than the inertia of the finger tissue displaced by the device in order to establish a robust causal relationship between an input force signal and perceived motions. This requirement is, to date, the most difficult to achieve.

A. J. Lenarčič and B. B. Ravani (eds.), Advances in Robot Kinematics and Computationed Geometry, 359–368.

Although the response can be improved by feedback or feedforward compensation techniques, these techniques have only limited applicability. It is well known that apparent inertia is not easily reduced by feedback [3]. Moreover, compensation, whether it is applied via feedback or feedforward, by principle will make up for the lack of uniformity of the transfer function from the actuators to the end-effector. We will propose a device which is based on linkages. Thus, its transfer function will vary with its posture from a best case to a worst case. Since ultimately, the performance is limited by the actuators, it follows that a superior design will result from the minimization of the "distance" between the worst and the best case. In this condition, the actuators can optimally be used.

There are many haptic devices which were designed with various goals in mind. Cadoz and colleagues describe an electromechanical system for the simulation of musical instruments which consists of a collection of one degree of freedom mechanisms to address the question of precise rendition of mechanical phenomena [4]. Millman, Stanley and Colgate report on the design of four degree-of-freedom haptic interface which can deliver a large output force [5]. Minsky and co-workers describe possible applications of such devices [13]. Howe built a high-fidelity two degree of freedom device for the study of tactile sensing in precise manipulation [7]. Kelley and Salcudean also designed a two degree of freedom device which avoids the use of linkages [8]. Matshuhira and colleagues used linkages to achieve large workspace [12]. Yokokoji and Yoshikawa looked at interaction of operator dynamics with hand controllers [22].

Previous work in dynamics optimization of linkages can be found in [1, 9].

II. Method

The device, because of its function [15], must have two degrees of freedom, and permit the displacement of a knob inside a rectangular area.

A design method derived from the pioneering work of Vertut and colleagues is illustrated [21, 20]. We will proceed similarly. From Section I, we have:

Size: The device should fit on a table top. A guideline for its size, in terms of its footprint, is provided by the size of an ordinary book.

Workspace: Translational motions should occur in the largest possible area within the footprint.

Intrusion: This is difficult to quantify. However, qualitatively we seek a low profile compatible with the human hand.

Inertia: For all practical purposes, given the state of actuator technology, an ideal inertia of a fraction of gram throughout a workspace of the order 0.01 m^2 is deemed unachieveable. Therefore out of several candidate designs, we will select those for which inertia is the smallest.

Response: Below its first natural frequency, the response of the device will be governed by its multi-body dynamics. In was noted that the Weber fraction $\Delta I/I$ of vibrotactile sensations is better than 0.4 for many kinds of stimuli [2]. For design purposes it was decided that the open loop response

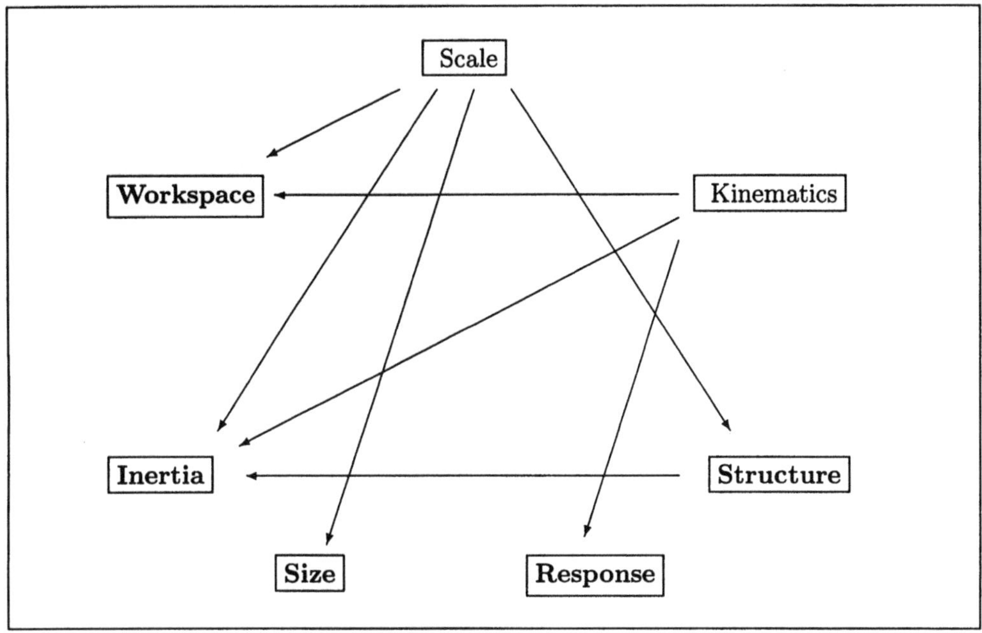

Figure 1: As seen above, many of the properties of the device depend on its scale and on the kinematic parameters. See text.

of the device in terms of torque-to-acceleration should not vary by more than a factor 0.5 which is a commonly accepted figure (3db) used for display devices.

Structure: The lowest natural frequency should be higher than the needed frequency response. Since the prototype is to be manufactured from metallic links, its response in the vicinity of the first natural frequency will be extremely undamped. Using the actuators to extend the frequency response beyond this point would require large amounts of actuator energy, which we saw before is in short supply. The use of fibrous or composite materials could provide the necessary damping (as for loudspeakers which are routinely used beyond their natural frequency), but this is beyond the scope of this paper. Strength is also an issue, since the device is likely to be exposed to abuse.

For any mechanism, the displacement workspace will increase with the square of its scale in the planar case, and by the cube in the spatial case; while the orientation workspace remains invariant in both cases. Except for the structural properties, all the criteria are governed by the kinematic parameters. Except for the change in acceleration response, all the criteria, depend on the scale. The Figure 1 summarizes the design dependencies.

When the scale is increased, the size (footprint) is obviously increased by its square and so is the workspace, since we are dealing with a planar mechanism. All other things being equal, the inertia will grow rapidly: with the cube of the scale. Evidently, the inertia will be a major limiting factor. Hence, the scale and everything that depend on it will be limited.

When the kinematics is changed, both in structure (the type of mechanism) and geometry (the various normalized link lengths), many things are affected. A parallel kinematic chain will almost certainly have better structural properties than a serial one. This is not represented on the Figure 1 because there is only a discrete number of choices. However, when the kinematic parameters are continuously changed for a given mechanism all of the criteria will vary: The inertial properties will change both in magnitude (Inertia) and smoothness (Response) and the workspace shape will be affected. Finally, for a given design, changes in materials and link shapes for the improvement of the structural properties, both in strength and resonance frequency may result in a rapidly rising inertia.

Inspection of Figure 1 suggest two methods of design: Search through the space of kinematic parameters that will satisfy the Response, Inertia, and Workspace requirements, and then adjust the scale, attempting to raise it as much as possible, or vice versa. Here we selected to fix the scale and then to search for kinematic parameters.

III. Carrying Out the Optimization

A five-bar mechanism was selected as a candidate kinematic chain to achieved the goals stated in Section I. When it is symmetrical with respect to its ground link, it is described by three lengths, or two ratios and the scale. The ground link length d is taken as a fixed parameter and the optimization search will be performed on the link lengths a (inner link) and b (outer link).

A. Multi-Objective Search.

The well known concept of singular values decomposition is applied to the optimization of the link lengths. Any matrix $\boldsymbol{A}$ can be factored into $\boldsymbol{A} = \boldsymbol{Q}_1 \boldsymbol{\Sigma} \boldsymbol{Q}_2$, where $\boldsymbol{Q}_1$ and $\boldsymbol{Q}_2$ are orthogonal ($\boldsymbol{Q}^{-1} = \boldsymbol{Q}^T$) and $\boldsymbol{\Sigma}$ is diagonal with the n singular values of $\boldsymbol{A}$ on its diagonal. The columns of $\boldsymbol{Q}_1$ are the eigenvectors of $\boldsymbol{A}\boldsymbol{A}^T$, and the columns of $\boldsymbol{Q}_2$ are the eigenvalues of $\boldsymbol{A}^T\boldsymbol{A}$ [18].

Here, because velocities are low in the desired operating conditions,[1] the transformation from actuator torques to tip acceleration can be locally approximated by a linear map $\ddot{x} = \boldsymbol{R}\,\tau$, where is x designates the coordinates of the tip and τ the vector of actuator torques. The determination of $\boldsymbol{R}$ is straightforward: the inertia tensor was derived using Lagrange's method and the Jacobian matrix was derived from the statics. This mechanism also admits a closed-form position analysis in the two directions.

This transformation $\boldsymbol{R}$ can be regarded as a *deformation.* The singular decomposition separates it into a stretching component, the $\boldsymbol{\Sigma}$ matrix and a rotation component, the $\boldsymbol{Q} = \boldsymbol{Q}_1 = \boldsymbol{Q}_2$ matrix with immediate physical interpretation: the ratio of singular

---

[1] Note that this observation was made by also made by Vertut and Liégeois who, like us, were concerned by the acceleration capabilities of manipulators [20].

values will measure the skew of the response. It is often called the dexterity [17] and its value is 1 when there is no skew. Their product will measure the overall gain of the system as a function of the operating point and is often called the manipulability [23].

Size: This is driven by two factors: the shape and size of the workspace, a connected 10 X 16 cm rectangular region; and its location with respect to the ground link. The distance of the region to the ground link should be minimized to improve the footprint.

Workspace: The workspace will be optimized using the hierarchical approach outlined in [10] with three criteria described in the following two paragraphs.

Inertia: The inertial properties of the device are rather intricate to describe. However, they have the general form $\sum_{\mathcal{V}} dm \,.\, d^2$. Because of its dependence on the structural properties (Figure 1), they will grow at least with the cube of the scale, if not faster. This criterion is consequently very sensitive and is selected as the primary objective.

Response: There are two secondary objectives which together capture this criterion. The ratio σ_2/σ_1 of the singular values of the acceleration map which should never be lower than 0.5. The geometric mean of the singular value has the same dimension as the dexterity: define $(\sigma_2\sigma_1)^{1/2}$ as the manipulability measure.

Structure: Since no finite element code was available in this project, this criterion was determined experimentally. The link shapes were approximated by simple shapes for the purpose of calculating an approximation of their moment of inertia as a function of scale. After determination of the link sizes, several models were built and their structural properties investigated. It can be readily observed that the outer links undergo in-plane stress in two directions: one due to the action of the actuators (principal direction), and the other due to the operator's hand resting on the knob. A triangular shape was selected. The inner links undergo bending on two direction, as well as torsion. A tapering H crossection beam structure was selected for this link.

B. Results.

It is obvious that any mechanism of this kind can be made as isotropic as desired (less skew) as the link lengths go to infinity. Of course, this is at the expense of a rapidly growing inertia. Consequently, of all the acceptable designs, we shall choose the smallest and this will improve inertia, size (footprint), and structural properties.

Both the dexterity and the manipulability of the acceleration map were calculated for rectangular ranges of the tip positions as a function of the link lengths a and b. Since they are related, the dexterity was examined first (higher in the hierarchy). The manipulability should be as large as possible, but should also remain constant, thus, it was normalized to its with respect to its lowest value within the workspace of interest.

The central observation made before searching for a and b is that the ground link length d must not be zero. The reason for this is twofold. For practical reasons, space must be provided for the actuators. But from a kinematic optimization view point, a zero length would causes the design to only depend on one parameter (besides the scale) which would considerably reduce the space of possible designs! One other way to see that is to observe that a zero ground link length would cause all kinematic and dynamic properties to depend on the radial extension of the linkage with their profile in a given direction dependent only on the ratio a/b.

The range of possible combinations of a and b is vast but this range could be reduced considerably by an early pruning. It was observed with the help of computer simulations that the performance severely deteriorated in two broad cases: when $a > b$, and when $a << b$. As a result, the search was narrowed down to regions where a and b were close and $b \geq a$, and several locally optimal designs were found.

For example, with $d = 4$, Figure 2(a) for $a = 13$ and $b = 15$ shows a wide vertical range, but it is a conservative design because the allowable range exceeds the specifications. Figure 2(b) ($a = 15$ and $b = 13$: violation of condition $a < b$) and 2(c) ($a = 10$ and $b = 18$: violation of condition a close to b) are examples of unacceptable performance. Figure 2(c) with $a = 10$ and $b = 18$, shows that the range of allowable dexterity is considerably reduced.

Since the example of Figure 2(a) does not fully exploit the allowable range of dexterity, the link lengths should be made smaller. A search in this direction was performed. With $a = 9$ and $b = 11$, Figure 2(d) reveals that this choice no longer yields a sufficient range. The final result of the search is shown on Figure 2(e) with $a = 10$ and $b = 12$. There are other combinations of link lengths which also achieve a sufficient vertical range, for example $a = b = 14$ on Figure 2(f), but they all correspond to longer link lengths and therefore a larger inertia. Figure 2(g) is the same as Figure 2(e), except that the full workspace has been zoomed in.

Before the lengths $a = 10$ and $b = 12$ can be deemed nearly optimal, the variation of the manipulability must be investigated. Figure 2(h) displays the normalized measure for this workspace and it does not vary by more than 30%, which is excellent.

C. Note on the Significance of Norms: Diamonds and Ellipses

So far, the discussion could be carried out without any reference to a particular norm.

An alternative (and more ornate) method for displaying the sets of achievable acceleration consists of mapping the sets $\|\boldsymbol{\tau}\|_\infty = \max(|\tau_1|, |\tau_2|) \leq \tau_{max}$, which may graphically be represented as diamonds. The infinity norm is the norm that really counts **instantaneously** since the actuators have their torques bounded by the demagnetization current. The 2-norm $\|\boldsymbol{\tau}\|_2 = (\tau_1^2 + \tau_2^2)^{1/2} \leq \tau_{max}$, which yields portraits made of ellipses, can be important too **on average** (i.e. RMS) since it is related to the dissipation in the windings. However, even its weighted version to account for dissimilar motors and torque amplification transmissions does not have a clear physical significance since one actuator could be going up in smoke while the others remain cold; the torques still being inside the unit ball.

At any rate, it is would be very difficult to appreciate the results of the optimization using these graphical techniques because from these representations it is hard to perceptually appreciate the various merits of a particular design [19]. It was found that the isoline method employed in this paper conveyed more concisely and accurately the crucial information.

When a manipulator at isotropic points, the norm does not matter, but globally speaking, consideration of various norms averaged over the operating conditions—in space and time—would allow to further optimize the design, see [11] for additional such global considerations.

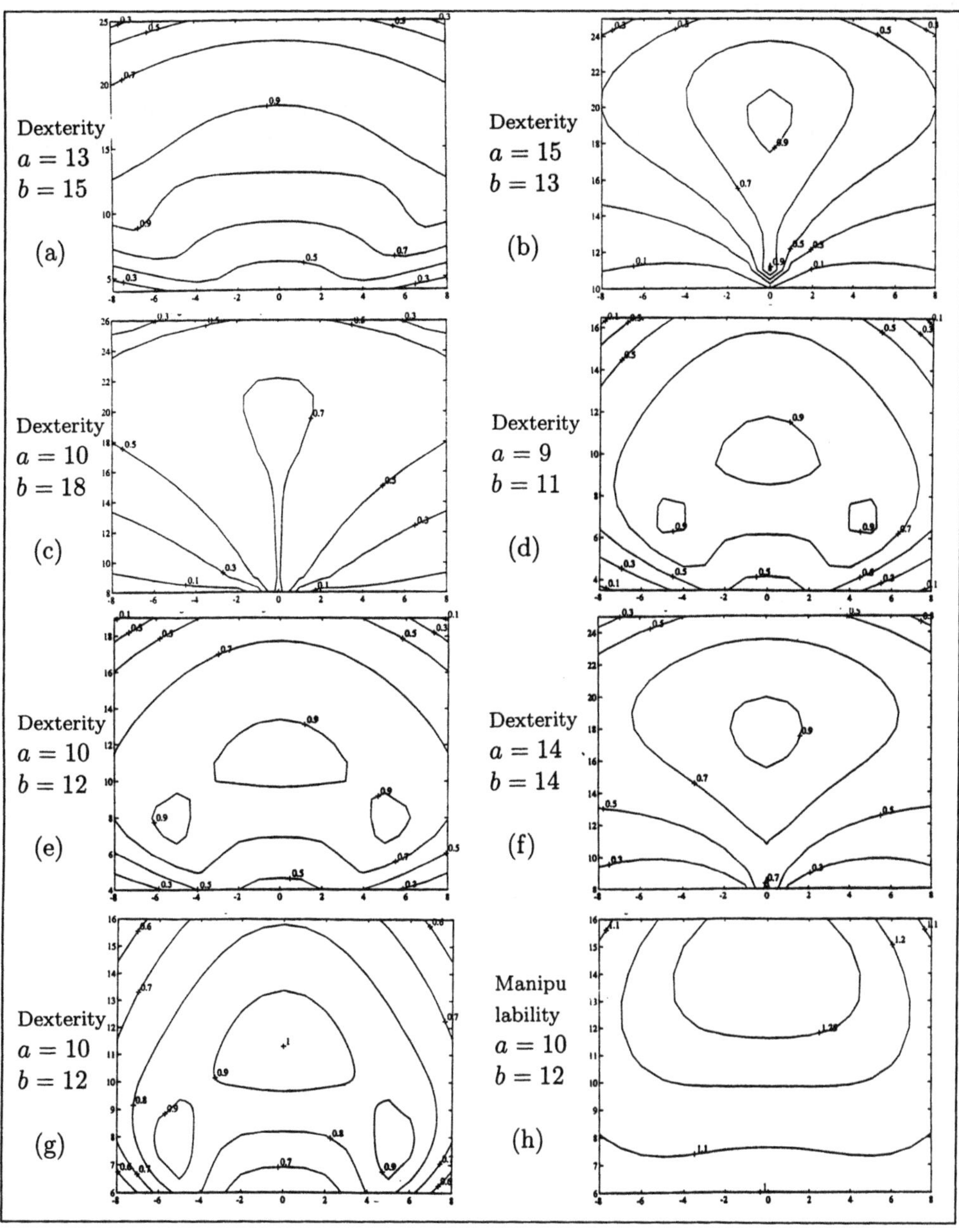

Figure 2: Isolines of the dexterity and the manipulability for various designs. See text.

IV. Conclusion

The choice of a structure, of its kinematic parameters and of a scale has impact on most requirements: workspace, inertial dynamics, structural dynamics, and uniformity of response. We selected for this project a very simple structure: a five-bar linkage with two grounded actuators driving a single knob and examined the effect of the kinematic parameters selection on the requirements in order to determine a preferred design in the presence of many conflicting objectives.

The most surprising result of this study is the **high sensitivity** of the dynamic performance of the linkage with respect to its kinematic parameters, whereas a **low sensitivity** to kinematic performance was observed in an earlier project [6]. The other surprising result was indeed that such a device could be made within the specifications outlined in the introduction.

One of the several existing prototypes was exhibited at the Conference on Human Factors in Computing Systems (ACM-SIGCHI'94) held in Boston in May 1994. Figure 3 shows a picture of this prototype.

Figure 3: Resulting device.

V. Acknowledgements

Many thanks to the people who helped turn an idea into a prototype thanks to their various contributions: Gilles Pepin and Francois Leroux (Visuaid 2000 Inc.), Steve Yeung and Tony Zandbelt (Canadian Space Agency), Pierre Billon, Raymond Descout and Robert Dupuy (Center for Information Technologies Innovation), Eric Bonneton who studied the control of the device, and Anthony Topper for following it up.

Partial funding for this project was provided by the Center for Information Technologies Innovation, and by the project "High-Performance Manipulators" (C3) funded by IRIS, the Institute for Robotics and Intelligent Systems part of Canada's National Centers of Excellence program (NCE). Additional funding was provided by a team grant from FCAR, le Fond pour les Chercheurs et l'Aide à la Recherche, Québec, and an operating grant from NSERC, the National Science and Engineering Council of Canada.

VI. References

[1] Asada, H. "A Geometrical Representation of Manipulator Dynamics and its Application to Arm Design. *ASME J. of Dynamical Systems, Measurement, and Control* 105(3), 131–135 (1983).

[2] K. R. Boff, L. Kaufman, and J. P. Thomas, *Handbook of Perception and Human Performance.* John Wiley and Sons (1986).

[3] T. L. Brooks, "Telerobotic Response Requirements." STX Publication ROB 90-03.

[4] Cadoz, C., Luciani, A., Florenz. "Responsive Input Devices and Sound Synthesis by Simulation of Musical Instruments: The Cordis System". *Computer Music Journal,* 8(3), 60–73 (1984).

[5] P. A. Millman, M. Stanley, J. E. Colgate "Design of a High-Performance Haptic Interface to Virtual Environments", *Proc. IEEE Virtual Reality Annual International Symposium VRAIS'93,* 216–221, Seattle, WA (1993).

[6] V. Hayward. "Design of a Hydraulic Robot Shoulder Based on a Combinatorial Mechanism." Preprints *Third International Symposium on Experimental Robotics,* to be published by Springer Verlag.

[7] R. D. Howe, "A Force Reflecting Teleoperated Hand System for the Study of Tactile Sensing in Precision Manipulation. *Proc. IEEE International Conference on Robotics and Automation,* 1321–1326, Nice, France (1991).

[8] A. J. Kelley, S. E. Salcudean, "MagicMouse: Tactile and Kinesthetic Feedback in the Human-Computer Interface Using an Electromagnetically Actuated Input/Output Device. University of British Columbia, Dept. of Electrical Engineering Tech. Report. (1993).

[9] O. Khatib, S. Agrawal, "Isotropic and Uniform Inertial and Acceleration Characteristics: Issues in the Design of Manipulators." *Dynamics of Controlled Mechanical Systems,* (G. Schweitzer and M. Mansour, Eds.), Springer Verlag. pp. 258–270. (1988).

[10] R. Kurtz, V. Hayward, "Multi-Goal Optimization of a Parallel Mechanism with Actuator Redundancy", *IEEE Transactions on Robotics and Automation.* 8(5) 633–651 (1992).

[11] J. Lenarčič, and L. Žlajpah, "Control Considerations on Minimum Joint Torque Motion", Preprints *Third International Symposium on Experimental Robotics*, to be published by Springer Verlag.

[12] M. Matsuhira, H. Banba, and M. Asakura. "Robot Hand Controller using a Twin Pantograph Mechanism", Proc. *IFToMM-jc International Symposium on the Theory of Machines and Mechanisms.* 167–171. Nagoya, Japan, (1992).

[13] M. Minsky, M. Ouh-young, O. Steele, F. P. Brooks, Jr., M. Behensky, "Feeling and Seeing: Issues in Force Display", *Computer Graphics.* 24(2) 235–243 (1990).

[14] Nevins *et al.* 1974 (August 1974). "A Scientific Approach to the Design of Computer Controlled Manipulators. C. S. Draper Lab. Report No. R-837.

[15] C. Ramstein, V. Hayward, "The Pantograph: a Large Workspace Haptic Interface Device for a Multi-Model Human Computer Interaction" *Proc. Conference on Human Factors in Computing Systems ACM-SIGCHI,* Boston, MA (1994).

[16] F. Reynier, and V. Hayward, "Summary of the Kinesthetic and Tactile Function of the Human Upper Extremity". McGill Center for Intelligent Machines Techical Report CIM-93-4, (1993).

[17] J. K. Salisbury and J. Craig, "Articulated Hands: Force Control and Kinematic Issues". *The International Journal of Robotics Research,* 1(1), 4-17. (1982).

[18] G. Strang. *Linear Algebra and its Applications,* Harcourt Brace Jovanovish, San Diego (1988).

[19] E. R. Tufte, *The Visual Display of Quantitative Information.* Graphics Press, Cheshire Connectitut (1983).

[20] J. Vertut, A. Liégeois. "General Design Criteria for Manipulators." *Mechanisms and Machine Theory,* 16, 65–70. (1981).

[21] J. Vertut, *et al.* "Contribution to Analyze Manipulator Coverage and Dexterity." *Proc. 1st ROMANSY.* Udine, Italy, (1973).

[22] Y. Yokokoji, T. Yoshikawa, "Design of Master Arms Considering Operator Dynamics. *Proc. 1990 Japan-U.S.A. Symposium on Flexible Automation—A Pacific Rim Conference—,* 35–40, Kyoto, Japan (1990).

[23] T. Yoshikawa, "Analysis and Design of Articulated Robot Arms from the Viewpoint of Dynamic Manipulabity" *Robotics Research: The 3rd Int. Symp.*, (O. D. Faugeras and G. Giralt Eds.), 273–279, MIT Press, (1986).

Design of a Three-dof Tendon-driven Manipulator with the Characteristics of Equal Maximum Tensions

Yeong-Jeong Ou and Lung-Wen Tsai

Mechanical Engineering Department and Institute for Systems Research
The University of Maryland
College Park, Maryland

Abstract - Through static force analysis, a design methodology is developed to determine the tendon routings and pulley sizes of a particular three-dof tendon-driven manipulator. This design methodology ensures that all tendons subject to equal maximum tensions in its entire workspace when an external force is applied at the end-effector in all possible orientations. The design is further enhanced when the criteria of isotropic transmission are imposed. An example is presented to demonstrate the features and to compare with the Stanford/JPL finger.

Introduction

Tendon-driven manipulators use tendons and pulleys to transmit power from actuators to joints. Depending on the number of actuators employed, different tendon transmission structures were reported in literature. Examples were the Okada's multi-jointed finger system [1], the anthropomorphic two-handed manipulator designed by Morecki, et al. [2], the Stanford/JPL three-fingered robotic hand [3], and the Utah/MIT dextrous hand [4]. The above mentioned designs partially mimic the functions of a human's hand and focus mostly on the control of the hands; few reveal the methodology which optimizes their structure designs.

For a manipulator designed without any structural optimization, tension in each tendon against an externally applied force at the end-effector will vary as a function of the direction of applied force and the position of the end-effector. As a result, the maximum tension for each tendon in the whole workspace would be different from one another. If equal strength tendons are used, the largest maximum tension among all tendons ought to be used for the sizing of tendons. This, however, will be inefficient. If different strength tendons are adopted, the transmission structure will be degraded due to diverse attributes of tendons. Therefore, a manipulator with the feature of equal maximum tensions in its whole workspace is crucial in the design of a manipulator with equal strength of actuators and of tendons. A manipulator with this feature is also expected to simplify the process of design and to reduce the cost of maintenance.

Lee and Tsai [5] developed a methodology to implement all possible tendon routings for an n-dof manipulator controlled by more than n actuators. Ou and Tsai [6] later developed a methodology to optimize the tendon routings and pulley sizes for an n-dof

A. J. Lenarčič and B. B. Ravani (eds.), Advances in Robot Kinematics and Computationed Geometry, 369–378.

manipulator with $n+1$ actuators based on the concept of local isotropic transmission characteristics. An isotropic transmission structure ensures a unity condition number of the static force transformation matrix and equal maximum tensions in all tendons against a unity force applied at the end-effector, when the end-effector is located at an isotropic point. However, as the end-effector moves away from the isotropic point, maximum tension in each tendon may become different.

Generally speaking, the feature of equal maximum tensions cannot be achieved simply by designing its transmission structure except for some particular manipulator configurations. For example, a manipulator with all perpendicular joint axes. In this paper, we present the design of a particular three-dof manipulator which possesses the feature of equal maximum tensions in its entire workspace. The description of the manipulator follows.

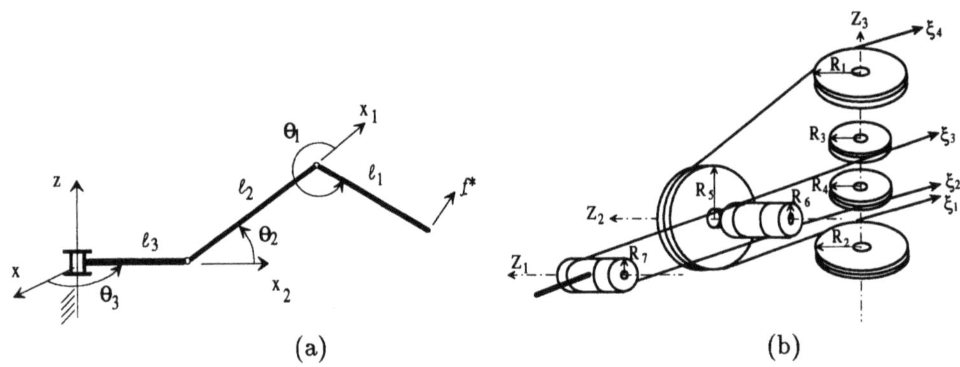

Figure 1: (a) The three-dof linkage structure and (b) the tendon routings and pulleys arrangement of the Stanford/JPL finger

Description of the Three-dof Manipulator

To improve static performance of existing devices with the aforementioned features and to illustrate the methodology, the linkage structure and the four tendons (actuators) of a three-dof manipulator (taken from the Stanford/JPL hand) was adopted. Figure 1-(a) shows its linkage structure; the first and the second joint axes are parallel to each other, and the third joint axis is perpendicular to the second. Note that the links and joints as shown in Fig. 1-(a) are numbered from the outest to the proximal, where ℓ_1, ℓ_2, ℓ_3, and θ_1, θ_2, θ_3 are the corresponding link lengths and joint angles. Although, the end-effector of this manipulator travels in a complex three dimensional space, its linkage structure and tendon routings are simple enough to be applied by the methodology.

By letting θ_3 be equal to 90°, without loosing generality, the Jacobian matrix of the manipulator, which represents the transformation from the joint space to the end-effector space, can be written as

$$\mathbf{J} = \ell_2 \begin{bmatrix} 0 & 0 & -(C_2 + l_{32} + C_{12}l_{12}) \\ -S_{12}l_{12} & -S_2 - S_{12}l_{12} & 0 \\ C_{12}l_{12} & C_2 + C_{12}l_{12} & 0 \end{bmatrix} \tag{1}$$

where $C_2 = Cos(\theta_2), S_2 = Sin(\theta_2), C_{12} = Cos(\theta_1 + \theta_2), S_{12} = Sin(\theta_1 + \theta_2)$, $l_{12} = \ell_1/\ell_2$ and $l_{32} = \ell_3/\ell_2$. Hence, l_{12} and l_{32} are two non-dimensional link length ratios.

Figure 1-(b) shows the tendon routings and pulley arrangement of the Stanford/JPL finger. The power is transmitted from the actuators to the joints through four tendons. The kinematic relationship between the joint angles and the tendon displacements can be expressed as [7]

$$\underline{S} = \mathbf{A}\ \underline{\theta} \tag{2}$$

where $\underline{S} = [S_1, S_2, S_3, S_4]^T$ denotes a 4×1 tendon displacement vector, $\underline{\theta} = [\theta_1, \theta_2, \theta_3]^T$ denotes a 3×1 joint angular displacement vector, and $\mathbf{A}$ is a 4×3 matrix determined by the tendon routings and pulley sizes. The transpose of $\mathbf{A}$ is called the structure matrix.

Most of researchers [8,9,10] studied only the Jacobian matrix which maps the joint space to the end-effector space; yet the transmission matrix which maps the joint space to the actuator space was not considered. It is clear that their results are valid only for direct-drive manipulators. To design a tendon-driven manipulator, the overall transformation from the actuator space to the end-effector space should be examined. Note that, for a tendon-driven manipulator, the transformation from actuator displacements to joint angles is not a one-to-one mapping.

Kinematic and Static Force Analysis

From the principle of conservation of energy, the fundamental force/torque relation between the end-effector space and the joint space can be written as

$$\underline{\tau} = -\mathbf{J}^T \underline{f}^* \tag{3}$$

where $\underline{\tau} = [\tau_1, \tau_2, \tau_3]^T$ denotes a 3×1 joint torque vector, and $\underline{f}^* = [f_x^*, f_y^*, f_z^*]^T$ denotes a 3×1 force vector acting on the end-effector.

Based on the same principle, the fundamental force/torque relation between the joint space and the actuator space can be derived from eq. (2) as

$$\underline{\tau} = \mathbf{A}^T\ \underline{\xi} \tag{4}$$

where $\underline{\xi} = [\xi_1, \xi_2, \xi_3, \xi_4]^T$ denotes a 4×1 tendon force vector.

The inverse transformation of eq. (4) can be written as

$$\underline{\xi} = \mathbf{A}^{+T} \underline{\tau} + \lambda\ \underline{h} \tag{5}$$

where $\mathbf{A}^{+T}$ is the pseudoinverse of $\mathbf{A}^T$ [11,12], $\underline{h}=[h_1, h_2, h_3, h_4]^T$ is a one-dimensional null vector of $\mathbf{A}^T$, and λ is an arbitrary constant. Note that all elements of $\underline{h}$ should have the same sign to keep all tendons under tension at all times [5].

Substituting eq. (3) into eq. (5) leads to

$$\underline{\xi} = -\mathbf{A}^{+T}\mathbf{J}^T\underline{f}^* + \lambda\,\underline{h} \tag{6}$$

Equation (6) represents the relationship between an external force applied at the end-effector and the tendon forces generated by input actuators. It is a underdetermined problem for the tendon forces. To maintain minimum positive values of the underdetermined tendon fcrces, λ should be determined by the following equation:

$$\lambda = \max_j\{\frac{(\mathbf{A}^{+T}\mathbf{J}^T)_j\underline{f}^*}{h_j}\} \tag{7}$$

where $(\ )_j$ denotes the j-th row of the matrix in the parentheses and where $\max\{\ \}_j$ denotes the maximum value among all the possible choices of j.

Substituting eq. (7) into eq. (6), we obtain scalar forms of eq. (6) as

$$\xi_i = \max_j\{[-(\mathbf{A}^{+T}\mathbf{J}^T)_i + (\mathbf{A}^{+T}\mathbf{J}^T)_j\frac{h_i}{h_j}]\underline{f}^*\}, \qquad i = 1,\cdots,4 \tag{8}$$

Structure Matrix with Equal Maximum Tensions

Confining the externally applied force on a unit sphere and applying Cauchy-Schwarz inequality to eq. (8) result in

$$\max(\xi_i) = \max_j\{\|\ (\mathbf{A}^{+T}\mathbf{J}^T)_i - (\mathbf{A}^{+T}\mathbf{J}^T)_j\frac{h_i}{h_j}\ \|\}, \qquad i = 1,\cdots,4 \tag{9}$$

When all four maximum tendon forces are equal, it is clear that the elements of $\underline{h}$ should satisfy

$$\frac{h_i}{h_j} = 1; \qquad i,j = 1,\cdots,4 \tag{10}$$

i.e.,

$$\underline{h} = [1,1,1,1]^T \tag{11}$$

Note that eq. (11) is one of the criteria for an isotropic transmission structure [6]. Using eq. (11), the condition of equal maximum tensions can be stated as

$$max(g_{12},g_{13},g_{14}) = max(g_{21},g_{23},g_{24}) = max(g_{31},g_{32},g_{34}) = max(g_{41},g_{42},g_{43}) \tag{12}$$

where

$$g_{ij} = g_{ji} = \|\ [(\mathbf{A}^{+T})_i - (\mathbf{A}^{+T})_j]\ \mathbf{J}^T\ \|; \quad i,j = 1,\cdots,4 \tag{13}$$

There are a total of six norms involved in eq. (12), namely: g_{12}, g_{13}, g_{14}, g_{23}, g_{24}, and g_{34}. Since the Jacobian matrix $\mathbf{J}$ depends on the posture of a manipulator, so are the six norms. It is clear that, for eq. (12) to be valid, the six norms must satisfy one of the following two conditions:

1. Three of the six norms are equal to each other and the other three are always less than or equal to the first three norms in the whole workspace.

2. Each maximum tendon force is determined from an equivalent set of norms, (g_{12}, g_{13}, g_{14}), provided the following conditions are satisfied:

$$g_{12} = g_{34}, \tag{14}$$

$$g_{13} = g_{24}, \quad \text{and} \tag{15}$$

$$g_{14} = g_{23} \tag{16}$$

Condition 1 results in rank-deficient transmission structure matrices which are not admissible. In what follows, we will only consider condition 2.

Equation (14) can be rewritten as

$$\| [(\mathbf{A}^{+T})_1 - (\mathbf{A}^{+T})_2]\, \mathbf{J}^T \| = \| [(\mathbf{A}^{+T})_3 - (\mathbf{A}^{+T})_4]\, \mathbf{J}^T \| \tag{17}$$

To derive matrix $\mathbf{A}$, we first express matrix $\mathbf{A}^{+T}$ as:

$$\mathbf{A}^{+T} = \begin{bmatrix} a_{11} & a_{12} & a_{13} \\ a_{21} & a_{22} & a_{23} \\ a_{31} & a_{32} & a_{33} \\ a_{41} & a_{42} & a_{43} \end{bmatrix} \tag{18}$$

where a_{ij} $(i, j = 1, \cdots, 4)$ are the unknown elements of $\mathbf{A}^{+T}$.

Substituting eqs. (1) and (18) into eq. (17), we obtain

$$\begin{aligned} &[(a_{12} - a_{22})^2 - (a_{32} - a_{42})^2](1 + 2l_{12}C_1) + \\ &[(a_{13} - a_{23})^2 - (a_{33} - a_{43})^2](C_2 + l_{32} + C_{12}l_{12})^2 + \\ &[(a_{12} - a_{22})(a_{11} - a_{21}) - (a_{32} - a_{42})(a_{31} - a_{41})]2l_{12}C_1 + \\ &[(a_{12} - a_{22} + a_{11} - a_{21})^2 - (a_{32} - a_{42} + a_{31} - a_{41})^2]l_{12}^2 = 0 \end{aligned} \tag{19}$$

Since θ_1 and θ_2 can take any two arbitrary angles, we conclude that

$$(a_{13} - a_{23})^2 - (a_{33} - a_{43})^2 = 0 \tag{20}$$

$$\begin{aligned} &(a_{12} - a_{22})^2 - (a_{32} - a_{42})^2 + \\ &\quad (a_{12} - a_{22})(a_{11} - a_{21}) - (a_{31} - a_{41})(a_{32} - a_{42}) = 0 \end{aligned} \tag{21}$$

$$\begin{aligned} &(a_{32} - a_{42})^2 - (a_{12} - a_{22})^2 = \\ &\quad l_{12}^2[(a_{12} - a_{22} - a_{11} - a_{21})^2 - (a_{32} - a_{42} + a_{31} - a_{41})^2] \end{aligned} \tag{22}$$

Similarly, from eqs. (15) and (16), we obtain

$$(a_{13} - a_{33})^2 - (a_{23} - a_{43})^2 = 0 \tag{23}$$

$$\begin{aligned} &(a_{12} - a_{32})^2 - (a_{22} - a_{42})^2 + \\ &\quad (a_{11} - a_{31})(a_{12} - a_{32}) - (a_{21} - a_{41})(a_{22} - a_{42}) = 0 \end{aligned} \tag{24}$$

$$\begin{aligned} &(a_{22} - a_{42})^2 - (a_{12} - a_{32})^2 = \\ &\quad l_{12}^2[(a_{12} - a_{32} - a_{11} - a_{31})^2 - (a_{22} - a_{42} + a_{21} - a_{41})^2] \end{aligned} \tag{25}$$

and

$$(a_{13} - a_{43})^2 - (a_{23} - a_{33})^2 = 0 \tag{26}$$

$$(a_{12} - a_{42})^2 - (a_{22} - a_{32})^2 + (a_{12} - a_{42})(a_{11} - a_{41}) - (a_{21} - a_{31})(a_{22} - a_{32}) = 0 \tag{27}$$

$$(a_{22} - a_{32})^2 - (a_{12} - a_{42})^2 = l_{12}^2[(a_{12} - a_{42} - a_{11} - a_{41})^2 - (a_{22} - a_{32} + a_{21} - a_{31})^2] \tag{28}$$

Since eq. (11) is a null vector of $\mathbf{A}^T$, $\mathbf{A}^{+T}$ should satisfy the following constraints:

$$a_{11} + a_{21} + a_{31} + a_{41} = 0 \tag{29}$$

$$a_{12} + a_{22} + a_{32} + a_{42} = 0 \tag{30}$$

$$a_{13} + a_{23} + a_{33} + a_{43} = 0 \tag{31}$$

Equations (20) through (31) are a set of twelve equations in twelve unknowns. However, only nine of them are independent. By solving eqs. (20) to (31) and eliminating those solutions which yield rank-deficient matrices, we obtain

$$\mathbf{A}^{+T} = \begin{bmatrix} -a_{12} \pm \frac{a_{22}}{l_{12}} & a_{12} & a_{13} \\ \pm\frac{a_{12}}{l_{12}} - a_{22} & a_{22} & -a_{13} \\ a_{12} \mp \frac{a_{22}}{l_{12}} & -a_{12} & a_{13} \\ \mp\frac{a_{12}}{l_{12}} + a_{22} & -a_{22} & -a_{13} \end{bmatrix} \tag{32}$$

Since

$$(\mathbf{A}^{+T})^+ = \mathbf{A}^T \tag{33}$$

we have

$$\mathbf{A}^T = (\mathbf{A}^+\mathbf{A}^{+T})^{-1}\mathbf{A}^+ \tag{34}$$

By introducing new variables,

$$a = \frac{1}{4a_{11}}, \quad b = \frac{a_{12}}{2(a_{12}^2 - a_{22}^2)}, \quad \text{and} \quad c = \frac{a_{22}}{2(a_{12}^2 - a_{22}^2)} \tag{35}$$

we obtain $\mathbf{A}^T$ as

$$\mathbf{A}^T = \begin{bmatrix} \mp cl_{12} & \pm bl_{12} & \pm cl_{12} & \mp bl_{12} \\ b \mp cl_{12} & -c \pm bl_{12} & -b \pm cl_{12} & c \mp bl_{12} \\ a & -a & a & -a \end{bmatrix} \tag{36}$$

Therefore, the three-dof manipulator with its structure matrix in the form of eq. (36) possesses the feature of equal maximum tensions in its whole workspace. Note that the link ratio l_{32} has no effect on the structure matrix.

Addition of Isotropic Transmission Characteristics

When the end-effector of a manipulator locates at an isotropic point, the maximum tensions on the four tendons are equal to one another. This feature is compatible with that of equal maximum tensions. The link ratio l_{32} and the four free variables in eq. (36) leave enough room for introducing more design constraints. Therefore, we can impose the concept of the isotropic transmission to the design of this manipulator.

To achieve isotropic transmission characteristics, two criteria should be satisfied: the first is eq. (11), and the second is [6]

$$\mathbf{A}^T\mathbf{A} = \frac{1}{\mu^2}\mathbf{J}^T\mathbf{J} \tag{37}$$

where μ is an arbitrary constant.

Since eq. (11) is automatically satisfied for eq. (36), we need only to consider eq. (37). Substituting eq. (36) into eq. (37) and after simplifying the result, we obtain the following two independent equations:

$$\frac{-2bc}{b^2+c^2} = \pm C_1 \tag{38}$$

and

$$\frac{2a^2}{b^2+c^2} = (l_{32} + C_2 + l_{12}C_{12})^2 \tag{39}$$

From eqs. (38) and (39), it is clear that the link ratios l_{12} and l_{32} and the location of isotropic point can be chosen arbitrary as long as the Jacobian matrix is not singular. After selecting the link ratios and the location of the isotropic point, eqs. (38) and (39) become two constraints for the matrix in eq. (36). Therefore, there is one free variable which serves as an amplification factor for the whole transmission structure.

The following section provides a numerical example to illustrate the characteristics of this type of manipulators. The results are compared with that of the Stanford/JPL finger.

Numerical Examples

The link proportion of the Stanford/JPL finger from the distal link to proximal link is 1 : 1 : 0.685. The radii of pulleys as shown in Fig. 1-(b) are R_1=1.0795cm, R_2=.8255cm, R_3=R_4=R_6=.5969cm, R_5=1.1862cm, and R_7=.6350cm. Hence, the transmission structure matrix of the manipulator can be written as

$$\mathbf{A}^T = \begin{bmatrix} 0 & -.6350 & .6350 & 0 \\ -1.1862 & -.5969 & .5969 & 1.1862 \\ -.8255 & .5969 & .5969 & -1.0795 \end{bmatrix} \tag{40}$$

where a positive element is assumed if a positive rotation of the corresponding pulley is produced by a positive displacement of the routing tendon, otherwise it is negative.

To demonstrate the effect of equal maximum tensions and the isotropic transmission characteristics, a new transmission structure is developed. The link arrangement and link proportion are the same as that of the Stanford/JPL finger. The variable c in eq. (36) is set to zero to maintain the same tendon routings as that of the Stanford/JPL finger for comparison. Thus, the isotropic points can be chosen only at the locations where the joint angle θ_1 is equal to $\pm 90°$. The new transmission structure is designed so that the manipulator possesses isotropic transmission characteristics when $\theta_1 = 90°$ and $\theta_2 = 0°$, and the point (x=0, y=$\ell_3 + \ell_2$, z=ℓ_1) is one of the isotropic points on the locus. The transmission structure matrix can be derived from eqs. (36) and (39) by letting $b = -1$. After discarding the solutions which involve different routings from that of the Stanford/JPL finger and after reordering columns of the transmission structure, such that the new transmission structure matrix has the same form as the matrix in eq. (40), we obtain

$$\mathbf{A}^T = \begin{bmatrix} 0 & -1 & 1 & 0 \\ -1 & -1 & 1 & 1 \\ -1.192 & 1.192 & 1.192 & -1.192 \end{bmatrix} \tag{41}$$

In what follows, we let $\ell_2 = 1$ unit for simplicity. The maximum tendon force on each tendon of the two transmission structures is computed by using eq. (9) at three different end-effector positions: the first at x=0, y=$\ell_3 + \ell_2$, z=ℓ_1, the second at x=0, y=$\ell_3 + \sqrt{2}\ell_2$, z=0, and the third at x=0, y=ℓ_3, z=$\sqrt{2}\ell_2$. The first position is an isotropic point for the new transmission structure.

Structures		(a)	(b)
κ		1.899	0.5942
position 1	max. tensions	$[1.516, 2.364, 2.418, 1.454]$	$[1.683, 1.683, 1.683, 1.683]$
	ratio	$1.042 : 1.626 : 1.663 : 1$	$1 : 1 : 1 : 1$
position 2	max. tensions	$[1.752, 2.749, 2.796, 1.699]$	$[1.901, 1.901, 1.901, 1.901]$
	ratio	$1.031 : 1.618 : 1.645 : 1$	$1 : 1 : 1 : 1$
position 3	max. tensions	$[1.13, 2.107, 2.107, 1.13]$	$[1.683, 1.683, 1.683, 1.683]$
	ratio	$1 : 1.865 : 1.865 : 1$	$1 : 1 : 1 : 1$

Table 1: List of κ, maximum tensions, their ratios at three end-effector positions

To achieve a fair comparison, each transmission structure matrix is multiplied by a constant κ. The value of κ is chosen so that the product of the three singular values of each $\mathbf{A}^T$ is equal to one. Table 1 lists the values of κ, the maximum tensions and their ratios at the three positions. In Table 1, structure (a) represents the Stanford/JPL finger and (b) represents the structure with the transmission structure matrix derived in eq. (41). Structure (a) has different maximum tensions at the three representative positions, and the ratios of its maximum tensions are never equal to 1:1:1:1. Structure (b) has equal maximum tensions at all three positions, and the ratios of its maximum tensions are always equal to 1:1:1:1.

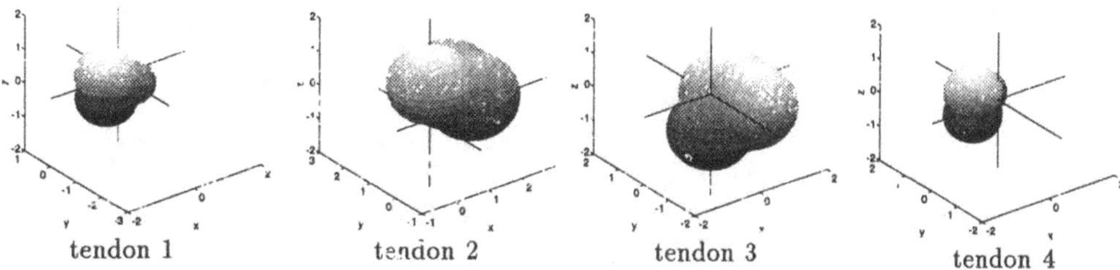

Figure 2: Spherical plots of the four tendon forces versus direction of applied force for structure (a) evaluated at position 1

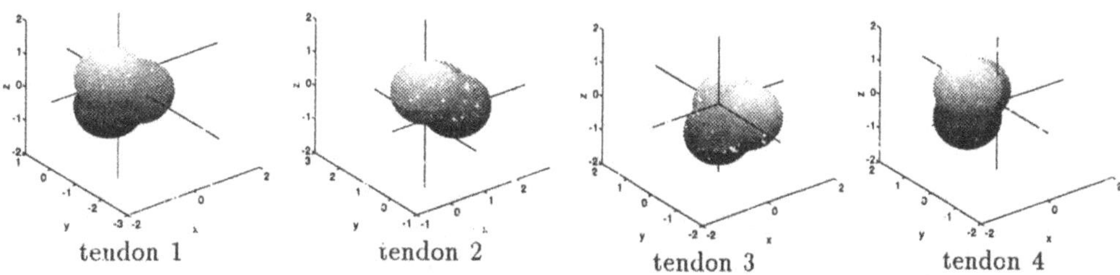

Figure 3: Spherical plots of the four tendon forces versus direction of applied force for structure (b) evaluated at position 1

Figures 2 and 3 show the spherical plots of four tendon forces for structures (a) and (b), respectively, evaluated at position 1. In a spherical plot, the radial distance represents the tendon force and the phase angle represents the direction of applied force. Except for a change in the orientation, the four spherical plots shown in Fig. 3 are identical in shape with each other while those shown in Fig. 2 are different from one another.

Conclusions

Through static force analysis, a design methodology for determining tendon routings and pulley sizes of a particular three-dof tendon-driven manipulator is developed. The manipulator features the characteristics of equal maximum tensions and isotropic transmission. The characteristics of equal maximum tensions ensure that all tendons subject to equal maximum tensions in its whole workspace when an external force is applied at the end-effector in all possible orientations. The isotropic transmission with appropriately selected isotropic points helps to improve the static performance of the manipulator.

Acknowledgment

The authors would like to thank Dr. Kenneth Salisbury for providing dimensions of the Stanford/JPL hand. This work was supported in part by the U.S. Department of Energy under Grant DEF05-88ER13977, and in part by the NSF Engineering Research Centers Program NSFD CDR 8803012. Such support does not constitute an endorsement by the supporting agencies of the views expressed in the paper.

References

[1] T. Okada,"On a Versatile Finger System", *Proc. 7th Int'l. Symposium on Industrial Robots*, 345-352, Tokyo, Japan (1977).

[2] A. Morecki, Z. Busko, H. Gasztold, and K. Jaworek,"Synthesis and Control of the Anthropomorphic Two-Handed Manipulator", *Proc. 10th Int'l. Symposium on Industrial Robots*, 461-474, Milan, Italy (1980).

[3] J. K. Salisbury,"Kinematic and Force Analysis of Articulated Hands," Ph.D. Dissertation, Mech. Eng. Dept., Stanford University, Stanford, CA. (1982).

[4] S. C. Jacobsen, J. E. Wood, D. F. Knutti, and K. B. Biggers,"The Utah/MIT Dexterous Hand: Work in Progress", The Int'l. J. of Robotics Research, 3(4), 21-50 (1984).

[5] J.-J. Lee, and L.-W. Tsai,"Topological Analysis of Tendon-Driven Manipulators", *8th World Congress on International Federation of the Theory of Machines and Mechanisms*, 479-482, Czechoslovakia (1991).

[6] Y.-J. Ou, and L.-W. Tsai,"Kinematic Synthesis of Tendon-Driven Manipulators with Isotropic Transmission Characteristics", ASME J. of Mechanical Design, 115(4), 884-891 (1993).

[7] J.-J. Lee, and L.-W. Tsai,"Kinematic Analysis of Tendon-Driven Robotic Mechanisms Using Graph Theory", ASME J. of Mechanisms, Transmissions, and Automation in Design, 111(1), 59-65 (1989).

[8] J. K. Salisbury and J. J. Craig,"Articulated Hands: Force Control and Kinematic Issues", The Int'l. J. of Robotics Research, 1(1), 4-17 (1982).

[9] H. Asada and J. A. Cro Granito,"Kinematic and Static Characterization of Wrist Joints and Their Optimal Design", *Proc. of IEEE Int'l. Conf. on Robotics and Automation*, 244-250 (1985).

[10] C. Gosselin and J. Angeles,"A New Performance Index for the Kinematic Optimization of Robotic Manipulators", ASME Trends and Developments in Mechanisms, Machines and Robotics, 15(3), 441-447 (1988).

[11] A. Ben-Israel and T. N. E. Greville,"*Generalized Inverses: Theory and Applications*", Wiley, New York (1974).

[12] G. Strang,*Linear Algebra and Its Applications*, 2nd ed., Academic Press, New York, N.Y. (1980).

Spherical 3 d.o.f. Geared Wrist with no Aligned Singularity

Giuseppe Quaglia and Massimo Sorli

Politecnico di Torino
C.so Duca degli Abruzzi 24
10129 Torino, Italy

Abstract - A new design of spherical 3 d.o.f. wrist for robot is presented. The mechanical transmission between motors and end-effector is performed via bevel gears.
The main feature of the wrist is the absence of singularity in the aligned configuration. In the article the kinematic scheme of the wrist is shown and the kinematic analysis (direct and inverse) is developed. Further the degeneration conditions are obtained.
The high stiffness and the compact overall dimension of the design, coupled with degeneration conditions outside the work space, make this type of wrist kinematically suitable for performing complex handling operations.

I. Introduction

Two basic aspects must be considered in the study and design of a robot wrist: the work space and degeneration.
The former defines the region in space where the end-effector can be positioned: clearly, the larger the work space, the more versatile the wrist will be.
As for the second aspect, degeneration is the inability to rotate the end-effector around a certain axis positioned in space. For 3 d.o.f. wrists, this occurs when all of the axes corresponding to the various degrees of freedom are located in the same plane.
In addition, the degeneracy configuration is not alone in being to some extent negative: the configurations adjacent to it are also unfavourable in that high motor rotational speeds may be required for small end-effector displacements.
Clearly, then, it is desirable that the wrist be free from degeneracy in the positions in which it must operate, and particularly in the configuration where the end-effector is in line with the robot arm on which the wrist is installed. This will be referred to below as the aligned configuration.
In previous researches, the authors investigated several wrist designs, both in Roll-Pitch-Roll [1], and in Pitch-Yaw-Roll configuration [2,3]. The main goal of these studies was the evaluation of the wrist degeneracy. These previous solution show two different kinds of drawbacks, because they present degeneracy inside the workspace [1], or they show low stiffness and structure complexity [2]. However, in both cases workspace was quite limited, with respect to usual industrial applications. In [4] and [5] the authors present different solution of wrist without degeneracy.
In the present work, the aim was to design and study a spherical wrist without degeneracy in the aligned configuration, with high stiffness and compact overall dimensions.

A. J. Lenarčič and B. B. Ravani (eds.), Advances in Robot Kinematics and Computationed Geometry, 379–388.

These considerations were the basis for developing a wrist with three degrees of freedom oriented in a set of three mutually perpendicular axes in the aligned configuration. Consequently, this wrist has no degeneration and can reach a situation of degeneracy only for a ±90° pitch motion. In addition, the degeneration positions were outside the wrist's work space, given the constraints deriving from the spatial arrangement of the gearing.
The wrist shows no degeneration for roll, yaw and pitch motions in the entire work space.
This article describes the kinematic layout of the wrist, indicating the relationships between joint coordinates and gear angles of rotation, the relationships between joint coordinates and cartesian coordinates, and the relationships between motor speeds and the end-effector's absolute rotational speeds in a cartesian space.
The latter relationship is then used to identify the wrist's degeneration configurations.

II. Description of wrist

A schematic representation of the wrist in the aligned configuration is shown in Fig. (1).

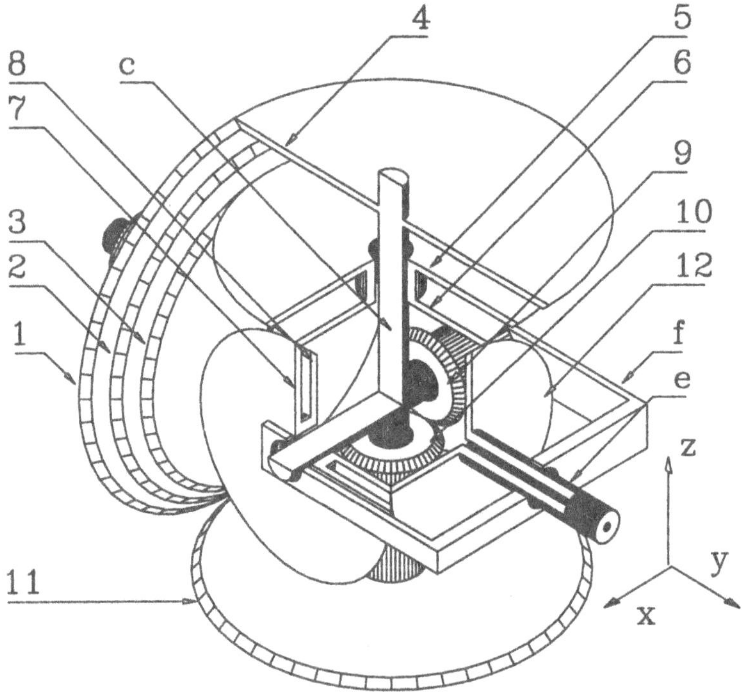

Fig. 1: Schematic view of wrist

The wrist consists of twelve bevel gears which are used to move end-effector *e* in space.

The wrist has three degrees of freedom with axes which are concurrent at the centre, and is thus spherical. A major feature is that the three DOFs are arranged in a set of three mutually perpendicular axes with the wrist in the aligned configuration, thus ensuring that there is no degeneration. This result is accomplished by means of a gear-driven cross *c* used to guide yoke *f*, which carries the end-effector output shaft ***e***.
Movement reaches the wrist via three concentric shafts, coaxial with the last robot arm, connected to bevel gears *1*, *2* and *3*. A fixed set of three cartesian reference axes *xyz* is located as indicated in Fig. (1) centred on the centre of cross *c*, with *x* lying on the horizontal cross axis and *y* lying on the end-effector output shaft axis.

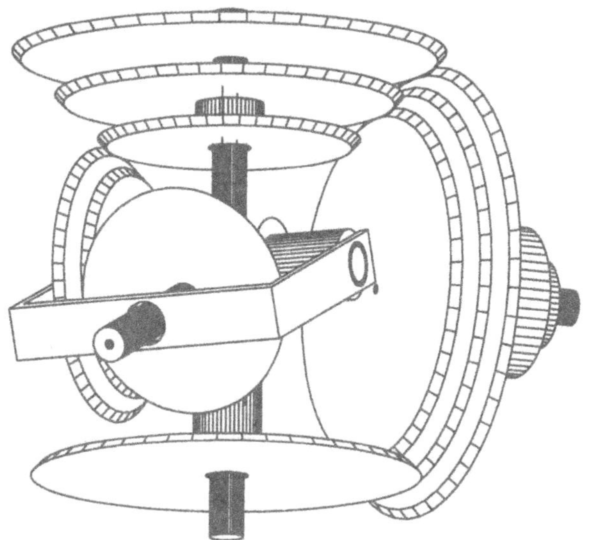

Fig. 2: View of the wrist rotated relative to the yaw and roll axes

Gear *1* meshes with gear *4*, which is positively connected to the vertical axis of the cross and thus controls its orientation (jaw). Gear *2* meshes with gear *11*, which is positively connected to gear *10* via a sleeve supported by the vertical cross shaft. Gear *10* in turn meshes with gear *9*, which is located at right angles to it and supported by the horizontal cross shaft. Gear *9* is positively connected to yoke *f*. Thus, rotating gear *2* causes yoke *f* to rotate relative to the horizontal axis of the cross, thereby controlling the pitch motion.
Finally, gear *3* meshes with gear *5*, which is positively connected to *6* and free to rotate relative to the vertical axis of the cross. Gear *6* meshes with gear *7*, which is positively connected to gear *8*; both are free to rotate relative to the horizontal axis of the cross. Gear *8* meshes with gear *12* which causes the end-effector to rotate relative to its axis and thus controls the roll motion.
It should be noted that two gears are required for the yaw motion inasmuch as the drive line is diverted a single time by 90°. Four gears corresponding to two 90° diversions are used for the pitch motion, while for the roll motion it is necessary to use six gears, corresponding to three

90° diversions.
For a clearer understanding of the wrist's possible spatial configurations, Fig. (2) shows a view of the wrist rotated relative to the yaw axis and the pitch axis.

III. Kinematic analysis of the wrist

Wrist kinematics can be studied with reference to Fig. (3), which indicates the symbols and conventions used.

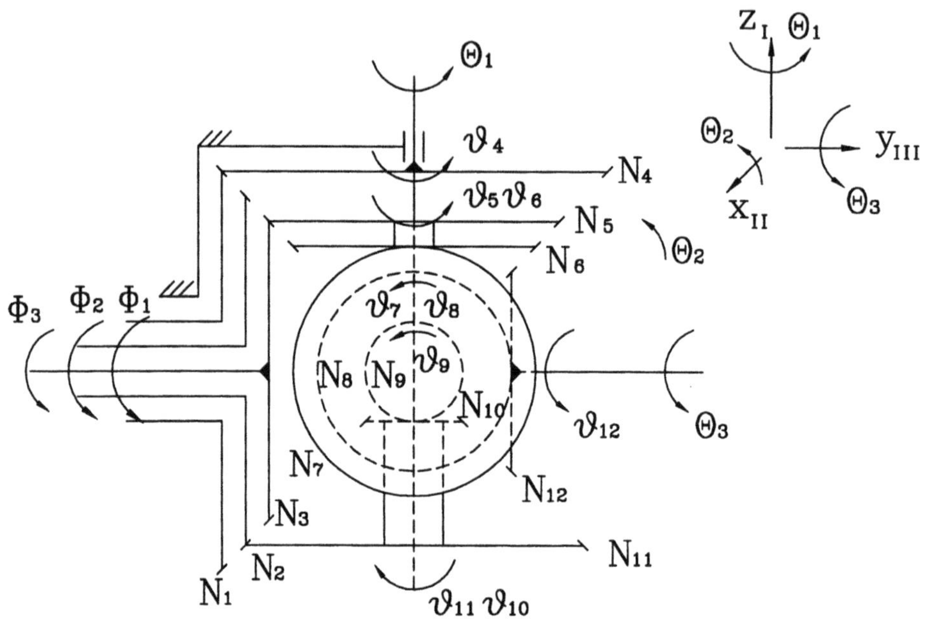

Fig. 3: Symbols and conventions for kinematic study of wrist

The symbol Φ_i denotes input shaft rotations, Θ_i are the rotations of the wrist's three degree of freedom, and N_j denotes the number of teeth of the j-nth gear.
The fixed reference system of mutually perpendicular cartesian axes (x,y,z) is assumed together with three mobile reference systems (x_n,y_n,z_n, with $n=I,II,III$) located on the three members making up the wrist (i.e. cross, yoke and end-effector output shaft) and originating at the centre of the wrist. In the aligned configuration, the four reference systems coincide, being y and y_n directed along the end effector axis.
Kinematic analysis of the wrist can be broken down into three points, that are the relationships between joint coordinates and gear angles of rotation, the relationships between joint coordinates and cartesian coordinates, and the relationships between motor speeds and the end-effector's absolute rotational speeds in a cartesian space.

A. Relationships between the angular velocities of the joints ($\dot{\Theta}_i$) and the angular velocities of the motors ($\dot{\Phi}_i$).

These relationships can be conveniently determined by using the expressions for the drive ratios, thus providing an explicit statement of the influence of wrist geometry and proportioning. Indicating the j-nth gear's angular velocity around its own axis with $\dot{\theta}_j$ and the gear radius with R_j, the kinematic relations can be formulated using sign convention as in Fig. (3).
Gears *5* and *6*, *7* and *8*, *10* and *11* are positively connected. One has:

$$\dot{\theta}_5 = \dot{\theta}_6 \tag{1}$$

$$\dot{\theta}_8 = \dot{\theta}_7 \tag{2}$$

$$\dot{\theta}_{10} = \dot{\theta}_{11} \tag{3}$$

The input transmission ratio are thus written:

$$\dot{\Phi}_1 = \frac{N_4}{N_1}\dot{\theta}_4 \tag{4}$$

$$\dot{\Phi}_2 = \frac{N_{11}}{N_2}\dot{\theta}_{11} = \frac{N_{11}}{N_2}\dot{\theta}_{10} \tag{5}$$

$$\dot{\Phi}_3 = \frac{N_5}{N_3}\dot{\theta}_5 = \frac{N_5}{N_3}\dot{\theta}_6 \tag{6}$$

By examining the motion of gears *7*, *9*, *12*, one obtains:

$$\dot{\theta}_7 = (\dot{\theta}_4 - \dot{\theta}_6)\cdot\frac{R_6}{R_7} = (\dot{\theta}_4 - \dot{\theta}_6)\cdot\frac{N_6}{N_7} \tag{7}$$

$$\dot{\theta}_9 = (\dot{\theta}_{10} + \dot{\theta}_4)\cdot\frac{R_{10}}{R_9} = (\dot{\theta}_{10} + \dot{\theta}_4)\cdot\frac{N_{10}}{N_9} \tag{8}$$

$$\dot{\theta}_{12} = (\dot{\theta}_9 - \dot{\theta}_8)\cdot\frac{R_8}{R_{12}} = (\dot{\theta}_9 - \dot{\theta}_8)\cdot\frac{N_8}{N_{12}} \tag{9}$$

The three DOFs are the rotation Θ_1 around z_{I}, the rotation Θ_2 around x_{II} and the rotation Θ_3 around y_{III}. Then we have:

$$\dot{\Theta}_1 = \dot{\theta}_4 \tag{10}$$

$$\dot{\Theta}_2 = \dot{\theta}_9 \tag{11}$$

$$\dot{\Theta}_3 = \dot{\theta}_{12} \tag{12}$$

By taking into account equations (7), (8), (9), introducing (4), (5), (6) in (10), (11), (12) we can write:

$$\dot{\Theta}_1 = \frac{N_1}{N_4}\dot{\Phi}_1 \tag{13}$$

$$\dot{\Theta}_2 = \frac{N_{10}}{N_9}\left(\frac{N_1}{N_4}\dot{\Phi}_1 + \frac{N_2}{N_{11}}\dot{\Phi}_2\right) \tag{14}$$

$$\dot{\Theta}_3 = \frac{N_8}{N_{12}}\left[\frac{N_1}{N_4}\left(\frac{N_{10}}{N_9} - \frac{N_6}{N_7}\right)\dot{\Phi}_1 + \frac{N_{10}}{N_9}\frac{N_2}{N_{11}}\dot{\Phi}_2 + \frac{N_6 N_3}{N_7 N_5}\dot{\Phi}_3\right] \tag{15}$$

In matrix form, (13) through (15) can be written thus:

$$\begin{vmatrix}\dot{\Theta}_1 \\ \dot{\Theta}_2 \\ \dot{\Theta}_3\end{vmatrix} = \begin{vmatrix} 1 & 0 & 0 \\ \frac{N_{10}}{N_9} & \frac{N_{10}}{N_9} & 0 \\ \frac{N_8}{N_{12}}\left(\frac{N_{10}}{N_9} - \frac{N_6}{N_7}\right) & \frac{N_8}{N_{12}}\frac{N_{10}}{N_9} & \frac{N_8}{N_{12}}\frac{N_6}{N_7} \end{vmatrix} \begin{vmatrix} \frac{N_1}{N_4} & 0 & 0 \\ 0 & \frac{N_2}{N_{11}} & 0 \\ 0 & 0 & \frac{N_3}{N_5} \end{vmatrix} \begin{vmatrix}\dot{\Phi}_1 \\ \dot{\Phi}_2 \\ \dot{\Phi}_3\end{vmatrix} \tag{16}$$

Or:

$$\vec{\dot{\Theta}} = |D|\,|E|\vec{\dot{\Phi}} = |F|\vec{\dot{\Phi}} \tag{17}$$

This division into two matrices |D| and |E| can be convenient in order to indicate the input gear drive ratios.
Reversing thus gives:

$$\vec{\dot{\Phi}} = |F|^{-1}\vec{\dot{\Theta}} \tag{18}$$

In analyzing (16), it will be noted that $\dot{\Phi}_3$ controls roll motions only ($\dot{\Theta}_3$), $\dot{\Phi}_2$ generates pitch motions ($\dot{\Theta}_2$) coupled with roll motions, and $\dot{\Phi}_1$ causes yaw motions ($\dot{\Theta}_1$) coupled with pitch and roll.
Expression (18) can be used to calculate the motor rotations needed to bring the wrist to the desired configuration, defined in the joint space.

B. Relationships between the angular velocities of the joints ($\dot{\Theta}_i$) and the angular velocities in cartesian space ($\vec{\omega}$).

The orientation of the set of reference axes located on the end-effector relative to the system located on the arm carrying the wrist can be expressed by the following rotation matrix:

$$A_3 = Rot(z,\Theta_1)Rot(x,\Theta_2)Rot(y,\Theta_3) = A_1\,{}^1A_2\,{}^2A_3 \tag{19}$$

given that:

$$A_1 = Rot(z,\Theta_1) = \begin{vmatrix} C_1 & -S_1 & 0 \\ S_1 & C_1 & 0 \\ 0 & 0 & 1 \end{vmatrix} \tag{20}$$

$$A_2 = A_1 Rot(x,\Theta_2) = \begin{vmatrix} C_1 & -S_1C_2 & S_1S_2 \\ S_1 & C_1C_2 & -C_1S_2 \\ 0 & S_2 & C_2 \end{vmatrix} \tag{21}$$

$$A_3 = A_1\,{}^1A_2 Rot(y,\Theta_3) = \begin{vmatrix} C_1C_3 - S_1S_2S_3 & -S_1C_2 & C_1S_3 + S_1S_2C_3 \\ S_1C_3 + C_1S_2S_3 & C_1C_2 & S_1S_3 - C_1S_2C_3 \\ -C_2S_3 & S_2 & C_2C_3 \end{vmatrix} \tag{22}$$

where $C_i = \cos(\Theta_i)$, $S_i = \sin(\Theta_i)$.

Indicating the versors of the three sets of axes rotated in transformation (19) with $\vec{i}_n, \vec{j}_n, \vec{k}_n$, the axes of the wrist's three DOFs Θ_1, Θ_2 and Θ_3 are the versors $\vec{k}_I, \vec{i}_{II}, \vec{j}_{III}$ respectively. The directing cosines of these three versors are indicated respectively in the third column of (20), the first column of (21) and the second column of (22).

Expressing the end-effector's absolute angular velocity $\vec{\omega}$ relative to axes *xyz*, we have:

$$\vec{\omega} = \omega_x\vec{i} + \omega_y\vec{j} + \omega_z\vec{k} = \dot{\Theta}_1\vec{k}_I + \dot{\Theta}_2\vec{i}_{II} + \dot{\Theta}_3\vec{j}_{III} \tag{23}$$

In matrix form, this becomes:

$$\vec{\omega} = \begin{vmatrix} \omega_x \\ \omega_y \\ \omega_z \end{vmatrix} = \begin{vmatrix} 0 & C_1 & -S_1C_2 \\ 0 & S_1 & C_1C_2 \\ 1 & 0 & S_2 \end{vmatrix} \begin{vmatrix} \dot{\Theta}_1 \\ \dot{\Theta}_2 \\ \dot{\Theta}_3 \end{vmatrix} = |J|\,\dot{\vec{\Theta}} \tag{24}$$

where $|J|$ is the Jacobi matrix.

C. Relationships between the angular velocities of the motors ($\vec{\Phi}$) and the angular velocities of the end-effector in cartesian space ($\vec{\omega}$).

Introducing (17) in (24) gives:

$$\vec{\omega} = |J|\vec{\Theta} = |J|\,|F|\vec{\Phi} = |G|\vec{\Phi} \tag{25}$$

where:

$$|G| = \begin{vmatrix} \dfrac{N_1(C_1N_{10}N_{12}N_7 - S_1C_2N_8N_{10}N_7 + S_1C_2N_8N_6N_9)}{N_{12}N_9N_7N_4} & \dfrac{N_{10}N_2(C_1N_{12} - S_1C_2N_8)}{N_{12}N_9N_{11}} & -\dfrac{S_1C_2N_8N_6N_3}{N_{12}N_7N_5} \\ -\dfrac{N_1(-S_1N_{10}N_{12}N_7 - C_1C_2N_8N_{10}N_7 + C_1C_2N_8N_6N_9)}{N_{12}N_9N_7N_4} & \dfrac{N_{10}N_2(S_1N_{12} + C_1C_2N_8)}{N_{12}N_9N_{11}} & \dfrac{C_1C_2N_8N_6N_3}{N_{12}N_7N_5} \\ -\dfrac{N_1(-N_{12}N_9N_7 - S_2N_8N_{10}N_7 + S_2N_8N_6N_9)}{N_{12}N_9N_7N_4} & \dfrac{S_2N_8N_{10}N_2}{N_{12}N_9N_{11}} & \dfrac{S_2N_8N_6N_3}{N_{12}N_7N_5} \end{vmatrix}$$

(26)

Expression (25) makes it possible to analyze the wrist's direct kinematics in cartesian space.

IV. Degeneration study

The wrist degeneration study is conducted by analyzing the reverse Jacobi matrix. In reality, the reverse of matrix (26) is analyzed so that the effect of gearing geometry on the motor speeds can also be taken into account.
We have:

$$\vec{\Phi} = |G|^{-1}\vec{\omega} \tag{27}$$

given that:

$$|G|^{-1} = \begin{vmatrix} \dfrac{S_2S_1N_4}{N_1C_2} & -\dfrac{S_2C_1N_4}{N_1C_2} & \dfrac{N_4}{N_1} \\ \dfrac{(-S_2S_1N_{10} + C_1C_2N_9)N_{11}}{N_{10}N_2C_2} & \dfrac{(S_2C_1N_{10} + S_1C_2N_9)N_{11}}{N_{10}N_2C_2} & -\dfrac{N_{11}}{N_2} \\ \dfrac{(-S_1N_{12}N_7 + S_1S_2N_8N_6 - C_1C_2N_8N_7)N_5}{N_8N_6N_3C_2} & -\dfrac{(-C_1N_{12}N_7 + C_1S_2N_8N_6 + S_1C_2N_8N_7)N_5}{N_8N_6N_3C_2} & \dfrac{N_5}{N_3} \end{vmatrix}$$

(28)

Degeneration configurations occur when there are coefficients of matrix (28) which tend to infinity for particular values assumed by the DOFs Θ_1 and Θ_2.
In this connection, indicating the corresponding x and y axes, following a rotation of Θ_1 around the z axis, with x_I and y_I, we have:

$$\omega_{xI} = \cos\Theta_1\omega_x + \sin\Theta_1\omega_y \tag{29}$$

$$\omega_{yI} = -\sin\Theta_1\omega_x + \cos\Theta_1\omega_y \tag{30}$$

$$\omega_{zI} = \omega_z \tag{31}$$

Introducing (29), (30), (31) in (27) and assuming for the sake of simplicity that the bevel gear drive ratios are all unitary yields:

$$\dot{\Phi}_1 = -\mathrm{tg}\,\Theta_2\omega_{yI} + \omega_{zI} \tag{32}$$

$$\dot{\Phi}_2 = \omega_{xI} + \mathrm{tg}\,\Theta_2\omega_{yI} - \omega_{zI} \tag{33}$$

$$\dot{\Phi}_3 = -\omega_{xI} + \left(\frac{1}{\cos\Theta_2} - \mathrm{tg}\,\Theta_2\right)\omega_{yI} + \omega_{zI} \tag{34}$$

Bearing in mind that degeneration occurs when motor velocities tend to infinity for assigned components of absolute velocity, (32)..(34) indicate that this takes place only when Θ_2 approximates $\pm\pi/2$, i.e. when the end-effector axis is perpendicular to the aligned configuration. In addition, it will be noted that the only component of absolute velocity which can cause degeneration is the one along the y_I axis, which is instantaneously the axis perpendicular to the cross plane.
In other words, for $\Theta_2 = \pm\pi/2$, degeneration conditions exist only if the composition of ω_x and ω_y is not directed along the x_I axis.

V. Conclusions

The wrist layout presented in this paper is of particular interest.
Kinematic analysis indicated that the wrist is not subject to degeneration in the aligned configuration, which is the condition in which it normally operates.
The only configuration in which degeneration occurs is that with the wrist rotated by 90° with respect to the pitch axis. However, this condition is outside the work space, given the geometric dimensions of the gears. The wrist is thus kinematically suitable for performing complex handling operations without particular kinematic constraints.
The basic relationships for wrist control and use were presented, specifying the links between the angular velocities of the joints and of the drive gears, between the angular velocities of the joints and the absolute velocities in a cartesian space, and between the motor velocities and the absolute velocities of the end-effector.

The kinematic configuration here presented has to be optimized in order to increase the workspace, choosing appropriate gear ratios and gear dimensions.

VI. Symbols

Θ_i	Rotations of the three wrist DOFs
Φ_i	Input shaft rotations
$\dot{\theta}_j$	Angular velocity of the j-nth gear relative to its axis
N_j	Number of teeth on the j-nth gear
R_j	Radius of j-nth gear
xyz	Fixed reference system of mutually perpendicular cartesian axes
$x_n y_n z_n$	Mobile reference system located on the i-nth DOF
$\vec{i}_n, \vec{j}_n, \vec{k}_n$	Reference system versors
ω	Absolute angular velocity of the end-effector in a fixed reference system

VII. Acknowledgments

This research was supported by the Italian National Research Council (C.N.R.) as part of the "Progetto Finalizzato Robotica".

VIII. References

[1] Romiti A., Sorli M, "Wrist analysis and degeneracy evaluation and comparison", *Proc. of 2nd Int. Symp. on Measurement and Control in Robotics ISMCR*, 437-444, Tsukuba, Japan, (1992).

[2] Romiti A., Raparelli T, "A new spherical robot wrist with remote degeneration regions", *Proc. of 6th Int. Conf. on Advanced Robotics ICAR*, 643-647, Tokyo, Japan.(1993).

[3] Romiti A., Raparelli T., Sorli M., "Robot wrist configurations, mechanism and kinematics", *Robotics in Alpe-Adria Region* (P. Kopacek ed.), 44-48, Springer-Verlag, Wien, (1994).

[4] B. Huang, V. Milenkovic, "On an algorithm negotiating wrist singularities", *Proc. of 17th int. Symposium on Industrial Robots ISIR,* 13-1, 13-6, (1987).

[5] J.P. Trevelyan, P.D. Kovesi, M. Ong, D. Elford, ET,"a wrist mechanism without singular position", *Int. J. of Robotic Research*, vol. 4, n. 4, 71-85, (1986).

[6] R.P. Paul, C.N. Stevenson, "Kinematics of robot wrist", *Int. J. of Robotic Research*, vol. 2, n.1, 31-38 (1983).

10. Kinematic Analysis

Is There a Most General Kinematic Chain?

Enric Celaya

Institut de Cibernètica (UPC – CSIC)
Diagonal 647, 08028-Barcelona, SPAIN

Abstract - The well known concept of kinematic equivalence of two kinematic chains is revised and precisely defined using the concept of liaison. With the aid of this definition a chain C_1 is said to be more general than C_2 if by freezing one or more joints of C_1 a chain kinematically equivalent to C_2 can be obtained. A question discussed in this paper is whether, by progressively increasing the complexity of a chain, new classes of liaisons can be obtained indefinitely or, on the contrary, chains above a certain length are always equivalent to simpler chains. This last situation would imply the existence of a most general kinematic chain, i.e., a single chain from which one could obtain chains kinematically equivalent to any other chain. It is shown that for the special cases of planar and spherical chains, such a most general kinematic chain effectively exists.

I. Introduction

The concept of *kinematic equivalence* is not new in kinematics (see for example [1] or [2]), and refers to the fact that the same coupling between two links can be obtained with structurally different kinematic chains. Perhaps the most simple example of kinematic equivalence is that existing between a C pair and its usual decomposition as an RP chain. The relevance of kinematic equivalence is emphasized by Hunt, who refers to it using the terms "substitute linkages" or "replacements" and says ([3]):

> *"... such substitutions are useful, both as actual linkages and as means helping us to analyse the behaviour better."* [p. 20]

implying that not only the use of a different but equivalent kinematic chain may find applications in industry and engineering, but also that the kinematic analysis of mechanisms and manipulators can be simplified through the formal substitution of a subchain by an equivalent but simpler one.

Despite its potential interest in kinematics, the concept of kinematic equivalence has been poorly formalized, and its practical use has been restricted to the observation that lower pairs with more than one d.o.f. (i.e., Cylindrical, Planar and Spherical pairs) can be replaced by appropriate chains of R and P pairs, thus reducing the analysis of kinematic chains of lower pairs to that of kinematic chains of R and P pairs (the Helical pair is not considered in many approaches).

A classification of kinematic chains based on mobility criteria using Group Theory has been done in [4] and later reelaborated in [5]. To my knowledge, no attempt has been made so far to achieve a classification of kinematic chains based on kinematic equivalence. This is, indeed, the main purpose of this paper: introducing a precise definition of kinematic equivalence and proposing a classification scheme based on what we call the n-bar model for kinematic chains. According to that, a complete

A. J. Lenarčič and B. B. Ravani (eds.), Advances in Robot Kinematics and Computationed Geometry, 391–400.

classification of planar and spherical chains is obtained, while the more challenging task of classifying all spatial chains remains to be done. As a consequence of the classification of the planar and spherical cases, the most significant result of the paper is derived: the existence of a most general planar chain and a most general spherical chain. A similar result could be obtained for general chains if a complete classification of spatial chains were achieved.

II. Liaisons

In the same way each lower pair defines a coupling between two adjacent links constraining their relative positions, an arbitrary kinematic chain defines a certain coupling between two distant links, constituting a generalization of the concept of pair that we call *liaison*. The term liaison was introduced in French in [4], and was subsequently used by some authors (cf. [6],[7]), but other terminologies have also been used, such as *bond* [5], or *connection* [3].

In order to compare the liaisons generated by different chains, we need a way to characterize them that refers directly to the coupling they define, rather than to the underlying chain. To this end, we take two reference frames: $\mathcal{F}_0$ attached to the base link and $\mathcal{F}_n$ attached to the last link. For each position of the last link allowed by the liaison, there is a corresponding displacement $d \in SE(3)$ transforming $\mathcal{F}_0$ into $\mathcal{F}_n$. The set of all such displacements d is a subset or *complex* $\{L\} \subset SE(3)$ that characterizes the liaison. This definition of $\{L\}$ is slightly different from that given in [4], which takes $\mathcal{F}_n$ coincident with $\mathcal{F}_0$ for some initial configuration, in which case the corresponding complex must always include the identity transformation I.

If we take a different base frame $\mathcal{F}'_0$ related with $\mathcal{F}_0$ by a displacement d_0, the corresponding complex is the result of the product $d_0^{-1} \cdot \{L\}$ and, similarly, taking a different $\mathcal{F}'_n$ obtained as $\mathcal{F}_n$ displaced by d_n, the corresponding complex is the result of the product $\{L\} \cdot d_n$. Since the selection of the reference frames $\mathcal{F}_0$ and $\mathcal{F}_n$ is arbitrary, we say that two complexes $\{L\}$ and $\{L'\}$ represent the same liaison if they are related in the form

$$\{L'\} = d_1 \cdot \{L\} \cdot d_2 ; \quad d_1, d_2 \in SE(3).$$

In practice, it will be desirable to choose reference frames $\mathcal{F}_0$ and $\mathcal{F}_n$ in such a way that the complex takes its simplest form. Below we show how a standard form for the complex associated with a planar liaison can be defined, that intuitively corresponds to take as z-axes of $\mathcal{F}_0$ and $\mathcal{F}_n$ the rotation axes of the first and last R pairs of the chain, respectively.

III. Kinematic Equivalence and Subsumption

Having defined a liaison as the coupling generated by a chain, the definition of kinematic equivalence follows in a natural way: *Two chains are equivalent if they provide the same liaison.* According to that, the equivalence of two chains implies that with an appropriate choice of frames $\mathcal{F}_0$ and $\mathcal{F}_n$ their associated complexes coincide. Actually, this requirement for equivalence is stronger than that implicitly used by most authors, who, avoiding a formal definition of kinematic equivalence, consider that a planar pair is equivalent, for example, to an RRR chain with three parallel rotation axes. While it is clear that both provide the same motion capabilities on a bounded region of the space, in the case of an RRR chain, points beyond a certain distance in the plane can not be reached and, according to our definition of equivalence, an RRR chain is not equivalent to a planar pair.

Now we introduce a new definition: we say that chain C_1 *subsumes* chain C_2, or also that chain C_1 *is more general* than chain C_2, if by freezing one or more pairs of C_1 at appropriate constant values, a chain equivalent to C_2 is obtained. Intuitively, if chain C_1 subsumes C_2, C_1 can mimic (or can be used to replace) C_2 just by fixing some of its pairs. We say that it is more general in the sense that C_1 can mimic many other different chains in addition to C_2. Subsumption implies the inclusion of the complex of the less general chain into that of the more general one (given an appropriate selection of frames $\mathcal{F}_0$ and $\mathcal{F}_n$). The converse is not true: The inclusion of one complex into another does not imply the subsumption of one chain by the other.

Note that subsumption does not define an ordering on chains (not even a partial ordering) since, in general, transitivity does not hold for subsumption.

IV. Structure of a Chain: The n-bar Model

In this work, we deal only with liaisons obtained from simple open chains of R and P pairs. (In a simple open chain each link is connected to two other links, except for the first and last links that are connected to only one link.) A convenient way to represent a simple kinematic chain is by the use of what we call the n-bar model. The n-bar model is based on the well-known Denavit-Hartenberg notation, where links are associated to common normals between consecutive pair-axes. Link parameters denote distance and twist angle between two consecutive pair-axes, and joint parameters denote distance and twist angle between two common normals. In the n-bar model pair-axes and common normals play equivalent roles, and both are called simply normal axes. One bar is associated with each intersection of two normal axes, that are taken as the x and z axes of a reference frame attached to the bar. The z-axis of one bar is aligned with the x-axis of the next, and they can be considered as joined by a C pair, in which one, none, or both parameters may be fixed. The n-bar model should be regarded as a theoretical tool rather than as a practically realizable implementation of actual manipulators for two reasons, at least: Bars in real manipulators will interfere in some configurations, and only finite range displacements along translation axes will be possible.

Depending on what parameters are fixed or variable in the C pair connecting one bar to the next, a bar can be of four different types. If we distinguish also the case of fixed null and fixed non-null parameters, there are eight possible types of bars: A bar for which both parameters are variable will be called a free bar, and will be represented by C. A bar with fixed rotation and variable translation will be represented by P_α, or simply P when the rotation is 0, and a bar with fixed translation and variable rotation will be represented by R_s, or R when the translation is 0. A bar with no variable rotation nor translation is a fixed bar and will be represented by c, α or s depending on whether both parameters are non-null, or the translation or rotation parameter is 0, respectively.

We define the *structure* of a chain as the sequence of bars used to define it, specifying the type of each bar but without referring to the particular values taken by constant non-null parameters. Thus the structure of a chain can be represented by a string of symbols from the set {C,R_s,R,P_α,P,c,α,s}. Chains with the same structure differ only in the values of their constant non-null parameters.

The task of classification of chains consists, mainly, in comparing the complexes generated by different chains in order to identify equivalences. The complex $\{L\}$ of the liaison generated by a kinematic chain can be expressed in a way that directly reflects its n-bar model in the form

$$\{L\} = Tz(z_1)Rz(\theta_1)\, Q\, Tz(z_2)Rz(\theta_2)\, Q \ldots Q\, Tz(z_n)Rz(\theta_n) \tag{1}$$

where Tz is a translation along the z-axis, Rz is a rotation around the z-axis and $Q \equiv Ry(\pi/2)Rz(\pi)$. Each Q operator makes the transition from the reference frame attached to one bar to the reference frame attached to the next, whose x-axis coincides with the z-axis of the previous bar. Arguments z_i and θ_i may be either constant or variable, depending on the type of the corresponding bar. In particular, if two chains differ only in the existence of a fixed bar at the beginning or at the end of the chain, they are equivalent, since the only effect of such fixed bars is just that of modifying the choice of frames $\mathcal{F}_0$ and $\mathcal{F}_n$. Accordingly, for the classification task only chains whose structures neither begin nor end by fixed bars need to be considered.

Clearly, a given chain subsumes all its subchains. Also, if in a given chain a fixed parameter is made variable, i.e., the corresponding bar is replaced by a less constrained one, then the new chain subsumes the old one. An n-bar chain in which all its bars are free is represented by a string of n C's and will be called the general n-bar. The general n-bar subsumes all chains with m-bar structures with $m \leq n$. In fact, the general n-bar can subsume also m-bar chains with $m \geq n$ due to kinematic equivalence. It is clear that if a given chain is subsumed by an n-bar chain, it is also subsumed by the general n-bar. But note that, as a consequence of the lack of transitivity of subsumption, the fact that a chain subsumes the general n-bar does not imply that it subsumes all chains with n-bar structure.

V. Classification of Planar Chains

When a planar chain is described according to the n-bar model, not all types of bars can be used. Since in a planar chain all rotation axes are parallel and all translation axes are perpendicular to them, only type R, P, s and α bars can appear. Furthermore, in a planar chain, translational (P and s) and rotational (R and α) bars must appear in an alternate form. Despite this, some freedom remains in the choice of the structure used to represent a given planar chain. It is easy to see that the following rules can always be met:

1. Between two same-type variable bars, there is always a single fixed bar (RsR or PαP).

2. If the first variable bar after a type R bar is a type P bar, no fixed bar appears between them (RP).

3. If the first variable bar after a type P bar is a type R bar, no fixed bar appears between them (PR), except when they are preceded by another R bar, in which case one can have RPαsR.

We say that a planar chain represented following these rules is in *standard form*. Using the n-bar model in standard form permits avoiding the analysis of many equivalent chains.

In the planar case, the expression of the complex given in (1) takes the form:

$$\begin{aligned}\{L\} \;=\; & Q\,Tz(x_0)\,Q\,Rz(\theta_1)\,Q\,Tz(x_1)\,Q\;\ldots\;Rz(\theta_n)\,Q\,Tz(x_n)\,Q\;\equiv \\ & Tx(x_0)\,Rz(\theta_1)\,Tx(x_1)\;\ldots\;Rz(\theta_n)\,Tx(x_n)\end{aligned} \tag{2}$$

where extra Q operators have been added at the beginning and the end just by cosmetic reasons. The first (last) translation in (2) may be omited if the chain begins (ends) with a rotation.

The complex of a planar chain can be explicitly represented using parametric expressions for variables x, y and θ, where (x, y) correspond to the Cartesian coordinates of the origin of $\mathcal{F}_n$ with respect to $\mathcal{F}_0$, and θ corresponds to the angle formed by their x-axes. Using (2), these parametric expressions take the form:

$$x = \sum_{i=0}^{n} x_i \cos(\sum_{j=1}^{i} \theta_j), \qquad y = \sum_{i=1}^{n} x_i \sin(\sum_{j=1}^{i} \theta_j), \qquad \theta = \sum_{i=1}^{n} \theta_i \tag{3}$$

in which only those x_i and θ_i that are variable act as free parameters.

For the classification task we will analyze the complexes of all planar chains according to the number of bars needed to model them in standard form. As noted before, only chains that neither begin nor end with a fixed bar need to be considered. The general procedure we follow is explained next: For each standard structure, we write the corresponding complex using (2) and obtain the expressions for x, y and θ given by (3). These parametric expressions are then used, through the elimination of parameters, to derive one or more constraints relating x, y and θ, that characterize the complex. Detailed derivations of such constraints in each case are not explicited due to space limitations, but they can be reproduced without difficulty. Chains from which the same constraints are obtained are equivalent. A 3-dimensional visualization of the complexes thus obtained is shown in Fig. 1 (except that generated by the RPR chain corresponding to the whole $SE(2)$ set). In the following, x_i and θ_i are used for variable parameters, while s_i and α_i are used for constant ones.

A. One-bar Chains

The only two possible planar chains with a single bar are R and P:

- $\{R\} = Rz(\theta_1) \implies \{x = 0,\, y = 0\}$.
- $\{P\} = Tx(x_1) \implies \{y = 0,\, \theta = 0\}$.

B. Two-bar Chains

The only two planar chains with standard two-bar structure are RP and PR:

- $\{RP\} = Rz(\theta_1)Tx(x_1) \implies y = x \tan\theta$.
- $\{PR\} = Tx(x_0)Rz(\theta_1) \implies y = 0$.

C. Three-bar Chains

Chains with four different standard 3-bar structures have to be considered:

- $\{Rs_1R\} = Rz(\theta_1)Tx(s_1)Rz(\theta_2) \implies x^2 + y^2 = s_1^2$.
  Different values of s_1 provide different complexes: there is a one-dimensional family of different (non-equivalent) planar chains with this structure.
- $\{RPR\} = Rz(\theta_1)Tx(x_1)Rz(\theta_2) \implies$ no constraint.
  The RPR chain is the shortest one that generates all $SE(2)$.
- $\{P\alpha_1P\} = Tx(x_0)Rz(\alpha_1)Tx(x_1) \implies \theta = \alpha_1$.
  While different values of α_1 provide different complexes, all of them represent the same liaison since they are related by $\{P\alpha_1P\} = \{P\alpha_2P\} \cdot Rz(\alpha_1 - \alpha_2)$.
- $\{PRP\} = Tx(x_0)Rz(\theta_1)Tx(x_1) \implies y = 0$, if $\sin\theta = 0$.

D. Four-bar Chains

There are 6 different standard 4-bar structures to be considered. Two of them, RPRP and PRPR, are equivalent to an RPR chain because they contain it as a subchain (whose generated complex is all $SE(2)$) and therefore they also generate all $SE(2)$. For the remaining four structures we have:

- $\{Rs_1RP\} = Rz(\theta_1)Tx(s_1)Rz(\theta_2)Tx(x_2) \implies y = x\tan\theta + s_1(\sin\phi - \tan\theta\cos\phi)$, where ϕ is a free parameter. Chains with the structure Rs_1RP define a one-dimensional family of liaisons parameterized by s_1.
- $\{RP\alpha_1P\} = Rz(\theta_1)Tx(x_1)Rz(\alpha_2)Tx(x_2) \implies$ no constraint.
  Equivalent to RPR.
- $\{P\alpha_1PR\} = Tx(x_0)Rz(\alpha_1)Tx(x_1)Rz(\theta_2) \implies$ no constraint.
  Equivalent to RPR.
- $\{PRs_1R\} = Tx(x_0)Rz(\theta_1)Tx(s_1)Rz(\theta_2) \implies -s_1 \leq y \leq s_1$.
  Chains with the structure PRs_1R define a one-dimensional family of liaisons parameterized by s_1.

E. Five-bar Chains

There are 11 standard five-bar structures of chains to consider. Seven of them are equivalent to RPR because they contain one of the 3 or 4-bar generators of $SE(2)$. For the remaining four structures we have:

- $\{Rs_1Rs_2R\} = Rz(\theta_1)Tx(s_1)Rz(\theta_2)Tx(s_2)Rz(\theta_3) \implies (x-a)^2 + (y-b)^2 = s_2^2$, with $a^2 + b^2 = s_1^2$. Chains with the structure Rs_1Rs_2R define a two-dimensional family of liaisons parameterized by s_1 and s_2.
- $\{RP\alpha_2s_2R\} = Rz(\theta_1)Tx(x_1)Rz(\alpha_2)Tx(s_2)Rz(\theta_3) \implies y = x\tan\phi + (t/\cos\phi)$, where ϕ is a free parameter and $t = s_2\sin\alpha_2$. Chains with the structure RPα_2s_2R define a one-dimensional family of liaisons parameterized by the product $s_2\sin\alpha_2$.

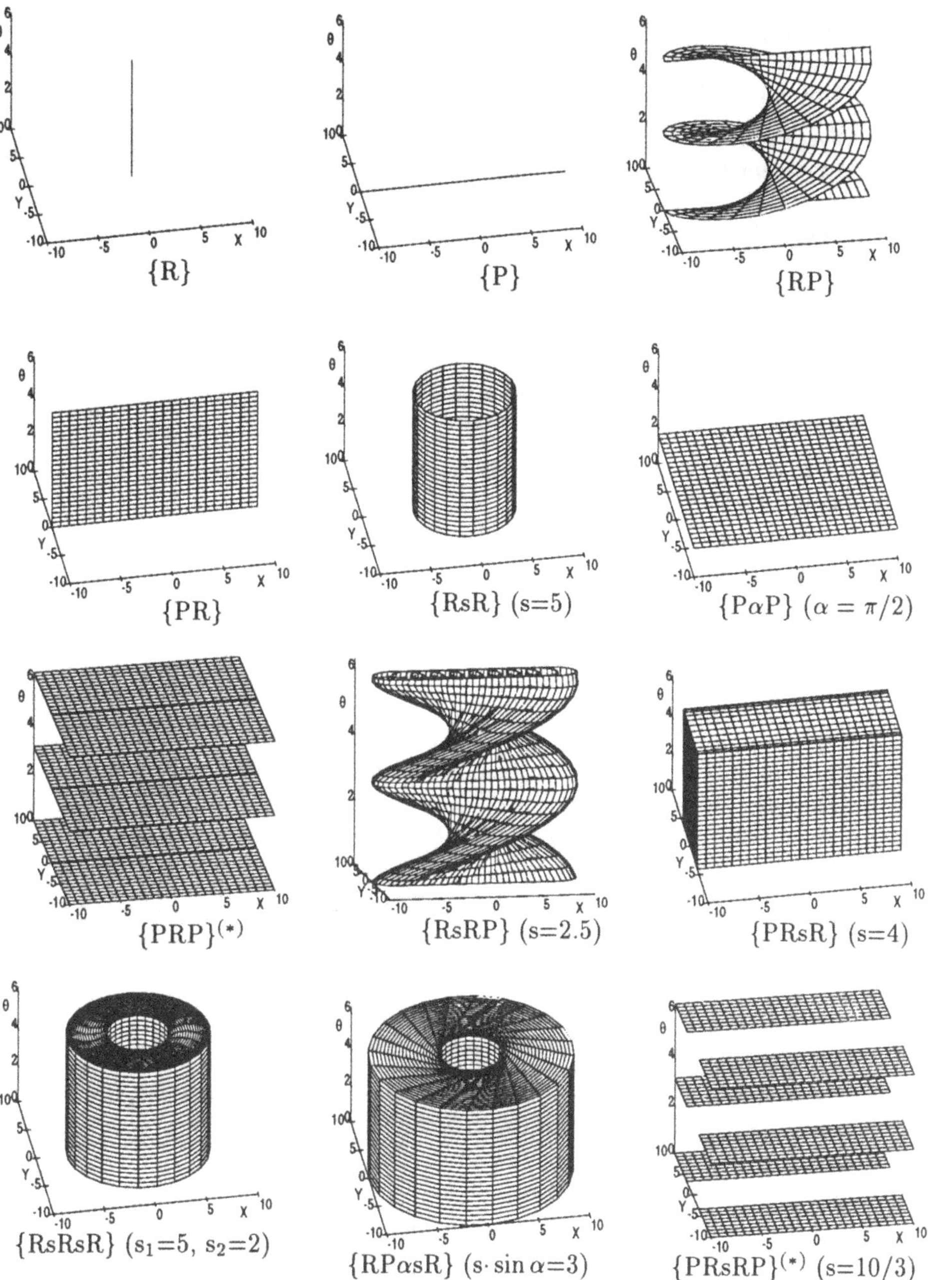

Figure 1: Complexes of planar chains. (*): Complementary of the complex.

- $\{PRs_1RP\} = Tx(x_0)Rz(\theta_1)Tx(s_1)Rz(\theta_2)Tx(x_2) \implies -s_1 \leq y \leq s_1$, if $\sin\theta = 0$. Chains with the structure PRs_1RP define a one-dimensional family of liaisons parameterized by s_1.
- $\{P\alpha_1P\alpha_2P\} = Tx(x_0)Rz(\alpha_1)Tx(x_1)Rz(\alpha_2)Tx(x_2) \implies \theta = \alpha_1 + \alpha_2$. Equivalent to $P\alpha P$ with $\alpha = \alpha_1 + \alpha_2$.

F. Six-bar Chains

There are 18 standard structures of six-bar chains among which 14 contain one of the generators of $SE(2)$ and are equivalent to RPR. Next we see that chains with each of the remaining four structures are equivalent to some shorter chain:

- $\{Rs_1Rs_2RP\} = Rz(\theta_1)Tx(s_1)Rz(\theta_2)Tx(s_2)Rz(\theta_3)Tx(x_3)$. Equivalent to $\{RsRP\}$ with $s = s_1 + s_2$.
- $\{RP\alpha_2s_2RP\} = Rz(\theta_1)Tx(x_1)Rz(\alpha_2)Tx(s_2)Rz(\theta_3)Tx(x_3)$. Equivalent to $\{RPR\}$.
- $\{PRP\alpha_2s_2R\} = Tx(x_0)Rz(\theta_1)Tx(x_1)Rz(\alpha_2)Tx(s_2)Rz(\theta_3)$. Equivalent to $\{RPR\}$.
- $\{PRs_1Rs_2R\} = Tx(x_0)Rz(\theta_1)Tx(s_1)Rz(\theta_2)Tx(s_2)Rz(\theta_3)$. Equivalent to $\{PRsR\}$ with $s = s_1 + s_2$.

G. Seven-bar Chains

By inspection of the 31 possible standard structures of seven-bar chains we find that 26 contain some of the generators of $SE(2)$, and two other (PαPαPαP and PRsRsRP) contain a subchain that is equivalent to a shorter one. The remaining three structures are:

- $\{Rs_1Rs_2Rs_3R\} = Rz(\theta_1)Tx(s_1)Rz(\theta_2)Tx(s_2)Rz(\theta_3)Tx(s_3)Rz(\theta_4)$. Equivalent to $\{Rs_aRs_bR\}$, where $s_a = s_1 + s_2 + s_3$ and $s_b = \max(0, 2s_m - s_a)$, with $s_m = \max(s_1, s_2, s_3)$.
- $\{Rs_1RP\alpha_3s_3R\} = Rz(\theta_1)Tx(s_1)Rz(\theta_2)Tx(x_2)Rz(\alpha_3)Tx(s_3)Rz(\theta_4)$. Equivalent to $\{RP\alpha sR\}$ with $s\sin\alpha = \max(0, s_3\sin\alpha_3 - s_1)$.
- $\{RP\alpha_2s_2Rs_3R\} = Rz(\theta_1)Tx(x_1)Rz(\alpha_2)Tx(s_2)Rz(\theta_3)Tx(s_3)Rz(\theta_4)$. Equivalent to $\{RP\alpha sR\}$ with $s\sin\alpha = \max(0, s_2\sin\alpha_2 - s_3)$.

H. Eight-bar and Longer Chains

Planar chains with eight or more bars necessarily contain as a subchain one of the six or seven-bar chains, all of which are equivalent to a shorter one. Thus the classification of planar chains is complete. As a result we conclude that any planar chain is equivalent to a chain with one of the 13 different structures summarized in Table 1.

To obtain the shortest planar chain equivalent to a given one the following rules should be applied:

Type	1-bar	2-bar	3-bar	4-bar	5-bar
Planar	R P	RP PR	RsR$^{(*)}$ PαP RPR PRP	RsRP$^{(*)}$ PRsR$^{(*)}$	RsRsR$^{(**)}$ RPαsR$^{(*)}$ PRsRP$^{(*)}$
Spherical	R		RαR$^{(*)}$		RαRαR$^{(**)}$

Table 1: Classification of planar and spherical chains.
(*): one-dimensional family; (**): two-dimensional family.

1. Write the chain in standard form.
2. If the chain contains as a subchain one of the following: RPR, RPαP, PαPR, RPαsRP, PRPαsR, then it is equivalent to RPR.
3. Use the following rewriting rules while possible:

PαPαP $\longrightarrow$ PαP	RsRsRsR $\longrightarrow$ RsRsR
RsRsRP $\longrightarrow$ RsRP	RsRPαsR $\longrightarrow$ RPαsR
PRsRsR $\longrightarrow$ PRsR	RPαsRsR $\longrightarrow$ RPαsR

After these rules are applied, the resulting chain will have one of the structures given in Table 1.

I. The Most General Planar Chain

Looking at the classification obtained for planar chains, we see that any planar chain is equivalent to some other with at most 5 bars. However, no particular 5-bar planar chain subsumes all of them. The shortest planar chains subsuming all planar chains are the 6-bar RPRPRP and PRPRPR, and both of them can be considered *most general planar chains.* On the other hand, if we consider arbitrary spatial chains, we find that the general 5-bar CCCCC subsumes all chains in Table 1, and therefore subsumes all planar chains. The general 5-bar is the shortest chain that is more general than all planar chains.

VI. Classification of Spherical Chains

In a spherical kinematic chain all translations are 0, and when the n-bar model is used to represent a spherical kinematic chain, only R and α type bars are required. The classification of spherical chains is, therefore, much simpler than that of planar chains.

Proceeding as in the planar case, we find that any spherical chain is equivalent to a chain with one of the three following structures: R, RαR and RαRαR (Table 1). Note that the chain (or subchain) RR is equivalent to Rα_1R with $\alpha_1 = \pi/2$, and is not included in the table. The simplest chain generating all the subgroup of rotations $SO(3)$ is Rα_1Rα_1R with $\alpha_1 = \pi/2$. It can be shown (see [8]) that any spherical chain with the structure RαRαRαR is equivalent to a chain with the structure RαRαR.

We conclude that the shortest *most general spherical chain* is the 5-bar RRRRR. Clearly, the general 5-bar CCCCC also subsumes all spherical chains, and therefore, it is more general than any planar and spherical chain.

VII. Conclusions

The definition of kinematic equivalence given here corresponds to a *global* equivalence, in contrast with the local equivalence implicitly used in previous approaches. With this definition, two equivalent chains can be interchanged without modifying in any way the (ideal) mobility capabilities of a mechanism containing them.

A classification of chains based on kinematic equivalence has been undertaken here for the first time. To this end, the n-bar model has been introduced, playing a fundamental role in the classification task. A complete classification of planar and spherical chains has been explicitly worked out, and as a result we have established that any arbitrarily long planar or spherical chain is equivalent to a chain with at most five bars.

The task of classifying all spatial chains remains to be done. The insights gained from the study of the planar and spherical cases compel us to believe that a complete classification of spatial chains is possible, though expectedly much more involved. The theoretically important question of the existence or not of a most general chain, or what is equivalent, the existence of an upper bound for the length of the chains that are worth to consider, is by now unanswered.

Acknowledgements: This work has been partially supported by the ESPRIT III Basic Research Action Program of the EC under contract No. 6546 (PROMotion).

References

[1] D.L. Pieper, *The Kinematics of Manipulators under Computer Control*, Ph.D. dissertation, Artificial Intelligence Laboratory, Stanford University (1968).

[2] J. Duffy, *Analysis of mechanisms and robot manipulators*, Edward Arnold, London (1980).

[3] K.H. Hunt, *Kinematic geometry of mechanisms*, Clarendon Press, Oxford (1978).

[4] J.M. Hervé, "Analyse Structurelle des Mécanismes par Groupe des Déplacements", *Mechanism and Machine Theory*, Vol. 13, pp. 437-450 (1978).

[5] J. Angeles, *Rational Kinematics*, Springer Tracts in Natural Philosophy, vol. 34, Springer-Verlag, New York (1988).

[6] J. Angeles, *Spatial Kinematic Chains*, Springer-Verlag, Berlin (1982).

[7] P. Fanghella "Kinematics of spatial linkages by group algebra: A structure-based approach", *Mechanism and Machine Theory*, Vol. 23, No. 3, pp. 171-183 (1988).

[8] E. Celaya, *Geometric reasoning for the determination of the position of objects linked by spatial relationships*, Ph.D. dissertation, Universitat Politècnica de Catalunya (1992).

Zero-Magnitude Screws in a 3-System of Finite Displacement Screws

I.A. Parkin

Basser Department of Computer Science
University of Sydney
Sydney, Australia 2006

Abstract. Consider that a rigid body is subject to angular displacements applied, always in the same sequence, at one then the other of two fixed revolute joints. It has recently been shown, when all possible angular displacements at the joints are considered, that the totality of screws of the finite displacements which are available to the rigid body constitute a 3-system of screws. This 3-system is not *special* in form; but it is unusual in not containing its own orthogonal basis screws, and in lacking an infinite number of other screws which might be expected to lie in a certain central 2-system.

To aid understanding of this practically important system, these peculiarities are examined. It is found that the unexpected absences of certain screws are localised, the "missing" screws occuring as limits, with zero magnitude, of spatial sequences of surrounding screws which do properly exist, and which populate the continuum of the 3-system.

1. Introduction

Consider a mechanism which contains revolute joints on two fixed but generally disposed lines S_1 and S_2 ; and that a body — it may be some link of the mechanism — is displaced by means, firstly, of a rotation through angle θ_1 about S_1 and then, secondly, of a rotation through angle θ_2 about S_2 . Under all possible variations of the angles θ_1 and θ_2 , applied in that sequence, the body suffers a variety of finite displacements which typically entail both translation and rotation. In fact it can be shown [1, 2] — the manner is outlined later — that the *finite displacement screws* which specify those displacements constitute a *real linear 3-system of screws* of general form, which contains the lines S_1 and S_2 as members.

This remark contains the seeds of some confusion: for one normally infers that such a 3-system will contain, *inter alia*, all linear combinations of the lines S_1 and S_2 ; that is, that an infinite number of these screws will exist to populate the contained 2-system in which S_1 and S_2 are generators, each such combination screw intersecting at right angles the common perpendicular line of S_1 and S_2 , which is the *nodal line* of the 2-system.

But, in fact, on evaluation and in practice, there are no such screws. As will be shown, apart from the screws of the individual revolute joints when either acts alone, the generally disposed revolute pair S_1 , S_2 cannot produce a displacement for which the screw axis — and the finite displacement screw upon it — perpendicularly intersects the common perpendicular line of S_1 and S_2 .

In resolution of this apparent discrepancy it is found that the entire infinity of screws in the 2-system of S_1 and S_2 does exist, but with a zero magnitude for every one of its

A. J. Lenarčič and B. B. Ravani (eds.), *Advances in Robot Kinematics and Computationed Geometry*, 401–410.

screws. Each of these screws occurs as the limit of a sequence of spatially adjacent screws of non-zero magnitude which do not quite intersect the nodal line. Thus, in review, the 3-system of finite displacement screws associated with a pair of revolute joints, though not *special* in the technical sense of Hunt [3, 4], is decidedly peculiar in the distribution of its screw magnitudes.

While, in essence, this finding simply illustrates a possible — but uncommon and often overlooked — configuration inherent in the 3-system of screws as a mathematical structure, it is brought to notice by the importance of the revolute-pair configuration in practical mechanism.

2. Specification of a Screw

We use familar conventions in writing the general *screw* as a 3-vector of dual numbers

$$\hat{\mathbf{S}} = \mathbf{S} + \varepsilon\, \mathbf{S}_p \,, \tag{2.1}$$

in which the real 3-vectors $\mathbf{S}$ and $\mathbf{S}_p$ are respectively the *direction* and *moment* parts of the screw, and ε is the *quasi-scalar* with the property $\varepsilon^2 = 0$. The *magnitude*, *pitch* and *origin radius* of the screw are

$$|\,\mathbf{S}\,| = \sqrt{\mathbf{S}^2}\,, \quad p = \frac{\mathbf{S}\cdot\mathbf{S}_p}{\mathbf{S}^2}\,, \quad \mathbf{R} = \frac{\mathbf{S}\times\mathbf{S}_p}{\mathbf{S}^2}\,, \tag{2.2}$$

respectively and, in terms of these, the screw may alternatively be written

$$\hat{\mathbf{S}} = |\,\mathbf{S}\,|\,(\,1 + \varepsilon\, p\,)\,\hat{\mathbf{s}}\,, \tag{2.3}$$

where $\hat{\mathbf{s}}$ is the *unit line*, of unit magnitude and zero pitch, of the screw. Whenever, throughout this paper, we refer to the *normalised* instance of a screw such as $\hat{\mathbf{S}}$, we shall mean the unit *line*, written with the corresponding lower case letter, like $\hat{\mathbf{s}}$; and we shall write $\mathbf{s}$ for the unit direction 3-vector of that line and of that screw.

Appeal is made to the following 3-vector properties: for any two screws

$$\hat{\mathbf{S}}_1 = |\,\mathbf{S}_1\,|\,(\,1 + \varepsilon\, p_1\,)\,\hat{\mathbf{s}}_1\,, \quad \hat{\mathbf{S}}_2 = |\,\mathbf{S}_2\,|\,(\,1 + \varepsilon\, p_2\,)\,\hat{\mathbf{s}}_2\,, \tag{2.4}$$

of respective pitches p_1 and p_2, their *cross product* screw $\hat{\mathbf{S}}_1 \times \hat{\mathbf{S}}_2$ is sited in the *common perpendicular line* of $\hat{\mathbf{S}}_1$ and $\hat{\mathbf{S}}_2$, and has pitch $p_1 + p_2 + d\, \cot\theta$ where d is the distance and θ is the angle from $\mathbf{S}_1$ to $\mathbf{S}_2$ as measured along and about that common perpendicular line. We shall say that two screws $\hat{\mathbf{S}}_1$ and $\hat{\mathbf{S}}_2$ are *orthogonal* if $\hat{\mathbf{S}}_1 \cdot \hat{\mathbf{S}}_2 = 0$, which implies that they intersect at right angles.

3. Chasles's Axis for a Finite Displacement

Consider a rigid body of general shape to undergo an arbitrary finite displacement between an initial and a final location in space. Then there is a directed line $\hat{\mathbf{s}}$ — Chasles's axis for the displacement [5] — such that the displacement could have been achieved by means of a translation parallel to that line and by a rotation about it. The line $\hat{\mathbf{s}}$ is unique except in the special case that the displacement is comprised by a pure translation, in which situation *any* line parallel with the translation will serve. It is a matter of experience that the axial motions of translation and rotation about the line $\hat{\mathbf{s}}$ may be applied in arbitrary and piecemeal sequence to achieve the given displacement; and that to any rotation which (with some associated translation) achieves the displacement we may add a further rotation of 2π radians to achieve the same displacement.

We follow Hunt [6] in defining the *cardinal motion* for the given displacement as those axial motion components which achieve the final location from the initial location by means of translation through a distance 2σ in the direction of $\hat{\mathbf{s}}$ and by rotation through an angle $2\theta, -\pi < 2\theta \leq \pi$, in a right-handed sense about that direction. The elements of this cardinal motion, namely the *axis line* $\hat{\mathbf{s}}$, the *half-translation* σ, and the *half-rotation* $\theta, -\frac{1}{2}\pi < \theta \leq \frac{1}{2}\pi$, are used to define the specifying screws of the displacement which are described later.

4. Sin-Screw Specification of a Finite Displacement

From the elements $\hat{\mathbf{s}}$, σ and $\theta, -\frac{1}{2}\pi < \theta \leq \frac{1}{2}\pi$, of the cardinal motion, we can readily construct a *finite displacement screw* $\hat{\mathbf{S}}$ which characterises the cardinal motion of the displacement [7]. Adopting the standard interpretation of a *dual angle*, namely

$$\hat{\theta} \equiv \theta + \varepsilon\,\sigma\,, \qquad \sin\hat{\theta} \equiv \sin\theta + \varepsilon\,\sigma\cos\theta\,, \qquad \cos\hat{\theta} \equiv \cos\theta - \varepsilon\,\sigma\sin\theta$$

we simply write

$$\hat{\mathbf{S}} \equiv \sin\hat{\theta}\;\hat{\mathbf{s}} = \sin\theta\,(\,1 + \varepsilon\,P_S\,)\;\hat{\mathbf{s}}\,, \qquad P_S \equiv \frac{\sigma}{\tan\theta}\,, \tag{4.1}$$

which is a directed screw of amplitude $\sin\theta$ and pitch P_S whose line is $\hat{\mathbf{s}}$. We shall refer to a such finite displacement screw $\hat{\mathbf{S}}$ as a *sin-screw*.

We note, for contrast and comparison, that with the same elements of the cardinal half-motion we may instead adopt the *dual tangent* function and, following Yang [8], write the finite displacement screw

$$\hat{\mathbf{T}} \equiv \tan\hat{\theta}\;\hat{\mathbf{s}} = \tan\theta\,(\,1 + \varepsilon\,P_T\,)\;\hat{\mathbf{s}}\,, \qquad P_T \equiv \sigma\,(\,\tan\theta + \frac{1}{\tan\theta}\,) = \frac{2\sigma}{\sin 2\theta}\,,$$

which we shall refer to as the *tan-screw* form of finite displacement screw.

5. Sin-Screw Formulation of the Screw Triangle

For a sequence of finite displacements applied to a rigid body we may represent both the sequence and the *resultant* screw of the aggregate displacement in the programmatic form

$$\hat{\mathbf{S}} \equiv \{\,\hat{\mathbf{S}}_1\,;\;\hat{\mathbf{S}}_2\,;\;\hat{\mathbf{S}}_3\,;\;\cdots\,\}\,.$$

The operations of such a sequence cannot, in general, be *commuted* without causing change to the resultant; but they may be freely *associated* in pairs. Following [7], we derive an expression for the sin-screw $\hat{\mathbf{S}}$ which is the resultant of applying such a pair of sin-screws, first $\hat{\mathbf{S}}_1 \equiv \sin\hat{\theta}_1\ \hat{\mathbf{s}}_1$ and then $\hat{\mathbf{S}}_2 \equiv \sin\hat{\theta}_2\ \hat{\mathbf{s}}_2$. Considered two at a time, each pair of the screws $\hat{\mathbf{S}}_1$, $\hat{\mathbf{S}}_2$ and $\hat{\mathbf{S}}$ has a common perpendicular line which we denote as follows: $\hat{\mathbf{j}}_1$ is the normalised common perpendicular line of $\hat{\mathbf{S}}$ and $\hat{\mathbf{S}}_1$ which is directed from $\hat{\mathbf{S}}$ to $\hat{\mathbf{S}}_1$; correspondingly, $\hat{\mathbf{j}}_2$ and $\hat{\mathbf{j}}_3$ are the normalised common perpendicular lines of the respective screw pairs $\hat{\mathbf{S}}_1$ and $\hat{\mathbf{S}}_2$, and $\hat{\mathbf{S}}_2$ and $\hat{\mathbf{S}}$.

We use without proof the geometry established for the *screw triangle* by Bottema and Roth [9]. Considering any one, say $\hat{\mathbf{S}}_1$, of the characterising screws, this shows that the distance and the angle between the lines $\hat{\mathbf{j}}_1$ and $\hat{\mathbf{j}}_2$, as measured along and about their common perpendicular on the screw $\hat{\mathbf{S}}_1$ as axis, are respectively equal to the half distance σ_1, and the half angle θ_1, of the displacement which is characterised by that screw. Corresponding remarks apply to the other screws. Thus we may write

$$\hat{\mathbf{S}}_1 = \hat{\mathbf{j}}_1 \times \hat{\mathbf{j}}_2\,, \quad \hat{\mathbf{S}}_2 = \hat{\mathbf{j}}_2 \times \hat{\mathbf{j}}_3\,, \quad \hat{\mathbf{S}} = \hat{\mathbf{j}}_1 \times \hat{\mathbf{j}}_3\,. \tag{5.1}$$

Since the lines $\hat{\mathbf{j}}_1$, $\hat{\mathbf{j}}_2$ and $\hat{\mathbf{j}}_3$ are normalised, the first two of these yield

$$\hat{\mathbf{j}}_1 = \hat{\mathbf{j}}_1 \cdot \hat{\mathbf{j}}_2\ \hat{\mathbf{j}}_2 + \hat{\mathbf{j}}_2 \times \hat{\mathbf{S}}_1 \quad \textit{and} \quad \hat{\mathbf{j}}_3 = \hat{\mathbf{j}}_2 \cdot \hat{\mathbf{j}}_3\ \hat{\mathbf{j}}_2 - \hat{\mathbf{j}}_2 \times \hat{\mathbf{S}}_2\,.$$

So by use of the third expression at (5.1) we obtain for the resultant screw:

$$\hat{\mathbf{S}} = -\hat{\mathbf{j}}_2 \cdot \hat{\mathbf{j}}_3\ \hat{\mathbf{j}}_2 \times (\hat{\mathbf{j}}_2 \times \hat{\mathbf{S}}_1) - \hat{\mathbf{j}}_1 \cdot \hat{\mathbf{j}}_2\ \hat{\mathbf{j}}_2 \times (\hat{\mathbf{j}}_2 \times \hat{\mathbf{S}}_2) - (\hat{\mathbf{j}}_2 \times \hat{\mathbf{S}}_1) \times (\hat{\mathbf{j}}_2 \times \hat{\mathbf{S}}_2)$$

which, since $\hat{\mathbf{j}}_2$ is normalised and parallel to the cross product $\hat{\mathbf{S}}_1 \times \hat{\mathbf{S}}_2$,

$$= \hat{\mathbf{j}}_2 \cdot \hat{\mathbf{j}}_3\ \hat{\mathbf{S}}_1 + \hat{\mathbf{j}}_1 \cdot \hat{\mathbf{j}}_2\ \hat{\mathbf{S}}_2 - \hat{\mathbf{S}}_1 \times \hat{\mathbf{S}}_2\ ;$$

so, on observing that $\hat{\mathbf{j}}_1 \cdot \hat{\mathbf{j}}_2 = \cos\hat{\theta}_1$ and $\hat{\mathbf{j}}_2 \cdot \hat{\mathbf{j}}_3 = \cos\hat{\theta}_2$, we derive

$$\hat{\mathbf{S}} \equiv \{\hat{\mathbf{S}}_1\,;\,\hat{\mathbf{S}}_2\} = \cos\hat{\theta}_2\ \hat{\mathbf{S}}_1 + \cos\hat{\theta}_1\ \hat{\mathbf{S}}_2 - \hat{\mathbf{S}}_1 \times \hat{\mathbf{S}}_2\,, \tag{5.2}$$

as the sin-screw formulation of the *screw triangle rule* for composition of finite displacements. Division by $\cos\hat{\theta}_1\ \cos\hat{\theta}_2$ followed by a little manipulation yields, for comparison, the tan-screw formulation of Yang [8], thus

$$\hat{\mathbf{T}} = \frac{\hat{\mathbf{T}}_1 + \hat{\mathbf{T}}_2 - \hat{\mathbf{T}}_1 \times \hat{\mathbf{T}}_2}{1 - \hat{\mathbf{T}}_1 \cdot \hat{\mathbf{T}}_2}\,.$$

6. Location Geometry of the Given Screws

For two finite displacements $\hat{\mathbf{S}}_1 \equiv \sin\hat{\theta}_1\ \hat{\mathbf{s}}_1$ and $\hat{\mathbf{S}}_2 \equiv \sin\hat{\theta}_2\ \hat{\mathbf{s}}_2$ applied in sequence, we now proceed to examine the spatial distribution of resultant screws described by Eq. (5.2). Given the large degree of variability available in this problem, we shall consider the lines of the given screws, $\hat{s}_1$ and $\hat{s}_2$, to be substantially fixed while variations are considered in the translations and rotations made about them, as comprised in the parameters θ_1 and θ_2. It will be convenient to recognise certain symmetries which derive from the location geometry of the given lines $\hat{s}_1$ and $\hat{s}_2$. Let $\hat{\phi}_{12} = \phi_{12} + \varepsilon\, d_{12}$ be half of the dual angle measured from the line $\hat{s}_1$ to the line $\hat{s}_2$. Then we may write

$$\hat{s}_1 + \hat{s}_2 = 2\cos\hat{\phi}_{12}\ \hat{s}_X\ , \quad \hat{s}_1 - \hat{s}_2 = 2\sin\hat{\phi}_{12}\ \hat{s}_Y\ , \quad \hat{s}_X \times \hat{s}_Y = \hat{s}_Z\ , \tag{6.1}$$

as definitions of the mutually orthogonal lines $\hat{s}_X$, $\hat{s}_Y$ and $\hat{s}_Z$ of the screws

$$\left.\begin{array}{lllll} \hat{S}_X & = & \cos\hat{\phi}_{12}\ \hat{s}_X & = & (\hat{s}_1 + \hat{s}_2)/2\ , \\ \hat{S}_Y & = & \sin\hat{\phi}_{12}\ \hat{s}_Y & = & (\hat{s}_1 - \hat{s}_2)/2\ , \\ \hat{S}_Z & = & \sin\hat{\phi}_{12}\ \cos\hat{\phi}_{12}\ \hat{s}_Z & = & \hat{S}_X \times \hat{S}_Y\ , \end{array}\right\} \tag{6.2}$$

which are axes of mirror-symmetry for the given lines. We shall refer later results to the reference frame whose xyz-axes are the mutually orthogonal lines $\hat{s}_X$, $\hat{s}_Y$ and $\hat{s}_Z$. These lines intersect in an origin point at the mid-point between the given screws S_1 and S_2 on their common perpendicular line $\hat{s}_Z$. We note the useful derived identity

$$-\hat{s}_1 \times \hat{s}_2 = \frac{(\hat{s}_1 + \hat{s}_2) \times (\hat{s}_1 - \hat{s}_2)}{2} = 2\sin\hat{\phi}_{12}\ \cos\hat{\phi}_{12}\ \hat{s}_Z = 2\,\hat{S}_Z\ . \tag{6.3}$$

7. A Dual 3-System, Observed as the Sum of Two Dual 2-Systems

Writing $\hat{S}_1 \equiv \sin\hat{\theta}_1\ \hat{s}_1$ and $\hat{S}_2 \equiv \sin\hat{\theta}_2\ \hat{s}_2$ we may, as in [2], re-express the screw triangle rule of Eq. (5.2) as

$$\begin{aligned} \hat{S} &= \sin\hat{\theta}_1\ \cos\hat{\theta}_2\ \hat{s}_1 + \sin\hat{\theta}_2\ \cos\hat{\theta}_1\ \hat{s}_2 - \sin\hat{\theta}_1\ \sin\hat{\theta}_2\ \hat{s}_1 \times \hat{s}_2 \\ &= \sin(\hat{\theta}_1 + \hat{\theta}_2)\ \frac{\hat{s}_1 + \hat{s}_2}{2} + \sin(\hat{\theta}_1 - \hat{\theta}_2)\ \frac{\hat{s}_1 - \hat{s}_2}{2} - \sin\hat{\theta}_1\ \sin\hat{\theta}_2\ \hat{s}_1 \times \hat{s}_2 \end{aligned}$$

which, through the definitions at Eqs. (6.2) and (6.3),

$$= \sin(\hat{\theta}_1 + \hat{\theta}_2)\ \hat{S}_X + \sin(\hat{\theta}_1 - \hat{\theta}_2)\ \hat{S}_Y + 2\sin\hat{\theta}_1\ \sin\hat{\theta}_2\ \hat{S}_Z\ , \tag{7.1}$$

which describes a *dual 3-system* whose orthogonal basis is the set of mirror-symmetry screws S_X, S_Y and S_Z. An alternative representation of this structure is obtained by defining new independent parameters

$$\hat{\psi}_y \equiv \hat{\theta}_1 + \hat{\theta}_2\ , \quad \hat{\psi}_x \equiv \hat{\theta}_1 - \hat{\theta}_2\ , \tag{7.2}$$

so that, on re-writing Eq. (7.1), we have

$$\hat{\mathbf{S}} = \sin\hat{\psi}_y\ \hat{\mathbf{S}}_X + \sin\hat{\psi}_x\ \hat{\mathbf{S}}_Y - (\cos\hat{\psi}_y - \cos\hat{\psi}_x)\ \hat{\mathbf{S}}_Z \tag{7.3}$$

which, on re-arrangement,

$$= (\sin\hat{\psi}_x\ \hat{\mathbf{S}}_Y + \cos\hat{\psi}_x\ \hat{\mathbf{S}}_Z) + (\sin\hat{\psi}_y\ \hat{\mathbf{S}}_X - \cos\hat{\psi}_y\ \hat{\mathbf{S}}_Z)\ , \tag{7.4}$$

in which each term has the form of a *dual 2-system*, combined by complementary dual sinusoids, whose structure has been studied in [7]. These dual 2-systems in the independent parameters $\hat{\psi}_x$ and $\hat{\psi}_y$, have orthogonal *nodal lines* on the axes $\hat{\mathbf{S}}_X$ and $\hat{\mathbf{S}}_Y$ respectively, and each finds its orthogonal basis in the remainder of the set $\mathbf{S}_X$, $\mathbf{S}_Y$ and $\mathbf{S}_Z$.

8. Discussion

As proposed in the introduction to this paper, let us now restrict attention to purely rotational displacements arising at *revolute* joints on the lines of the screws $\mathbf{S}_1$ and $\mathbf{S}_2$. We achieve specification of all available resultant displacement screws by setting the translation components of θ_1 and θ_2 to zero. The resulting expressions, obtained by "removing the hats" of the displacement angle parameters in Eqs. (7.1) and (7.3), are

$$\hat{\mathbf{S}} = \sin(\theta_1 + \theta_2)\ \hat{\mathbf{S}}_X + \sin(\theta_1 - \theta_2)\ \hat{\mathbf{S}}_Y + 2\sin\theta_1\ \sin\theta_2\ \hat{\mathbf{S}}_Z \tag{8.1}$$

or, alternatively,

$$= \sin\psi_y\ \hat{\mathbf{S}}_X + \sin\psi_x\ \hat{\mathbf{S}}_Y - (\cos\psi_y - \cos\psi_x)\ \hat{\mathbf{S}}_Z \tag{8.2}$$

which describe a *general real 3-system* whose orthogonal basis screws are $\hat{\mathbf{S}}_X$, $\hat{\mathbf{S}}_Y$ and $\hat{\mathbf{S}}_Z$.

When familiarising ourselves with this 3-system structure we expect to find, *inter alia*, an infinite number of screws which intersect the $\mathbf{S}_Z$-axis and lie parallel with the $\mathbf{S}_X\mathbf{S}_Y$-plane, and which constitute the 2-system in which $\mathbf{S}_1$ and $\mathbf{S}_2$ are generators. But when examined for such screws — in which the $\mathbf{S}_Z$-component is zero-valued — Eq. (8.2) yields only two solutions, namely

$$\cos\psi_x - \cos\psi_y = 0\ , \quad i.e.\quad \psi_x = \pm\,\psi_y\ .$$

These are equivalent, under the definitions (7.2), to the alternate solutions $\theta_1 = 0$ or $\theta_2 = 0$, in which only one or the other of the contributing revolutes is active. In terms of expectations of 3-system structure, it comes as a surprise that there are no other finite displacement screws which lie parallel with the plane of $\mathbf{S}_X$ and $\mathbf{S}_Y$ and, in particular, that these orthogonal basis screws are themselves excluded. (We can show, similarly, that no resultant screw is sited on the line of the basis screw $\mathbf{S}_Z$.)

These observations necessarily question our identification of the structure with a *general* 3-system, there being the possibility that it is some *special* 3-system [3, 4] or, indeed, some other more distantly related form. Put briefly, if a zero of screw magnitude — that is, the absence of a screw — is found to be restricted to an individual point of the ψ_x, ψ_y - parameter space and continuously related to magnitudes in the neighbourhood, we may retain the identification; however, if the absence applies throughout an extended volume of that space, and is therefore substantially structural, the nature of this particular specialisation of the structure should be identified and brought to notice. In the following Section 9 we perform an analysis of the typical screw $\hat{\mathbf{S}}$ so that in Section 10 we may determine in detail if these effects are local or extended.

9. Analysis of the Typical Displacement Screw

Consider the origin O to be placed at the point of intersection of the screws $\hat{\mathbf{S}}_X$, $\hat{\mathbf{S}}_Y$, and $\hat{\mathbf{S}}_Z$. Then the lines of those screws take the particularly simple forms $\hat{\mathbf{s}}_X = \mathbf{s}_X + \varepsilon\ 0$, $\hat{\mathbf{s}}_Y = \mathbf{s}_Y + \varepsilon\ 0$ and $\hat{\mathbf{s}}_Z = \mathbf{s}_Z + \varepsilon\ 0$ in which the moment components have vanished and — using definitions given at Eqs. (2.2) — we can reveal the characteristics of the screw $\hat{\mathbf{S}}$ in terms, simply, of the unit direction 3-vectors $\mathbf{s}_X$, $\mathbf{s}_Y$ and $\mathbf{s}_Z$. On substituting from Eq. (6.2) into Eq. (8.2) and separating $\hat{\mathbf{S}} = \mathbf{S} + \varepsilon\ \mathbf{S}_p$ into real and dual parts we obtain

$$\hat{\mathbf{S}} = \sin\psi_y\ \cos\phi_{12}\ \mathbf{s}_X + \sin\psi_x\ \sin\phi_{12}\ \mathbf{s}_Y - (\cos\psi_y - \cos\psi_x)\ \sin\phi_{12}\ \cos\phi_{12}\ \mathbf{s}_Z$$
$$-\varepsilon\ d_{12}\ \{\sin\psi_y\ \sin\phi_{12}\ \mathbf{s}_X - \sin\psi_x\ \cos\phi_{12}\ \mathbf{s}_Y + (\cos\psi_y - \cos\psi_x)\ \cos 2\phi_{12}\ \mathbf{s}_Z\}\ .$$

The Orientation of $\hat{\mathbf{S}}$.

The azimuth angle ϕ of $\hat{\mathbf{S}}$, as measured around the axis $\hat{\mathbf{s}}_Z$ from the axis $\hat{\mathbf{s}}_X$, is determined from the ratio of the $\hat{\mathbf{s}}_Y$- and $\hat{\mathbf{s}}_X$-components of the direction vector $\mathbf{S}$, thus

$$\frac{\sin\phi}{\cos\phi} = \frac{\sin\psi_x}{\sin\psi_y}\ \frac{\sin\phi_{12}}{\cos\phi_{12}}\ . \tag{9.1}$$

The Magnitude and Rotational Component of $\hat{\mathbf{S}}$.

Observing in Eq. (4.1) that the direction 3-vector $\mathbf{S}$ of the resultant $\hat{\mathbf{S}}$ has amplitude $\sin\theta$, in which θ is the angular component of the resultant displacement, we achieve the double purpose of isolating this component and determining a normalising factor for other measures by evaluating

$$\mathbf{S}^2 \equiv \sin^2\theta = \sin^2\psi_y\ \cos^2\phi_{12} + \sin^2\psi_x\ \sin^2\phi_{12} + (\cos\psi_y - \cos\psi_x)^2\ \sin^2\phi_{12}\ \cos^2\phi_{12}$$
$$= 1 - (\cos\psi_y\ \cos^2\phi_{12} + \cos\psi_x\ \sin^2\phi_{12})^2\ . \tag{9.2}$$

On taking square roots and choosing signs such that $\cos\theta \to \cos\psi_y$ as $\cos\psi_y \to \cos\psi_x$ (or, equivalently, such that $\theta \to \theta_1$ *or* θ_2 as θ_2 *or* $\theta_1 \to 0$ respectively), we find that

$$\cos\theta = \cos\psi_y\ \cos^2\phi_{12} + \cos\psi_x\ \sin^2\phi_{12}\ . \tag{9.3}$$

The Origin Radius $\mathbf{R}$ *of* $\hat{\mathbf{S}}$.

For the origin radius $\mathbf{R}$ we derive, on expansion and rearrangement:

$$\mathbf{S}^2\ \mathbf{R} = \mathbf{S} \times \mathbf{S}_p = d_{12}\ (\cos\psi_y - \cos\psi_x)\ (\sin\psi_x\ \sin^3\phi_{12}\ \mathbf{s}_X + \sin\psi_y\ \cos^3\phi_{12}\ \mathbf{s}_Y)$$
$$+ d_{12}\ \sin\psi_x\ \sin\psi_y\ \mathbf{s}_Z\ . \tag{9.4}$$

The Pitch P *of* $\hat{\mathbf{S}}$.

We evaluate $\mathbf{S}^2 \; P \equiv \sin^2\theta \; \dfrac{\sigma}{\tan\theta}$ (by Eq. 4.1) which yields, on rearrangement:

$$\mathbf{S}^2 \; P \;=\; \mathbf{S}\cdot\mathbf{S}_p \;=\; d_{12}\;(\cos\psi_y - \cos\psi_x)\;(\cos\psi_y\;\cos^2\phi_{12} + \cos\psi_x\;\sin^2\phi_{12})\;\sin 2\phi_{12}$$

$$= d_{12}\;\cos\theta\;(\cos\psi_y - \cos\psi_x)\;\sin 2\phi_{12}\;, \qquad (9.5)$$

by Eq. (9.3). It follows that, for the translational component of the resultant displacement,

$$\sigma \;=\; d_{12}\;\sin 2\phi_{12}\;\frac{\cos\psi_y - \cos\psi_x}{\sin\theta}\;. \qquad (9.6)$$

10. Limit Screws of Zero Magnitude

It is sufficient for our purpose to examine the resultant screw $\hat{\mathbf{S}}$ in the limit situation $\psi_x \to 0$ and $\psi_y \to 0$. Applying appropriate approximations, namely

$$\sin\psi_x \;=\; \psi_x - \cdots\;, \qquad \cos\psi_x \;=\; 1 - \frac{\psi_x^{\;2}}{2} + \cdots\;, \qquad \sin\psi_y \;=\; \psi_y - \cdots\;, \qquad etc.\;,$$

we proceed to evaluate key values for the resultant screw $\hat{\mathbf{S}}$; thus, by Eq. (9.3),

$$\cos\theta \;=\; 1 \;-\; \frac{\psi_y^{\;2}}{2}\;\cos^2\phi_{12} \;-\; \frac{\psi_x^{\;2}}{2}\;\sin^2\phi_{12} \;+\; \cdots \qquad (10.1)$$

so that, by Eq. (9.2), the squared magnitude of $\hat{\mathbf{S}}$, which passes through diminishing values to vanish in the limit, is given by

$$\mathbf{S}^2 \;\equiv\; \sin^2\theta \;=\; \psi_y^{\;2}\;\cos^2\phi_{12} \;+\; \psi_x^{\;2}\;\sin^2\phi_{12} \;-\; \cdots\;. \qquad (10.2)$$

Now let us additionally constrain the variables ψ_x and ψ_y in such a way that, while $\psi_x \to 0$ and $\psi_y \to 0$, the ratio $\psi_y / \psi_x = t$ maintains a constant value t, $-\infty \le t \le \infty$. We observe that, in doing so, we constrain the azimuth angle ϕ of the screw $\hat{\mathbf{S}}$ which, on adapting Eq. (9.1) to the limit situation, is given by

$$\frac{\sin\phi}{\cos\phi} \;=\; \frac{\psi_x}{\psi_y}\;\frac{\sin\phi_{12}}{\cos\phi_{12}} \;=\; \frac{1}{t}\;\frac{\sin\phi_{12}}{\cos\phi_{12}}\;, \quad i.e. \quad t \;=\; \frac{\sin\phi_{12}}{\cos\phi_{12}}\;\frac{\cos\phi}{\sin\phi}\;. \qquad (10.3)$$

The Origin Radius $\mathbf{R}$ *of the Limit Screw* $\hat{\mathbf{S}}$.

Making use of Eq. (10.2) and ignoring higher order terms, we adapt Eq. (9.4) to

$$\psi_x^{\;2}\;\{\,1 \;-\; \cos^2\phi_{12}\;(1 - t^2)\,\}\;\mathbf{R}$$

$$= \psi_x^{\;3}\;\frac{d_{12}}{2}\;(1 - t^2)\;(\sin^3\phi_{12}\;\mathbf{s}_X + t\;\cos^3\phi_{12}\;\mathbf{s}_Y) \;+\; \psi_x^{\;2}\;d_{12}\;t\;\mathbf{s}_Z\;.$$

Thus, in the limit as $\psi_x \rightarrow 0$ and $\psi_y \rightarrow 0$, the terms providing s_X and s_Y vanish, and

$$\mathbf{R} = d_{12} \frac{t}{1 - \cos^2\phi_{12}\,(1-t^2)}\, s_Z = d_{12} \frac{1/t}{1 - \sin^2\phi_{12}\,(1-1/t^2)}\, s_Z \quad (10.4)$$

This shows that the limit screw $\hat{\mathbf{S}}$, already observed to be of vanishing magnitude, is always parallel to the $\hat{s}_X\hat{s}_Y$-plane and intersects the $\hat{s}_Z$-axis. Substitution from Eq. (10.3) yields

$$\mathbf{R} = \frac{d_{12}}{\sin 2\phi_{12}} \sin 2\phi\; s_Z\,, \quad (10.5)$$

which in its form shows the limit screw $\hat{\mathbf{S}}$ to be a generator of a *cylindroid* which is symmetrically disposed about the $\hat{s}_X\hat{s}_Y$-plane with the $\hat{s}_Z$-axis as nodal line. The origin-distance of $\mathbf{S}$ obviously adopts the values

$$\mathbf{R} = \pm d_{12}\, s_Z \quad for \quad \sin 2\phi = \pm \sin 2\phi_{12}\,, \quad (10.6)$$

which shows that the limit screw occupies the position of the revolute screw, either $\hat{\mathbf{S}}_1$ or $\mathbf{S}_2$, for those respective values. The revolute screws are, therefore, particular generators of the cylindroid. From the association

$$\mathbf{R} = \pm 0 \quad for \quad \sin 2\phi = \pm 0\,, \quad (10.7)$$

it is equally clear that the axial lines $\hat{s}_X$ and $\hat{s}_Y$ are locations adopted by the limit screw; they are also, therefore, generators of the cylindroid.

The Pitch P of the Limit Screw $\hat{\mathbf{S}}$.

In similar terms we adapt Eq. (9.5) to provide

$$P = \frac{d_{12}}{2} \sin 2\phi_{12} \frac{\psi_x^2\,(1-t^2)}{\psi_x^2\,(\sin^2\phi_{12} + t^2\cos^2\phi_{12})} - \frac{d_{12}}{4} \sin 2\phi_{12}\, \psi_x^2\,(1-t^2)\,.$$

So, from Eq. (10.3), on proceeding to the limit and simplifying,

$$P = \frac{d_{12}}{\sin 2\phi_{12}} (\cos 2\phi_{12} - \cos 2\phi)\,, \quad (10.8)$$

which we recognise, in its form, to be the pitch distribution of a 2-system which can exist on the cylindroid identified above. The pitch of a revolute generator, $\mathbf{S}_1$ or $\mathbf{S}_2$, where $\phi = \pm\phi_{12}$ respectively, is clearly zero as is appropriate; and for the generators $\hat{s}_x$ and $\hat{s}_y$, for which $\phi = 0$ or $\phi = \frac{1}{2}\pi$, is respectively

$$P = \frac{d_{12}}{\sin 2\phi_{12}} (\cos 2\phi_{12} - \pm 1) = \frac{-d_{12}}{\cot\phi_{12}} \quad or \quad = \frac{d_{12}}{\tan\phi_{12}}\,,$$

which are respectively the pitches of the screws $\hat{\mathbf{S}}_x$ or $\hat{\mathbf{S}}_y$ themselves.

11. Conclusion

Thus we see that the limit screws $\hat{\mathbf{S}}$ for $\psi_x \to 0$ and $\psi_y \to 0$, while being themselves of zero magnitude, occur in sequences of screws in the continuum of the 3-system which are not of zero magnitude. The region of their absence is not spatially extended in the ψ_x, ψ_y -parameter space but, for all values of the parameter t, is located in the origin point of that space.

We have seen that, under variation of the parameter t, the limit screw $\hat{\mathbf{S}}$ generates a 2-system of screws, all of which have zero magnitude. Since it contains the given revolute screws $\mathbf{S}_1$ and $\mathbf{S}_2$ as zero-pitched generators, we recognise this as the 2-system whose absence from the 3-system was first remarked upon in the introductory section. This 2-system has the common perpendicular of $\mathbf{S}_1$ and $\mathbf{S}_2$, that is, the $\hat{\mathbf{s}}_Z$-axis, as nodal line; and it contains the orthogonal axes $\hat{\mathbf{s}}_X$ and $\hat{\mathbf{s}}_Y$, both with zero magnitude, as generators.

12. Acknowledgements

The author acknowledges the assistance of the Basser Department of Computer Science, University of Sydney, in providing facilities used in preparation of this paper.

13. References

[1] C. Huang, "On the finite screw system of the third order associated with a revolute-revolute chain", (submitted to *Trans. ASME* (*J. Mech. Design*), Jan. 1993) Private communication, March 1993.

[2] I.A. Parkin, "Dual systems of finite displacement screws in the screw triangle". *Technical Report TR 470*, Basser Dept. of Computer Science, University of Sydney, Jun (1993), (submitted to *Mechanism and Machine Theory*, Jul. 1993).

[3] K.H. Hunt, *Kinematic geometry of mechanisms*, Clarendon Press, Oxford (1990).

[4] C.G. Gibson and K.H. Hunt, "Geometry of screw systems - 2. Classification of screw systems", *Mechanism and Machine Theory*, *25*, 11-27 (1990).

[5] M. Chasles, "Note sur les proprietes ... et sur le deplacement fini ou infiniment petit d'un corps solide libre", *Correspondance mathematique et physique* (A. Quetelet, Ed.), 352-357 (1832).

[6] K.H. Hunt, "Manipulating a body through a finite displacement". *Proc. Seventh World Congress on the Theory of Machines and Mechanisms*, 187-191, Sevilla, Spain, (1987).

[7] I.A. Parkin, "Composition and additive combination of finite displacement screws - Part 1", *Technical Report TR 454*, Basser Dept. of Computer Science, University of Sydney, Dec (1992), (submitted to *Mechanism and Machine Theory*, Mar. 1993).

[8] A.T. Yang, "Calculus of screws", *Basic Questions of Design Theory* (W.R. Spillers, Ed.), Elsevier, New York, pp. 265-281 (1974).

[9] O. Bottema and B. Roth, *Theoretical kinematics*, Dover Publications, New York (1990).

True Straight-line Linkages Having a Rectilinear Translating Bar

E.A. Dijksman

Precision Engineering
Faculty of Mechanical Engineering
Eindhoven University of Technology
The Netherlands

Summary - Eight- and ten-bar straight-line mechanisms with a rectilinear translating bar, have been derived from two basic types, namely from Watt's - and from Stephenson's type of Hart's true straight-line mechanism(s). Though the generalization of Watt's type, representing Hart's inversor, was easily done, the other one, namely Stephenson's type, representing Hart's 2nd straight-line mechanism, required a new design circumventing Burmester's design of focal linkages. In the end spectacular results emerged based on a new but easy going design of these true straight-line linkages. Because of the simplification obtained here, the applicability of these "lifting devices" now comes within easy reach of the designer.

1. Introduction

Planar linkage mechanisms producing an exact straight-line are very rare. The more so when the elementary pairs contained in the mechanism, are restricted to revolute pairs (i.e.turning-joints) only. Four-bar linkages for instance, do not produce a straight-line at all. But, **six**-bar linkages may produce them, provided the dimensions are specifically chosen. It appears that principally only *two* solutions exist. Both have been found by H.**Hart**, ref.[1], and are known as respectively, the *first* and the *second* straight-line mechanism of Hart. The first one is a six-bar of **Watt**'s type, whereas the second one may be recognized as a six-bar of type **Stephenson**, ref.[2].

The first one represents the *inversor* mechanism (figure **1**), while the 2nd one represents a special case of **Kempe**'s *focal* mechanism (figures **2** and **9**), ref.[3,4].

Hart's inversor has further been generalized, which lead to the so-called *quadruplane inversor* of **Sylvester** and **Kempe** (figure **3**). The result though, may still be recognized as a six-bar linkage mechanism of Watt's form.(ref.[5])

The generalization of the focal type, however, does lead to an *eight*-bar but then containing a rectilinear translating bar. (figure **4**, see also figure 10 of ref.[4])

A. J. Lenarčič and B. B. Ravani (eds.), Advances in Robot Kinematics and Computationed Geometry, 411–420.

Eight-bar linkages with such a rectilinear translating bar, may also be obtained from a design based on Harts' inversor. (figure **5**, see also figure 10 of ref.[6].)

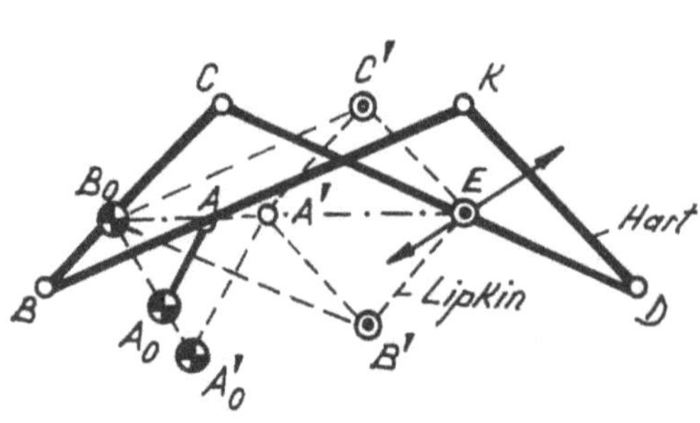

1st straight-line mechanism of Hart mechanically interconnected with the *Peaucellier-Lipkin inversor* of 1864.
Figure **1**

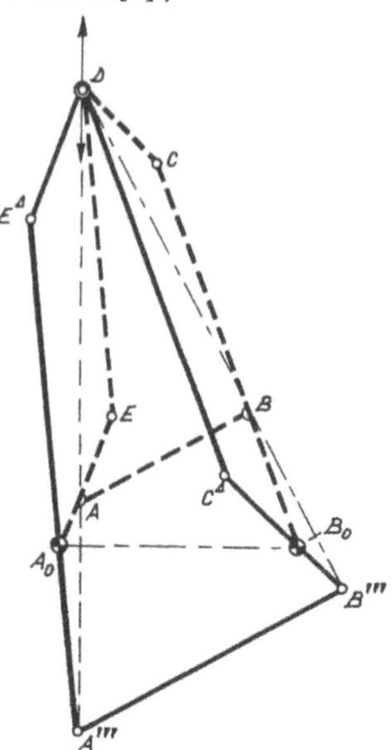

Two *curve-cognates* each representing the *2nd straight-line mechanism of Hart*(1877)
Figure **2**

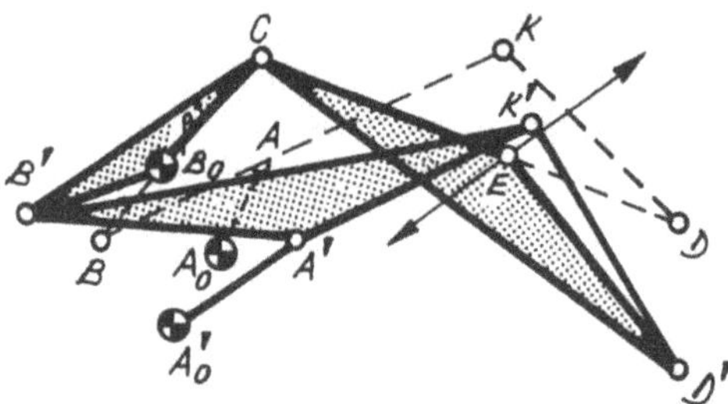

Quadruplane inversor of Sylvester and Kempe (a generalization of Hart's inversor)
Figure **3**

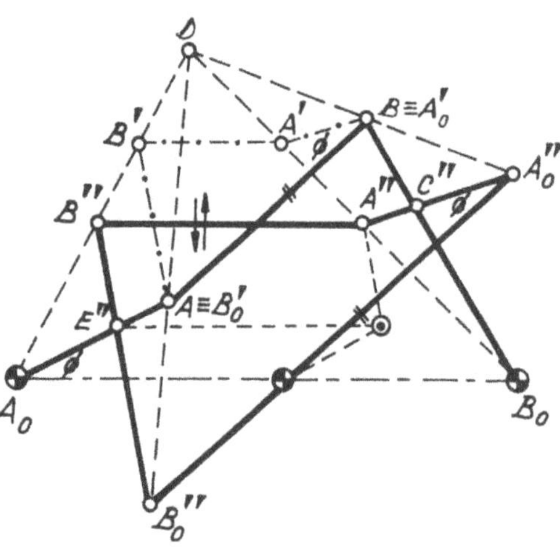

8-bar linkage mechanism with bar A''B'' moving perpendicular to the frame
Figure **4**

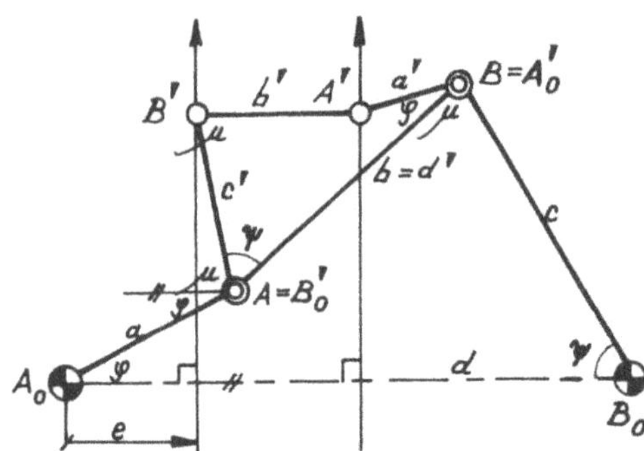

A random 4-bar and a reflected similar one, built on top Figure **7**

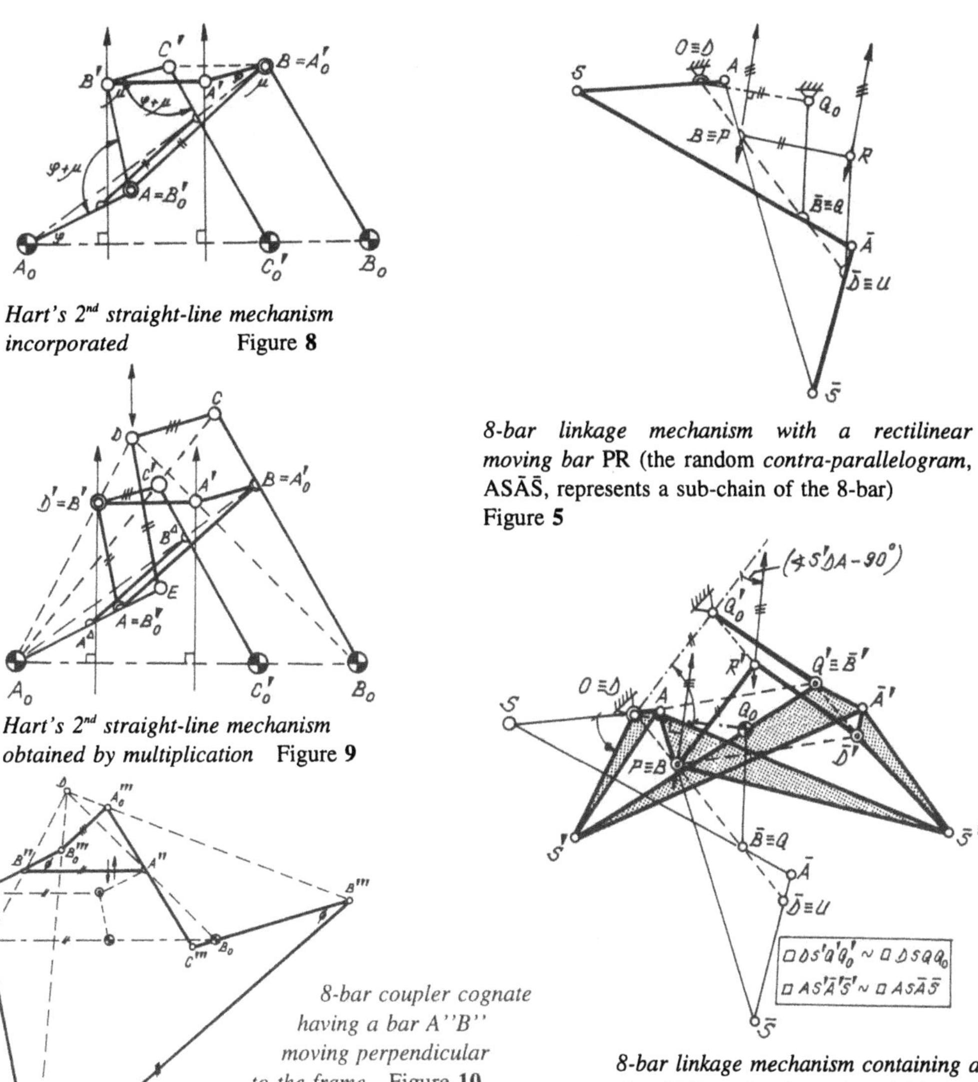

Hart's 2nd straight-line mechanism incorporated Figure **8**

8-bar linkage mechanism with a rectilinear moving bar PR (the random *contra-parallelogram*, ASĀS̄, represents a sub-chain of the 8-bar) Figure **5**

Hart's 2nd straight-line mechanism obtained by multiplication Figure **9**

8-bar coupler cognate having a bar A''B'' moving perpendicular to the frame Figure **10**

8-bar linkage mechanism containing a bar PR' moving in an invariable but oblique direction with the frame Figure **6**

2. Eight-bar Inversor with rectilinear translating bar

In practice, only one equilateral dyad-linkage is needed to adjoin Hart's inversor in order to obtain an 8-bar with a translating link moving perpendicular to the frame, (figure **5**). The mechanism contains a contra-parallelogram and two identical dyad chains, remaining parallel during the motion.
The same procedure may be applied at the quadruplane inversor of Sylvester and Kempe.

Then, the equilateral dyad PR'$\bar{D}$' adjoined to the 6-bar, remains parallel to the identical dyad OQ_0'Q' giving a rectilinear translating bar PR' moving straight with respect to the fixed link OQ_0'. (figure **6**) The angle enclosed between the fixed link and the direction of the straight-lines, traced by points of PR' , then differs from 90^0 and equals
($\measuredangle$ S'DA - 90^0).
We conclude that this kind of generalization practically results into a change of direction of the straight motion of the translating bar. (See also figure 27 of ref.[7].)

3. Six-, eight- and ten-bar linkages derived from the focal type

Hart's 2nd straight-line mechanism, which is a six-bar of type Stephenson, is to be generalized in the way Kempe did with his focal linkage.(ref.[8] and [4])
However, it is much easier to follow a different road: Doing that, one obtains Hart's 2nd straight-line mechanism as a besides. In order to design the mechanism as successively drawn in the figures **7** and **8**,

1. one starts with a completely *random* four-bar A_0ABB_0 ,
2. then, one adjoins the chain A_0'A'B'B_0' in such a way that the 4-bar A_0'A'B'B_0' is *reflected similar* to the initial four-bar A_0ABB_0 , whereas
   $A_0' \equiv B$ and $B_0' \equiv A$.
3. One further adjoins the linkage-dyad B'C'B in such a way that B'C'BA' forms a linkage parallelogram.
4. One finally adjoins the bar C'C_0' , such that B_0BC'C_0' too forms a linkage parallelogram, of which B_0C_0' represents the frame-link.

The result is a **10-bar** linkage mechanism of which the bar A'B' moves perpendicular to the frame.

Considering the stage in which the three bars C'B , C'B' and C'C_0' are not yet installed and assuming that the 4-bar A_0'A'B'B_0', built on top of the random 4-bar A_0ABB_0 , *remains* reflected similar with the latter, it is still easy to prove that A'B' *remains parallel* to the frame A_0B_0:

$$\measuredangle\,(\mathbf{B'A'},\mathbf{A_0B_0}) \equiv \measuredangle\, AA_0B_0 - \measuredangle\, A_0AB' + \measuredangle\, A'B'B_0' \equiv$$

$$\equiv \measuredangle\, AA_0B_0 - \measuredangle\, A_0AB + \measuredangle\, B'B_0'A_0' + \measuredangle\, A'B'B_0' \equiv$$

$$\equiv \measuredangle\, AA_0B_0 - \measuredangle\, A_0AB + \measuredangle\, A_0B_0B + \measuredangle\, B_0BA \equiv 0^0 , \qquad (1)$$

practically using the fact that the four angles of the random four-bar are all to be indicated at the turning-joint $A \equiv B_0$'. (The sum of them being 360^0.) Thus, if we draw a horizontal passing through A, the bar B_0'B' encloses the same angle μ with the horizontal as well as

with the bar A'B'.
Further note that because of the reflected similarity the bar AB encloses the same angle with the fixed link as with the bar A'B'. (Actually, this remark represents the shortest proof for the parallel motion of A'B')
The reflected similarity of the four-bars may be enforced in different ways. One of them is shown in figure **8** through the adjoining of the three bars having C' as their common joint. Then, the parallelism of the bar A'B' is enforced. But other methods are allowed too. (Think, for instance of the later to be explained possibility to replace the bar C'B by a bar interconnecting A_0A and $C_0'C'$, simultaneously avoiding the appearance of linkage parallelograms.)
Notwithstanding the shown parallel motion, it still necessary to prove that a singular point of A'B' moves perpendicular to the frame. Choosing the point B', it suffices to demonstrate that the dyad-chain A_0AB' represents an eccentric crank-and-slider mechanism. Whence, it remains to prove that the *eccentricity* "*e*" of that mechanism shows to be a constant.
Clearly, the distance *e* between the fixed center A_0 and the perpendicular let down from B' on to the fixed link may be determined by projection of the dyad A_0AB' on to the frame:

$$e = a.\cos\varphi - c'.\cos\mu = a.\cos\varphi - (bc/d)\cos\mu \qquad (2)$$

in which $\varphi = \sphericalangle AA_0B_0$ and $\mu = \sphericalangle B_0BA$ (3)
(thus, φ and μ are opposite angles in the initial four-bar)
Further note, that the subsequent dimensions of the reflected similar four-bar are respectively: $a' = ab/d$; $b' = b^2/d$; $c' = bc/d$ and $d' = b$, (4)
the initial dimensions being a, b, c and d respectively.
The *Rule of Cosine* twice applied in the random four-bar (abcd) on top of which the reflected similar four-bar (a'b'c'd') has been built as shown in figure **7**, gives

$$a^2 + d^2 - 2ad.\cos\varphi = \overline{AB_0}^2 = b^2 + c^2 - 2bc.\cos\mu \qquad (5)$$

Whence after substitution apparently

$$e = \tfrac{1}{2}(a^2 + d^2 - b^2 - c^2)/d = \text{constant}. \qquad (6)$$

Thus, since the eccentricity remains a constant, point B' has to trace a straight-line running normal to the frame, that is presuming the reflected similar four-bar on top of the initial one, *remains* reflected similar. So, all in all we proved that the motion of the bar B'A' is in fact a *rectilinear translation* in the mentioned direction.

At this point, it is easily seen that $\sphericalangle A_0AB' \equiv \sphericalangle C_0'C'B'$,whence the five-bar $A_0AB'C'C_0'$ alone represents a constrained linkage-pentagon. It is further possible to find a singular bar interconnecting the sides A_0A and $C_0'C'$ of that pentagon. To obtain that bar, we observe the pantograph $B_0BC'C_0'$, simultaneously taking A_0A as the fixed link. Point B of that pantograph then traces a circle about A with respect to A_0A, whence equally the intersection (A_0B x $C_0'C'$) of A_0B and $C_0'C'$ traces a circle about the intersection of A_0A and a line running parallel to AB but meeting the point (A_0B x $C_0'C'$). The resulting six-bar containing the pentagon and the adjoined bar appears to be Hart's 2nd straight-line linkage mechanism since among others B' traces a straight-line.

Multiplying the six-bar with the factor $\overline{A_0D}/\overline{A_0B'}$ in which D = A_0B' x B_0A' then results into an enlarged Hart's 2nd straight-line mechanism, also shown in figure **9**. This mechanism then consists of the pentagonal linkage A_0EDCB_0 and the bar AB. Naturally, turning-joint D of the pentagon similarly traces a straight-line like B' did in the initial six-bar.
The six-bar obtained this way, is to be set up by the initial four-bar A_0ABB_0 and the additional linkage-dyad CDE. (Note that *any* four-bar contains such a unique point D. The point plays a center-rôle in further generalization.)

It is quite possible to turn the 10-bar into an 8-bar linkage. To carry this out, one multiplies the reflected four-bar $A_0'A'B'B_0'$ geometrically from Hart's dyad-joint D. (figure **4**) The multiplication-factor is then decided by the choice of, for instance, B" at A_0D. Whence the factor equals **DB"/DB'**.
Thus, □ A_0"A"B"B_0" = **(DB"/DB')**.□ $A_0'A'B'B_0'$, (7)
Note that corresponding vertices of these two four-bars remain at the same ray joining the ray-center D.
The initial four-bar (A_0ABB_0) and the obtained one (A_0"A"B"B_0") are further interlinked at the common turning-joints E" and C", being the intersections of corresponding sides. Thus, E" = A_0A x B"B_0" and C" = B_0B x A_0"A". (The connectivity at E" for example is due to the fact that the eccentric slider-crank A_0AB' has been turned into the similar one A_0E"B" through the similarity center A_0 of these sub-chains.) In fact, the former common joint A has now moved to E" while former common joint B moved to C".
Clearly, the bar A"B" moves perpendicular to the frame. Now, the parallel motion of A"B" doesn't have to be sustained by linkage parallelograms, needed earlier for the motion of A'B'. Thus, an eight-bar is obtained with a rectilinear motion of A"B" moving normal to the frame.
However, in case the rectilinear motion is badly transmitted - as may be the case sometimes - one still has the possibility to adjoin linkage parallelograms to sustain the motion. To avoid the then appearing overconstrainedness, one may simultaneously omit for example the superfluous bar A_0"B_0". The result is a 10-bar linkage to be recognized in figure **4**.

4. The cognate 8-bar

Hart's 2nd straight-line mechanism occurs in pairs. That is to say, for each mechanism of this type, a *curve* cognate exists producing the same part of a straight-line. (figure **2** and also figure 9 of ref.[2].)
Both cognates may further be generalized in a way as demonstrated in the figures **4** and **10**. However, it is quite possible to let them also have the **same** (rectilinear) moving bar A"B". To attain this, one takes for both generalizations the same point B" at the ray A_0D (and/or the same point A" at DB_0.). The two generalizations then appear to be each others' *coupler* cognate producing the identical (rectilinear) motion for the common bar A"B". Figure **11** demonstrates the two of them in one figure, whereas figures **4** and **10** show them separately.
So, for each point B" at A_0D, **two** 8-bar coupler cognates exist, producing the same straight motion perpendicular to the fixed link they have in common.

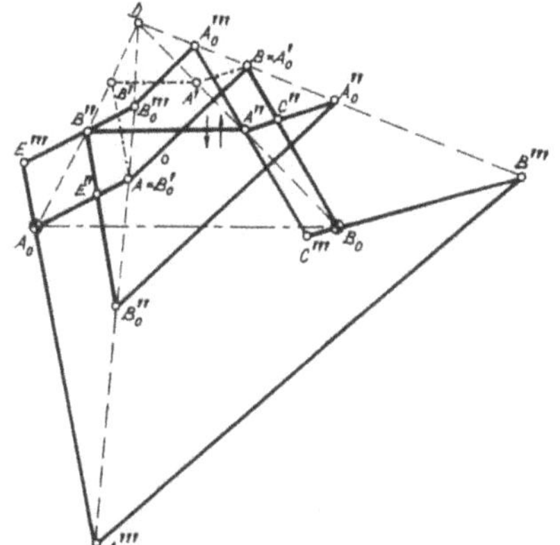

Two 8-bar coupler cognates having a common bar A''B'' moving perpendicular to the common frame A_0B_0 Figure **11**

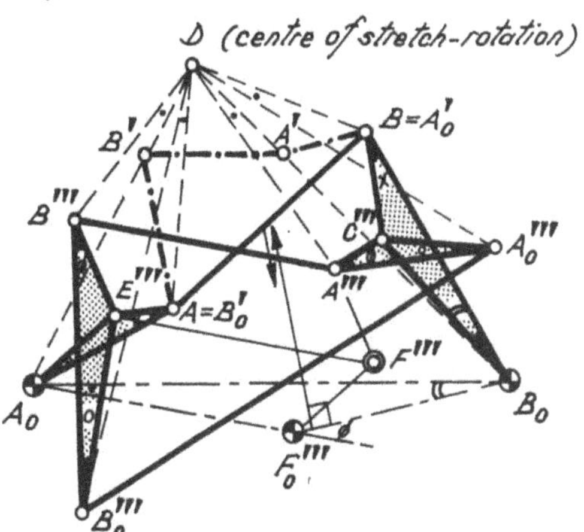

random choice for E''' (or C''')
Generalized 8-bar with link A'''B''' moving normal to B_0F_0''' Figure **12**

5. Special Cases

All 6-, 8- and 10-bar straight-line mechanisms of the focal type, have the advantage of a *random* choice for the dimensions of the initial four-bar A_0ABB_0 contained in them. Naturally, particular examples are found by taking specific four-bars to start the design with. If, for instance, the initial four-bar resembles a *deltoid*- linkage, possessing two pairs of adjacent equal sides, the well-known 10-bar linkage of A.B.**Kempe** is obtained. (ref.[7]). Note that in this case AB resembles the angle-bisector of the angle A_0AB'.
In case the initial four-bar A_0ABB_0 represents a contra-parallelogram, the adjoined four-bar

$A_0'A'B'B_0'$ coincides with the initial one. So, the design collapses. Section 2 then shows how to proceed in that case.

6. Generalization through stretch-rotation

In figure **4** point E" joined the input-crank A_0A. It is still possible, however, to generalize the design by a completely random choice of this point. See, for instance figure **12** in which we took the point E''' as a random point attached to the input-crank A_0A. The corresponding design of the generalized 8-bar straight-line mechanism having a rectilinear moving bar would then read something like:

a. Start with the *random* choice of the *4-bar* A_0ABB_0,
b. Make the four-bar □ BA'B'A reflected similar with the initial 4-bar A_0ABB_0,
c. Determine the intersection point D = A_0B'x B_0A',
d. choose a *random point* E''' in the moving plane of the input-crank A_0A,
e. Stretch-rotate A_0AB' about A_0 into $A_0E'''B'''$ determining point B''',
f. Stretch-rotate □ BA'B'A about D into the 4-bar $A_0'''A'''B'''B_0'''$,
g. Form the similar and rigid triangles B_0BC''' and A_0AE''',
h. Finally, form the similar and rigid triangles $B_0'''B'''E'''$ and $A_0'''A'''C'''$,
i. Note, that ∡ $B_0'''E'''B'''$ = ∡ $AE'''A_0$ = ∡ $A_0'''C'''A'''$ = ∡ $BC'''B_0$,

The choice of E''' has the advantage of choosing the motion-direction of the translating bar A'''B'''. This bar moves in a fixed direction under an angle of ($\pi/2$ + ∡ $E'''A_0A$) with A_0B_0. The length of the straight-line is governed by the utmost positions of the eccentric slider-crank $A_0E'''B'''$. Whence, the *length* L of the straight-line, traced by the point B''', is determined by the values of $(\overline{A_0E'''} + \overline{E'''B'''})$, $|\overline{A_0E'''} - \overline{E'''B'''}|$ and by the eccentricity *e'*

In order to sustain the rectilinear motion it is allowed to adjoin the three bars A'''F''', E'''F''' and $F_0'''F'''$ such that E'''B'''A'''F''' as well as $A_0E'''F'''F_0'''$ form linkage-parallelograms. After omitting either one of the bars AB or $A_0'''B_0'''$, one obtains a ten-bar linkage-mechanism still containing a bar always moving perpendicular to a fixed line ($F_0'''B_0$) in the frame, as proved hereafter:

Since for the 5-bar $F_0'''F'''A'''C'''B_0$, the ∡ $F_0'''F'''A'''$ ≡ ∡ $B_0C'''A'''$, the rectilinear motion of A''' will be perpendicular to $F_0'''B_0$, the frame of the 5-bar. (See Wunderlich's proof with isotropic coordinates in ref.[9])

Further, as the 5-bar $F_0'''F'''A'''C'''B_0$ represents part of the focal mechanism, it is possible to find a bar interlinking the cranks $F_0'''F'''$ and B_0C'''. This bar too may replace the then superfluous bars AB and $A_0'''B_0'''$, again creating another 10-bar linkage with a rectilinear moving bar. The mechanism so obtained then represents Hart's 2nd straight-line mechanism adjoined with two linkage parallelograms, $A_0E'''F'''F_0'''$ and E'''B'''A'''F''', in order to produce the parallel motion for the rectilinear moving bar.

Thus, from the 8-bar, many alternative 10-bar linkages are to be derived producing the same rectilinear motion with sometimes better transmission.
The designer may choose which one he prefers.

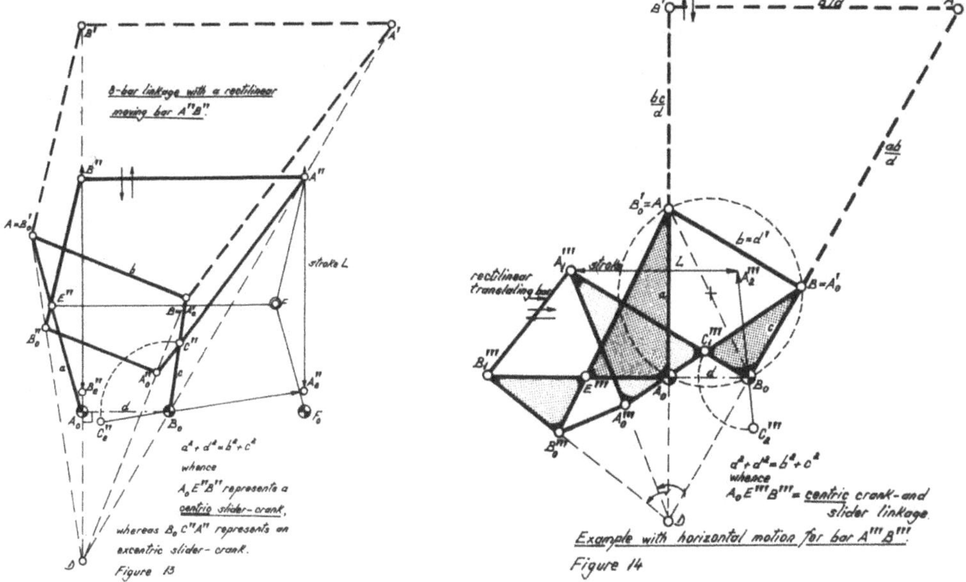

Figure 13

Figure 14

7. Zero eccentricity

In the particular case for which B''' traces a straight-line passing through A_0, the eccentricity e' of the slider-crank $A_0E'''B'''$ equals zero, whence e = 0 and so
$a^2 + d^2 = b^2 + c^2$, since

$$e' = (\overline{A_0E'''}/\overline{A_0A}).e = \overline{A_0E'''}\,(a^2+d^2-b^2-c^2)/(2ad) \qquad (8)$$

(See figure **13** for vertical translation and figure **14** for a horizontal one.)
Then, the initial four-bar has equal transmission angles ($\mu_1 = \mu_2$) in the two aligned positions of the crank A_0A with the frame A_0B_0.
And, since e = 0, we find that

$$\overline{E'''B'''}/\overline{A_0E'''} = \cos\varphi / \cos\mu = (bc)/(ad) \qquad (9)$$

If, *additionally*, $\overline{E'''B'''} = \overline{A_0E'''}$, we find that $\mu \equiv \pm\varphi$ and bc = ad (10)
Then, the initial four-bar has to be a **kite**, whereas the bar B'''E''' generates the *elliptic motion*.

8. Historical Note on Linearizers (i.e. *linkages with a rectilinear moving link*)

Apart from the mentioned literature of Hart and Kempe, a booklet of the russian L.D. **Ruzinov** (ref.[10]) has been devoted to this subject. He mentioned more inventors, such as **Gagarin** and others, but all linkages of this kind seem to have been restricted to those containing either deltoids (i.e. kites) or contra-parallelograms.
Ruzinov, for example, wrote, quote: "It is not possible to produce a linearizer based on a four-bar chain", unquote, (page 61 of ref.[10]). Though he searched for a more general approach, he didn't succeed and really thought that such was not possible and only very particular linkages were able to meet the rectilinear motion. That in reality, a *random* four-bar can be used as a base to start the design of a linearizer with, he never uncovered.
Remarkable is also that even Kempe, who founded the generalization of his *compound linkage*, later named the focal linkage by L.Burmester, did not saw the connection between his three kite- or *deltoid-linearizers* and his own interconnected, so-called *conjugate four-piece linkages*. Special cases of the latter were either restricted to Hart's 2nd straight-line mechanism or to other more general linkages, but not to linearizers, having a complete link moving rectilinear.
We conclude that the general approach as developed in the underlying manuscript, really represents a generalized method simultaneously interconnecting all linearizers known to mankind.

References

[1] Hart,H.: *A parallel motion*, Proc.London Math.Soc.**6**(1875),p.137-9

[2] Dijksman,E.A.: *Six-Bar Cognates of a Stephenson Mechanism*, Journal of Mechanisms,Vol.6(1970),Nr.1,p.31-57,(page 40)

[3] Hart,H.: *On some cases of parallel motion*, Proc.London Math.Soc.**8**(1877),p.286-289.

[4] Dijksman,E.A.: *Kempe's (Focal) Linkage Generalized, particularly in connection with Hart's second straight-line mechanism*,Mechanism and Machine Theory,Vol.10(1974),Nr.6,p.445-460,(figure 10)

[5] Kempe,A.B.: *On a general method of producing exact rectilinear motion by linkworks*, Proc. Royal Soc.London,**23**(1875),p.565-77.

[6] Dijksman,E.A.: *Overconstrained Linkages to be derived from perspectivity and reflection*, Proc.7th World Congress on TMM,Sevilla, Spain,Vol.1(1987),p.69-73(figure 10)

[7] Kempe,A.B.: *How to draw a straight-line*-III,Nature,16(1877)p.127

[8] Kempe,A.B.: *On conjugate Four-piece Linkages,*Proceedings of the London Mathematical Society,9(1878)p.133-147

[9] Wunderlich,W.: *On Burmester's focal mechanism and Hart's straight-line motion*, Journal of Mechanisms, Vol.3 (1968), Nr.2, p.79-86.

[10] Ruzinov,L.D.: **Design of Mechanisms by Geometric Transformations**, (translated from the Russian into English), Iliffe Books Ltd., London,(1968).

On the Reduction of Parameters in Kinematic Equations

K. Dobrovodský

Institute of Control Theory and Robotics
Slovak Academy of Sciences
Dúbravská cesta 9, 842 37 Bratislava, Slovakia

Abstract - This paper presents a new systematic approach to the representation and classification of kinematic equations. Equivalent substitutions of joint variables and link parameters are developed and the equivalence of open serial kinematic chains is defined. Algebraic representation is based on the free semigroup over the alphabet consisting of 12 letters. Equivalent substitutions are defined and the parameters reduction procedure is derived. The main result is a reduction of the number of parameters of the given kinematic equation.

I. Introduction

The problem of positioning a serial open kinematic chain in a manner that places the end effector in a specified position and orientation is a problem of fundamental importance in the robotics field. In this problem, we are given an open kinematic chain and some cartesian goal specification, and we are to obtain a configuration for the chain which satisfies that goal. The complexity of actual physical situations calls for simplified descriptions in terms of symbolic models which abstract suitably chosen essential properties of physical objects and relations between these objects. A mathematical model will reproduce suitably chosen features of a physical situation if it is possible to establish rules of correspondence relating specific physical objects and relationships to corresponding mathematical objects and relations.

II. Representation of serial kinematic chains

Suppose that we have a fixed coordinate system (reference) and a moving coordinate system, and let both coordinate systems be identical at the instant. Denote $i(\delta)$, $j(\delta)$, $k(\delta)$ elementary translations and $I(\delta)$, $J(\delta)$, $K(\delta)$ elementary rotations in respect to rectangular coordinate axes i, j, k, respectively. Parametric operators

$$i(\delta), j(\delta), k(\delta) \tag{1}$$

$$I(\delta), J(\delta), K(\delta) \tag{2}$$

may be represented by the well known 4.4 homogeneous transformation matrices. In a concise symbolic notation parameters δ will be omitted. We emphasize quality (constant or variable) of the parameter only. Let's introduce the following symbols:

A. J. Lenarčič and B. B. Ravani (eds.), Advances in Robot Kinematics and Computationed Geometry, 421–426.

a, b, c	constant translations
A, B, C	constant rotations
x, y, z	variable translations
X, Y, Z	variable rotations

Under this notation a sequence $cZcAcXcXcZcaXcZcA$ defines the kinematic structure of a serial kinematic chain with seven links: c, cAc, c, c, ca, c, cA which are connected by six joints: Z, X, X, Z, X, Z. An implicit indexing of link parameters $\lambda = (\lambda_1, \lambda_2, \ldots)$ and joint variables $\vartheta = (\vartheta_1, \vartheta_2, \ldots)$ assumed from left to right uniquely defines the following parametric operator

$$\begin{aligned}\Omega(\vartheta, \lambda) \quad = \quad & k(\lambda_1) \cdot K(\vartheta_1) \cdot k(\lambda_2) \cdot I(\lambda_3) \cdot k(\lambda_4) \cdot I(\vartheta_2) \cdot \\ & \cdot k(\lambda_5) \cdot I(\vartheta_3) \cdot k(\lambda_6) \cdot K(\vartheta_4) \cdot k(\lambda_7) \cdot i(\lambda_8) \cdot \\ & \cdot I(\vartheta_5) \cdot k(\lambda_9) \cdot K(\vartheta_6) \cdot k(\lambda_{10}) \cdot I(\lambda_{11})\end{aligned} \tag{3}$$

We will use a power for repeated subsequences, e.g. $\Omega = ZaXcXcZXZ = Za(Xc)^2ZXZ$. It doesn't mean a power in the corresponding parametric operator

$$\Omega(\vartheta, \lambda) = K(\vartheta_1) \cdot i(\lambda_1) \cdot I(\vartheta_2) \cdot k(\lambda_2) \cdot I(\vartheta_3) \cdot k(\lambda_3) \cdot K(\vartheta_4) \cdot I(\vartheta_5) \cdot K(\vartheta_6) \tag{4}$$

Any string Ω over the alphabet $V = \{a, b, c, x, y, z, A, B, C, X, Y, Z\}$ defines an operator $\Omega(\vartheta, \lambda)$. Two strings are identical if and only if they are composed of the same sequence of letters. An empty string consists of no letters. The product of two strings is simply the concatenation of the strings, e.g. $aBx \cdot cZ = aBxcZ$. This product is associative. Hence, the set of all finite strings over the alphabet V (including the empty string) is the free semigroup V^* over the alphabet V as defined in [1]. It should be emphasized at this point that the string Ω is an element of a semigroup, but the corresponding parametric operator $\Omega(\vartheta, \lambda)$ is an element of a group. In the concise "semigroup notation" the Denavit–Hartenberg notation yields substrings $ZcaA$ (revolute joint) and $CzaA$ (prismatic joint). For example Stanford manipulator according to [2] may be expressed in the form $(ZcaA)^2 \cdot CzaA \cdot (ZcaA)^3$. Operators i, j, k, I, J, K with zero parameters simply vanish and operators I, J, K with parameters being nonzero integer multiples of $\pi/2$ may be eliminated by similarity transformations. Thus we can search for the shortest (reduced) string corresponding with the given kinematic structure.

III. Kinematic equations

The specified representation of the serial kinematic chains is not unique. Two different strings from V^* may represent "conceptually equivalent" kinematic chains. The problem arises, how to specify such equivalence and how to decide whether or not two strings are equivalent. We will specify the equivalence of the strings over the alphabet V in accordance with the equivalence of the corresponding kinematic equations. We will write the kinematic equation in the form

$$\Omega(\vartheta, \lambda) = T \tag{5}$$

where Ω is an operator illustrated in Eqs. (3,4), and T is a constant general position transformation (translation and rotation). From the physical point of view quantities $\lambda = (\lambda_1, \lambda_2, \ldots)$ represent metrical properties of the links, $\vartheta = (\vartheta_1, \vartheta_2, \ldots)$ are unknown joint variables and T are the coordinates of the desired position of the last link at the end of the open kinematic chain in respect to the reference frame.

Suppose we have Eq. (5). We can say that we know how to solve this equation if and only if we know how to evaluate the inverse of $T = \Omega(\vartheta, \lambda)$ in respect to ϑ. There is a great variety of different serial kinematic chains (kinematic structures Ω) and thus we usually don't know directly how to "invert" the given structure Ω. It is desirable to make some substitutions of variables ϑ and parameters λ :

$$\begin{aligned} \vartheta' &= F(\lambda) \cdot \vartheta + f(\lambda) \\ \lambda' &= g(\lambda) \end{aligned} \tag{6}$$

and (or) some constant homogeneous transformations of both sides of the kinematic equation leading to an "equivalent" kinematic structure Ω' and the corresponding equation

$$\Omega'(\vartheta', \lambda') = T' \tag{7}$$

Having solved Eq. (7) in respect to the variables ϑ', we must return back to the original variables ϑ :

$$\vartheta = F^{-1}(\vartheta' - f) \tag{8}$$

where $F = F(\lambda)$ and $f = f(\lambda)$.

IV. Similarity transformations

We can eliminate any right angle rotation in the given kinematic equation and thus to reduce the number of operators characterizing a serial kinematic chain. Suppose Eq. (5) involves the right angle rotation $P(\pi/2)$, $P \in \{I, J, K\}$:

$$\Phi(\vartheta_a, \lambda_a) \cdot P(\pi/2) \cdot \Psi(\vartheta_b, \lambda_b) = T \tag{9}$$

We postmultiply Eq. (9) by $P(-\pi/2)$

$$\Phi(\vartheta_a, \lambda_a) \cdot P(\pi/2) \cdot \Psi(\vartheta_b, \lambda_b) \cdot P(-\pi/2) = T \cdot P(-\pi/2) \tag{10}$$

A similarity transformation with $P = I(\pi/2)$ converts operators $j(\delta)$, $J(\delta)$ to operators $k(\delta)$, $K(\delta)$ and operators $k(\delta)$, $K(\delta)$ to operators $j(-\delta)$, $J(-\delta)$, respectively. Analogously for $P \in \{J, K\}$. Thus we can write

$$\Phi(\vartheta_a, \lambda_a) \cdot \Psi^P(\vartheta'_b, \lambda'_b) = T' \tag{11}$$

where $\Psi^P(\vartheta'_b, \lambda'_b) = P(\pi/2) \cdot \Psi(\vartheta_b, \lambda_b) \cdot P(-\pi/2)$ and $T' = T \cdot P(-\pi/2)$.

V. Local substitutions

A substitution of variables and parameters of some part (a substring) in the kinematic structure like Eq. (3) provides the most important way of a context-free "string manipulation". It is obvious that any occurrence of two translations along (or two rotations about) the same coordinate axis may be reduced to one by the substitution $\delta' = \delta_1 + \delta_2$ and, in the case of a constant λ and a variable ϑ, by the substitution $\vartheta' = \vartheta + \lambda$. We can easily verify that operators Eqs. (1,2) concerned with the same coordinate axis commute and all translation operators Eq. (1) commute.

Generally, we can "replace" a substring Ω by Ω' and vice versa in any context if there exists a substitution of variables and parameters preserving equality $\Omega(\vartheta, \lambda) = \Omega'(\vartheta', \lambda')$. For example $ZabZ$ may be replaced by ZbZ according to the following substitution:

$$K(\vartheta_1) \cdot i(\lambda_1) \cdot j(\lambda_2) \cdot K(\vartheta_2) = K(\vartheta') \cdot j(\lambda') \cdot K(\vartheta') \tag{12}$$

$$\begin{aligned} \vartheta'_1 &= \vartheta_1 - \operatorname{atan2}(\lambda_1, \lambda_2) \\ \vartheta'_2 &= \vartheta_2 + \operatorname{atan2}(\lambda_1, \lambda_2) \\ \lambda' &= \sqrt{(\lambda_1^2 + \lambda_2^2)} \end{aligned} \tag{13}$$

We will specify all basic "replaceable" substrings more exactly in the following text.

VI. Definition of equivalence

We will define two equivalencies of strings over the alphabet V. The local equivalence of strings Ω, Ω' is associated with the existence of the mutual substitutions of variables ϑ, ϑ' and parameters λ, λ' preserving the equality $\Omega(\vartheta, \lambda) = \Omega'(\vartheta', \lambda')$. The global equivalence of strings is associated with the existence of the regular transformations leading from the kinematic equation $\Omega(\vartheta, \lambda) = T$ to the equation $\Omega'(\vartheta', \lambda') = T'$. Let's introduce the following notation:

Ω^I mutual renaming of letters associated with j and k
Ω^J mutual renaming of letters associated with i and k
Ω^K mutual renaming of letters associated with i and j

Example: $\Omega = AYcy$; $\Omega^I = AZbz$; $\Omega^J = CYay$; $\Omega^K = BXcx$
In particular, $(AYcy)^I = A^I Y^I c^I y^I = AZbz$

Definition 1: Let $\Omega, \Omega' \in V^*$ are the strings associated with parametric operators $\Omega(\vartheta, \lambda)$ and $\Omega'(\vartheta', \lambda')$. We will call two strings Ω, Ω' locally equivalent (i.e. $\Omega \equiv \Omega'$) if and only if a surjection $\lambda' = g(\lambda)$ and a λ - parametric linear surjection $\vartheta' = F(\lambda) \cdot \vartheta + f(\lambda)$ are given implicitly by the relation $\Omega(\vartheta, \lambda) = \Omega'(\vartheta', \lambda')$, and vice versa.

Basic local equivalence relations provided by the above definition:
$gg \equiv g$; $g \in V$ (idempotency)

$\{a,x,A,X\}^*$, $\{b,y,B,Y\}^*$, $\{c,z,C,Z\}^*$ and $\{a,b,c,x,y,z\}^*$ are commutative
$A\Omega A \equiv A\Omega^I A$; $B\Omega B \equiv B\Omega^J B$; $C\Omega C \equiv C\Omega^K C$; $\Omega \in V^*$
$ax \equiv x$; $AX \equiv X$; $by \equiv y$; $BY \equiv Y$; $cz \equiv z$; $CZ \equiv Z$;
$Abc \equiv bcA \equiv AbA \equiv AcA \equiv cAc \equiv cAb \equiv bAc \equiv bAb$;
$Ayz \equiv yzA \equiv yAy \equiv yAz \equiv zAz \equiv zAy$;
$Bca \equiv caB \equiv BcB \equiv BaB \equiv aBa \equiv aBc \equiv cBa \equiv cBc$;
$Bxz \equiv xzB \equiv xBx \equiv xBz \equiv zBz \equiv zBx$;
$Cab \equiv abC \equiv CaC \equiv CbC \equiv bCb \equiv bCa \equiv aCb \equiv aCa$;
$Cxy \equiv xyC \equiv xCx \equiv xCy \equiv yCy \equiv yCx$;
$ABA \equiv ACA \equiv ABC \equiv ACB \equiv BAB \equiv BCB \equiv BAC \equiv BCA \equiv CAC \equiv CBC \equiv CAB \equiv CBA$;

Definition 2: We will call two strings Ω, Ω' from V^* equivalent (i.e. $\Omega \cong \Omega'$) if and only if they are mutually derivable by local equivalencies and/or by adding, truncating constant prefixes and/or postfixes.

Example: x $\not\equiv y$ but $x \cong y$
since $x \cong CxC \equiv CyC \cong y$

Definition 3: We call a string Ω from V^* minimal if and only if no equivalent string is shorter.

It follows directly from the Definition 3: If two strings are minimal and equivalent, then they are of the same length.

VII. Equivalent substitutions procedure

(1) Eliminate right angle rotations by similarity transformations
(2) Eliminate all Y by similarity transformations
(3) Commute constant translations to the right
(4) Remove the constant postfix by the regular transformation
(5) Commute constant translations to the left
(6) Remove the constant prefix by the regular transformation
(7) Reduce constant parameters by local substitutions

In the following table, we can compare original and reduced kinematic structures of several representative robot manipulators:

	original	# par	reduced	# par
PUMA [3]	$cZabc(Xac)^2ZcXcZc$	11	$ZabXcXcZXZ$	4
SCARA	$czc(Zcb)^2YbZbYb$	9	$zZbZaXZX$	2
Stanford [2]	$cZcaXbzZcXcZc$	7	$ZaXbzZXZ$	2
HITACHI	$cZc(Xac)^3ZcXa$	10	$Za(Xc)^3ZX$	4
ASEA	$cZc(Xac)^3ZcA$	10	$Za(Xb)^2XcZ$	4

VIII. Conclusion

In accordance with the above approach an expert system for the analysis and solving kinematic equations was realized at the Institute of Control Theory and Robotics. The goal of the system is to point out special classes of structures having closed form solutions of the inverse kinematics problem. The system provides an automatic analysis of the given kinematic structure and generates the direct and the inverse kinematics algorithms for structures solvable in a closed form.

References

[1] J.E. Hopcroft and J.D. Ullman, *Formal languages and their relation to automata*, Addison-Wesley Publishing Company, Reading, Massachusetts, USA (1969)

[2] R.P. Paul, *Robot manipulators: mathematics, programming, and control*, MIT Press, Cambridge, Massachusetts, London, England (1981)

[3] C.S.G. Lee, "Robot arm kinematics, dynamics and control", *IEEE Computer*, 62-79, December (1982)

[4] K. Dobrovodský, "On the Equivalence of Kinematic Equations", *Proc. 1st Int. Meeting on Robotics in Alpe-Adria Region*, Portorož , Slovenia (1992)

[5] H. Rieseler, H. Schrake and F.M. Wahl, "Symbolic Computation of Closed Form Solutions with Prototype Equations", *Proc. 2-nd Int. Workshop on Advances in Robot Kinematics*, Linz, Austria (1990)

[6] H. Rieseler, H. Schrake and F.M. Wahl, "Symbolic Kinematics Inversion of Redundant Robots", *Proc. ISFR'90, 4-th Int. Symposium on Foundations of Robotics*, Holzhau, G.D.R. (1990)

[7] P. Andris and K. Dobrovodský, "Inverse Kinematics of the 3P3R Manipulator of General Geometry", *Kybernetika a informatika*, *5*(2/3), (1992)

[8] M. Raghavan and B. Roth, "Kinematic Analysis of the 6R Manipulator of General Geometry", *Proc. 5-th Int. Symposium on Robotics Research*, (H. Miura and S. Arimoto, Eds.), 314-320 (1989)

11. Parallel Manipulators

Symbolic-Form Forward Kinematics of a 5-4 Fully-Parallel Manipulator

Carlo Innocenti
Dipartimento di Ingegneria delle Costruzioni Meccaniche
Università di Bologna - Italy

Vincenzo Parenti-Castelli
Istituto di Ingegneria - Facoltà di Ingegneria
Università di Ferrara - Italy

Abstract - Fully parallel manipulators feature two rigid bodies, base and platform, each other connected by six binary links with spherical pairs at the extremities. Links have adjustable lengths to control position and orientation of the platform with respect to the base. The forward position analysis solves for the location of the platform once the link lengths are given. The paper presents the forward position analysis in symbolic form of a 5-4 fully-parallel manipulator having general geometry. By a new procedure a kinematic model is devised that leads to a system of three closure equations in three unknowns. As a result of a single-step elimination process, a final 32nd order polynomial equation in only one unknown is obtained. Thirty-two configurations of the manipulator are hence possible in the complex field. A case study validates the new theoretical results.

I. Introduction

Recently, increasing attention has been focused on parallel manipulators whose technical relevance in many fields has been spotted by several authors [1,2]. Parallel manipulators are closed chains with one or more loops where only some kinematic pairs are actively controlled. The most common arrangement comprises a frame (base) connected via serial, parallel, or serial-parallel (hybrid) chains to the output link (platform). Actuated pairs provide the platform with the desired degrees of freedom (dof) relative to the base. Fully-parallel manipulators (FPM) are characterized by six binary links (legs), each joined to base and platform through spherical pairs. Up to six legs may have controllable variable length, providing the platform with a corresponding number of dofs.

Different manipulators arise when two or three leg extremities coalesce in a multiple spherical pair centered on the base and/or platform. An established notation labels these manipulators as M-m fully-parallel manipulators, where M and m are the number of distinct connection points on the base and platform. The 6-6 FPM, shown in Fig. (1), often referred to as the generalized Stewart platform, has the most general leg arrangement. For any FPM, base and platform can have general, nearly general (planar shape) or special (symmetric planar shape) geometry.

A significant research topic for FPMs is their position analysis. The forward position analysis (FPA) solves for the position and orientation of the platform once the leg lengths are given. Conversely the inverse position analysis solves for the leg lengths when the location of the platform is given. Only the first is a difficult and challenging problem since non-linear equations are involved. A bibliography of the research activity on forward kinematics can be constructed from [1,3,4,5].

A. J. Lenarčič and B. B. Ravani (eds.), Advances in Robot Kinematics and Computationed Geometry, 429–438.

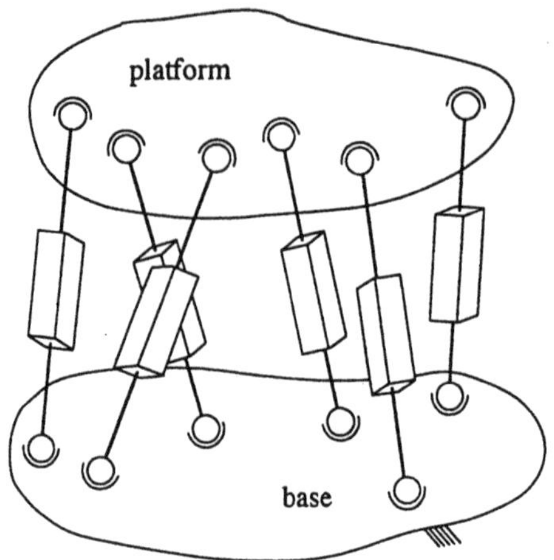

Fig. 1: The generalized 6-6 Stewart platform

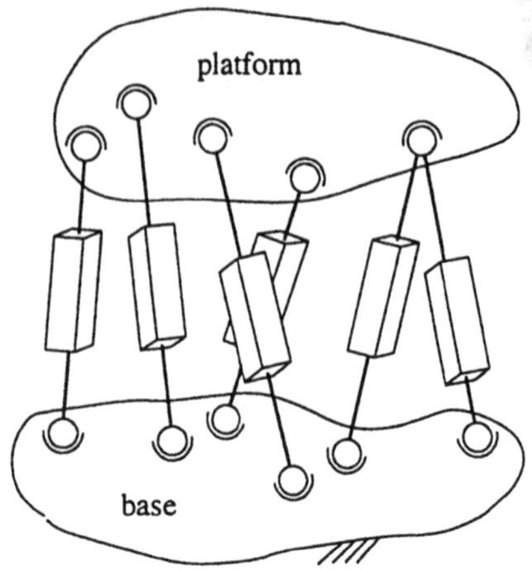

Fig. 2: The 6-5 fully-parallel manipulator

The main difficulty of the FPA is to find all the possible solutions. Standard numerical methods provide only a few of them and do not guarantee exhaustiveness. Numerical methods such as the continuation method - introduced in kinematics by Freudenstein and Roth [6] and recently developed in a reliable procedure by Morgan and Sommese [7] - could be used for finding all solutions, but they prove to be efficient when the number of solutions is known in advance. On the other hand, as emphasized by many authors, a symbolic-form solution would provide all possible solutions of the FPA and contribute to enlightening the kinematic behaviour of the mechanism.

Since no general procedure exists for the FPA in symbolic-form of FPMs, customized strategies must be devised [4]. Moreover, it is worth noting that *ad hoc* procedures for manipulators with special or nearly general geometry cannot be extended to general geometry manipulators even if they feature the same topology.

Symbolic-form solution, when feasible, may be obtained by the elimination of some unknowns from the closure equations of the manipulator. The elimination, based on either the dyalitic method [8] or customized procedures, aims at transforming the original system into an equivalent system in echelon form, i.e., a system where each equation can be regarded as univariate. When, for instance, the equations are algebraic, the first equation to be solved is frequently a polynomial equation of high degree, whereas the remaining are linear. Once the roots of the first polynomial equation are found, all solutions of the system can then be easily determined. Each root provides one solution of the FPA. While for degrees not greater than four a closed form solution is possible, roots of higher degree polynomial equations must be determined numerically. However, because numerical solution of a polynomial equation is, in practice, a routine task, even in this case all possible FPA solutions can be found and the FPA is said to be solvable in symbolic form.

Recently it has been shown [5] that the FPA symbolic form solution of almost any FPMs with general geometry has been solved in symbolic form. Yet to be solved are three FPMs, namely, the 6-6 FPM, the 6-5 FPM and one type of a 5-4 FPM, which are schematically shown in Fig. (1), Fig. (2) and Fig. (3a) respectively.

The purpose of this paper is to solve in symbolic form the FPA of the general geometry 5-4 FPM shown in Fig. (3a). It is the only 5-4 arrangement still to be solved in symbolic form; the other three existing 5-4 arrangements, represented in Figs. (3b,3c,3d), have been

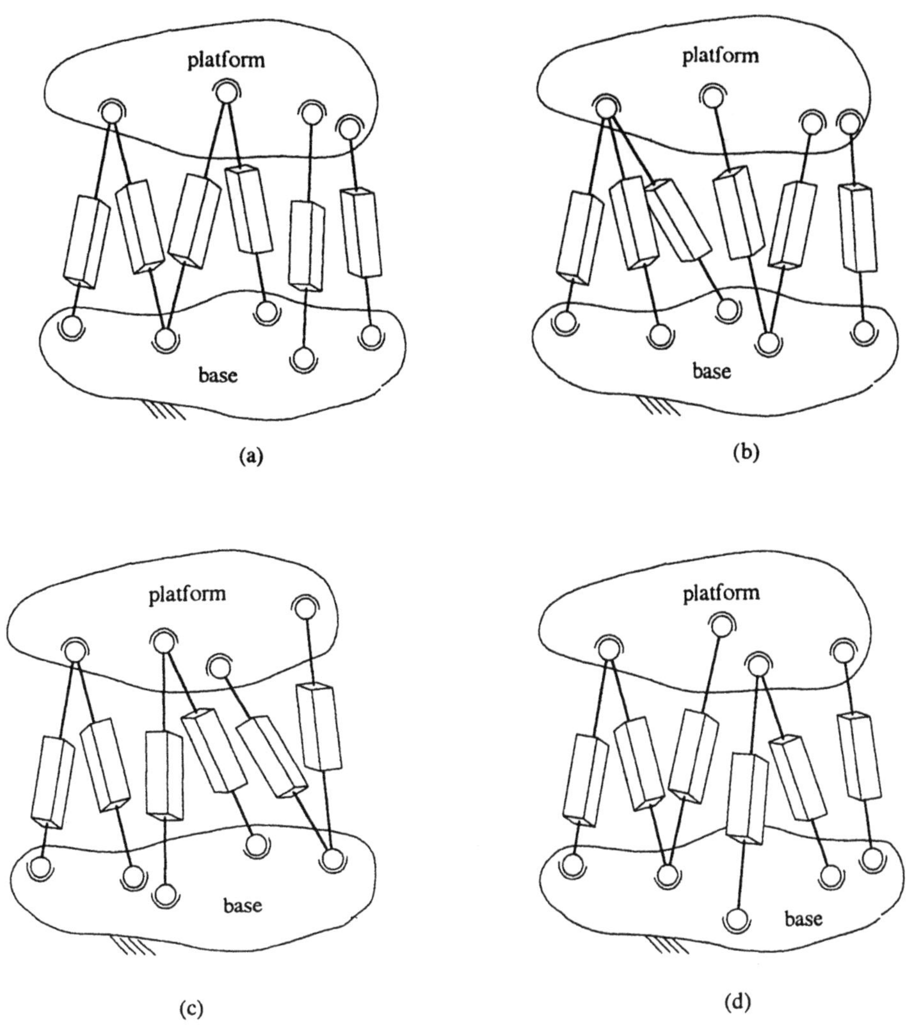

Fig. 3: The four possible leg arrangements for a 5-4 FPM

solved in symbolic-form in [9,10,11] respectively. A symbolic form solution for the arrangements (a), (c), and (d) of Fig. (3) have been presented in [12] for base and platform both planar.

The symbolic form solution, of the 5-4 FPM of Fig. (3a), is obtained in two phases. First a system of three closure equations in three unknowns is obtained, then a suitable elimination process leads to a polynomial equation of order thirty-two in only one unknown whose solutions correspond to the manipulator configurations. Hence thirty-two configurations of the 5-4 FPM are possible in the complex field. Finally a numerical example validates the new theoretical results.

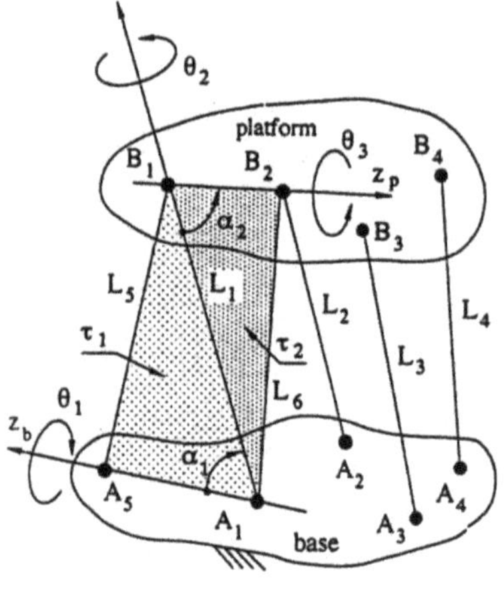

Fig. 4: Schematic of the 5-4 structure related to the 5-4 fully-parallel manipulator of Fig. 3a

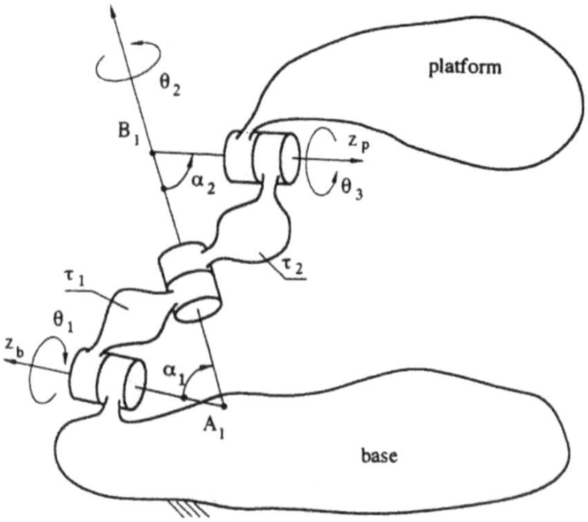

Fig. 5: The R-R-R serial chain

II. Forward Kinematics

A. Kinematic Model

For a given set of leg lengths, a fully-parallel manipulator becomes a statically determined structure, hence the solution of the manipulator FPA can be regarded as the determination of all possible ways of assembly (configurations) of the related structure. The structure pertaining to the 5-4 fully parallel manipulator shown in Fig. (3a) is schematically represented in Fig. (4). As shown in the following, the position and orientation of the platform with respect to the base can be univocally defined by a reduced set of parameters. Three equations in three unknowns are devised that represent the closure equation system from which symbolic form solution of the FPA will be obtained.

With reference to Fig. (4) it can be noted that triangles $\tau_1 \equiv A_1B_1A_5$ and $\tau_2 \equiv A_1B_2B_1$, can be considered as rigid bodies. If legs A_2B_2, A_3B_3, and A_4B_4 are momentarily removed from the structure, the remaining links (base, triangles τ_1 and τ_2, and platform) form a serial chain. Triangles τ_1 and τ_2 may be considered as binary links connected to each other and, respectively, to the base and platform by revolute pairs. The resulting spatial R-R-R serial chain is diagrammatically shown in Fig. (5). Axes of revolute pairs are defined by couples of points (A_1, A_5), (A_1, B_1), and (B_1, B_2) (see Fig. (4)). The configuration of the R-R-R chain is univocally defined by angles θ_1, θ_2, and θ_3.

B. Closure Equations

The closure equations of the structure can be found by imposing that the distance between points A_j and B_j, (j=2,3,4), equals the corresponding leg length:

$$(B_j - A_j)^2 = L_j^2 \qquad (j=2,3,4) \qquad (1)$$

A reference system S_b, embedded in the base, with origin in A_1 is chosen: its z axis, z_b, is directed from A_1 to A_5, while the x axis, perpendicular to z_b, can be chosen arbitrarily. A reference system S_p fixed to the platform is chosen with origin in B_1. It has z axis, z_p, directed from B_1 to B_2 and x axis arbitrarily chosen in a plane perpendicular to z_p. According to the D-H notation [13], reference systems fixed to triangles τ_1 and τ_2 are also chosen.

In order to impose the constraints given by Eq. (1), the 4×4 matrix for the coordinate transformation from system S_p to system S_b must be determined. Its 3×3 rotational part **R** is given by:

$$\mathbf{R} = \begin{vmatrix} c_1c_2c_3-u_1s_1s_2c_3+u_2c_1s_2s_3 & -c_1c_2s_3+u_1s_1s_2s_3+u_2c_1s_2c_3 & v_2c_1s_2+u_1v_2s_1c_2 \\ +u_1u_2s_1c_2s_3+v_1v_2s_1s_3 & +u_1u_2s_1c_2c_3+v_1v_2s_1c_3 & -v_1u_2s_1 \\ s_1c_2c_3+u_1c_1s_2c_3+u_2s_1s_2s_3 & -s_1c_2s_3-u_1c_1s_2s_3+u_2s_1s_2c_3 & v_2s_1s_2-u_1v_2c_1c_2 \\ -u_1u_2c_1c_2s_3-v_1v_2c_1s_3 & -u_1u_2c_1c_2c_3-v_1v_2c_1c_3 & +v_1u_2c_1 \\ v_1s_2c_3-v_1u_2c_2s_3+u_1v_2s_3 & -v_1s_2s_3-v_1u_2c_2c_3+u_1v_2c_3 & -v_1v_2c_2-u_1u_2 \end{vmatrix} \tag{2}$$

while its displacement vector, **d**, that represents the position of point B_1 in reference system S_b, is:

$$\mathbf{d} = L_1(v_1s_1, -v_1c_1, u_1)^T \tag{3}$$

where T is for transpose, $L_1=\|B_1 - A_1\|$, and the positions:

$$c_j=\cos\theta_j; \qquad s_j=\sin\theta_j \qquad (j=1,2,3)$$
$$u_j=\cos\alpha_j; \qquad v_j=\sin\alpha_j \qquad (j=1,2) \tag{4}$$

are adopted. Cosine and sine of angles α_1 and α_2, with reference to triangles τ_1 and τ_2, are given by:

$$\cos\alpha_1 = \frac{L_1^2 - L_5^2 + (A_5-A_1)^2}{2\,L_1\,\|A_5-A_1\|}; \qquad \sin\alpha_1 = +(1-\cos^2\alpha_1)^{1/2}$$
$$\cos\alpha_2 = \frac{L_1^2 - L_6^2 + (B_2-B_1)^2}{2\,L_1\,\|B_2-B_1\|}; \qquad \sin\alpha_2 = +(1-\cos^2\alpha_2)^{1/2} \tag{5}$$

The coordinates in reference system S_b of any point B_j of the platform can be obtained by:

$$(B_j - A_1)_b = \mathbf{R}\,(B_j - B_1)_p + \mathbf{d} \tag{6}$$

(a vector subscript refers to the coordinate system where the vector is measured).

According to Eqs. (1), by also considering Eq. (6), the three closure equations of the 5-4 structure can be written as:

$$[(\mathbf{R}\,(B_j - B_1)_p + \mathbf{d} - (A_j - A_1)_b]^2 = L_j^{\,2} \qquad (j=2,3,4) \tag{7}$$

Expansion of Eqs. (7) yields:

$$(B_j - B_1)_p^2 + d^2 + (A_j - A_1)_b^2 + 2\,(B_j - B_1)_p^T\, \mathbf{R}^T\, \mathbf{d}$$

$$- 2\,(A_j - A_1)_b^T\, \mathbf{R}\,(B_j - B_1)_p - 2\,(A_j - A_1)_b^T\, \mathbf{d} = L_j^{\,2} \qquad (j=2,3,4) \tag{8}$$

where $\mathbf{R}^T\,\mathbf{d}$, computed from Eqs. (2,3), is:

$$\mathbf{R}^T\,\mathbf{d} = L_1(v_2 s_3,\ v_2 c_3,\ -u_2)^T \tag{9}$$

By inspection, Eqs. (8) result to be cosine-sine (c-s for brevity) linear in θ_j, j=1,2,3. Cosine-sine linearity has the following meaning: if a term of Eqs. (8) depends on θ_j, it contains either the cosine or sine of θ_j.

By noting that the expressions for $\mathbf{R}$, $\mathbf{d}$, and $\mathbf{R}^T\mathbf{d}$ are c-s linear in θ_1, θ_2, and θ_3, Eqs. (8) can be given the following form:

$$\sum_{i,j=0,2} a_{ij}\, s_1^{p(i)}\, c_1^{q(i)}\, s_2^{p(j)}\, c_2^{q(j)} = 0 \tag{10}$$

$$\sum_{i,j,k=0,2} m_{ijk}\, s_1^{p(i)}\, c_1^{q(i)}\, s_2^{p(j)}\, c_2^{q(j)}\, s_3^{p(k)}\, c_3^{q(k)} = 0 \tag{11}$$

$$\sum_{i,j,k=0,2} n_{ijk}\, s_1^{p(i)}\, c_1^{q(i)}\, s_2^{p(j)}\, c_2^{q(j)}\, s_3^{p(k)}\, c_3^{q(k)} = 0 \tag{12}$$

where coefficients a_{ij}, m_{ijk}, and n_{ijk} depend only on the geometry of the 5-4 structure.

C. Elimination of θ_2 and θ_3

In order to obtain a final polynomial equation in only one unknown, elimination of two unknowns from the three closure equations must be accomplished. Although a two-steps dyalitic method could be employed [8], the single-step elimination procedure presented in [12,14] proves more efficient in the case being studied.

First, closure Eqs. (10-12) can be written in algebraic form after substituting, for cosine and sine of θ_j, j=1,2,3, the well known expressions:

$$c_j = (1 - t_j^2)\,/\,(1 + t_j^2); \qquad s_j = 2\,t_j\,/\,(1 + t_j^2) \tag{13}$$

where $t_j=\tan(\theta_j/2)$. Substitution allows Eqs. (10-12) to be respectively written in the following form:

$$\sum_{j=0,2} F_j\, t_2^j = 0 \tag{14.1}$$

$$\sum_{j,k=0,2} G_{jk}\, t_2^j\, t_3^k = 0 \tag{14.2}$$

$$\sum_{j,k=0,2} H_{jk}\, t_2^j\, t_3^k = 0 \tag{14.3}$$

where:

$$F_j = \sum_{i=0,2} f_{ij}\, t_1^{\,i} \qquad (j=0,1,2) \tag{15.1}$$

$$G_{jk} = \sum_{i=0,2} g_{ijk}\, t_1^{\,i} \qquad (j,k=0,1,2) \tag{15.2}$$

$$H_{jk} = \sum_{i=0,2} h_{ijk}\, t_1^{\,i} \qquad (j,k=0,1,2) \tag{15.3}$$

Coefficients f_{ij}, g_{ijk}, h_{ijk}, (i,j,k=0,1,2), which result by substituting positions (13) into Eqs. (10-12), are functions of quantities a_{ij}, m_{ijk}, and n_{ijk}: thus they depend only on the geometry of the 5-4 structure.

A system of sixteen equations can be devised by multiplying Eqs. (14) by suitable factors. The resulting system can be regarded as a homogeneous linear system in sixteen unknowns. In particular, the first eight equations are obtained by multiplying Eq. (14.1) by factors:

$$1,\ t_2,\ t_3,\ t_2\, t_3,\ t_3^2,\ t_2\, t_3^2,\ t_3^3,\ t_2\, t_3^3 \tag{16}$$

and the remaining eight equations, grouped in fours, are obtained by multiplying Eqs. (14.2) and (14.3) by factors:

$$1,\ t_2,\ t_3,\ t_2\, t_3 \tag{17}$$

The sixteen unknowns of the homogeneous linear system, apart from an arbitrary scaling factor, are:

$$1,\ t_2,\ t_2^2,\ t_2^3,\ t_3,\ t_2\, t_3,\ t_2^2\, t_3,\ t_2^3\, t_3,\ t_3^2,\ t_2\, t_3^2,\ t_2^2\, t_3^2,\ t_2^3\, t_3^2,\ t_3^3,\ t_2\, t_3^3,\ t_2^2\, t_3^3,\ t_2^3\, t_3^3 \tag{18}$$

The homogeneous linear system will provide a non-trivial solution if the determinant of its coefficient matrix:

$$\mathbf{K} = \begin{vmatrix}
F_0 & F_1 & F_2 & 0 & 0 & 0 & 0 & 0 & 0 & 0 & 0 & 0 & 0 & 0 & 0 & 0 \\
0 & F_0 & F_1 & F_2 & 0 & 0 & 0 & 0 & 0 & 0 & 0 & 0 & 0 & 0 & 0 & 0 \\
0 & 0 & 0 & 0 & F_0 & F_1 & F_2 & 0 & 0 & 0 & 0 & 0 & 0 & 0 & 0 & 0 \\
0 & 0 & 0 & 0 & 0 & F_0 & F_1 & F_2 & 0 & 0 & 0 & 0 & 0 & 0 & 0 & 0 \\
0 & 0 & 0 & 0 & 0 & 0 & 0 & 0 & F_0 & F_1 & F_2 & 0 & 0 & 0 & 0 & 0 \\
0 & 0 & 0 & 0 & 0 & 0 & 0 & 0 & 0 & F_0 & F_1 & F_2 & 0 & 0 & 0 & 0 \\
0 & 0 & 0 & 0 & 0 & 0 & 0 & 0 & 0 & 0 & 0 & 0 & F_0 & F_1 & F_2 & 0 \\
0 & 0 & 0 & 0 & 0 & 0 & 0 & 0 & 0 & 0 & 0 & 0 & 0 & F_0 & F_1 & F_2 \\
G_{00} & G_{10} & G_{20} & 0 & G_{01} & G_{11} & G_{21} & 0 & G_{02} & G_{12} & G_{22} & 0 & 0 & 0 & 0 & 0 \\
0 & G_{00} & G_{10} & G_{20} & 0 & G_{01} & G_{11} & G_{21} & 0 & G_{02} & G_{12} & G_{22} & 0 & 0 & 0 & 0 \\
0 & 0 & 0 & 0 & G_{00} & G_{10} & G_{20} & 0 & G_{01} & G_{11} & G_{21} & 0 & G_{02} & G_{12} & G_{22} & 0 \\
0 & 0 & 0 & 0 & 0 & G_{00} & G_{10} & G_{20} & 0 & G_{01} & G_{11} & G_{21} & 0 & G_{02} & G_{12} & G_{22} \\
H_{00} & H_{10} & H_{20} & 0 & H_{01} & H_{11} & H_{21} & 0 & H_{02} & H_{12} & H_{22} & 0 & 0 & 0 & 0 & 0 \\
0 & H_{00} & H_{10} & H_{20} & 0 & H_{01} & H_{11} & H_{21} & 0 & H_{02} & H_{12} & H_{22} & 0 & 0 & 0 & 0 \\
0 & 0 & 0 & 0 & H_{00} & H_{10} & H_{20} & 0 & H_{01} & H_{11} & H_{21} & 0 & H_{02} & H_{12} & H_{22} & 0 \\
0 & 0 & 0 & 0 & 0 & H_{00} & H_{10} & H_{20} & 0 & H_{01} & H_{11} & H_{21} & 0 & H_{02} & H_{12} & H_{22}
\end{vmatrix} \tag{19}$$

vanishes. Then:

$$\det \mathbf{K} = 0 \qquad (20)$$

represents the necessary and sufficient condition so that a couple of values for t_2 and t_3 simultaneously satisfy Eqs. (14). In words, Eq. (20) is the final result of the elimination of unknowns t_2 and t_3 from Eqs. (14).

Based on the degrees of terms F_j, G_{jk}, and H_{jk}, j,k=0,1,2, the left hand side of Eq. (20) is a polynomial in t_1 with degree not greater than thirty-two. Direct computation shows, indeed, that Eq. (20) can be rewritten as:

$$\sum_{i=0,32} b_i \, t_1{}^i = 0 \qquad (21)$$

where coefficients b_i, i=0,1,...,32, depend only on the geometry of the 5-4 structure. Equation (21) provides thirty-two solutions for t_1 in the complex field.

D. Back Substitution

For a given 5-4 structure, Eq. (21) can be solved for t_1. Every root t_{1r}, r = 1,2,...,32, through Eqs. (13), allows computation of values s_{1r} and c_{1r} for s_1 and c_1, then angle θ_{1r} can be determined. For every value θ_{1r} of θ_1, all elements of matrix $\mathbf{K}$ can now be computed by Eqs. (15). Since Eq. (20) is satisfied, the homogeneous linear system of sixteen equations, whose coefficient matrix is $\mathbf{K}$, provides a non-trivial solution for the sixteen unknowns (18). In particular, solutions t_{2r} and t_{3r} for the unknowns t_2 and t_3 can be determined (they are the second and fifth quantities of (18) when the first one is made unitary). From Eqs. (13), angles θ_{2r}, and θ_{3r} can be determined. Since every $(\theta_{1r},\theta_{2r},\theta_{3r})$ triplet univocally defines one configuration of the 5-4 structure, thirty-two configurations are possible for the 5-4 structure in the complex field.

III. Case Study

The following geometrical data of a 5-4 structure have been considered (see Fig. (4)). Coordinates of points A_i, (i=1,...,5) are given in an arbitrary reference W_b system fixed to the base; they are in the order: (2.1, −1.4, 1.4), (1.5, 2.1, 0.2), (−1.7, 1.8, −2.2), (1.3, −1.9, −0.9), (0.2, −0.8, −1.6). Similarly, coordinates of points B_j, i=1,2,3,4, are given in an arbitrary reference system W_p fixed to the platform; they are in the order (−0.6, 0.4, −1.2), (−1.5, 1.1, 2.1), (−0.6, −1.3, −1.6), (1.4, −1.5, 1.8). The leg lengths L_i, i=1,...,6, are in the order 4.16, 3.69, 3.34, 2.73, 1.5, 4.44. (Data are expressed in arbitrary length unit.)

The coordinates of points A_i and B_i are first transformed in reference systems S_b and S_p respectively. By adopting the presented procedure for the solution of the FPA, thirty-two configurations in terms of triplets $(\theta_1,\theta_2,\theta_3)$ are obtained in the complex field. For each triplet, by Eq. (6), coordinates of points B_i are determined in reference system S_b, then transformed in reference system W_b. In such a form, points B_i are reported in Tb. (1) for every configuration (only one configuration out of a couple of complex conjugate ones is reported). The results reported in Tb. (1) can be straightforwardly validated by inverse position analysis.

IV. Conclusions

The paper presents the forward position analysis in symbolic form of a 5-4 fully-parallel manipulator with general geometry. The 5-4 structure obtained from the manipulator once

Table 1. x, y, z coordinates in reference system W_b of points B_1, B_2, B_3, and B_4 for every configuration (only one out of two complex conjugate configurations is reported).

B_1	B_2	B_3	B_4
(-0.20652392, 0.00000000)	(-1.64334597, 0.00000000)	(0.06266091, 0.00000000)	(1.57236903, 0.00000000)
(0.63050194, 0.00000000)	(0.96023541, 0.00000000)	(-1.02139950, 0.00000000)	(-1.25758137, 0.00000000)
(-1.40403446, 0.00000000)	(1.76090128, 0.00000000)	(-1.90279407, 0.00000000)	(1.73932063, 0.00000000)
(0.80046434, 0.00000000)	(4.27381641, 0.00000000)	(0.10369529, 0.00000000)	(3.04379016, 0.00000000)
(0.49189729, 0.00000000)	(0.74613196, 0.00000000)	(-0.79239107, 0.00000000)	(-2.83552834, 0.00000000)
(-2.06951462, 0.00000000)	(-1.82214832, 0.00000000)	(-1.11289750, 0.00000000)	(-2.78066015, 0.00000000)
(-0.67717277, 0.00000000)	(-1.48046330, 0.00000000)	(-0.82450742, 0.00000000)	(1.18700352, 0.00000000)
(0.33629542, 0.00000000)	(1.11869900, 0.00000000)	(-1.35875608, 0.00000000)	(-1.71716849, 0.00000000)
(-1.16479816, 0.00000000)	(2.14164539, 0.00000000)	(-1.55861652, 0.00000000)	(1.82152612, 0.00000000)
(-0.90183770, 0.00000000)	(1.77367216, 0.00000000)	(-1.83144240, 0.00000000)	(1.78343023, 0.00000000)
(-1.80687420, 0.00000000)	(-0.02032106, 0.00000000)	(-0.92098164, 0.00000000)	(-0.10349815, 0.00000000)
(-1.45114430, 0.00000000)	(-2.80756416, 0.00000000)	(-0.26749339, 0.00000000)	(1.09794302, 0.00000000)
(0.88522771, 0.00000000)	(-1.76503244, 0.00000000)	(0.48975721, 0.00000000)	(-1.07817722, 0.00000000)
(0.42138042, 0.00000000)	(0.38201486, 0.00000000)	(-0.21699534, 0.00000000)	(-3.09432183, 0.00000000)
(-2.13730146, 0.00000000)	(0.13526790, 0.00000000)	(-3.71403209, 0.00000000)	(-1.50890762, 0.00000000)
(0.40353138, 0.00000000)	(2.97858419, 0.00000000)	(-1.08666224, 0.00000000)	(1.40698120, 0.00000000)
(-2.09173182, 0.00000000)	(0.24771795, 0.00000000)	(-1.44407622, 0.00000000)	(0.10897932, 0.00000000)
(-2.33484957, 0.00000000)	(-2.62823974, 0.00000000)	(-1.69464241, 0.00000000)	(0.94538806, 0.00000000)
(-0.60391338, 0.00000000)	(-1.50002735, 0.00000000)	(-0.60850872, 0.00000000)	(1.39313238, 0.00000000)
(0.40097000, 0.00000000)	(1.10195395, 0.00000000)	(-1.29886531, 0.00000000)	(-1.50532324, 0.00000000)
(-1.19826086, 0.00000000)	(2.10258770, 0.00000000)	(-1.59893379, 0.00000000)	(1.79971417, 0.00000000)
(-0.41487379, 0.00000000)	(2.76473754, 0.00000000)	(-0.88692524, 0.00000000)	(0.46002619, 0.00000000)
(0.53297070, 0.00000000)	(0.18911522, 0.00000000)	(-0.78507416, 0.00000000)	(-2.90695509, 0.00000000)
(-1.29158579, 0.00000000)	(-2.69224105, 0.00000000)	(-0.24759097, 0.00000000)	(-3.29444888, 0.00000000)
(-0.05278453, 1.20612486)	(3.10493481, 2.18085581)	(0.46012853, 0.32766390)	(2.78955813, -0.16679525)
(1.23961146, 0.22712692)	(0.03610320, 0.59573912)	(2.18909484, -0.14578324)	(-0.96614132, -0.36210628)
(-1.37958084, -0.71845370)	(-3.30862473, 0.64714454)	(0.35569283, -0.25475386)	(1.24379103, 0.27363083)
(0.39439194, 1.70880252)	(-6.76725665, 4.82347593)	(0.62151387, 0.42523973)	(-2.95425635, 7.94642679)
(1.72330002, -0.11677084)	(-1.87848488, -1.85389008)	(3.87895691, -0.21854672)	(0.70485370, 0.56924279)
(-1.56605489, -1.10559577)	(-3.95674425, -7.81891736)	(-0.84795574, -0.39410745)	(-8.81843934, -4.08203448)
(0.42893833, -1.66155569)	(2.55039874, -2.26488334)	(-2.17679984, -2.19819100)	(-2.53035405, 0.25228125)
(-3.09118174, -0.55654048)	(-0.10063203, -0.50748112)	(-3.03512357, -0.42410528)	(0.46072004, -0.75351745)
(-2.55083062, 0.94101051)	(-3.43016778, -0.34771159)	(-3.25911499, 2.92574327)	(-0.13123695, 3.57089254)
(-2.23303687, 1.56761699)	(-3.37966818, -0.87093844)	(0.04247084, 0.65261243)	(-3.27366472, -4.07353943)
(-3.03986968, -2.17814772)	(0.72171623, -1.36093665)	(-1.66904264, -0.60791448)	(5.33914046, -3.55690668)
(-0.85465059, -1.42845364)	(1.93338144, -3.53392934)	(-0.77115056, -2.27178547)	(-2.05529035, -6.16117284)
(1.21032560, 1.94404160)	(-1.02965064, 0.29546082)	(11.98595906, 11.16669085)	(-2.26163702, 31.58170815)
(-3.19380441, 1.40279452)	(-0.04630320, -0.07076465)	(2.85965809, 13.32574218)	(-13.31078992, 28.22009842)
(-3.06623376, -0.95066744)	(-1.48168401, -0.35412732)	(-17.71649319, 10.75936439)	(-42.91528285, -10.34138452)
(-0.68908699, 2.67672173)	(0.44476666, 1.77487986)	(0.37070634, 1.51684002)	(-1.74144053, -1.49062379)
(2.55063154, 1.10018516)	(-0.31455710, 0.08237535)	(1.77637083, 0.68246641)	(-2.91055199, 2.02890922)
(-0.71438527, -1.47522007)	(-3.00129987, -0.64717848)	(1.04939467, -0.96165731)	(1.29170537, -1.13305620)
(1.61035185, 0.15989579)	(4.56481717, 0.46259864)	(0.37321730, 0.39287626)	(3.58282458, 0.72672631)
(-0.86179311, 1.12798492)	(0.86355803, 0.28048626)	(-0.78479368, 0.42659468)	(-2.33930078, -1.26010282)
(-2.85318146, 0.12432965)	(-1.62515598, 0.58678561)	(-1.39881872, 0.35964567)	(1.36288096, -0.97775927)
(1.28302667, -5.34308359)	(3.79595230, -1.03409407)	(0.91304955, -1.19089683)	(12.86746172, 2.88905687)
(5.72487439, 0.73992676)	(2.43449856, 0.07898763)	(0.76396114, -0.34105850)	(-6.06686385, 8.85974320)
(-1.32854201, 3.53193829)	(3.34118631, 0.74742762)	(0.25908766, 1.12176667)	(-1.36299960, -7.55570608)
(-0.29452875, -8.24276457)	(17.80719598, 14.94898766)	(-17.10724607, -11.82691975)	(2.39311798, -16.28493962)
(-10.41658744, -0.61388195)	(8.55876306, -2.94267417)	(-19.08511581, 8.56337570)	(-18.03329261, -4.16382382)
(-3.55771595, 5.09764117)	(14.19800260,-16.05729350)	(-2.35914833, 21.19513123)	(-9.00230927, 6.09392039)
(3.03239866, 6.62456781)	(14.05848334, -0.53347871)	(0.96822628, 11.40083299)	(2.42875103, -0.17021116)
(-8.14778908, 4.20243658)	(3.19251375, 3.82421687)	(-10.17335892, 6.94836068)	(-1.45994703, 0.14013809)
(-5.21101030, -3.35507230)	(0.42079844, 11.42070522)	(-10.90159807, -6.06500251)	(1.55695022, 0.05309747)
(-2.16646505, -13.92216879)	(-0.13088881, -0.65918396)	(-2.88959174, -0.76267018)	(29.26721726, 14.83976424)
(15.11195479, -4.24091387)	(-0.18001948, 0.02534282)	(1.57827264, -0.31650840)	(-6.56135874, 37.07676210)
(2.73355215, 7.96919079)	(-2.32107076, 0.40350855)	(-5.43488041, 0.30215792)	(-30.33105017, 8.22936389)
(-37.56019715, -40.53531278)	(4.46182260, -4.62098911)	(-6.49290277, -20.30221582)	(-18.27836933, 96.06633498)
(41.46936649, -48.33582992)	(3.36958918, 0.05148290)	(20.00118561, -14.38476030)	(-107.80936206, -23.36838157)
(30.42106482, 16.00519878)	(5.73559881, 2.46065301)	(16.58112270, 8.75949475)	(12.40204978, -44.66290598)

the leg lengths are frozen has been considered. Based on an original kinematic model, the rigid body position of the platform with respect to the base has been expressed as a function of three parameters that, by considering all existing constraints between base and platform, have to satisfy three equations. Through a single-step elimination procedure two unknown parameters have been suppressed and a final polynomial equation of order thirty-two in the remaining parameter is obtained. Since each root of the polynomial equation corresponds to a solution of the forward position analysis, thirty-two configurations are possible in the complex field for the considered 5-4 fully-parallel manipulator. Numerical results validate, through inverse position analysis, the new theoretical findings.

V. Acknowledgments

The financial support of Ministry of University, Scientific and Technological Research (MURST) and Research National Council (CNR) is gratefully acknowledged.

VI. References

[1] M. Griffis and J. Duffy, "A Forward Displacement Analysis of a Class of Stewart Platform", *J. Robot. Sys.*, 6(6), 703-720 (1989).
[2] C.C. Nguyen and M. Jamshidi, "Guest Editorial: Parallel Closed-Kinematic Chain Manipulators and Devices", *J. Robot. Sys.*, 10(5), 557-560 (1993).
[3] K.J. Hunt, "Structural Kinematics of In-Parallel Actuated Robot-Arms", *ASME, J. Mech. Trans. Auto. in Design*, 105, 705-712 (1983).
[4] C. Innocenti and V. Parenti-Castelli, "Basic Ideas and Recent Techniques for the Analytical Form Solution of the Direct Position Analysis of Fully-Parallel Mechanisms", *Int. J. Laboratory and Automation*, 4, 107-113 (1992).
[5] C. Innocenti and V. Parenti-Castelli, "Exhaustive Enumeration of Fully-Parallel Kinematic Chains", submitted to *ASME J. Mech. Design* (1994).
[6] F. Freudenstein and B. Roth, "Numerical Solution of Systems of Nonlinear Equations", *J. of Association of Computing Machinery*, 10, 550-556 (1963).
[7] A. Morgan and A. Sommese, "Computing All Solution to Polynomial Systems Using Homotopy Continuation", *Appl. Math. Comput.*, 24, 115-138 (1987).
[8] G.D.D. Salmon, *Modern Higher Algebra*, Hodges, Figgis, and Co., Dublin (1885).
[9] P. Nanua and K.J. Waldron, "Direct Kinematic Solution of a Special Parallel Robot Structure", *Proc. 8th CISM-IFToMM Symposium on Theory and Practice of Robots and Manipulators*, 134-142, Cracow, Poland (1990).
[10] C. Innocenti and V. Parenti-Castelli, "Direct Kinematics of the Reverse Stewart Platform Mechanism", *IFAC'91-SYROCO'91*, 75-80, Wien, Austria (1991).
[11] C. Innocenti and V. Parenti-Castelli, "Direct Kinematics in Analytical Form of a General Geometry 5-4 Fully-Parallel Manipulator", *Computational Kinematics*, (J. Angeles, P. Hommel, and P. Kovács Eds.), Kluwer Academic Publisher, Dordrecht (1993).
[12] W. Lin, C.D. Crane, and J. Duffy, "Closed-Form Forward Displacement Analysis of the 4-5 In-Parallel Platform", *Proc. 22nd ASME Biennial Mechanisms Conference*, 521-527, Scottsdale, USA (1992).
[13] J. Denavit and R.S. Hartenberg, "A Kinematic Notation for Lower-Pair Mechanisms Based on Matrices", *ASME J. Appl. Mech.*, E, 22, 215-221 (1955).
[14] C. Innocenti, "Analytical-Form Position Analysis of the 7-Link Assur Kinematic Chain with Four Serially-Connected Ternary Links", *ASME J. Mech. Design* (to appear) (1993).

Evaluation of the Errors When Solving the Direct Kinematics of Parallel Manipulators With Extra Sensors

Luc TANCREDI   Jean-Pierre MERLET
INRIA - Sophia-Antipolis
2004, route des Lucioles B.P. 93
06902 Sophia-Antipolis - FRANCE

Abstract - The measurement of the links lengths of a general six d.o.f parallel manipulator is not sufficient to determine the actual unique posture of its platform. It has been shown that adding four extra sensors leads in general to an unique solution. We derive an analytical form of this solution and then study its robustness with respect to the sensors errors. As example, maps of the errors for a particular manipulator with a given orientation are exhibited.

I. Introduction

When studying robots kinematics two main problems appear. They are called direct kinematics and inverse kinematics problems. By direct kinematics we mean the problem of computing the position and the orientation[1] of the end effector of the manipulator from the values of the articular variables. By inverse kinematics we mean the dual problem, *i.e.* determining the values of articular coordinates for a given posture. The problem of direct kinematics for parallel manipulators is, by far, the more difficult one ([1], [2]).

A general 6 d.o.f. parallel manipulator is shown in Fig. (1), which has six linear adjustable actuators connecting a mobile platform and a base platform. As the length of the actuators change an end-effector attached to the mobile platform can be moved in six d.o.f. space. Each link is connected to the base platform through an universal joint and to the mobile platform through a ball-and-socket joint.

Assuming that the length of each of the six links is given, the problem is to determine the manipulator posture. Firstly, it is known that this problem has not a unique solution. For the general parallel manipulator the maximum number of *real* solutions is not known but an upper bound is 40 ([3], [4]) and at this time examples with at most 16 real solutions have been found ([5]). For the special case where the mobile platform

[1] In all this paper we will use *posture* to refer to position AND orientation

A. J. Lenarčič and B. B. Ravani (eds.), Advances in Robot Kinematics and Computational Geometry, 439–448.

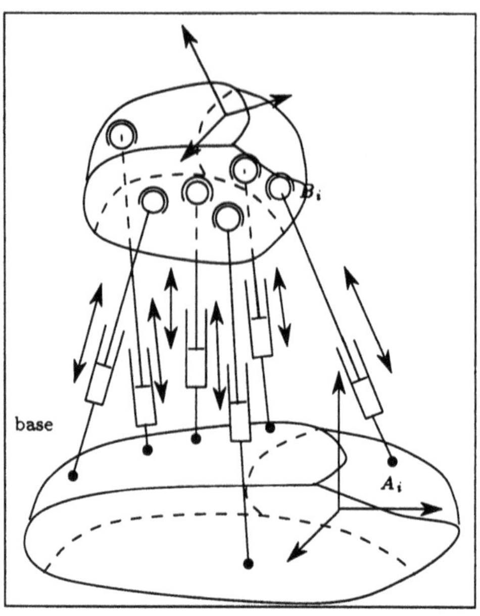

Fig. 1: A 6 d.o.f. parallel manipulator.

is a triangle it has been shown that there can be up to 16 solutions and examples with 16 solutions have been found ([6],[7]).

In any case the numerical methods which determine all the solutions, are not usable in real time as their computation time is by far too great. Furthermore sorting out the unique solution (*i.e.* the actual posture of the end-effector) among the set of solutions remains an open and difficult problem. Therefore it may be interesting to add sensors in order to solve the direct kinematics problem. Stoughton and Arai [8] have followed this approach. In their article they discuss about sensors placement and sensibility to sensors errors and loss. Although their study deals with a particular class of parallel manipulator and so is too restricted for our purpose, we shall remember that optimal placement is on the base.

We will denote A_i the joint centre of link i near the base, B_i the joint center near the platform and ρ_i the link length. We define a reference frame $\mathcal{R}(O, \boldsymbol{x}, \boldsymbol{y}, \boldsymbol{z})$ attached to the base and a moving frame $\mathcal{R}_r(C, \boldsymbol{x}_r, \boldsymbol{y}_r, \boldsymbol{z}_r)$ attached to the platform, where C is any reference point on the platform.

II. Adding four sensors

A first method for solving the direct kinematics problem will be to add passive legs [9]. But these legs will decrease the size of the workspace of the robot as more links interference will occur. Another method is to put rotary sensors on the passive joints of the existing legs. Two sensors are needed to measure the link direction, and with six extra sensors we may determine the position of three points of the platform and therefore its posture. But it has been shown in [10] that four sensors lead to a unique solution in general.

A. *Geometrical result*

We will assume that four sensors are used to equip two links with two sensors each. The sensors measurements give the direction of the link. On the other hand the length of each link is given by other sensors and therefore we are able to compute the position of the centres of the platform joints B_1 and B_2 for two legs. Thus the remaining unknown for the posture of the platform is its rotation angle around the axis (B_1B_2). Consequently any joint centre (*e.g.* B_3) moves on a circle C_3 lying in a plane P perpendicular to (B_1B_2) (see Fig. (2)). But B_3 lie also on the sphere S of centre A_3,

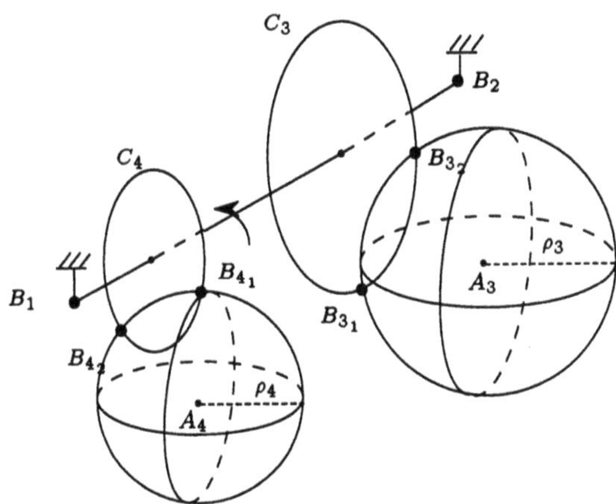

Fig. 2: Positions of B_3 and B_4 when two points of the platform are fixed

with radius equal to the link length ρ_3. Therefore B_3 is at the intersection of the plane P and the sphere *i.e.* a circle C_3^S. Consequently the possible positions of B_3 are the intersections of the circles C_3 and C_3^p *i.e.* there are two solutions for B_3 (see Fig. (3)). A similar reasoning can be made for B_4 and we project the circles C_4 and C_4^S on the plane P. Let B_{4_1p} and B_{4_2p} be the two possible solutions for B_4 and δ_1 (resp. δ_2) the angle between the lines $(B_{1,2p}B_{4_1p})$ (resp. $(B_{1,2p}B_{4_2p})$). According to Fig. (3) we have:

$$\delta_1 = \theta_1 + \lambda + \alpha \tag{1}$$

Purely geometrical considerations show that $\delta_1 = -\delta_2$. Hence

$$\delta_2 = -\delta_1 = -\theta_1 - \lambda + \alpha \tag{2}$$

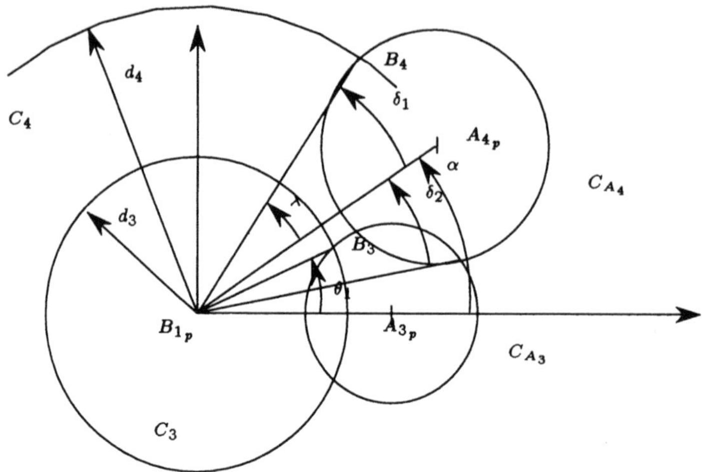

Fig. 3: Four extra sensors are sufficient to determine the posture.

From equations (1) and (2) we deduce

$$\delta_2 = \quad \theta_2 + \lambda - \alpha = \quad -\theta_1 + \lambda - \alpha \tag{3}$$

According to this equation the two solutions for B_3 will be valid if and only if $\lambda = \alpha$ (this has to be true for every joint B_3, B_4, B_5, B_6). Except for this special case, one unique solution can always be found with four extra sensors. Practically, one position among the two for B_4 is taken at random, so angle δ is fixed. This give four criterion to choose whether the solution for B_3 is B_{3_1} or B_{3_2}

$$B_3 = B_{3_1} \quad if \quad \delta = \theta_1 + \lambda - \alpha \quad or \quad -\delta = \theta_1 + \lambda - \alpha \tag{4}$$
$$B_3 = B_{3_2} \quad if \quad \delta = \theta_2 + \lambda - \alpha \quad or \quad -\delta = \theta_2 + \lambda - \alpha \tag{5}$$

B. Computing the location of three points of the platform

We define a joint frame $(A_i, \boldsymbol{x}, \boldsymbol{y}, \boldsymbol{z})$ attached to each joint near the base where $\boldsymbol{x}$ is coincident with the first rotation axis of the joint, $\boldsymbol{z}$ being vertical. Let $\boldsymbol{z}''$ be the unit vector of the link. This vector is obtained from the $\boldsymbol{z}$ vector by applying a rotation of angle α around x (rotation matrix R_α) followed by a rotation around $y' = R_\alpha y$ with the angle β (rotation matrix R_β) (see Fig. (4)). We get

$$AB \quad = \quad \| AB \| \, R_\alpha R_\beta \left[\begin{array}{ccc} 0 & 0 & 1 \end{array} \right]^T \tag{6}$$

As $\| AB \| = \rho$ is given by the prismatic sensor the measurement of α, β enables to compute the coordinates of B in the joint frame as:

$$B \quad = \quad \left[\begin{array}{ccc} -\rho \sin\beta & -\rho \sin\alpha \cos\beta & \rho \cos\alpha \cos\beta \end{array} \right]^T \tag{7}$$

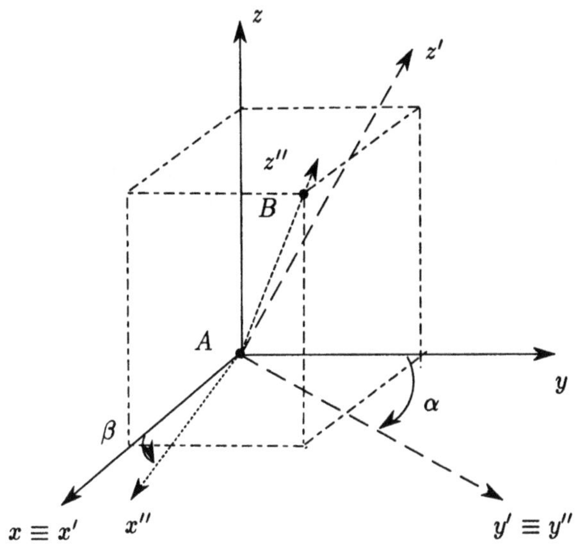

Fig. 4: Location and measures for angular sensors.

Let R_i be the rotation matrix between the joint frame and the reference frame $\mathcal{R}$, and $A_iB_{i\mathcal{C}}$ the leg vector in the joint frame. Therefore we have

$$OB_i = R_i \cdot A_iB_{i\mathcal{C}} + OA_i \tag{8}$$

Adding four sensors to the two links (e.g. links 1, 2) enable thus to compute the location of B_1 and B_2 in the reference frame. Using the geometrical results presented in the previous section we are able to compute the two possible locations of joint centres $B_{3..6}$ and using the selection criteria we deduce from them the location of B_3.

C. Retrieving the posture of the platform

Let CB_{i_r} be the vectors in the mobile frame defining the location of the joint centres B_i. Let $\widehat{v}$ denote the unitary vector corresponding to $\frac{\boldsymbol{v}}{\|\boldsymbol{v}\|}$. It is possible to find three real values μ_1, μ_2, μ_3 such that

$$B_1C_r = \mu_1 \cdot \widehat{B_1B_{2_r}} + \mu_2 \cdot \widehat{B_1B_{3_r}} + \mu_3 \cdot (\widehat{B_1B_{2_r} \wedge B_1B_{3_r}}) \tag{9}$$

Indeed, this equation is a linear system in the μ_i. Since the platform geometry is fixed Eq. (9) gives the values of μ_i which are independent of the current posture. Therefore after computing the location of B_1, B_2, B_3 from the sensors measurements we may compute $B_1C_{\mathcal{B}}$ as:

$$B_1C_{\mathcal{B}} = \mu_1 \cdot \widehat{B_1B_{2\mathcal{B}}} + \mu_2 \cdot \widehat{B_1B_{3\mathcal{B}}} + \mu_3 \cdot (\widehat{B_1B_{2\mathcal{B}} \wedge B_1B_{3\mathcal{B}}}) \tag{10}$$

For which we deduce the location of C in the reference frame as

$$OC_{\mathcal{B}} = \quad OB_1 + B_1C_{\mathcal{B}} = \quad \lambda_1 \cdot OB_{1\mathcal{B}} + \lambda_2 \cdot OB_{2\mathcal{B}} + \lambda_3 \cdot OB_{3\mathcal{B}} \tag{11}$$

To compute the complete posture of the platform we need to evaluate the rotation matrix $R_{(\psi,\theta,\phi)}$ from the moving frame to the reference frame. Using a similar method we can find a triplet of coefficients of linear combination for each base vector $(\boldsymbol{u}_{x_r}, \boldsymbol{u}_{y_r}, \boldsymbol{u}_{z_r})$ of the moving frame. Then as in Eq. (10) we can express each base vector of the moving frame in the reference frame, say $(\boldsymbol{u}_{x\mathcal{B}}, \boldsymbol{u}_{y\mathcal{B}}, \boldsymbol{u}_{z\mathcal{B}})$. We get the rotation matrix by

$$R \quad = \quad \left[\ \boldsymbol{u}_{x\mathcal{B}} \quad \boldsymbol{u}_{y\mathcal{B}} \quad \boldsymbol{u}_{z\mathcal{B}} \ \right] \tag{12}$$

III. Effect of measurement errors

After the calculus of the platform posture, we would like to know how this solution depends on sensors errors.

All our study is based on a first order derivation of Eq. (11).

$$dOC \quad = \quad \lambda_1 \cdot dOB_1 + \lambda_2 \cdot dOB_2 + \lambda_3 \cdot dOB_3 \tag{13}$$

We use the jacobian matrix related to extra angular sensors data.

$$dOC \quad = \quad \lambda_1 J_1 d\eta + \lambda_2 J_2 d\eta + \lambda_3 J_3 d\eta \tag{14}$$

where $J_i = \frac{\partial B_i}{\partial \eta}, i = 1..3, \eta = (\alpha_1, \beta_1, \alpha_2, \beta_2)^T$.
J_1, J_2 can be written as

$$J_1 = \left[\ J_1^{\alpha_1} \quad J_1^{\beta_1} \quad 0 \quad 0 \ \right] \qquad J_2 = \left[\ 0 \quad 0 \quad J_2^{\alpha_2} \quad J_2^{\beta_2} \ \right]$$

where $J_1^{\alpha_1}, J_1^{\beta_1}, J_2^{\alpha_2}$ and $J_2^{\beta_2}$are the (3×1) matrices:

$$J_1^{\alpha_1} = \left[\frac{\partial OB_1}{\partial \alpha_1} \right] \qquad J_1^{\beta_1} = \left[\frac{\partial OB_1}{\partial \beta_1} \right] \qquad J_2^{\alpha_2} = \left[\frac{\partial OB_2}{\partial \alpha_2} \right] \qquad J_2^{\beta_2} = \left[\frac{\partial OB_2}{\partial \beta_2} \right]$$

We have

$$J_3 = \frac{\partial B_3}{\partial \eta} \quad = \frac{\partial B_3}{\partial B_1}\frac{\partial B_1}{\partial \eta} + \frac{\partial B_3}{\partial B_2}\frac{\partial B_2}{\partial \eta} \quad = J_{3_1}\frac{\partial B_1}{\partial \eta} + J_{3_2}\frac{\partial B_2}{\partial \eta}$$

Consequently we have $dOC = J \cdot d\eta$ with

$$\begin{aligned} J \quad &= \quad \left[\ \lambda_1 J_1^{\alpha_1} + \lambda_3 J_{3_1} J_1^{\alpha_1} \quad \lambda_1 J_1^{\beta_1} + \lambda_3 J_{3_1} J_1^{\beta_1} \quad \lambda_2 J_2^{\alpha_2} + \lambda_3 J_{3_2} J_1^{\alpha_2} \quad \lambda_2 J_2^{\beta_2} + \lambda_3 J_{3_2} J_2^{\beta_2} \right. \\ &= \quad \left[\ J^{\alpha_1} \quad J^{\beta_1} \quad J^{\alpha_2} \quad J^{\beta_2} \ \right] \end{aligned}$$

	Base			Platform		
Joints	x_B	y_B	z_B	x_P	y_P	z_P
1	-9.7	9.1	0	-3	7.3	0
2	9.7	9.1	0	3	7.3	0
3	12.76	3.9	0	7.822	-1.052	0
4	3	-13	0	4.822	-6.248	0
5	-3	-13	0	-4.822	-6.248	0
6	-12.76	3.9	0	-7.822	-1.052	0

Tb. 1: Location of the joint centres in the reference frame and in the moving frame.

IV. Example on the *left hand* manipulator

To illustrate our purpose we ran our algorithms on a particular geometry of parallel manipulator. This robot is called the left hand, an INRIA's prototype, defined by the following table Tb. (1).

Each map in Fig. (5),(6) has the following features. The coordinates on x and y axes represent coordinates of the centre for a constant height and orientation. Value on the z axis is the norm of the error vector on platform centre for a unit value of the corresponding sensor error. These maps represent graphically numerical results that we can use like a reference table. Since we made a first order approximation on the error, we can linearly interpolate to obtain a value for a point not on the grid. Provided we use a grid fine enough, this approach needs to compute over the working space only once per robot, independantly of sensors quality. Thus for a given precision of sensors the global error on the centre position is

$$\Delta OC = J^{\alpha_1}\Delta\alpha_1 + J^{\beta_1}\Delta\beta_1 + J^{\alpha_2}\Delta\alpha_2 + J^{\beta_2}\Delta\beta_2 \qquad (15)$$

With the maps presented, we illustrate how each angle influence the final value of the position of the centre. For each curve of Fig. (7) we have a minimum error when x is about 8.00.

V. Conclusion

In this paper we have shown that adding four extra sensors leads to a unique solution for the direct kinematics problem. We have then studied the influence of the sensors measurement errors by producing maps enabling to compute the errors for any value of the errors. According to these maps the designer may determine the maximum sensors errors to obtain a given accuracy on the posture of the platform. Another method will be to use the result of the centre position as an initial guess for an iterative algorithm enabling to ensure a fast and safe convergence of the algorithm.

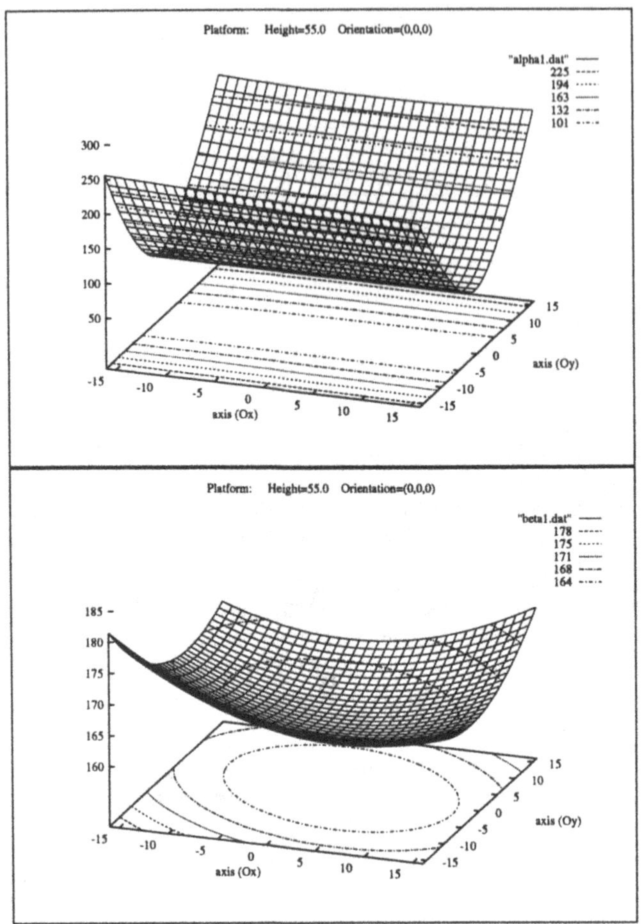

Fig. 5: Maps for J^{α_1} and J^{β_1}

References

[1] J. Angeles and K.E. Zanganeh "The semi-graphical solution of the direct kinematics of general platform manipulators." *ISRAM 92*, 45–52, Santa-Fe, USA (1992).

[2] J-P. Merlet "Direct kinematics and assembly modes of parallel manipulators." *International Journal of Robotics Research*, 11(2), 150–162 (1992).

[3] D. Lazard "Stewart platform and Gröbner basis." *Advances in Robot Kinematics*, 136–142, Ferrare, Italia (1992).

[4] D. Lazard "Generalized Stewart Platform: How to compute with rigid motions?" *IMACS*, 85–88, Lille, France (1993).

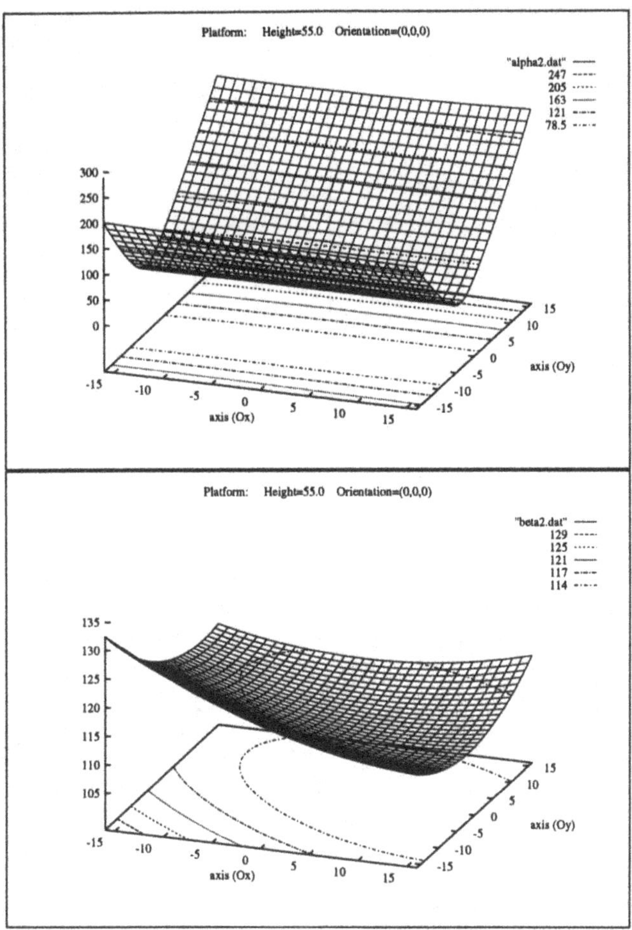

Fig. 6: Maps for J^{α_2} and J^{β_2}

[5] M. Ait-Ahmed *Contribution à la modélisation géométrique et dynamique des robots parallèles.* PhD thesis, Université Paul Sabatier, Toulouse, France (1993)

[6] C. Innocenti and V. Parenti-Castelli "Direct position analysis of the Stewart platform mechanism." *Mechanism and Machine Theory*, 25(6), 611–621 (1990).

[7] S. Charentus and M. Renaud "Calcul du modèle géométrique direct de la plateforme de Stewart." Research Report 89260, LAAS, Toulouse, France (1989).

[8] R. Stoughton and T. Arai "Optimal sensor placement for forward kinematics evaluation of a 6-dof parallel link manipulator." *IEEE/RSJ Int. Workshop on Intelligent Robots and Systems, IROS'91*, 785–790, Osaka, Japan (1991).

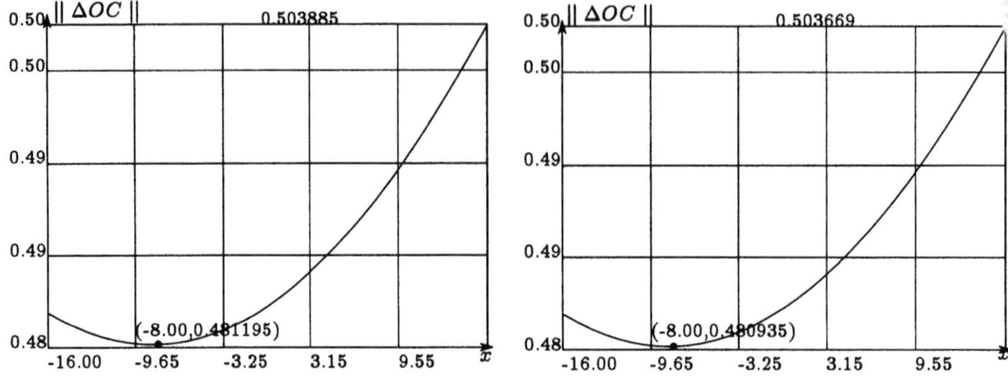

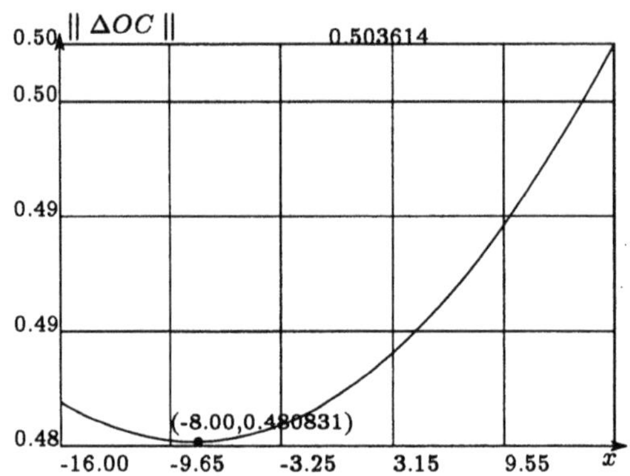

Fig. 7: Norm of error along x axis for $y = -1, 0$ and $1, z = 55.0, \psi = 0, \theta = 0, \phi = 0$.

[9] P. Nair "On the kinematics geometry of parallel robot manipulators." Master's thesis, University of Maryland, College Park, USA (1992).

[10] J-P. Merlet "Closed-form resolution of the direct kinematics of parallel manipulators using extra sensors data." *IEEE Int. Conf. on Robotics and Automation*, 200–304, Atlanta, USA (1993).

A Special Type of Singular Stewart-Gough Platform

M. L. Husty   P. Zsombor-Murray

Department of Mechanical Engineering &
McGill Centre for Intelligent Machines
McGill University
817 Sherbrooke St. West
Montréal, Québec, Canada, H3A 2K6
husty@cim.mcgill.ca   paul@cim.mcgill.ca

Abstract - A Stewart-Gough platform, whose base attachment points occupy a particular cubic surface, may exhibit continuous motion while all six prismatic actuators are locked. Line geometric analysis reveals that, during such motion, the six leg axes remain in a specific linear complex, congruence or hyperboloidal ruled surface. Furthermore the pose or direct kinematics of any platform, five of whose leg base attachment points lie in such a cubic surface, is readily obtained and admits no more than four real solutions.

I Introduction

Six-legged parallel manipulators(StGp), introduced by Gough and adapted to move flight simulators [1], continue to stimulate considerable research interest. Direct kinematics and singularities are of particular concern and have been investigated by [2], [3], [4] and [5] as well as others. These investigations consider solution of the so-called *assembly problem*, where six legs of given length, attached to six base points, are positioned to satisfy the rigid body constraint imposed by the six attachment points on the mobile platform.

Consider a different approach which deals with singularities in terms of a class of continuous motions where some points on a rigid body are each constrained to remain on the surface of as many given, fixed spheres. This class of motion was the topic of a competition conducted by L'Académie des Sciences de l'Institut National de France. Results contained in the prize-winning papers submitted by Borel [6] and Bricard [7] will be used. Our paper applies a rather remarkable member of this class to singular motion of a StGp. Except for fixed axis rotation it is the *only* way in which *all* points on a rigid body may be moved on non concentric spherical paths and it shows how to

A. J. Lenarčič and B. B. Ravani (eds.), Advances in Robot Kinematics and Computationed Geometry, 449–458.

construct nontrivial platforms, having any number of fixed length legs, which move in a finite, one degree-of-freedom(dof) workspace.

II Borel-Bricard Motion

The competition question was not completely answered, documentation of n-sphere rigid body motions is by no means complete. However the discovery of an ∞-sphere rigid body motion and the proof of its uniqueness were independently achieved by Borel and Bricard. This motion can be expressed as follows.

$$\begin{aligned} X &= x\cos 2t - y\sin 2t \\ Y &= x\sin 2t + y\cos 2t \\ Z &= z + 2\sqrt{\rho^2 - a^2\sin^2 t} \end{aligned} \tag{1}$$

t is the angular displacement of the mobile platform. (x, y, z) are the coordinates of any point on the platform's rigid body, in the moving frame, while (X, Y, Z) are the coordinates of this point measured in the fixed, reference system. Finally, ρ and a are design constants, chosen as described in [8], which characterize the platform. Elegant properties of the motion, reported by Borel and Bricard and later elaborated in [9], are listed below.

- All points on the platform move on different, fixed spheres except that points on the z-axis move along it; even this is an arc upon a sphere of ∞-radius. The centre coordinates (X_c, Y_c, Z_c), of the sphere whose radius is given by $r^2 = (X_c - X)^2 + (Y_c - Y)^2 + (Z_c - Z)^2$, are expressed below.

$$X_c = \frac{a^2 x}{x^2 + y^2},\ Y_c = \frac{a^2 y}{x^2 + y^2},\ Z_c = z \tag{2}$$

- It is a Schönflies motion [10] wherein any plane in the platform which is parallel to $Z = 0$, remains so.
- Eq. 2 is a cubic equation. It relates a point on the curve to its osculating sphere centre [10].
- $x^2 + y^2 = a^2$ is an inversion circle in the plane $z = 0$, obtained by setting $X_c = x$ and $Y_c = y$ in the first two Eqs. 2. All points on the platform, not on its circumference when viewed in this projection, may be related to their sphere centres by inversion with respect to it. In this motion, the ruled surface of axodes degenerates to coincide with the axis of this circle, called the axis of a Type-I Borel-Bricard motion(BBM).
- The relationship $X^2+Y^2 = x^2+y^2$, obtained by squaring and adding the first two Eqs. 1, states that paths of all points on the platform are curves of intersection

between cylinders, coaxial on the z-axis, and spheres. These fourth order curves are generally non rational except when $\rho = a$ and there is a double point.

- BBM is a line symmetric, one dof motion whose basic surface is a spherical conoid [9].

Legs of fixed length, connecting any n-points on a platform to points on its base, which satisfy the relation Eq. 2, create a continuously movable, BBM platform(BBMP). Figure 1 shows a BBMP with six points, chosen randomly, on the platform. Figure 2 shows a second position as it executes BBM. Such motion is physically limited by leg collisions. Notwithstanding, a real BBM platform would exhibit a substantial continuum of unimpeded, continuous movement.

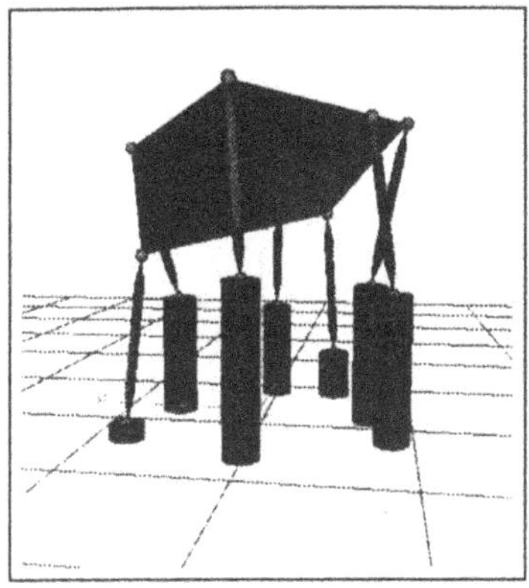

Figure 1: General Platform

Figure 2: Second Position

Some practical BBM mechanisms already exist. Jacobsen's LADD actuator is one [11]. Figure 3 shows three BBM positions of this mechanism also devised by Sir Christopher Wren to prove, in 1645, that the one sheeted hyperboloid of revolution contains straight lines.

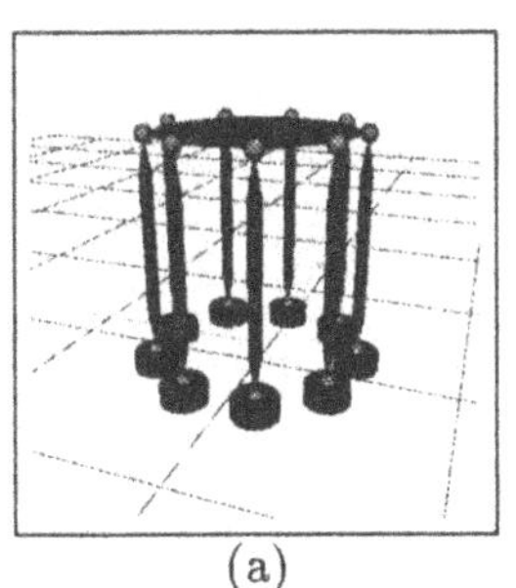

(a)

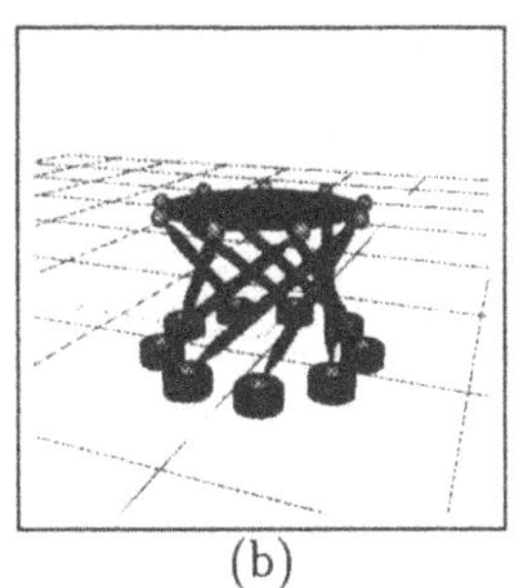

(b)

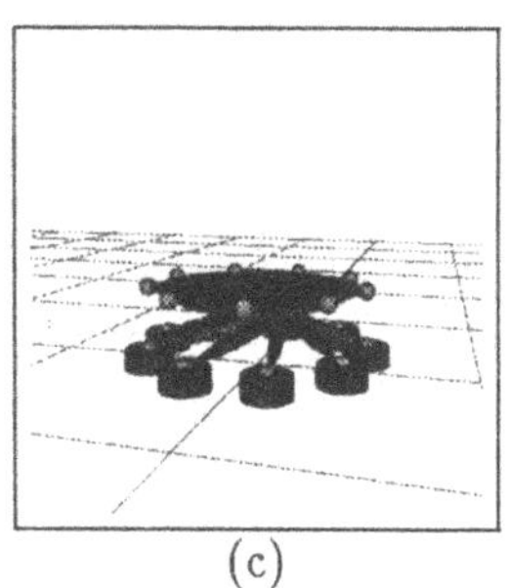

(c)

Figure 3: Platform Points on a Circle centered at the axis

Three positions of a BBMP with 18 leg attachment points in an elliptical array are presented in Figure 4.

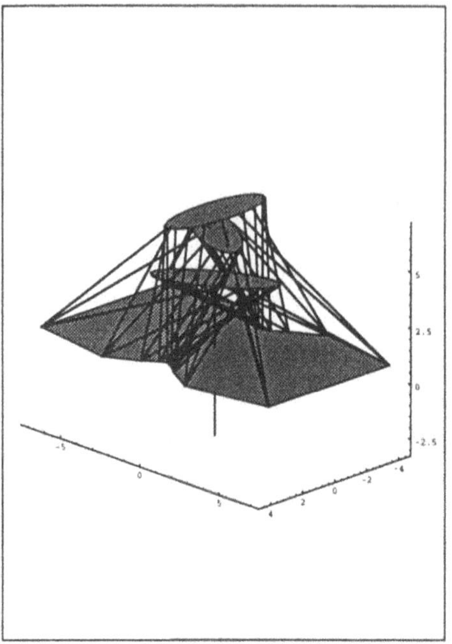

Figure 4: Eighteen Legged Movable Platform

III Line Geometry and Platforms

Merlet [3] pointed out that singularity of a StGp depends on the line geometric relation among its six legs. A platform with fixed length legs is in a rigid pose if the six axes comprise a general layout. If they are in a special line geometric configuration, then the platform is locally movable. The Plücker coordinates of these axes are written as $(p_1 : p_2 : p_3 : p_4 : p_5 : p_6) \neq (0:0:0:0:0:0)$ where $p_i, i = 1, \ldots, 6$ are homogeneous coordinates of points in a five dimensional projective space, P_5. A point in this space represents a line in ordinary, three dimensional Euclidean space, E_3, only if p_i fulfill the so-called Plücker condition.

$$\Omega = \sum_{i=1}^{3} p_i p_j = 0, j = i + 3 \tag{3}$$

These points in P_5 occupy the quadratic Plücker surface, M_4^2 (This notation stands for, " 4-dimensional {indicated by $_4$}, quadratic {indicated by ^{2}} manifold {M}"), sometimes called Klein quadric. Special line geometric configurations can be conveniently expressed in terms of P_5, whose coordinate space is a six dimensional vector space, V_6. With V_6 it is possible to establish linear dependency among lines in the corresponding three dimensional Euclidean space, E_3. Usually six arbitrary lines in E_3 correspond to six vectors in V_6 which span V_6 and constitute a basis. If one line is linearly dependent on the five others, which means geometrically that the corresponding point of P_5 is in the four dimensional subspace spanned by the other five, then the six lines belong to a special line geometric configuration called *linear complex* [12]. If this happens with the six leg axes of the StGp then it is singular and gains one infinitesimal, local dof. Note that points in P_5 represent lines in E_3 only if they are on M_4^2. This means that a linear complex in P_5 is represented by the intersection of a four dimensional subspace and M_4^2. But linear dependency can be compounded so that six points of P_5, corresponding to six lines, are in a three dimensional subspace of P_5. Then the six lines in E_3 belong to a more restricted line geometric configuration called *linear congruence* which is represented in P_5 by the intersection of a three dimensional subspace with M_4^2. A third special line geometric configuration, pertinent to BBMP/StGp's, occurs when the six corresponding points in P_5 are in a two dimensional subspace of P_5 and the lines belong to a *quadratic surface regulus*. According to [3] and [13], all three cases produce a singular StGp leg Jacobian matrix. The rank of the Jacobian depends on the dimension of the corresponding subspace in P_5. A search of the literature on StGp's revealed references only to singularities occurring at discrete platform positions. However it is seen that BBMP's produce a continuous set of singular positions which correspond to a continuous set of subspaces in P_5. With appropriate choice of six points on either base or platform, all three special types of line geometric leg configuration can be achieved and the BBMP moves without changing leg length. The line geometric reason for continuous movement is that at least one vector in the Jacobian null space is directionally invariant during motion. Expressed in another way, all instantaneous screws belonging to this null space vector, always share the same axis but have different, continuously changing pitch.

If a point $P(x, y, z)$ is chosen on the platform, the corresponding point on the base is determined by Eq. 2. Plücker coordinates, $\mathcal{P}_k$, of the leg axes l_k, $k = 1, .., 6$ are given by Eq. 4.

$$
\begin{aligned}
p_{1k} &= ax - (x^2 + y^2)(x\cos t - y\sin t) \\
p_{2k} &= ay - (x^2 + y^2)(y\cos t + x\sin t) \\
p_{3k} &= (x^2 + y^2)z - (x^2 + y^2)\left(z + 2\sqrt{r - a\sin[t/2]^2}\right) \\
p_{4k} &= -\left(ay(z + 2\sqrt{r - a\sin[t/2]^2})\right) + (x^2 + y^2)z(y\cos t + x\sin t) \\
p_{5k} &= ax\left(z + 2\sqrt{r - d\sin[t/2]^2}\right) - (x^2 + y^2)z(x\cos t - y\sin t) \\
p_{6k} &= -a(x^2 + y^2)\sin t
\end{aligned}
\tag{4}
$$

Normed Plücker coordinates are produced by dividing p_{ik} by n_k.

$$\begin{aligned} n_k &= \|(p_{1k}, p_{2k}, p_{3k})\| \\ &= \sqrt{(x^2+y^2)(a^2-2ax^2+4rx^2+x^4-2ay^2+4ry^2+2x^2y^2+y^4)} \end{aligned} \quad (5)$$

The six legs' normed Plücker coordinates, $\mathcal{P}_k$, may be written in a 6×6 matrix.

$$\mathbf{P} = [\mathcal{P}_1, \ldots, \mathcal{P}_6] \quad (6)$$

Then the Jacobian, $\mathbf{J}$, of the platform is given by

$$\mathbf{J} = \mathbf{P}^T \quad (7)$$

Fully expanding the Jacobian matrix, especially in normalized form, is a tedious process with an unwieldy result. But in the case of BBMP's the determinant of this matrix is independent of t, the motion parameter, and the choice of the points; $\det \mathbf{J} = 0$, always. This means that all positions are singular. Six randomly chosen points on the platform and their corresponding points define six fixed length leg axes which, during motion, remain in a common, varying linear complex of the form

$$Ap_3 + Bp_6 = 0 \quad (8)$$

where A and B are functions of the motion parameter, t, the points on the moving platform and the design constants, ρ and a. Eq. 8 shows that all complices have a common axis, *i.e.*, the axis of their reciprocal screws [14], which is the z-axis of the base coordinate system. So all the instantaneous screws have the same axis and their pitches change continuously, thereby producing continuous motion. It is easy to verify that all leg axes intersect the z-axis of the base when $t = 0$ and $t = \pi$. Therefore $A = 0$ in Eq. 8 which means that at $t = 0$ and π, the legs belong to a special line complex consisting of all lines intersecting the z-axis of the base.

Depending on the choice of points on the moving platform, all three types of special line geometric configurations can be obtained. *E.g.*, a circular array of points, on the platform, whose axis coincides with the axis of a BBM, Fig.3, is chosen. Then, during motion, all leg axes remain on various hyperboloids of revolution which share the z-axis of the base. At $t = 0$ and $t = \pi$ the surface of revolution becomes a cone or cylinder; essentially singular cases of the hyperboloid. Only the axes of legs attached on the cylinder, $x^2 + y^2 = a^2$, degenerate to a cylindrical surface of revolution. The infinitesimal dof in this example is always three. To demonstrate the possibility of leg axes which remain in a linear congruence, one point is moved from the circle to some other, arbitrary position and reanchored on the base at a new, compatible point. Computation will show that the legs move through continuously varying linear congruences. A "snap-shot" will reveal the six axes as partners in the same, instantaneous congruence. Some BBMP designs can change their line geometric behaviour during motion. In general they move in an one dof manifold with only one local, infinitesimal dof. This behaviour can change, at certain instants, to a one dof manifold with two or even three local, infinitesimal dof.

IV Direct Kinematics of Borel Bricard Platforms

Consider the direct kinematics of certain StGp's that have five BBMP legs. *I.e.*, only five "hip" points on the platform, chosen randomly, are given and the five "ankle" points on the base are obtained with Eq. 2. With only five constraints, the platform will naturally move with one dof and this will be BBM. Any sixth hip point, Q_p, will describe a fully cyclic fourth order curve because it moves on the sphere centered at the ankle of its "phantom", sixth limb. Now an arbitrary ankle point is chosen at Q_B on the base and connected to a leg of given length l. How may this leg reach Q_p? Consider the intersections of the sphere of radius l, centred at Q_B, with the trajectory of Q_p. The number of solutions is easily calculated; there are eight, four of them are in the plane at infinity, because the trajectory is fully cyclic. Only four finite solutions exist. This produces the following important result.

Theorem 1: *If five legs of a StGp platform belong to a BBMP, then the platform may have no more than four practical assembly positions.*

If we assume the sixth hip point to be at $Q_p(x,y,z)$, *i.e.*, at $Q_P(X,Y,Z)$ in the base system, and its ankle is arbitrarily chosen at $Q_B(L,M,N)$ so that for a valid solution the length of the sixth leg is $l^2=(L-X)^2+(M-Y)^2+(N-Z)^2$, then the direct kinematics are obtained as follows.

Using Eq. 1 and performing a half tangent substitution we get

$$\begin{aligned} -r^2 &+ \left(-L+\frac{[-4t^2+(1-t^2)^2]x}{(1+t^2)}-\frac{4t(1-t^2)y}{(1+t^2)^2}\right)^2+ \\ &+ \left(-M+\frac{4t(1-t^2)x}{(1+t^2)^2}+\frac{[-4t^2+(1-t^2)^2]y}{(1+t^2)^2}\right)^2+ \\ &+ \left(-N+2\sqrt{d^2-\frac{4\rho^2t^2}{(1+t^2)^2}}+z\right)^2=0 \end{aligned} \tag{9}$$

Squaring this equation produces an eighth order polynomial in t which yields eight solutions. But because of the squaring process, only four of them satisfy Eq. 9.

Example:

Choose five hip points, (x,y,z), as $P_1(1,0,3)$, $P_2(0,1,3), P_3(-1,0,3)$, $P_4(-1,-1,3)$, $P_5(0,-1,3)$, to move in a BBM with design parameters $\rho=a=1$. Then the ankle points, (X_m,Y_m,Z_m), are given by Eq. 2 $Q_1(1,0,3), Q_2(0,1,3), Q_3(-1,0,3), Q_4(-1/2, -1/2,3), Q_5(0,-1,3)$. Now choose a sixth hip point, (x,y,z), to be $P_6(1,2,3)$ and pair it with an arbitrary sixth ankle point, (X,Y,Z), at $Q_6(2,1,1)$. Furthermore choose the leg length to be $l=4$. Substituting in Eq. 9, we get

$$-7-72t+36t^2+504t^3+86t^4-504t^5+36t^6+72t^7-7t^8=0 \tag{10}$$

This equation yields eight solutions but only four of them, ($t_1\to-0.4441447342254855$, $t_2\to-0.0989246998462375$, $t_3\to2.251518306850657$, $t_4\to10.10869885432393$), produce valid positions of the platform as shown in Fig. 5 – 8. Fig. 9 shows all four positions in superposition.

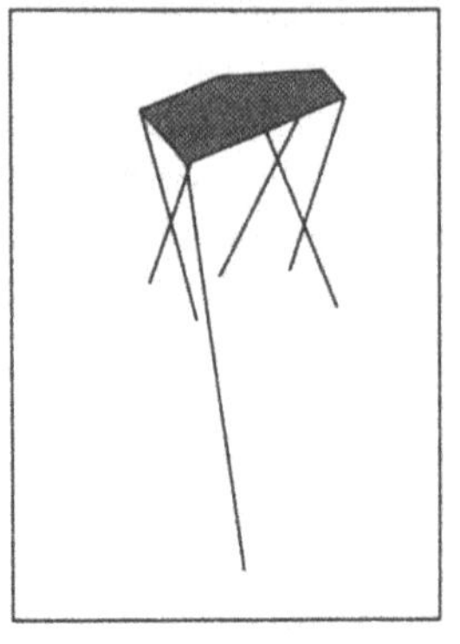

Figure 5: Position t_1

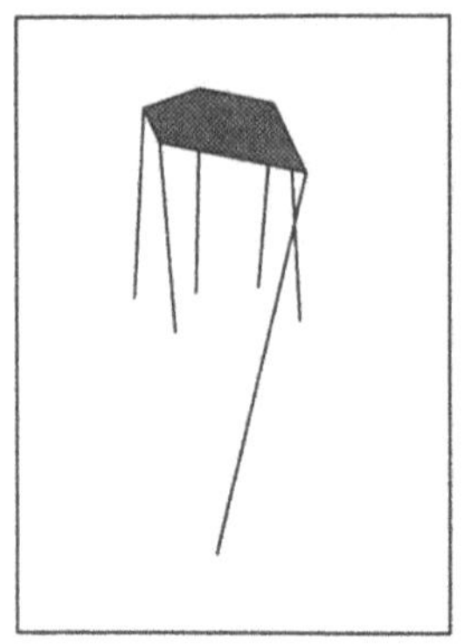

Figure 6: Position t_2

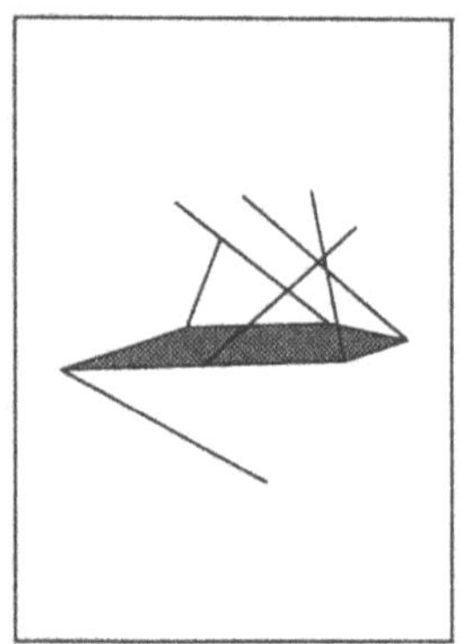

Figure 7: Position t_3

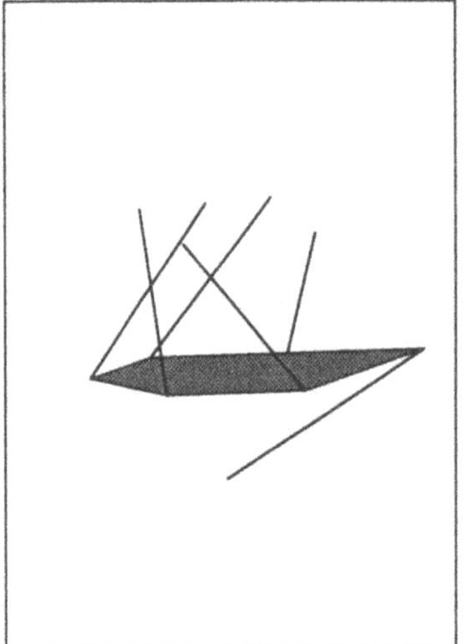

Figure 8: Position t_4

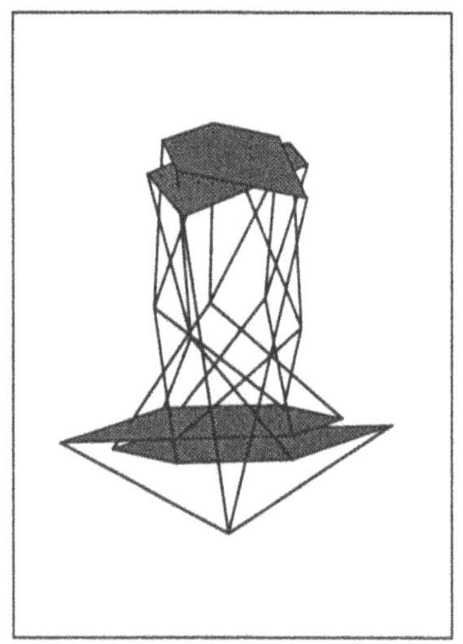

Figure 9: All Four Positions

V A Spherocylindrical Mechanism to Generate BBM

A simple mechanism, described in [15], can be devised to demonstrate BBM. It is shown in Fig. 10 and consists of an inverted "T" made of wooden dowels glued together. A third dowel is cylindrically jointed and perpendicular to and bisected by the T-leg. Strings of equal length are attached to each end of the movable crossbar and connected to respective ends of the fixed one. When the movable crossbar is turned, while holding the strings taut, its ends trace the curve of intersection between a sphere and cylinder. Rôles of the BBM design constants, a and ρ, and the motion parameter, t, as they pertain to the mechanism, are indicated.

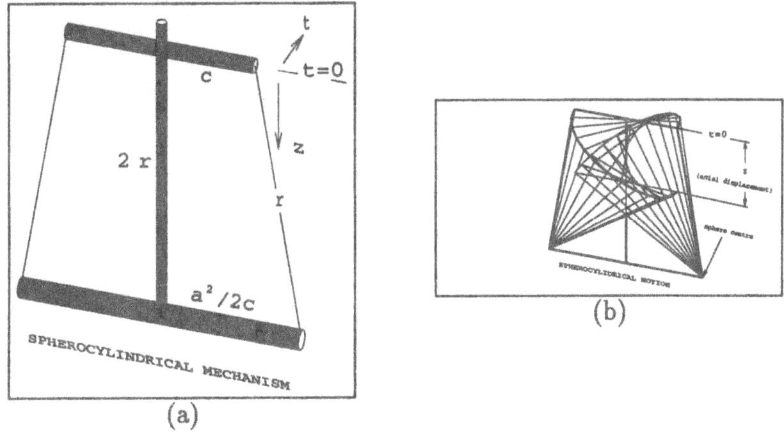

Figure 10: Spherocylindrical Mechanism & Swept Surfaces

VI Conclusions

A very curious StGp example has been introduced. It executes finite, one dof motion while all legs remain at constant length. This situation might jeopardize a flight simulator. On the other hand the property that permits one dof motion with any number of legs of fixed length provides possibilities to build very stiff, one dof platforms which are easy to actuate. Other types motions were discovered and described by Borel and Bricard. The application of these to the study of StGp singularity and direct kinematics is promising and will be published elsewhere.

VII Acknowledgements

The first author acknowledges the Austrian FWF (Fonds zur Förderung der wissenschaftlichen Forschung) and NSERC (Natural Science and Engineering Research Council of Canada) for their support under grants Nos. J0862-PHY and ISE0150033 respectively. The second author acknowledges the support of NSERC through grant No. OGP0139964. The contribution of Shadi Habib, who provided platform animation software & the resulting images reproduced in Figs. (1,2, & 3), is greatfully acknowledged.

References

[1] Stewart, D., 1965, "A platform with 6 degrees of freedom", *Proc. Institution of Mechanical Engineers*, Vol. 180, Part 1, No. 15, pp. 371-386.

[2] Fichter, E. F., 1986, "A Stewart Platform-Based Manipulator: General Theory and Practical Construction," *The International Journal of Robotics Research*, Vol 5., No. 2, pp. 157-182

[3] Merlet, J-P., 1992, "Singular Configurations of Parallel Manipulators and Grassmann Geometry", *The International Journal of Robotics Research* , Vol. 8, No. 5, pp. 45-56.

[4] Ma, O. & Angeles, J., 1992, "Architecture Singularities of Parallel Manipulator", *The International Journal of Robotics Research*, Vol. 7, No. 1, pp. 23-29.

[5] Hunt, K. H. & Primrose, E. J. F., 1993, "Assembly Configurations of Some In-Parallel-Actuated Manipulators", *Mech. Mach. Theory*, Vol 28, No 1, pp. 31-42

[6] Borel, E., 1908, "Mémoire sur les déplacements à trajectoires sphériques,"*Mém. présentés par divers savantes* , Paris (2) Vol. 33 Nr.1, pp. 1-128.

[7] Bricard, R., 1906, "Mémoire sur les déplacements à trajectiores sphériques", *J.Éc. Polyt.* (2) Vol. 11, pp. 1-96.

[8] Bottema, O. & Roth, B., 1979, *Theoretical Kinematics*, North-Holland, Amsterdam.

[9] Krames,J., 1939, "Über Symmetrische Schrotungen I-VI", *Monatsh. Math. Phys.*, Vol. 45, pp. 394-406, pp. 407-417, Vol. 46, pp. 38-50, pp. 172-195, *Sitz. Berichte. d. österr. Akademie. d. Wiss., Wien*, Vol. 146, pp. 145-158, pp. 159-173.

[10] Schönflies, A., 1886, *Geometrie der Bewegung in synthetischer Darstellung*, Leipzig, 194pp.

[11] Jacobsen, S. C., 1975, "1-Oz 'Rope Ladder' Actuator Pulls 60-Lb Loads", *Electro Mechanical Design*, Vol. 19, No. 5.

[12] Jessop, C. M., 1903, *A Treatise on the Line Complex*, Cambridge University Press

[13] Hunt, K. H., 1983, "Structural Kinematics of In-Parallel Actuated Robot-Arms", *Trans. ASME, Journal Mechanisms, Transmissions and Automation in Design*, Vol 105, pp. 705-712.

[14] Hunt, K. H., 1990, *Kinematic Geometry of mechanisms*, Clarendon Press, Oxford.

[15] Zsombor-Murray, P.J., 1992, "A Spherocylindrical Mechanism and Intersection among Other Quadratic Surfaces", *Proc. 5th Int. Conf. ECG&DG*, Vol. 2, pp. 370-376.

A Smart Kinestatic Interactive Platform

M. Griffis, C. Crane, and J. Duffy

Center for Intelligent Machines and Robotics
University of Florida
Gainesville, Florida 32611

Abstract - This paper presents a newly patented in-parallel platform which is geometrically simple. It has six distinct connecting points in both the base and top platforms. The forward solution can be determined by solving an eighth degree polynomial.

I. Introduction

Over the past few years, there has been an ever increasing interest in direct applications of parallel mechanisms to real-world industrial problems. In situations where the needs for accuracy and sturdiness dominate the requirement or a large workspace, parallel mechanisms present themselves as viable alternatives to their serial counterparts. This article is confined to the forward displacement analysis of Stewart Platform-type parallel mechanisms. In the general sense, each of these mechanisms consists of two platforms that are connected by six prismatic joints acting in-parallel. One of the platforms is defined as the "top platform". It has six degrees of freedom relative to the other platform, which is the "base". It is then required to compute all possible locations (positions and orientations) of the top platform measured relative to the base for arbitrary sets of six connecting prismatic leg lengths. Each prismatic pair is connected at each end to the base and top platforms by ball and socket joints.

Stewart [1] introduced his platform in 1965 as an aircraft simulator. Since then, many parallel mechanisms containing prismatic joints have been called Stewart Platforms, although Stewart originally suggested only two different arrangements. Hunt [2,3], Mohamed and Duffy [4], Fichter [5], Sugimoto [6,7], Rees-Jones [8], and Kerr [9] all suggest use of Stewart Platforms, with various applications ranging from manipulators to force/torque sensors. Reinholz and Gokhale [10] (as well as Miura, Furuya, and Suzuki [11], and Miura and Furuya [12]) investigated an interesting application in the form of "Variable Geometry Truss Robots (VGTs)". It is apparent that NASA's octahedral VGT is in fact founded on the simplest of the Stewart Platforms.

Much of the research in the literature has devoted extensive effort to the reverse displacement analysis that is inherently simple for parallel mechanisms (viz. it is required to compute a set of leg lengths given a desired location of the top platform relative to the base). Few, however, have investigated closed-form forward displacement analyses for parallel mechanisms. Instead, they depend on purely numerical solutions. Behi [13] investigated a forward displacement analysis of a parallel mechanism that closely resembles a Stewart Platform. Numerically, he was able to find eight solutions. Reinholz and Gokhale [10] used the Newton-Raphson technique to obtain an iterative solution for the forward displacement analysis of a Stewart Platform.

A. J. Lenarčič and B. B. Ravani (eds.), Advances in Robot Kinematics and Computationed Geometry, 459–464.

A closed-form forward displacement analysis (as opposed to an iterative one) will yield much important information on the geometry and kinematics of a parallel mechanism. For instance, a closed-form solution for a Stewart Platform will not only yield the exact number of real configurations of the top platform relative to the base for a specified set of leg lengths, but also quantify the effects of errors in leg lengths on the position and orientation of the top platform. Furthermore, a forward displacement analysis of a Stewart Platform manipulator will provide a Cartesian controller with necessary feedback information, namely the position and orientation of the top platform relative to the base. This is especially important when the actual position and orientation cannot be directly sensed, and when the manipulator's configuration is determined solely from lengths of the connecting prismatic legs. Near singularities, purely numerical solutions may experience difficulties, since they provide no way to determine changes in closure.

II. Forward Analysis

The forward analysis of in-parallel platforms has attracted much attention in academia in recent years. The most general in-parallel platform is called a 6-6 platform because there are six distinct connecting points in the base and six distinct connecting points in the top platform (see Fig. 1). In this paper all connecting points are considered to be coplanar in both the base and top platforms. The closed-form analysis for the 6-6 platform has proven to be a most challenging task.

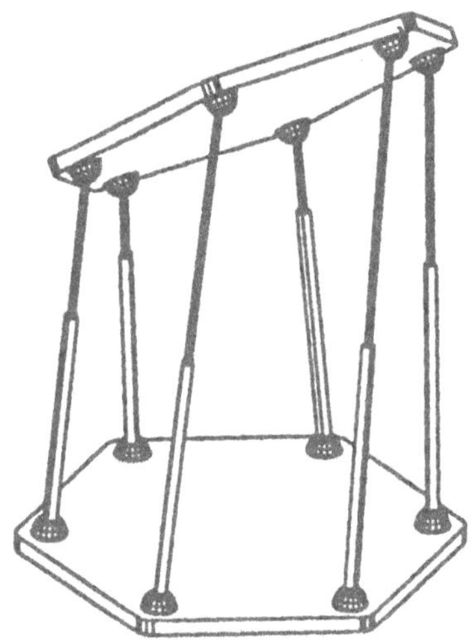

Fig. 1: General 6-6 Platform

The most simplified form of the mechanism contains six legs which meet in a pair-wise fashion at three points in both the top platform and base (see Fig. 2). This form of mechanism which was called the "3-3 Stewart Platform" was solved by Griffis and Duffy [14]. Their solution was easily extended to a "6-3 Platform" whose legs meet at six distinct points in a planar base (Fig. 3). Nanua, Waldron, and Murthy [15] also obtained a closed-form solution for the 6-3 Platform. A 6-3 Platform with a non-planar base was analysed by Innocenti and Parenti-Castelli [16]. The maximum number of assembly configurations for the 3-3 and 6-3 Platforms are sixteen. For the 3-3 and for the 6-3 Platform with six distinct connecting points lying in a plane it is necessary to solve an eighth degree in a variable x^2. This yields a maximum of 8 real assembly configurations above the base and 8 reflected assembly configurations through the base. For the more general 6-3 Platform with non-planar connecting points it is necessary to solve a general 16th degree polynomial.

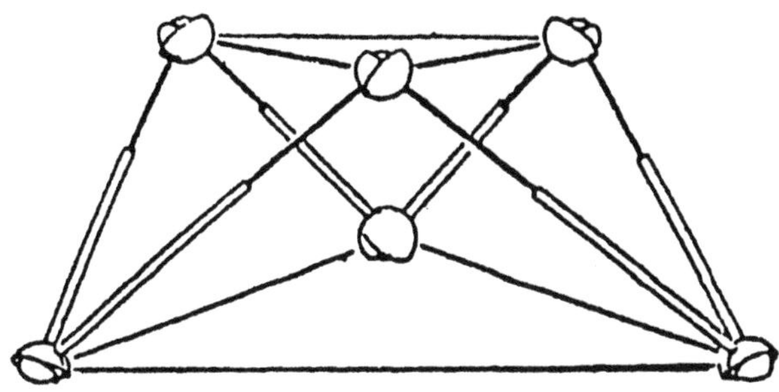

Fig. 2: 3-3 Platform

It is clear that the 3-3 Platform, an octahedron, has the simplest forward solution and it further became clear that apart from the 6-3 Platform, as more distinct connecting points were introduced the degree of complexity of the forward analysis increased. The most complex forward analysis is that for the 6-6 Platform. In order to gain insight into the problem formulation for the general 6-6 Platform, the forward analysis for a sequence of platforms was performed.

A number of 4-4 Platforms, for which the six legs connected either singly or pair-wise at four points in both the top and base platforms were solved by Lin, Griffis and Duffy [17]. This was accomplished by deriving and solving polynomials of degree four, eight, and twelve. About the same time, Innocenti and Parenti-Castelli [18] solved a 5-5 Platform, which contains three singly connected legs. This was accomplished by deriving and solving a fortieth degree polynomial. Later Lin, Crane and Duffy [19] solved a series of 4-5 Platforms. This was accomplished by deriving and solving various polynomials ranging from sixteenth degree to thirty second degree. Most recently, the solution to the general 6-6 Platform was obtained by Wen and Liang [20] and Zhang and Song [21]. It was necessary to derive and solve a fortieth degree polynomial.

III. The New In-Parallel Platform

As previously stated, there is a real need for a closed-form forward analysis. It is, of course, possible to perform numerical iterations (an optimisation using six independent variables) to obtain the position orientation of the platform. However, it is well known that such iterative solutions have a tendency to "jump" from one closure to another. From a practical viewpoint, this is undesirable.

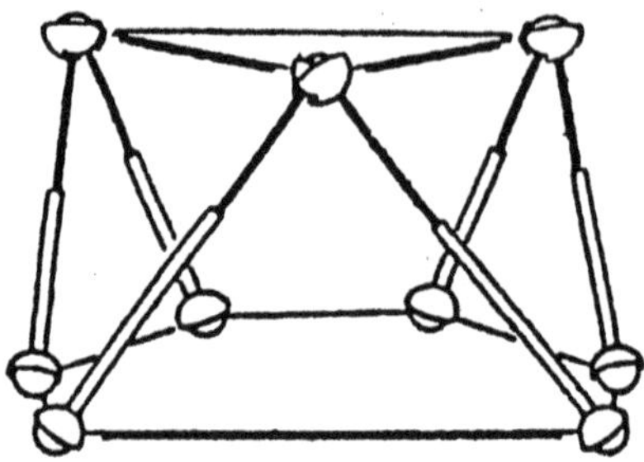

Fig. 3: 6-3 Platform

The necessity for a closed-form analysis really manifests itself whenever a platform is to be used to control force and motion simultaneously. Any in-parallel structure lends itself well to a static force analysis particularly when utilising the theory of screws [2]. The wrench applied to the top platform can be statically equated to the summation of the forces measured along the lines of the six prismatic legs. Thus far this particular application of an in-parallel platform has depended on relatively "small" leg deflections, resulting in a "constant" configuration. However, employing a closed form forward analysis provides the analytics to monitor gross deflections of the platform and thereby permit one to consider the design of a compliant force/torque sensor.

There is, however, a dilemma. The geometrically simplest parallel mechanism is clearly the octrahedron which contains six legs that meet in a pair-wise fashion at three points in both the top platform and base. The closed form forward analysis involves the solution of an eighth degree polynomial. It is practical to solve such a polynomial rapidly and in real time. This is necessary when employing the platform to control force and motion simultaneously. However, such a platform has a serious mechanical disadvantage. It is not possible to design the necessary concentric ball and socket joints at each of the double connection points without mechanical interference. Such a design is mechanically unacceptable. It is preferable to separate the double connection points in order to overcome the mechanical design problem. However, the closed-form forward solution for the general 6-6 Platform involves the solution of a fortieth degree polynomial. It is not practical to solve a fortieth degree polynomial in real time.

The newly patented platform, illustrated in Fig. 4 incorporates the benefits of both the octrahedron Platform and the general 6-6 Platform. There are six distinct connecting points in both the top platform and the base which avoids the mechanical interference problem. At the same time it is possible to control the position of the top platform employing the eighth degree polynomial for the forward solution of the octrahedron. This combination makes possible the simultaneous control of force and motion using the newly patented device.

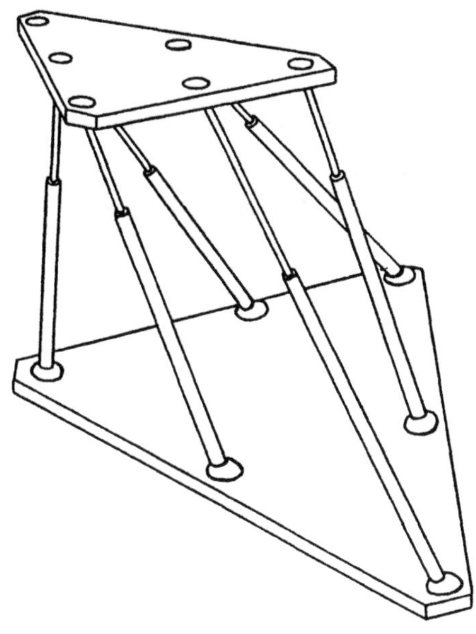

Fig. 4: Special 6-6 Platform (U.S. Patent # 5,179,525)

IV. Conclusions

This novel platform has practical applications. Initial results demonstrate a stable workspace with pitch, roll, and yaw of 30 degrees from a central position.

V. References

[1] D. Stewart, "A Platform With Six Degrees of Freedom", *Proc. Inst. Eng.*, London, ***180***, 371-386 (1965).

[2] K.H. Hunt, *Kinematic Geometry of Mechanisms*, Oxford University Press, London (1978).

[3] K.H. Hunt, "Structural Kinematics of In-Parallel-Actuated Robot-Arms", *Trans. ASME, J. of Mech., Trans., and Auto. in Design*, ***105***, 705-712 (1983).

[4] M.G. Mohammed and J. Duffy, "A Direct Determination of the Instantaneous Kinematics of Fully Parallel Robotic Manipulators," *ASME Journal of Mechanisms, Transmissions, and Automation in Design*, ***107***, 226-229 (1985).

[5] E.F. Fichter, "A Stewart Platform-Based Manipulator: General Theory and Practical Construction", *Int. J. Robotics Research*, ***5***, 157-182 (1986).

[6] K. Sugimoto, "Kinematic and Dynamic Analysis of Parallel Manipulators by Means of Motor Algebra", *ASME J. of Mech. Trans. and Auto. in Design*, ***109***, 3-7 (1987).

[7] K. Sugimoto, "Computational Scheme for Dynamic Analysis of Parallel Manipulators", Trends and Developments in Mechanisms, Machines, and Robotics-1988, *ASME Proceedings of the 20th Biennial Mechanisms Conference* (1988).

[8] J. Rees-Jones, "Cross Coordinate Control of Robotic Manipulators", *Proceedings of the International Workshop on Nuclear Robotic Technologies and Applications*, Present and Future, University of Lancaster, UK, June 29-July 1(1987).

[9] D.R. Kerr, "Analysis, Properties, and Design of a Stewart-Platform Transducer," Trends and Developments in Mechanisms, Machines, and Robotics-1988, *ASME Proceedings of the 20th Biennial Mechanisms Conference* (1988).

[10] C. Reinholz and D. Gokhale, "Design and Analysis of Variable Geometry Truss Robots", *Proceedings of the Ninth Annual Applied Mechanisms Conference*, Oklahoma State University, (1987).

[11] K. Miura, H. Furuya and K. Suzuki, "Variable Geometry Truss and its Application to Deployable Truss and Space Crane Arm", *35th Congress of the International Astronautical Federation*, Lausanne, Switzerland, Pergamon, New York (1984).

[12] K. Miura and H. Furuya, "An Adaptive Structure Concept for Future Space Applications", *36th Congress of the International Astronautical Federation*, Stockholm (1985).

[13] F. Behi, "Kinematic Analysis for a Six-Degree-of-Freedom 3-PRPS Parallel Mechanism", *IEEE Journal of Robotics and Automation, 4*, 561-565 (1988).

[14] M. Griffis and J. Duffy, "A Forward Displacement Analysis of a Class of Stewart Platforms", *J. of Robotic Systems*, John Wiley, *6*(6), 703-720 (1989).

[15] P. Nanua, K.J. Waldron, V. Murthy, "Direct Kinematic Solution of a Stewart Platform", *IEEE Trans. on Robotics and Automation, 6*(4), 438-444 (1990).

[16] C. Innocenti and V. Parenti-Castelli, "Direct Position Analysis of the Stewart Platform mechanism", *Mechanism and Machine Theory, 25*(6), 611-621 (1989).

[17] W. Lin, M. Griffis,J. Duffy, "Forward Displacement Analyses of the 4-4 Stewart Platforms", *Proceedings of the 21st ASME Mechanisms Conference*, Chicago, IL, Sept. 16-19 (1990).

[18] C. Innocenti and V. Parenti-Castelli, "Closed-Form Direct Position Analysis of a 5-5 Parallel Mechanism", Dipartimento di Ingegneria delle costruzioni meccaniche, nucleari, aeronautiche e di Metallurgia - Universita de Bologna, Italy (1990) - private communication.

[19] W. Lin, C.D. Crane and J. Duffy, "Closed-Form Forward Displacement Analyses of the 4-5 In-Parallel Platforms", *Proceedings of the 22nd ASME Mechanisms Conference*, Phoenix, AR, (1992).

[20] F.A. Wen and C.G. Liang, "Displacement Analysis for the General Stewart Platform-Type Mechanism", Beijing Univ. of Posts and Telecommunications, June (1992) - private communication

[21] Change-de Zhang and Shin-Min Song, "Forward Position Analysis of Nearly General Stewart Platforms", *Proceedings of the 22nd Biannual ASME Mechanisms Conf.*, Scottsdale, Arizona, (1992).

12. Task and Motion Planning

Strategic Collision Avoidance of Two Robot Arms in the Same Work Cell

P. Alison, M.J. Gilmartin

School of Engineering and Technology Management
Liverpool John Moores University, Liverpool, U.K.

P. Urwin

School of Electrical Engineering, De Monfort University
Leichester, U.K.

Abstract-A strategy is proposed whereby two robot arms can move together with independent motions in a common workspace without collision. This is achieved using the concept of a virtual arm and assessing the 'safety' of its position at discrete intervals. If unsafe, the path vector of either arm is redirected using a rotation function defined from a number of specified collision avoidance functions (CAFs). These are rated in performance according to the fastest and shortest paths required to avoid collisions. Graphical simulations are used to illustrate the effects of different CAFs.

I. Introduction

In practice when two robots are to operate in the same work cell some CAD and off-line or pendant programming system will produce both a specification for local driving hardware and sensor controllers together with trajectory planning. Collision-free motions could be obtained by keeping one arm motionless while the other completes its task and vice versa. Off-line arm location programs such as Grasp [1] could then be used to obtain collision-free trajectories. For a fully free operation however such pre-planning is not possible and some form of motion coordination between the robots needs to be established for a collision-free efficient operation.

The practical implementation of on-line collision avoidance techniques is clearly limited by the constraints of real-time computation. Obstacle avoidance by manipulators in real time has been investigated by Khatib [2] and Freund and Hoyer [3].

The objective in this work is to navigate either arm from its start or pick-up position to a destination or place position in the presence of a stationary or moving object which in both cases is the other arm. The use of right-of-way precedence as suggested by Freund [4] is not assumed here. Each manipulator is considered for its own find-path problem as a separate independent entity. It is shown that using the path control characteristics embodied in a standard rotation matrix and the concept of a collision avoidance function (CAF), a sub-optimal path may be found for each robot. This is as close to a shortest path solution that either arm can safely travel without risking a situation where it would be impossible to stop in time to avoid a collision with any other obstacle in the workspace.

A. J. Lenarčič and B. B. Ravani (eds.), Advances in Robot Kinematics and Computationed Geometry, 467–476.

II. The Generalised Model

The model used for simulation purposes is shown in Fig.1. Line representations of the arms are used here for simplicity although the simulation can also be used with polygonal representations for the shapes of the arms. Each arm is considered to move in the same plane with two degrees of freedom, a rotation about an axis perpendicular to the plane and a translation along the axis of the arm. End effector motion at the wrist is not considered.

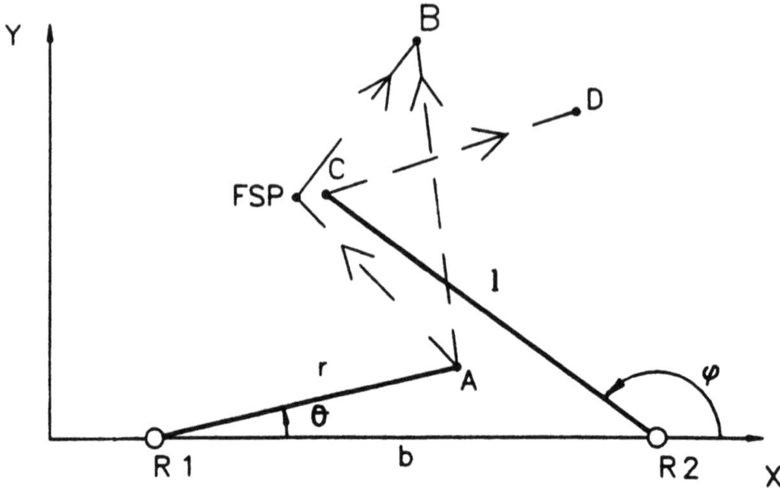

Fig.1. The kinematic model

Referring to Fig.1, points A and B are always the start and end of the path for robot R1 and similarly C and D for R2. A possible anti-collision strategy could be posed as follows: when arm R1 has to move its hand at A around robot R2 when travelling to its goal at point B then a free space point (FSP) is the initial motion destination until hand A can 'see' around the arm of R2 and thus complete its journey without collisions. Although the foregoing refers to the motion of R1, similar considerations apply to the moving arm R2. Moving either arm to its FSP may produce an inefficient trajectory since no account is taken of the other's motion in the interim. Either arm might move out of the way of a direct path to the other's goal while completing its own task. A more acceptable strategy is to examine the relative positions of each arm at discrete intervals during the overall motion period and enable each to modify its trajectory only when a collision is imminent. The judgement of when a collision is imminent and the evasive action to be taken forms the basis of the anti-collision strategies proposed in this paper.

Discrete analysis is used here because the moving robots constitute a continually changing dynamic environment. An optimisation of path trajectories using Euler-Lagrange equations [5] with an appropriate minimising function was considered. However, because of the non-linearities and the fact that the parameters of each arm cannot be known in advance, this approach was abandoned in favour of a system where global information about how the environment was changing was unavailable. This was achieved by including time as a state variable i.e. discretising the system. The resulting control strategy is now at best sub-optimal.

III. Collision Avoidance Strategy

At time t_{k-1} assume both arms are known in position and velocity. Arm R1 is located at A_{k-1} and arm R2 at C_{k-1}, Fig.2. The avoidance strategy is illustrated with respect to R1 although similar steps are carried out with respect to R2 also. The strategy is described in the following steps:

(i) Estimate the position of a virtual arm. This corresponds to the concept of a virtual object defined by Shen and Bien [6]. The position of the virtual arm is determined for the next sample time t_k by considering the end of R1 to move to A_k along A_{k-1} B.

(ii) Determine the nearest FSP with respect to C_k which provides a clear path from A_{k-1}.

(iii) Calculate the perpendicular distance d_k and the relative approach velocity between the virtual arm and R2.

(iv) Determine the deceleration required to stop the arm within the distance d_k and with the relative approach velocity calculated in (iii).

(v) If the arm can be safely stopped within distance d_k, then move it to this virtual position at the next sample time.

(vi) If the position is unsafe then rotate the path vector A_{k-1} A_k towards the FSP. The amount of rotation is governed by the choice made for the CAF.

(vii) Repeat the procedure until both arms reach their final destinations.

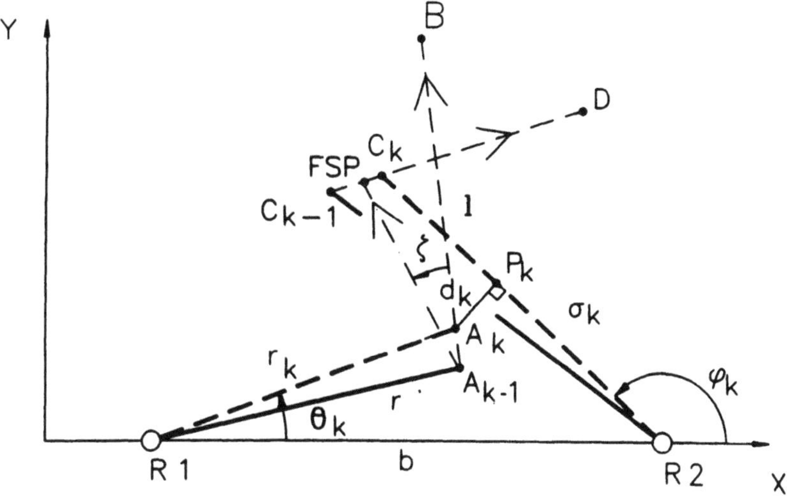

Fig.2. Notation for the mathematical model

The CAFs are formulated as functions of the perpendicular distance and the relative approach velocity as calculated for step (iii). Their various forms are listed in Table 1. The effects of the CAFs on the arms are examined later.

IV. Mathematical Background

The calculations are noted for arm R1 but are equally applicable to R2. For robot R1 with hand at $A(x_A, y_A)$ moving linearly towards destination $B(x_B, y_B)$, the world coordinates (r, θ) must satisfy

$$r = \frac{\gamma}{(\beta\cos\theta - \alpha\sin\theta)} \tag{1}$$

where

$$\alpha = (x_A - x_B)$$
$$\beta = (y_A - y_B)$$
$$\gamma = x_A y_B - x_B y_A$$

Differentiation of (1) with respect to time gives the velocity condition for straight-line motion, that is

$$\frac{\dot{r}}{\dot{\theta}} = \gamma \frac{(\beta\sin\theta + \alpha\cos\theta)}{(\beta\cos\theta - \alpha\sin\theta)^2} \tag{2}$$

For simulation purposes, it is assumed as in Shin and McKay [7] that the acceleration and deceleration bounds are quadratic functions in velocity so that for a generalised state variable q

$$\dot{q} = \begin{bmatrix} \ddot{\theta} \\ \ddot{r} \end{bmatrix} = \pm\left[\ddot{q}_{max} - \lambda \dot{q}^2\right] \tag{3}$$

In Eq.(3) the value of λ depends on the extreme values for velocity $\dot{q}$ .In practice these would depend on the saturation levels of the driving hardware. In this case we arbitrarily set $\dot{q} = \pm\frac{3}{2}\dot{q}_{max}$ when $\ddot{q} = \pm\frac{1}{2}\ddot{q}_{max}$ giving

$$\lambda = \frac{2\ddot{q}_{max}}{(3\dot{q}_{max})^2} \tag{4}$$

so that

$$\ddot{q} = \pm\ddot{q}_{max}\left[1 - \frac{2}{9}\left(\frac{\dot{q}}{\dot{q}_{max}}\right)^2\right] \tag{5}$$

Therefore by specifying absolute maximum values for both velocity and acceleration (or deceleration),the acceleration (or deceleration) for velocity $\dot{q}$ may be calculated. In practice a value much smaller or even zero may be used.

The motion of the arm then proceeds such that the velocity along the path is a maximum i.e. $\max\left[(\dot{\theta}_{max}, \dot{r}_c), (\dot{\theta}_c, \dot{r}_{max})\right]$ where $\dot{r}_c$ and $\dot{\theta}_c$ are the associated velocities for straight-line motion. Thus if the arm is driven at $\dot{\theta}_{max}$, $\dot{r}_c$ is the radial velocity as given by Eq. (2) while if driven at $\ddot{r}_{max}$, $\dot{\theta}_c$ is the angular velocity. The motion continues until it becomes necessary to initiate a deceleration procedure either due to a change in direction or because the arm is approaching a final value in one of its state variables, q.

With the discrete dynamic forces assumed piecewise constant over each time interval and the distance and velocity of the wrist along a shortest cartesian path taken as the output vectors, then with a sample time t_s, the position of a new virtual arm q_k with respect to the current arm position q_{k-1} is calculated using Euler's method [8]. That is

$$q_k = q_{k-1} + t_s \dot{q}_{k-1} + \frac{t_s^2 \ddot{q}_{k-1}}{2} \tag{6}$$

If the path calculated for the virtual arm is obstructed, the perpendicular is dropped on to the other arm. Referring to Fig.2, the path of the virtual arm from A_k to B is obstructed by R2 with its hand at C_k. The length of the perpendicular d_k is given by

$$d_k = b\sin\phi_k - r_k \sin(\phi_k - \theta_k) \tag{7}$$

The relative velocity and deceleration along d_k can be calculated as

$$v_k = \dot{r}_k \sin(\phi_k - \theta_k) - r_k \dot{\theta}_k \cos(\phi_k - \theta_k) - \sigma_k \dot{\phi}_k \tag{8}$$

$$a_k = (\ddot{r}_{max} - r_k \dot{\theta}^2)\sin(\phi_k - \theta_k) - (r_k \ddot{\theta}_{max} + 2\dot{r}_k \dot{\theta}_k)\cos(\phi_k - \theta_k) - (\sigma_k \ddot{\phi}_{max} - 2\dot{\sigma}_k \dot{\phi}_k) \tag{9}$$

V. Collision Avoidance

With the perpendicular distance d_k for the virtual position A_k known, together with the relative velocity and acceleration in this direction, a safe stopping distance d_s may be calculated assuming linear deceleration as

$$d_s = \frac{v_k^2}{2a_k} \tag{10}$$

If $d_k \le d_s$, the virtual arm position A_k is judged unsafe and the path must be changed. A simple corrective strategy is to rotate the current path vector $A_{k-1} A_k$ (Fig.2) through angle ς to pass through the nearest FSP. This rotation may be effected by the matrix R where

$$R = \begin{bmatrix} \cos\varsigma & -\sin\varsigma \\ \sin\varsigma & \cos\varsigma \end{bmatrix} \tag{11}$$

However with d_s only being an estimate based on a linear deceleration profile and not necessarily accurate over time, it was decided that this rotation R of the direction vector be carried out automatically at each sample time. Modulation of the rotation angle ς is achieved by using a collision avoidance function (CAF) such that

$$\psi = \text{CAF} \cdot \varsigma \qquad (0 \le \text{CAF} \le 1) \tag{12}$$

where ψ replaces ς in the rotation matrix given by Eq. (11).

CAF No.	Mathematical Representation
0	0 Clear path 1 Obstructed path
1	$\left\lvert \frac{v_k}{v_s} \right\rvert$
2	$\left(\frac{v_k}{v_s} \right)^2$
3	$\left\lvert \frac{(d_s \cdot v_k)}{(d_k \cdot v_s)} \right\rvert$
4	$\left\lvert \frac{d_s}{d_k} \right\rvert + \left\lvert \frac{v_k}{v_s} \right\rvert$
5	$\exp\left(1 - \left\lvert \frac{v_s}{v_k} \right\rvert \right)$
6	$\exp\left(1 - \left\lvert \frac{(d_k \cdot v_s)}{(d_s \cdot v_k)} \right\rvert \right)$
7	$\left(\frac{d_s}{d_k} \right)^2$
8	$\left\lvert \frac{d_s}{d_k} \right\rvert$
9	$\exp\left(1 - \left\lvert \frac{d_k}{d_s} \right\rvert \right)$

Table 1. CAF Functions

Thus when $\text{CAF} \to 0$ there is no impending collision and no change in path is required while when $\text{CAF} \to 1$, an imminent collision is detected and the maximum rotation is implemented. Intermediate CAF values vary the value of ψ and hence the degree of rotation.

The major part of this work was to investigate the effect of different CAFs on the resulting motion of the two arms. The CAFs, shown in Table 1, were f ormulated as functions of the distance d_k and d_s and the relative velocities v_k and v_s, where v_s is defined as a safe velocity, compatible with the arm stopping within the distance d_k and given by

$$v_s = \left|2 \cdot d_s \cdot a_k\right| \tag{13}$$

VI. Simulation Results

To assess the effectiveness of the CAFs in Table 1, a large number of simulations were carried out using a DEC-VAX minicomputer using GINO-F as a graphical interface. Three example outputs are shown in Figs.3,4 and 5.

Each figure shows the trajectories for the hands of the two arms R1 and R2 and the position of each arm at 100 ms intervals of simulation time. The nomenclature used is defined in Table 2.

A, B	start and finish position coordinates for R1
C, D	start and finish position coordinates for R2
ADMX	maximum angular velocity (rad/s)
PDMX	maximum prismatic velocity (cm/s)
ADDMX	maximum angular acceleration (rad/s^2)
PDDMX	maximum prismatic acceleration (cm/s^2)
s/s	samples per second

Table 2. Nomenclature for simulation figures

Fig.3 illustrates the action of CAF0. Both hands start initially at A and C and have clear line-of-sight to their respective destinations at B and D. It is not until some two seconds into the journey when both are close to maximum velocity that arm R1 suddenly crosses the path of R2. Arm R2 attempts to accelerate its hand away from the potential collision and around R1 at B. However its maximum acceleration is insufficient to clear the other arm and a collision occurs after 2.5 seconds into its journey.

Fig.4 shows the same position requirements but this time with CAF8 operating. This function depends on a safe distance which itself is dependent on the relative velocities of the two arms. R1's path to B deviates little from a straight line while that of R2 changes when a potential collision with R1 is detected. Evasive action is sufficient in this case to avoid a collision but with little clearance evident between the trajectories after 2.6 seconds.

In Fig.5 with CAF1 in operation, the function is now dependent on the ratio of the actual to safe velocity, collisions are again avoided but in this case the path of R1 is no longer along a straight line to B. Although R1's initial line of sight is clear because of the velocity ratio, avoidance action is still taken. R1's path is deviated away from B. The effective slowing of R1's movement ensures a smoother path for R2 which is noticeable by the savings in both time and distance. However these are clearly at the expense of R1.

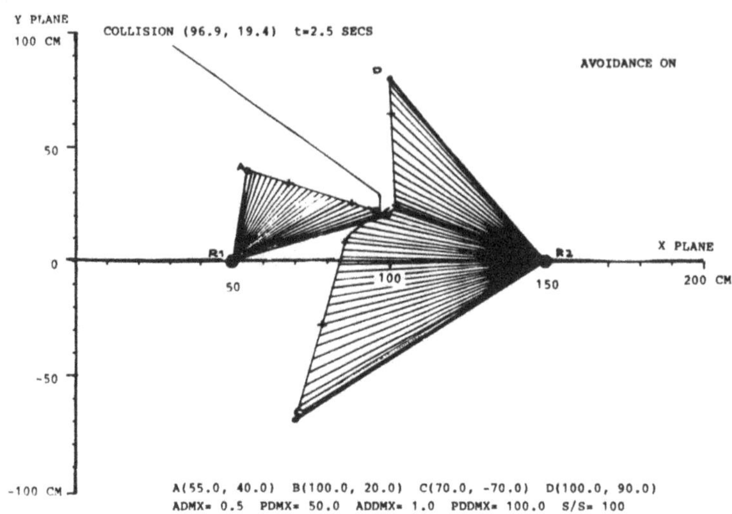

Fig.3 Collision avoidance using CAF0

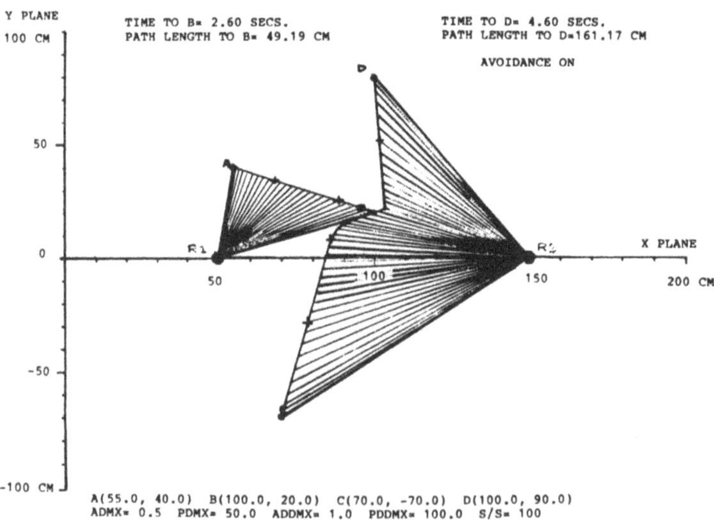

Fig.4 Collision avoidance using CAF8

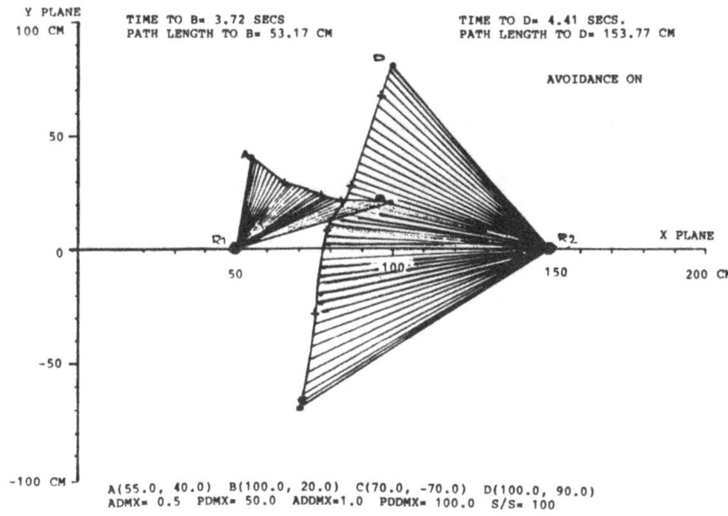

Fig.5 Collision avoidance using CAF1

VII. Conclusions

The strategy proposed for avoiding collisions between two robot arms was tested using the range of CAFs listed in Table 1. Their comparative effectiveness was found to be dependent on the path requirements of the two arms especially when potential collisions were apparent.

When the paths for each arm were clear of each other i.e. no collisions were possible, the logic CAF0 was found to be superior giving the best time and shortest path for each arm as well as being the quickest to calculate.

Where the paths of either or both arms were liable to result in collisions, the most general case, CAFs 1,2 and 5 were found to give better results than the others.However all the CAFs, 1-9, were effective in giving collision-free motions.

Overall it was found that CAF1, the ratio of actual velocity to safe velocity along the perpendicular, gave the best results in terms of shortest collision-free paths, lowest times and smooth jerk-free motions for all the cases examined [8].

The implementation of the collision avoidance strategy in an on-line application is dependent on the necessary calculations being performed in real time. In addition to this, the sensory information on which the calculations depend must also be available within the sample period. Although many of these problems still need to be addressed, the formulation of a list of CAFs as given here offers some flexibility for making a minimum cost assessment in terms of calculation time.

VIII. References

[1] K.K.D. Young et al., "Performance and Evaluation of Flexible Manufacturing Systems". *Proc. 4th Polytechnic Symposium on Manufacture*, Birmingham,(1984).

[2] O. Khatib, "Real-Time Obstacle Avoidance for Manipulators and Mobile Robots". *Int. J. Robotics Research,* 5(1), 90-98, (1986).

[3] E. Freund and H. Hoyer, "Real-Time Pathfinding in Multi-Robot Systems Including Obstacle Avoidance". *Int. J. Robotics Research*,7(1), 42-70, (1988).

[4] E. Freund, "Fast Non-Linear Control With Arbitrary Pole-Placement for Industrial Robots and Manipulators". *Int. J. Robotics Research*,1(1), (1982).

[5] L.S. Pontryagin, *The Mathematical Theory of Optimal Processes*. Interscience Press, N.Y. (1962).

[6] Y. Shen and Z. Bien, "Collision-Free Trajectory Planning for Two Robot Arms". *Robotics*, 7, 205-212, (1989).

[7] K.G. Shin and N.D. McKay, "A Dynamic Programming Approach to Trajectory Planning of Robotic Manipulators". *IEEE Trans. Auto. Cont.*, AC-31(6), Jun. (1986).

[8] R.L. Burden and J.D. Faires, *Numerical Analysis.* 3rd Ed.. Prindle, Weber and Schmidt, Boston, (1985).

[9] P. Alison, Optimal Strategic Deployment of Robot Arms in a Dynamically-Constrained Environment. *Ph.D. Thesis*. Liverpool Polytechnic, (1991).

A Collision-Detection Scheme Based on Convex-Hulls Concept for Generating Kinematically Feasible Robot Trajectories

A. C. Nearchou and N. A. Aspragathos

Department of Mechanical Engineering
University of Patras
26 110 Patras, Greece.

Abstract - In this paper a technique for collision detection between a robot and a collection of obstacles is presented. The technique is based on decomposing the problem of interference into two sub-problems: Firstly, the problem of determining the arm's links too closely located to obstacles so that they are candidate to collide, and secondly the problem of checking for interference between the links in question, and the associated near by obstacles. Well known concepts in computational geometry such as convex hulls and minimum spanning circles are used in robot trajectory control moving among obstacles. The efficiency of the method is tested by numerical experiments applied on an existing path planning algorithm.

I. Introduction

Controlling robot motion is one of the most significant problems in todays Robotics Science. The basic goal of this problem is to determine what actions must a robot do in order to perform a desired task. In its simplest form robot motion planning (RMP) problem is twofold: Firstly, is a problem of finding a feasible path for the robot from a start placement to a desired final placement, and secondly is a problem of planning the necessary actions needed for the robot to follow this path. Obstacle avoidance control is included in the first sub-problem.

There is a rather large literature on the development of collision-free path planning algorithms. An early work [1] proposed the use of Ternary algebra as a method for detection of interference, among planar shapes or among prismatic bodies. A good review of the work on the geometric interference between objects is also presented by Lozano-Perez [2,3], with algorithms to solve the collision avoidance. The Lozano-Perez's approach is based on characterising the position and orientation of an object as a single point in a configuration space (C-space) by building geometric objects, called configuration space obstacles (C-obstacles). These algorithms have the advantage that the intersection of a point relative to a set of objects is easier to deal with than the intersection of objects among themselves.

Many approaches have been proposed by the research community to solve RMP

A. J. Lenarčič and B. B. Ravani (eds.), Advances in Robot Kinematics and Computationed Geometry, 477–484.

problem using C-space. The most important of them and most extensively studied so far, reduce the problem to a shortest-path problem in a graph. The visibility graph approach[3,4], is based on the construction of a non-directed graph whose vertices are the initial and the goal configuration of the robot and the vertices of the C-obstacles. The problem of path planning is thereby converted to searching the graph for a path (usually the shortest) between the initial and the final configuration. Retraction method[5,6] uses the Voronoi diagram to solve the RMP problem. The edges of the Voronoi diagram represent paths that are equidistant from the closest pair of obstacles, and its vertices are points where three or more such path meet. A solution to the problem is found by searching this graph for shortest path. Brooks[7] proposed the freeway method for an explicit representation of the free space, based on overlapping generalised cones having straight spines and non increasing radii. Translations are performed along freeways and rotations at the intersections of freeways. A similar method based on the concept of good representation of the free space and adaptable on articulated robots appeared in [8]. This work determines a spine of the free space between the obstacles and proposes a kinematic control algorithm which guides the tip of the hand to a desired point of the free space, while the manipulator's body is kept close to the spine of the free space. Khatib [9] introduced the potential field approach. This approach treats the robot as a point moving in its C-space under the influence of an artificial potential field produced by the final desired configuration, and the C-obstacles. Actually, the idea with this approach is that, the final configuration produces an "attractive" potential which pulses the robot toward its goal, and the C-obstacles generate a "repulsive" potential which pushes the robot away from them. The negated gradient of the summation of these potentials is treated as an artificial force applied to the robot and control its motion. The decomposition of the free-space of the robot into cells is another important approach to solve RMP problem. The methods based on this approach include the Quad Tree and the Oct Tree [10,11] methods. Quad and Oct Trees are hierarchical data structures that recursively subdivides the work space of the robot into cells until a certain criterion is satisfied. In the case of the Quad Tree the workspace is represented by a rectangle and in each iteration this rectangle is subdivided into four smaller rectangles; where in the case of the Oct Trees the workspace is a cube which recursively is subdivided into eight smaller cubes (called octants). Again, a solution can be found by searching the trees for a valid path.

Minimum distances between objects is another important aspect for collision avoidance. Several path planning algorithms which require the computation of the minimum distances between the robot and the obstacles have been appeared in the literature; for example Hurteau and Stewart[12], Gilbert and Johnson[13], Bobrow[14]. The main idea with this approach is to use the minimum distance measure to determine if any components of the pair of objects are about to collide. A method for generating interference-free robot trajectories has been appeared in [15]. The method approximates the volume swept by a manipulator's link moving between adjacent positions, by the convex hull containing the two positions. Penetration is thus determined by testing for intersection between the set of connecting convex hulls and the stationary obstacles. The method was applied into point-to-point robot trajectories. Application of the method into continuous path problems seems to be very difficult because of its complexity.

This work introduces another technique to detect collisions between robot and obstacles based on the convex-hull concept. Moreover, using an hierarchical structure that decomposes the original problem into sub-problems, the time-complexity of the problem is significantly reduced and thereby can be effectively applied into continuous path control of robots.

The remaining of the paper is organised as in the following: In section II the collision-detection scheme is presented and analysed. Section III discusses the efficiency of the proposed method through some experiments in continuous path problems; while section IV summarises the contribution of this paper.

II. The Collision-Detection Scheme

Obstacle avoidance control is a basic part of RMP problem and deals with the problem of moving the robot among obstacles in such a way so that it reaches its goal with safety. The rapid increasing need for good path planning algorithms for obstacle avoidance in industrial automation and robotics, made the obstacle avoidance control a significant problem in its own right. The purpose of this work is to develop an efficient scheme of detection of interference between the robot and the obstacles, capable to cooperate with existing trajectory generation algorithms and therefore to be included in a real industrial path planner system.

A. *Basic Assumptions And Analysis.*

In our discussion we make the following assumptions:

- We consider 2-dimensional problems in the plane,
- the robot is a redundant or non-redundant planar manipulator,
- the links of the robot are straight line segments, whose endpoints are the corresponding link's joints,
- the obstacles are convex polygons, and their geometry is fixed and known.

In order to quarantee the collision-free motion of a robot between a collection of obstacles, an hierarchical structure that decomposes the original problem into two sub-problems is proposed. The structure consists of a two-level detection mechanism. In the higher-level the proposed mechanism determines those links of the robot arm that are too near the obstacles and thus, there is an increased possibility to collide. In the lower-level the mechanism checks for interference between the candidate for collision links (found in the previous level) and the corresponding near by obstacles. Thereby, with the proposed method there is no need to check for collisions between each one of the robot's links and every obstacle, but only between the candidate for collision links and the associated near by obstacles. The complexity of the original problem is therefore significantly reduced.

B. The High-Level Detection Mechanism

In this level each link and each obstacle are approximated by a circle. This circle is the minimum spanning circle i.e. the smallest circle that encloses a link or an obstacle. The detection mechanism in this level, is therefore consists of determining those obstacles that are too closed to a link and so that, a collision may be occurred. An obstacle is characterised as closed to a link if the circles enclosing the link and the obstacle are overlapped, or are in a contact (see Fig. 1).

C. The Low-Level Detection Mechanism

The lower-level of the proposed mechanism is responsible to detect for collisions between each one of the links and the corresponding closest obstacles found in the previous level. It must be underlined that links which do not have closed enough obstacles, meaning that they are in a safe distance from all the obstacles, do not included in this level, and therefore do not be examined.

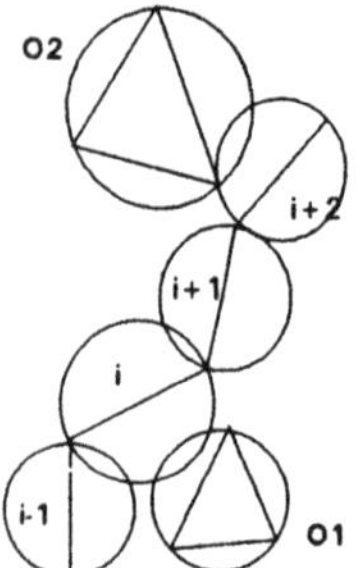

Fig. 1: Link--i is candidate to collide with obstacle O1, and link-(i+2) with O2.

To check whether there is a collision between a link and an obstacle it is suggested to build the convex-hull of the following set of points: the two joints of the link and the vertices of the obstacle. After the construction of the convex-hull, by simply locating the positions corresponding to the link's joints in the hull, the current level of the control mechanism can decide for a collision or not. This mechanism is given below in algorithmic form:

For each link of the robot arm .

1. Construct the convex-hull of the following set of points: the two joints of the link and the vertices of the polygon (Fig. 2.a)
2. Search this hull and locate the positions corresponding to the joints of the link.
3. If no one of the joints are contained in the hull, then we have a collision (Fig. 2.b).
4. If both of the joints are contained in the hull in successive positions say k and k+1, or

in the first and the last position, respectively, then there is no collision (Fig. 2.a), else there is a collision (Fig. 2.c).

5. If only one joint is contained in the hull check if the other joint lies outside the polygon. If this is the case, then there is no collision (Fig. 2.d), otherwise a collision occurs (Fig. 2.e)

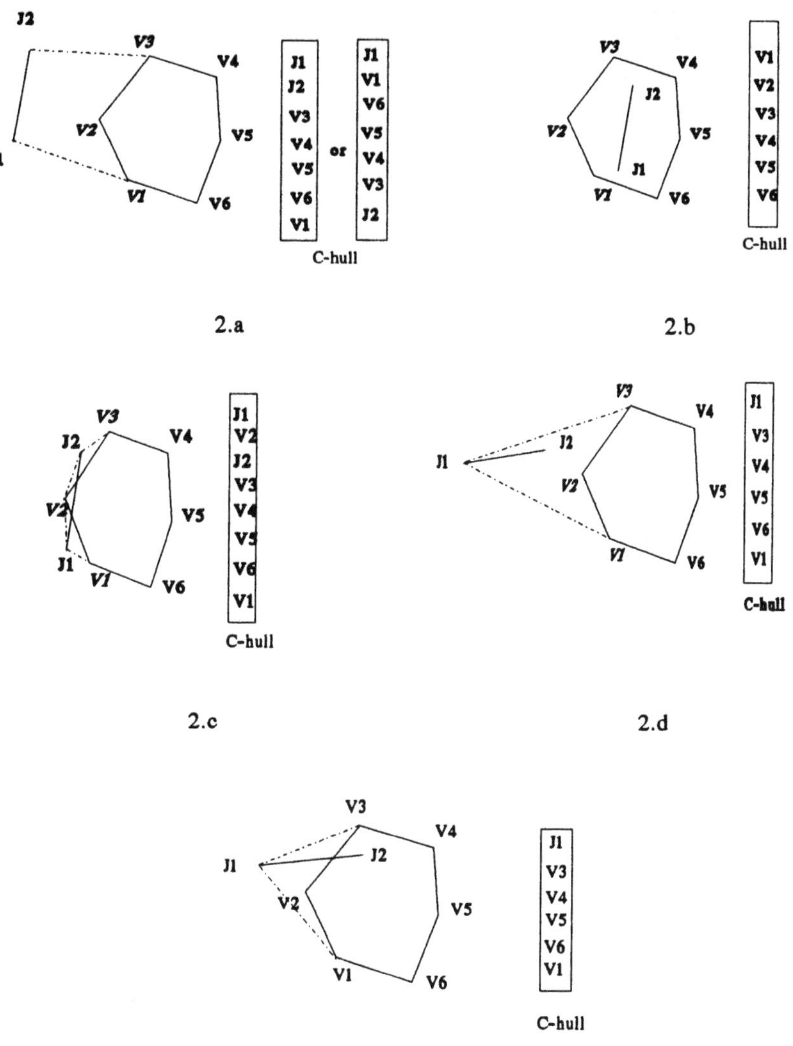

2.e

Fig. 2: Construction of the convex-hull of the link's joints and the obstacle's vertices.

III. Experiments and Results

The proposed method was simulated by a computer procedure and incorporated into an existing trajectory generation algorithm [16] in order to check and test its efficiency in generating collision-free robot trajectories. This algorithm [16] can be used in kinematics control of redundant or non-redundant manipulators along continuous path motion and in environments cluttered with obstacles. The algorithm drives the tip of the working hand of the robot along a desired path (whose parametric form is known a priori) until the tip reaches its goal, while the arm moves in a safe way among the obstacles. To do so, three appropriate criteria are used: A direction criterion which deals with the advancing of the tip of the hand towards its goal, a proximity criterion which is related to the proximity between the actual and the desired path, and a criterion of obstacle avoidance which is related to the collision-free motion of the arm. The problem is formulated as a constrained optimisation problem and solved using a technique based on genetic algorithms. The basic structure of this optimisation problem is:

Minimise { The Proximity Function }
Subject to,
a. {The Direction function}
b. {The Collision-free function}

It must be underlined that the problem is very important for applications such as welding, glueing, spray painting, etc. The analytical and mathematical formulation of this problem is presented in a previous work by one of the authors of this paper [17].

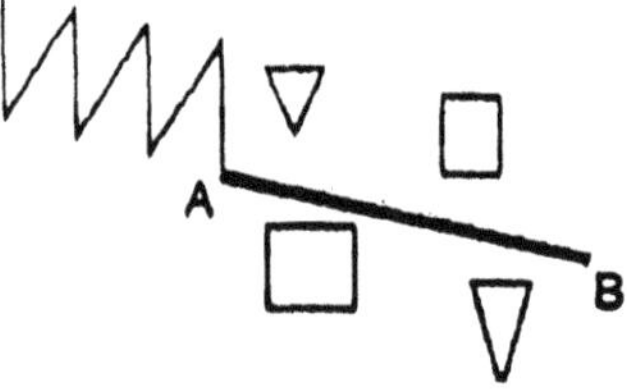

Fig. 3.a: Initial position of the 7-DOF Robot.

Fig. 3.b: Collision-free continuous motion of the 7-DOF robot on a specified straight line.

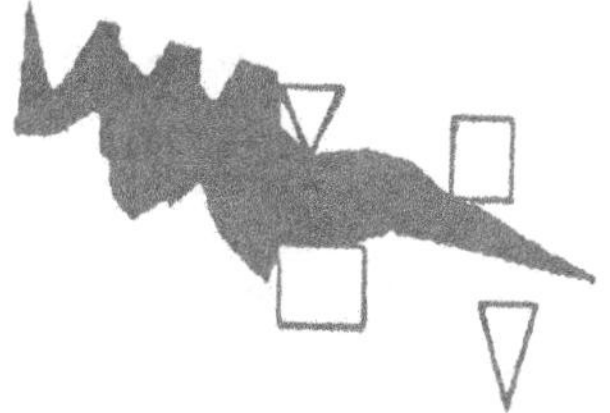

Fig. 3.c: The whole motion of the 7-DOF robot on the specified line.

The proposed collision-detection mechanism has been incorporated into the above problem in the place of the collision-free function. Thereby, in each step of the robot's motion, the algorithm guarantee the collision-free movement of the robot by first determining if a link is too near to an obstacle and then by computing the convex-hull of the set of the points as described above. Multiple experiments have been done on a number of planar manipulators with redundant degrees-of-freedom (DOF) moving in several environments cluttered with obstacles. Unfortunately, because of the limit in the allowed pages of the paper, only one of these experiments will be presented. In this experiment (see Fig. 3) the end-effector of a 7-DOF robot must traverse a straight line with endpoints A and B, while the robot's links must avoid the four obstacles in its environment. Initially, the robot's end-effector is posed on the point A (see Fig. 3.a). Fig. (3.b) displays some selected steps of robot's motion using the proposed method; while fig. (3.c) shows the whole motion of the robot with all the generated configurations. From these figures we can see the effectiveness of the proposed collision-detection scheme applied into continuous path planning problems.

IV. Conclusions

This paper presents a method for detection of interference between a robot and a collection of obstacles based on the convex-hull concept. The proposed method is simple and efficient and reduces the complexity of the original problem by taking into account only those links of the robot, which have increased possibilities to collide. The method can be effectively applied into continuous path planning problems. The feasibility of this method is demonstrated by experiments in 2-D. Work must be done to apply the method in problems in 3-D.

V. References

[1] Tai-Soku and Bahram Revani, "Interference detection in Computer-Aided Geometry Design of prismatic objects", ASME paper, no 84-DET-(1979).

[2] T. Lozano Perez,"Spatial Planning: A configuration space approach", IEEE Trans. on Comp., Vol. C-32, no.2, 108-120, February (1983).

[3] T. Lozano Perez and M..A. Wesley,"An algorithm for planning collision-free paths among polyhedral obstacles", Communications of the ACM, 22,(10), 560-570, (1979).

[4] H. Mitchell,"An algorithmic approach to some problems in terrain navigation", Artificial Intelligence, 37, 171-201, (1988).

[5] C. O'Dunlaing and K. Yap,"A retraction method for planning the motion of a disc", Journal of Algorithms, 6, 104-111, (1985).

[6] C. O'Dunlaing M. Sharir and K. Yap, "Retraction: A new approach to motion planning", Proc. of the 15th ACM Symposioum on the Theory of Computing, Boston, 207-220, (1983).

[7] R.A. Brooks, "Solving the find-path problem by good representation of free space", IEEE Trans. on Systems, Man and Cybernetics, SMC-13(3), 190-197, (1983).

[8] M. Anthimopoulou and N. Aspragathos,"Kinematic control of planar manipulators moving between obstacles", 2nd International Workshop on Advances in Robot Kinematics, Linz, September, 10-12, (1990).

[9] O. Khatib, "Real-time obstacle avoidance for manipulators and mobile robots", Int. Journal of Robotics Research, 5(1), 90-98, (1986).

[10] S. Arimoto and H. Noborio, "A 3D closest pair algorithm and its application to robot motion planning", IFAC Robot control, Karlsuhe, FRG, (1988).

[11] G. Garcia, Ph. Wenger and P. Chedmail,"Computing moveability areas of a robot among obstacles using Octrees", Proc. of the 4th Intern. Conf. on Advanced Robotics, Columbus, Ohio, June 13-15, (1989).

[12] G. Hurteau and N.F. Stewart, "Distance calculation for imminent collision indication in a robot system simulation", Robotica, Vol. 6, 47-51, (1988).

[13] E.G. Gilbert and D.W. Jonson, "Distance functions and their application to robot path planning in the presence of obstacles", IEEE Journal Robotics Automation. RA-1: 21-30.

[14] J. E. Bobrow, "A direct minimization approach for obtaining the distance between convex polyherda", The International Journal of Robotics Research, 8(3), June, (1989).

[15] R.O. Buchal and D.B. Cherchas, "An iterative method for generating kinematically feasible interference-free robot trajectories", Robotica, Vol. 7, 119-127, (1989).

[16] A. Nearchou and N. Aspragathos, "Collision-free continuous path control of manipulators using genetic algorithms", submitted for publication to Robotica.

[17] T. Lianos, D. Kiritsis and N. Aspragathos, "A Direct Algorithm for Continuous Path Control of Manipulators", Robotics & Computer Integrated Manufacturing, 8(2), (1991).

Collision-Avoidance Robot Path Planning Using Fully Cartesian Coordinates

F. Valero, J.I.Cuadrado and V. Mata

Department of Mechanical Engineering
Polytechnic University of Valencia
46022 Valencia, Spain

M. Ceccarelli

Department of Industrial Engineering
University of Cassino
03043 Cassino (FR), Italy.

Abstract - An algorithm for robot path planning in the presence of obstacles is proposed. The problem is formulated in the Cartesian space. Nonlinear programming techniques are used to obtain a map of feasible robot configurations. Then a weighted graph is associated to the map, and a search algorithm is used to get a sequence of collision-free robot configurations between two previously selected points. The algorithm has been successfully applied to redundant planar robotic manipulators with revolute and prismatic joints and spatial robotic manipulators.

I. Introduction

Algorithms for robot path planning among obstacles can be classified in two main classes: global algorithms and local algorithms. Most of the global algorithms for collision avoidance are based on the method of Configuration Spaces. This method has been introduced by Udupa in [1], and formalized by Lozano-Perez in [2], requires transforming the objects from the Cartesian space to the robot joint space, so that a robot can be represented as a single point and the free-space as a tree of cells. Search algorithms are then used to find collision-free paths between the initial and final robot configurations. The algorithms proposed by Lozano-Perez in [3] and by Faverjon in [4] can be considered as global algorithms. Further developments can be found in Chen [5], Laugier [6], and Dupont [7]. The main disadvantage of the global algorithms is that the computation of the free-space may be very consuming both in CPU time and storage requirements.

Local algorithms are attractive because of their simplicity. However, they do not guarantee the determination of a path even if it exists. The method of Potential Functions was proposed by Khatib, [8], and it considers the moving object in a field of potential, with an attractive potential (final position) and some repulsive potentials (obstacles). Another local

A. J. Lenarčič and B. B. Ravani (eds.), Advances in Robot Kinematics and Computationed Geometry, 485–494.

approach for the path planning problem is the Method of Constraints, introduced by Faverjon in [9]. The basic idea of this method is to impose constraints on the variation of the robot configuration parameters.

The algorithms for robot path planning in the presence of obstacles proposed until now have two aspects in common: they describe the robot placement through a system of independent coordinates, and they operate in an abstract space (robot joint space). However, in the Mechanism and Machine Theory a variety of sets of dependent coordinates has been successfully used in order to describe the mechanisms.

The approach, which is proposed in this paper for the path planning problem with obstacle avoidance, is based on two main ideas: we shall consider a set of dependent coordinates to describe the robot location, and the robot system will operate in a physical space (Cartesian space). Among the several types of dependent coordinates currently used, we have chosen fully Cartesian coordinates, that have been introduced by Garcia de Jalón in [10].

Particularly, this paper presents a global algorithm for planning collision-free paths for robots with any number and class of joints. The method proposed operates in two consecutive phases. In the first phase, a map of admissible robot configurations will be generated over a discrete robot environment. The concept of adjacent configurations will be introduced, so that two robot configurations are said to be adjacent if they are the nearest according to a suitable norm which has been previously defined. Based on the above definition, the problem of adjacent configurations will be established as a constrained nonlinear optimization problem. In a second phase of the method, a labelled graph will be associated to the map of configurations, by reducing the robot path planning problem to that of finding the shortest distance between points in a graph.

II. Problem Formulation

Three main subjects will be considered in this section: the geometric modelling of a robotic manipulator, the discrete representation of the robot environment and the definition of adjacent configurations. To better illustrate, the problem of moving a planar manipulator in a planar environment with obstacles is considered first, and in a following section, an example for a spatial robot will be considered.

A planar robotic manipulator can be modelled as a sequence of $n+1$ rigid links, numbered from 0 to n. The base link, which is fixed to the ground, is labeled 0 and the most distant link is numbered n. Each pair of consecutive links is joined together by a kinematic pair, either of rotational or prismatic type. In order to make easy the mathematical treatment of the problem, robot mobile links will be represented by a sequence of n segments with lengths l_1, $l_2,\ldots,l_n$; a growing obstacles technique can be then used in order to take into account real robot dimensions. Also we assume that links are allowed to cross over one another (each link could move in a separate plane parallel to the ground). The pair joining link 0 and link 1 is assumed to be a revolute joint. Figure 1 shows a planar robotic manipulator composed by n mobile links joined by rotational or prismatic joints.

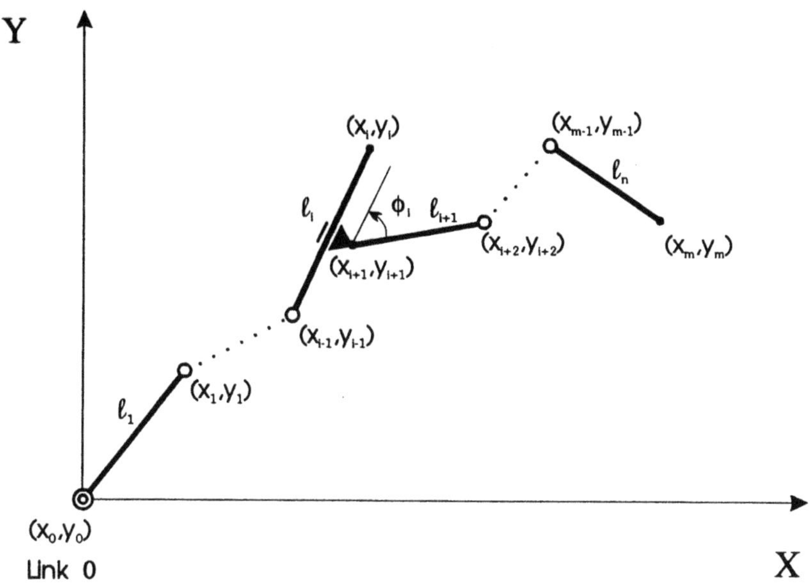

Fig. 1: A planar robotic manipulator with n degrees of freedom

Cartesian coordinates of some basic points are used in the fully Cartesian coordinates Method to mathematically model a mechanism. Once m reference points are established, a general robot configuration can be represented by the variables $(x_1,y_1,x_2,y_2,\ldots,x_m,y_m)$ excluding the coordinates (x_0,y_0) of the robot fixed point. These Cartesian coordinates are related among each other by constraint equations, see Garcia de Jalón [11] for an exhaustive description of the method. It is remarkable that the number of mobile links, n, is not necessarily the same as the number of basic points, m.

Then, the j-th robot configuration can be denoted synthetically by $C^j(x_1^j,y_1^j,x_2^j,y_2^j,\ldots,x_m^j,y_m^j)$ (using the absolute Cartesian coordinates (x_i^j,y_i^j) of m basic points).

In addition, in order to generate a Map of Configurations, it is necessary to transform the robot environment into discrete states. The discrete space can be defined to enclose all possible robot endpoint locations. In the planar case, a rectangular work area is considered, so that it necessarily includes initial and final robot endpoint locations. These points will have integer Cartesian coordinates in a previously established Cartesian Reference System.

The discrete transformation routine proposed in this paper is based on three main assumptions:

1) The procedure starts from a final (and safe) robot configuration given by the user;

2) The robot endpoint movements are constrained to be horizontal or vertical;

3) The robot endpoint locations must be points with integer coordinates in the Cartesian Reference System stated above.

The concept of adjacent configuration is now better illustrated. Let $C^k(x_1^k,y_1^k,x_2^k,y_2^k,....,x_m^k,y_m^k)$ and $C^p(x_1^p,y_1^p,x_2^p,y_2^p,....,x_m^p,y_m^p)$ be two robot configurations; they will be said to be adjacent configurations if the following two conditions are satisfied:

$$|x_m^k-x_m^p| + |y_m^k-y_m^p| = 1$$

$$||C^k-C^p|| \text{ is minimum} \qquad (1)$$

that is, the configuration C^k is the nearest to the C^p, according to the following norm:

$$||C^k-C^p|| = \sum_{i=1}^{m} |x_i^k-x_i^p| + \sum_{i=1}^{m} |y_i^k-y_i^p| \qquad (2)$$

The basic idea for a generation of a Map of Configurations is the following: given a final robot configuration, and taking into account Assumption 2, the endpoint of the preceding configuration (in a ideal sequence of robot configurations) will be located in one of the four nearest points in the Cartesian plane. From this, a backtracking routine is established in order to cover all the robot work area (or volume in the three-dimensional case).

III. Solving for a Collision-Free Adjacent Configuration

The determination of an adjacent configuration can be better stated by referring to a planar manipulator. Two different work areas are considered. The first of them, a rectangular area with dimensions $h \times v$ and left bottom corner with coordinates (x_A,y_A), where the complete robot could be operated. The second one, with dimensions $h' \times v'$ and left bottom corner with coordinates (x_w,y_w) , where the robot endpoint has to be located. A Cartesian reference system will be considered in the first work area. Let us consider that the above mentioned rectangles are parallel to the axes of the reference system. Some polyhedral obstacles are located inside the first work area. These obstacles can be modelled through p circles, where r_j will be the radius of the i-th circle and P_j its centre. Thus, these circles can be considered as the obstacles. The second work area, which defines allowable robot endpoint locations, can be discrete, being Δs the step length (both horizontal and vertical) of the mesh.

Let $C^k(x_1^k,y_1^k,x_2^k,y_2^k,....,x_m^k,y_m^k)$ and $C^p(x_1^p,y_1^p,x_2^p,y_2^p,....,x_m^p,y_m^p)$ be two adjacent robot configurations, where C^k is known and both of them are collision-free with any of the obstacles.

Let d_{ij}^{P} be the minimum distance from the i-th link of the robot in the C^{P} configuration to the centre of the j-th obstacle, Figure 2.

The problem of finding a safe robot configuration, named C^{P}, from its adjacent configuration, C^{k}, may be stated in a general form as follows:

$$Min \quad \{ \sum_{i=1}^{m} (x_i^{P}-x_i^{k})^2 + \sum_{i=1}^{m} (y_i^{P}-y_i^{k})^2 \}$$

subject to:

$$(x_i^{P}-x_{i-1}^{P})^2 + (y_i^{P}-y_{i-1}^{P})^2 = l_i^2 \qquad \text{Joint i rotational} \qquad (3a)$$

$$l_i^2 \leq (x_i^{P}-x_{i-1}^{P})^2 + (y_i^{P}-y_{i-1}^{P})^2 \leq L_i^2 \qquad \text{Joint i prismatic} \qquad (3b)$$

$$-h \leq x_i^{P} \leq h \qquad i=1,2,\ldots,m \qquad (4)$$

$$0 \leq y_i^{P} \leq v \qquad i=1,2,\ldots,m \qquad (5)$$

$$x_w \leq x_m^{P} \leq x_w + h' \qquad i=1,2,\ldots,m \qquad (6)$$

$$y_w \leq y_i^{P} \leq y_w + v' \qquad i=1,2,\ldots,m \qquad (7)$$

$$|x_m^{P}-x_m^{k}| + |y_m^{P}-y_m^{k}| = \Delta s \qquad (8)$$

$$d_{ij}^{P} > r_j \qquad i=1,2,\ldots,n, \quad j=1,2,\ldots,p \qquad (9)$$

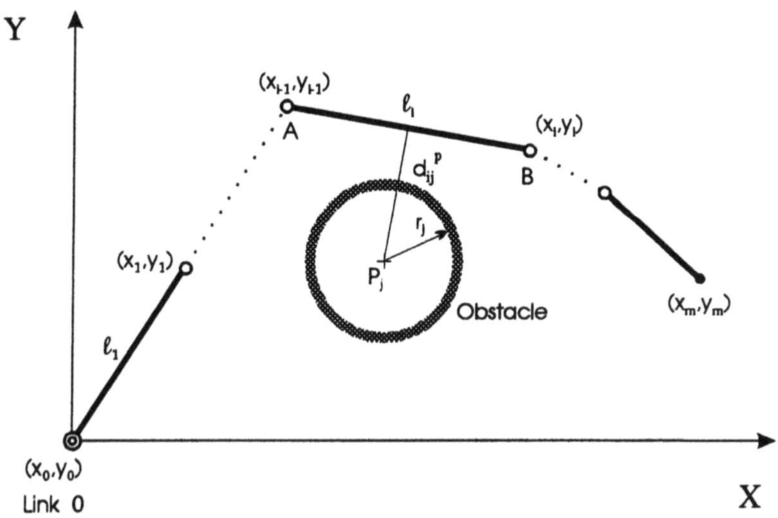

Fig. 2: Planar robot with revolute joints

Constraints (3a) and (3b) ensure the geometrical compatibility of the elements (links and joints) that the robot is composed of. Note that in the case of prismatic joints, l_i is the minimum length allowable for the link and L_i is the maximum one, also they are followed by a rotational pair ($l_{i+1}=0$, in Fig. 1). Constraints (4) and (5) keep all the robot links inside the predefined robot work area. Constraints (6), (7) and (8) are concerned with the movement of the robot endpoint within the workspace. Finally, the set of constraints (9) states that the robot links and the obstacles must not collide.

The above formulated problem can be recognized as a constrained nonlinear optimization problem. Algorithms based on the Sequential Quadratic Programming (SQP) method have been proved to be efficient in solving these kind of optimization problems (see Powell [12]). The basic idea of this class of algorithms is that, instead of applying the Newton method for solving the Kuhn-Tucker equations, they consider a quadratic approximation of the Lagrangian function and a linearization of the constraints. This quadratic subproblem can be solved by specific methods (Gill and Murray or Goldfarb and Idnani). Subsequently a line search can be performed with respect to a merit function and the Hessian aproximation is usually updated by the Broyden-Fletcher-Goldfarb-Shanno (BFGS) formula. The algorithm considered in this paper was proposed by Schittkowski in [13], and it contains some additional refinements: an active set strategy and one additional variable in the objective function in order to avoid the problem of an empty feasible region in the quadratic subproblem.

Two aspects of the optimization problem are remarkable: the objective function and the constraints have analytical derivatives, and a good initial guess (the adjacent configuration) is always available to start the optimization procedure.

Once a map of safe robot configurations has been generated, a labelled graph can be associated to it in order to find a path (sequence of robot configurations) between the initial location of the robot endpoint and the goal robot configuration. The nodes of this graph correspond to the robot configurations and the edges connect pairs of adjacent configurations. Dijkstra's shortest path algorithm is used in order to find a minimum distance path. If such a path exists, then it is easy to transform it into a sequence of robot configurations.

IV. A Numerical Example of a Collision-Avoidance Path Planning for a Planar Robot

Several planar manipulators have been considered in order to test the performance of the path planning procedure proposed in this paper. Particularly, an example for a planar redundant manipulator composed of 3 rotational joints and 1 prismatic joint (see Figure 3) is reported. The data of the problem are the following: $l_1= 5$, $l_2=1$, $L_2=5$, $l_3= 5$, $l_4= 5$, $x_A= -12$, $y_A= 0$, $h= 24$, $v= 12$, $x_W= -9$, $y_W= 0$, $h'= 18$, $v'= 10$, $p= 20$, $r_j= 1$ $j=1,...,20$, $\Delta s= 1$ and $C^f(0.0, 1.2, 4.9, 0.5, 6.5, -3.5, 9.5, -7.0, 6.0)$.

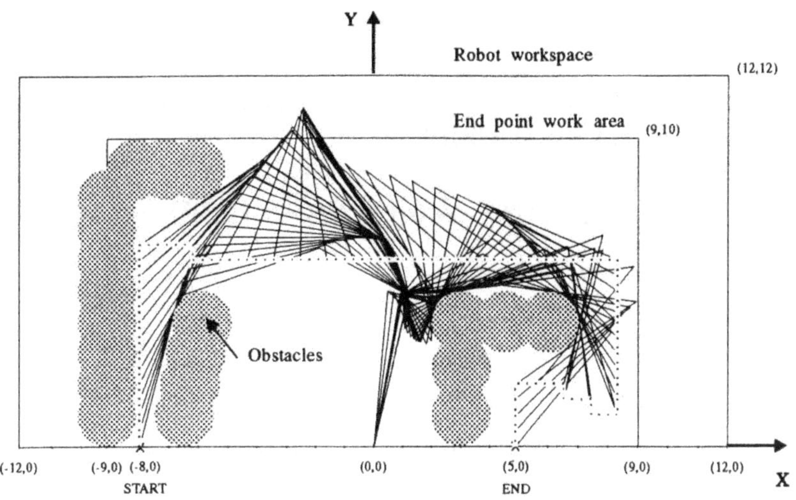

Fig. 3: An illustrative example of a proposed collision-avoidance path planning for a 4 d.o.f. planar manipulator

V. A Numerical Example of a Collision-Avoidance Path Planning for a Spatial Robot

One of the main advantages of the method proposed is that the extension of the problem formulation from the two-dimensional case to the three-dimensional one is straightforward. For instance, Figure 4 depicts a robot similar to the Unimation PUMA 560 industrial robot, for which no physical constraints in the joints have been considered. In this particular case, the robot has been modelled with five basic points, so that a p-th robot configuration can be represented by $C^p(x_1^p,y_1^p,x_2^p,y_2^p,z_2^p,x_3^p,y_3^p,z_3^p,x_4^p,y_4^p,z_4^p,x_5^p,y_5^p,z_5^p)$. Moreover, taking into account the specific geometry of the robot, trivial relations can be stated among some of the Cartesian coordinates:

$$z_1^p = z_0$$

$$y_3^p = y_2^p \pm \frac{l_3.(x_2^p - x_1^p)}{\sqrt{(x_2^p-x_1^p)^2+(y_2^p-y_1^p)^2}} \qquad \begin{cases} + \text{ if } y_2^p > 0 \\ - \text{ if } y_2^p \leq 0 \end{cases}$$

$$x_3^p = x_2^p - \frac{(y_2^p-y_1^p).(y_3^p-y_2^p)}{(x_2^p-x_1^p)}$$

$$z_3^p = z_2^p$$

So that, a robot configuration will be represented as $C^p(x_1^p,y_1^p,z_1^p,x_2^p,y_2^p,z_2^p,x_4^p,y_4^p,z_4^p,x_5^p,y_5^p,z_5^p)$. The data of the problem are the following: l_1= 255, l_2= 432, l_3= 145, l_4= 433, l_5= 420, x_A= -700, y_A= -700, z_A= -400, h=h'= 1400, v=v'= 1400, w=w'= 800, p= 20, r_j= 100 j= 1,..,25 (main obstacle), r_j= 150 j=26,..,30 (secondary obstacle), r_j= 150 j= 31,..,38 (fixed base of the robot), Δs= 10 and C^f(112, 230, 0, 8, 280, 435, 345, 7, 505, 570, -140, 210).

The problem of finding a safe robot configuration, assumed C^p, from its adjacent configuration C^k, can be stated as follows:

$$Min \quad \{ \sum_{i=1}^{5} (x_i^p - x_i^k)^2 + \sum_{i=1}^{5} (y_i^p - y_i^k)^2 + \sum_{i=1}^{5} (z_i^p - z_i^k)^2 \}$$

subject to:

$$(x_1^p - x_0)^2 + (y_1^p - y_0)^2 - l_1^2 = 0 \qquad (10)$$
$$(x_2^p - x_1^p)^2 + (y_2^p - y_1^p)^2 + (z_2^p - z_0)^2 - l_2^2 = 0 \qquad (11)$$
$$(x_1^p - x_0).(x_2^p - x_1^p) + (y_1^p - y_0).(y_2^p - y_2^p) = 0 \qquad (12)$$
$$(x_4^p - x_3^p)^2 + (y_4^p - y_3^p)^2 + (z_4^p - z_2^p)^2 - l_4^2 = 0 \qquad (13)$$
$$(x_4^p - x_3^p)^2.(x_3^p - x_2^p)^2 + (y_4^p - y_3^p).(y_3^p - y_2^p) = 0 \qquad (14)$$
$$(x_5^p - x_4^p)^2 + (y_5^p - y_4^p)^2 + (z_5^p - z_4^p)^2 - l_5^2 = 0 \qquad (15)$$
$$-h \leq x_i^p \leq h \qquad i=1,2,...,5 \qquad (16)$$
$$-v \leq y_i^p \leq v \qquad i=1,2,...,5 \qquad (17)$$
$$0 \leq z_i^p \leq w \qquad i=1,2,...,5 \qquad (18)$$
$$|x_m^p - x_m^k| + |y_m^p - y_m^k| + |z_m^p - z_m^k| = \Delta s \qquad (19)$$
$$d_{kj}^p > r_j \qquad j=1,2,...,33 \qquad k=1,2,...,5 \qquad (20)$$

Constraints (10), (11), (13) and (15) correspond to the rigid body condition for links 1, 2, 4 and 5. Constraints (12) and (14) keeps an angle of 90 degrees between links 1 and 2 and 3 and 4. It is remarkable that the spherical wrist of the robot is considered automatically in the formulation, with no need for constraint equation. The other constaints are analogous to a planar case. It is remarkable that constraints (20) avoids collision not only between robot links and the U-shaped obstacle considered, but also with robot's base.

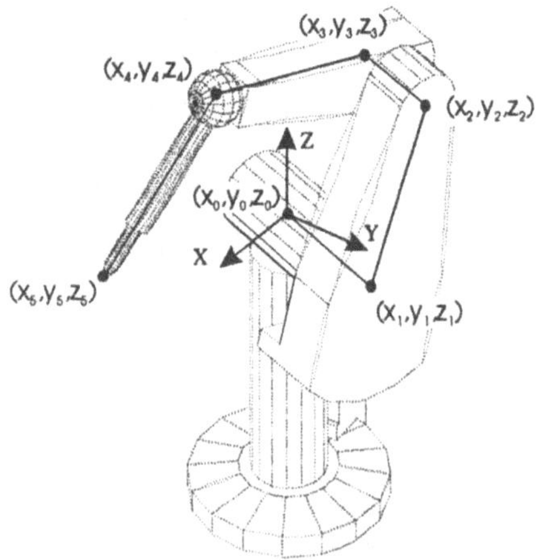

Fig. 4: Industrial robot and its basic points.

In the Figure 5 some configurations of a simulation of the results of a path plannig for an industrial robot are illustrated from two different perspectives. The dotted line shows the end-effector path. The obstacles which model the base of the robot are not included in the figure. The robot path between the two selected points consists of 50 robot configurations.

VI. Conclusions

In this paper, a new procedure for the collision-avoidance path planning problem for planar or spatial robotic manipulators has been developed. The problem has been formulated in the Cartesian Space through fully cartesian coordinates. The procedure has turned out to be very versatile as the reported examples prove, since it can handle robots with revolute and prismatic joints. It is also remarkable that a high degree of redundancy in the robots does not increase significantly the complexity of the proposed formulation.

The efficacy of the method is demonstrated through two and three dimensional examples. The complexity of the environments (number of obstacles) that have been considered is not essentially a restriction of the method but is a restriction of the computer that has been used in the simulation.

Among the drawbacks of this procedure, it is possible to point up the dependence of the computed path on the given final configuration. Moreover, having in mind that the proposed algorithm operates with via points, it is not guaranteed that the movement between two adjacent configurations has no collision. Nevertheless, it is possible to overcome this difficulty through an adequate discrete representation of the robot environment.

VII. References

[1] S.M. Udupa, "Collision detection and avoidance in computer controlled manipulators", *Proc. of the Int. Joint Conference on Artificial Intelligence*, 737-748 (1977).

[2] T. Lozano-Pérez, M.A. Wesley and T.J. Watson, "An Algorithm for Planning Collision-Free Paths Among Polyhedral Obstacles", *Communications of ACM*, 22, 560-570 (1979).

[3] T. Lozano-Pérez, "Spatial Planning: A Configuration Space Approach", *IEEE Transactions on Computers*, 32, 108-120 (1983).

[4] B. Faverjon, "Obstacle avoidance using an octree in the configuration space of a manipulator", *Proc. of IEEE Int. Conference on Robotics and Automation*, Atlanta (1984).

[5] R.T. Chien, L. Zhang and B. Zhang, "Planning Collision-Free Paths for Robotic Arm Among Obstacles", *IEEE Trans. on Pattern Analysis and Machine Intelligence*, 6(1), 91-96 (1984).

[6] C. Laugier and F. Germain, "An adaptive collision-free trajectory planner", *Proc. of the Int. Conference on Advance Robotics*, Tokyo (1985).

[7] P.E. Dupont and S. Derby, "Two-Phase Path Planning for Robots With Six or More Joints", *ASME Journal of Mechanical Design*, 112, 50-58 (1990).

[8] O. Khatib, "Real Time Obstacle Avoidance for Manipulators and Mobile Robots", *International Journal of Robotics Research*, 7(1), 90-98 (1986).

[9] B. Faverjon and P. Tournassoud, "A local based approach for path planning of manipulators with a high numer of degrees of freedom", *Proc. of IEEE Int. Conference on Robotics and Automation*, Raleigh, 1987.

[10] J. García de Jalón, M.A. Serna and R. Avilés, "Computer Method for Kinematic Analysis of Lowe-Pair Mechanisms. Part I: Velocities and Accelerations and Part II: Position Problems", *Mechanism and Machine Theory*, 16, 543-566 (1981).

[11] J. García de Jalón and E. Bayo, *Kinematic and Dynamic Simulation of Multibody Systems*, Springer-Verlag, New York (1994).

[12] M.J. Powell, "Algorithms for Nonlinear Constraints that Use Lagrangian Functions", *Mathematical Programming*, 14, 224-248 (1978).

[13] K. Schittkowski, "NLPQL: A FORTRAN subroutine solving constrained nonlinear programming problems", *Annals of Operations Research*, 5, 485-500 (1986).

Acknowledgments

This work was partially supported by the Italian and Spanish Governments "Integrated Actions 1993" program ; and by the Special Actions Nº TAP93-1296-E, TAP93-1324-E of Spanish. Their support is gratefully acknowledged. Government "

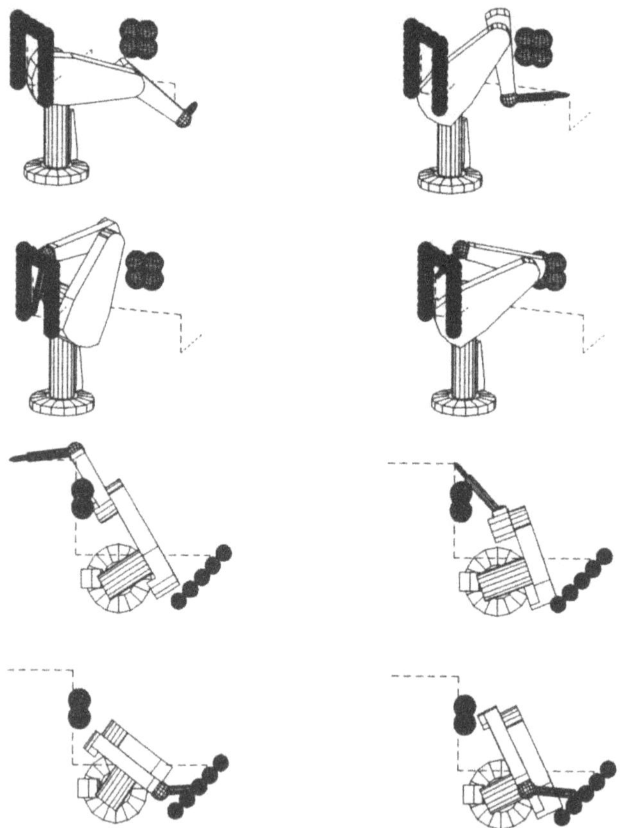

Fig. 5: An illustrative example of the proposed collision-avoidance path planning for an industrial robot.

Planning Grasp Points and Base Locations of Single Robots and Cooperating Robot Systems

P. Dietmaier

Department of Mechanical Engineering
Technical University, Graz
Graz, Austria

J. M. McCarthy

Department of Mechanical Engineering
University of California Irvine
Irvine, California 92717

Abstract - This paper describes a method for determining the set of available grasp points and the base locations for single robot and cooperating robot systems so a workpiece can be moved through a set of specified positions. The method uses a simple model for the individual robot workspaces and does not take into account collision avoidance. It is applicable to planar as well as spatial robots, robot systems, and multifingered hands. The focus is especially on the development of fast algorithms for interactive and real time planning systems.

I. Introduction

In order to move a workpiece through a set of specified positions, with a given robot (or a system of cooperating robots), it is necessary to know where the robot can grasp the object and where the base of the robot can be located. Depending on the task, the object, the robot, and the environment, there usually exist certain restrictions which require planning the object grasp point and the base location of the robot in advance. In the past, this has quite often be done by trial and error at the scene. For robots with a moving base or robots with frequently changing tasks, planing ahead becomes even more important. In certain applications planing might even have to be done in real time. Therefore fast algorithms are important in this context.

For our calculation, we assume that the workspace of each single robot is shaped like a disk (or a ball in the spatial case) and that the robot can arbitrarily orient an object at each point within this workspace, Fig. 1. For many robots this is a good approximation of their workspace, and forms an upper bound for all other robots. Any kind of collisions between the robot(s) and the moved object or the robot(s) are neglected. The analysis of robots under these simplified conditions provides a framework for the study of specific robotic system. Studies of the workspace of cooperating robots can found in McCarthy and Bodduluri [5], which considers the extreme reach of two 3R planar robots forming a planar 4R linkage, and Kerr and Roth [4], which describe the workspace of multifingered hands.

A. J. Lenarčič and B. B. Ravani (eds.), Advances in Robot Kinematics and Computationed Geometry, 495–504.

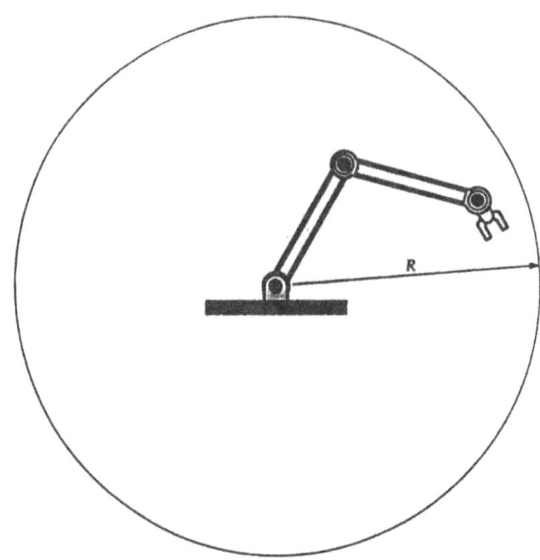

Figure 1: Simplified workspace of a planar 3R robot manipulator.

II. Finding all Possible Grasp Points for n Planar Positions

Our goal is to find possible grasp points on the object such that they do not leave the circular workspace of the robot while moving the object through n specified positions. McCarthy and Bodduluri considered the 3-position case [5]. Attach a Cartesian coordinate system to the moving object and describe the relation between coordinates of two relative positions, denoted by the 2×1 column vectors $\mathbf{X}_i$ and $\mathbf{x}$, with the matrix equation

$$\mathbf{X}_i = [\mathbf{A}_i]\mathbf{x} + \mathbf{d}_i, \qquad i = 1, ..., n. \tag{1}$$

The 2×2 rotation matrix $[\mathbf{A}_i]$ and the 2×1 column vector $\mathbf{d}_i$ represent the rotation and translation of the ith position relative to the fixed frame. For the inverse transformation we have

$$\mathbf{x} = [\mathbf{A}_i]^T(\mathbf{X}_i - \mathbf{d}_i). \tag{2}$$

For each grasp point G on the moving object we obtain a set of n, so called, homologous points G_i, $i = 1,...,n$ as the object coincides with the n specified positions. For a possible grasp point G, all homologous positions G_i of this grasp point have to lie within the circular workspace of the robot. For a given set of homologous points we can find a circular workspace with minimal radius $\overline{R}$ that encloses all these points. If we now vary the grasp point G in such a way that $\overline{R}$ becomes equal to the workspace radius R of the given robot, then G becomes a point on the grasp space boundary $\overline{G}$ for the robot. In the following we shall see that this approach allows us to compute the grasp space boundary in linear time.

A. *Finding the Smallest Circle*

The problem of finding the smallest circle enclosing n points has been studied for more than a century. However, Megiddo [7], was the first to show that the problem can be solved in linear time with a deterministic algorithm; but his algorithm is difficult to implement. Recently, Welzl [9], proposed a method based on Seidel's linear programming algorithm [8]. He also gives some experimental results on his algorithms. Although this seemed to be the fastest algorithm available in the literature, we found a different method, that at least for the two and three dimensional case, outperforms these algorithms by at least a factor of two. This algorithm is as follows (see Fig. 2):

1. Determine the circle C_1 with smallest radius enclosing the first three points.
2. Check if the next point (start with the fourth point, and if the last point is reached begin again with the first one) lies inside the circle C_i (the boundary is considered to belong to the interior). If the new point lies outside of C_i, then compute the smallest circle which encloses the points *determining* C_i (either 2 or 3) and the new point. Call the result C_{i+1}.
3. If all points have been checked to lie inside C_k, then C_k is the smallest circle enclosing all points, and we are done. Otherwise continue with step 2.

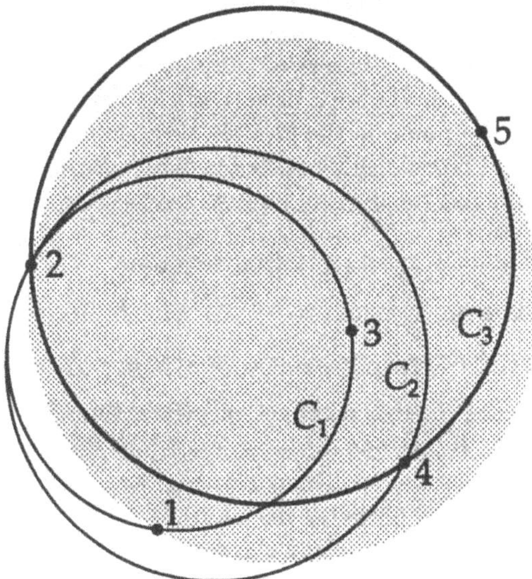

Figure 2: The first three steps in finding the smallest circle for a given 5 points.

Since in step 2 we only consider the points determining the circle C_i and not all points prior to the current point (which is the advantage of the method), it may happen that after having checked $n-2$ points (see C_3 in Fig. 2), there are still some points lying outside of the current circle C_k. So, we have to start again with the first point and have to continue this procedure until all points lie inside of C_k.

Using Welzl's examples, we found that in average just a little bit over n steps are necessary to determine the final circle, and it takes an additional n containment tests to ensure this result. Therefore the average number of containment tests lies between $2n$ and $2.2n$, while the number of computed intermediate circles C_i is just a fraction of n (e.g. 30 - 40 for 1000 points).

To give the reader a feeling for how quickly this algorithm runs, we mention that computing all homologous positions of a grasp point takes about the same number of operations as finding the smallest circle for the same set of points.

B. Properties of the Planar Grasp Space Boundary

Before we describe how to compute the grasp space boundary, we want to summarize what we know about it: (i). The grasp space boundary is the intersection of circles and tri-circular sextics (McCarthy and Bodduluri [5]). (ii). A grasp space associated with a certain workspace radius R encloses all grasp spaces with smaller workspace radius. (iii). This boundary is strictly convex. (iv). The grasp space is simply connected, and there exists a grasp point associated with a minimal workspace radius $R_{\min} \geq 0$. While property (i) is known and (ii) is obvious, since a robot with a larger workspace can always do the job of a smaller one (according to our workspace definition), we have to show (iii). For the proof of property (iii), we pick two possible grasp points (1, 2) on the grasp space boundary $\overline{G}$ for a workspace with radius R (see Fig. 3). Associated with these two grasp points are two sets of homologous positions denoted as G_{1i}, G_{2i} ($i = 1, ..., n$) and two workspaces with base locations C_1, C_2 and the same radius R. We now pick a third grasp point on the straight line connecting the two grasp points, and obtain a third set of homologous points $G_{\lambda i}$ which satisfies the condition $\overline{G_{1i}G_{\lambda i}} = \lambda\overline{G_{1i}G_{2i}}$ for all i, because the moving body is rigid. From Fig. 3 we can determine the vector relation

$$\mathbf{r}_{\lambda i} = \mathbf{r}_{1i} + (\Delta\mathbf{x}_i - \Delta\mathbf{x}_C)\lambda \tag{3}$$

and if $\Delta\mathbf{x}_i - \Delta\mathbf{x}_C \neq 0$, then

$$|\mathbf{r}_{\lambda i}|^2 = |\mathbf{r}_{1i}|^2 + 2\mathbf{r}_{1i} \cdot (\Delta\mathbf{x}_i - \Delta\mathbf{x}_C)\lambda + |\Delta\mathbf{x}_i - \Delta\mathbf{x}_C|^2\lambda^2 < R^2, \tag{4}$$

holds for $0 < \lambda < 1$, since $|\mathbf{r}_{\lambda i}|^2$ is a convex quadratic function of λ with $|\mathbf{r}_{\lambda i}|^2 \leq R^2$ for $\lambda = 0$ and for $\lambda = 1$. If just one of the points which determine the smallest circle enclosing the $G_{\lambda i}$'s, satisfies the condition $\Delta\mathbf{x}_i \neq \Delta\mathbf{x}_C$, then the radius of the circle C_λ is less than R; according to property (ii) all grasp points on the line between 1 and 2 (not including the endpoints) lie inside the grasp space associated with R. Now, if any two points on $\overline{G}$ result in such a configuration, then the grasp space has a strictly convex boundary. We now show that this is always the case.

Suppose all points determining the smallest circle in position λ satisfy $\Delta\mathbf{x}_i = \Delta\mathbf{x}_C$, then there is no relative rotation between the associated positions of these homologous points. This would be a special case of our three position analysis for which we found the whole plane to be the grasp space. Therefore there has to be a point with $\Delta\mathbf{x}_i \neq \Delta\mathbf{x}_C$ involved in determining the smallest circle, which had to be shown.

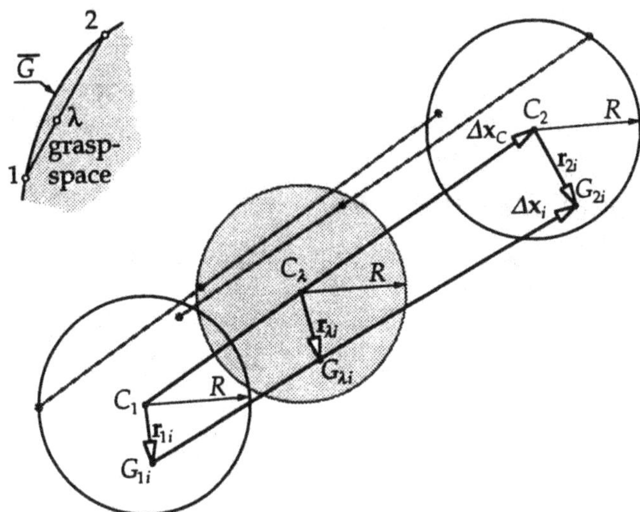

Figure 3: Configurations for three different grasp points on the boundary.

C. Determining the Planar Grasp Space

In order to compute the boundary of the grasp space for a given workspace with radius R, we use the following steps:

1. Compute the minimal workspace (radius R_{min}) required to move the object through all positions by optimal choice of the grasp point ($\mathbf{x}_{G\text{min}}$). This can be achieved by minimizing the square of the minimal radius for a given set of homologous points ($\overline{R}^2(\mathbf{x}_G)$) determined by a chosen grasp point denoted by $\mathbf{x}_G$. The function $\overline{R}^2(\mathbf{x}_G)$ is strictly convex for all grasp points, as we can see by substituting $\overline{R}^2(\mathbf{x}_{G1}) + (\overline{R}^2(\mathbf{x}_{G2})$ - $\overline{R}^2(\mathbf{x}_{G1}))\ \lambda$ into the right hand side of Eq. 4, and taking into account that $|\mathbf{r}_{1i}|^2 \leq \overline{R}^2(\mathbf{x}_{G1})$ for $\lambda = 0$ and $|\mathbf{r}_{2i}|^2 \leq \overline{R}^2(\mathbf{x}_{G2})$ for $\lambda = 1$. Because of this property of $\overline{R}^2(\mathbf{x}_G)$, any optimization algorithm which does not require derivatives will work (see e.g. Brent [3]). Associated with the minimal workspace size is the optimal grasp point denoted by $\mathbf{x}_{G\text{min}}$ and the optimal base location as the center of the minimal circle given by $\mathbf{x}_{B\text{min}}$.

2. A boundary point for a given radius R is found by moving $\mathbf{x}_G$ along a ray starting from $\mathbf{x}_{G\text{min}}$ until $\overline{R}(\mathbf{x}_G)$ is equal to R. So, the task is to find the root of a smooth but not analytically given function. Since we know that there is just one such root this is also an easy task. By sweeping this ray around and using the previous roots as estimations for the one we want to compute, we obtain an efficient way to determine the grasp space boundary.

The algorithm described above depends linearly on the number of positions, while behavior of the optimization and root finding algorithms just determine the constant of the linear term. Note: The minimal radius could also be found by determining the minimum of all minimal radii of all possible three positions problems (see Bottema and Roth [2] Ch. 8, §3), but this would yield an algorithm of $O(n^3 \log n)$ in time.

III. Finding all Base Locations for n Planar Positions

We defined the grasp space boundary by those grasp points where the associated homologous points were enclosed by a minimal disk with the radius of the workspace. The problem of finding the possible base locations can now be formulated in a similar way. If we pick a base location for our manipulator (the center of the workspace) in a fixed frame, and observe the relative positions of this point seen from a frame attached to the moving object, we obtain a set of homologous points for the inverse motion. Since all these points must be in reach of the manipulator, they have to lie inside a disk with the radius R of the workspace. So, we can use the same algorithm as the one for determining the grasp space boundary by using the inverse transformations $[\mathbf{A}_i]^T$ and $-[\mathbf{A}_i]^T\mathbf{d}_i$ instead of $[\mathbf{A}_i]$ and $\mathbf{d}_i$.

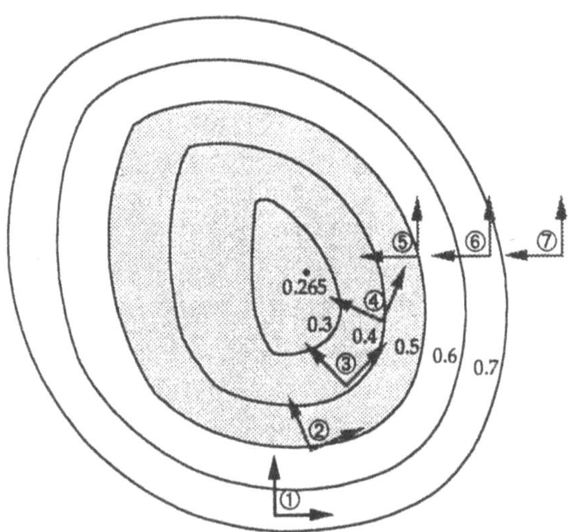

Figure 4: Set of available grasp points for 7 specified positions.

Figs. 4 and 5 show an example with the grasp and the base spaces for a planar movement determined by 7 positions, for different workspace radii. As the figure shows, the grasp and base spaces look very similar. This comes from the fact that parts of both curves are determined by the same sets of three positions which yields a tricircular sextic boundary arranged symmetrically with respect to a line through two of the displacement poles (see Bodduluri and McCarthy [5], and Alt [1]). The sets of three positions, determining these parts of the boundary, are in general changing and so are the sextic curves and the lines of symmetry. So, parts of the two curves may be the same, but the boundaries as a whole are different in general.

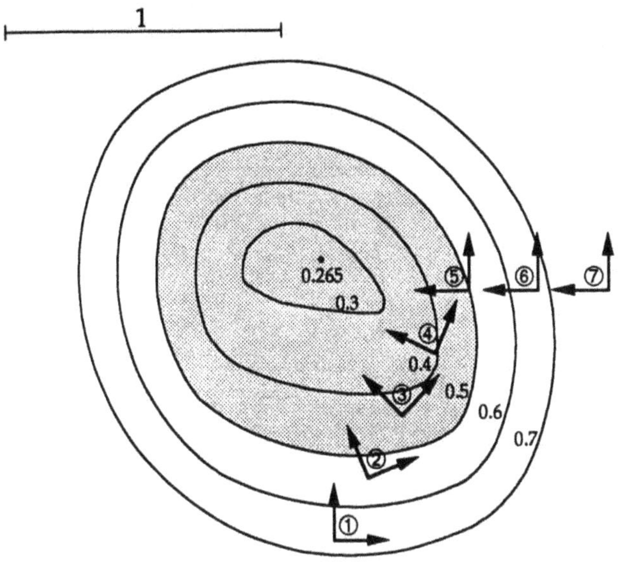

Figure 5: Base locations available for 7 specified positions.

IV. Finding Base Locations for a Given Grasp Point

So far we have considered how to find all available grasp points and base locations, independently. Now we determine the possible base locations, or grasp points, given either the grasp point or the base location, respectively, of a specified robot. In the case of a given grasp point, each base location must lie in a disk of radius R (the workspace) with the grasp point as its center. The intersection of the disks obtained for each homologous position of a grasp point is the desired space of base locations.

Fig. 6 shows the possible base locations for the specified grasp point G (shown in the first position), and the 7 position planar movement from Fig. 4. The solid lines represent the boundaries of the grasp spaces while the dashed lines denote the boundaries for the associated base locations. The point labeled $B_{\min}$ is the base location of the minimal size manipulator coupled to the homologous positions of G (the center of the minimal circle determined by the G_i's). Note that $B_{\min}$ lies inside the base space boundary although we picked G on the boundary of the grasp space for $R = 0.4$. Determining the exact intersection of these circles takes at least $O(n \log n)$ time, and the algorithms for this problem are usually complicated. In most cases, such as the graphical representation of the intersection, an approximation of this area may be sufficient. Since the point $B_{\min}$, which we can determine in linear time, always lies inside the intersection of the circles we can use this fact with an approximation techniques. For example, as described above, draw rays starting in $B_{\min}$ and intersect them with the circles which denote the workspace boundaries, the one with minimal distance to $B_{\min}$ is then a point of the boundary. By using adaptive methods and the knowledge that the boundary is formed by parts of circles with a known radius we can compute an approximation to the base space boundary in an efficient way with simple algorithms.

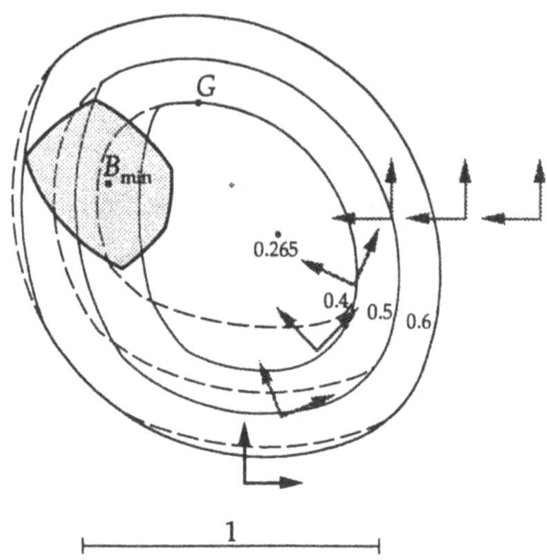

Figure 6: Base locations for a robot of radius $R = 0.6$ and a given grasp point.

V. Cooperating Robots

For a system of cooperating robots (or multifingered hands) the set of available grasp points and base locations of each robot can be determined in the way described above. Therefore many of the problems related to cooperating robots can be solved by simply treating them as a gathering of independent robots. However, there are tasks which need the interaction of the robots to be taken into account. One such problem, for example, is where to locate the bases of a system of robots if the relative positions of the bases to each other are specified, such as occurs in a mechanical hand. If we assume, for simplicity, that all the robots are the same size, we can compute the required minimal size of the robots and the loci of the bases and the grasp points that allow a certain movement. The algorithm for this computation is simply: For a given configuration of the bases, compute the smallest required robot size for each base, and then minimize the maximum over all robot sizes by moving the base configuration around.

For the 7 position planar movement of Fig. 4, and two cooperating robots with bases being a distance of 1 apart, we obtain for the smallest possible robots $R_{\min} = 0.407$ and grasp and base points as shown in Fig. 7. Because of the convexity of the base space there exist just this solution. From the picture we see that both robots touch the boundary of the workspace three times (robots in extreme reach position) as has to be the case for a minimal robot. Note that in this problem increasing the distance between the base locations of the robots will always force the size of the robots to increase due to the convexity of the base space.

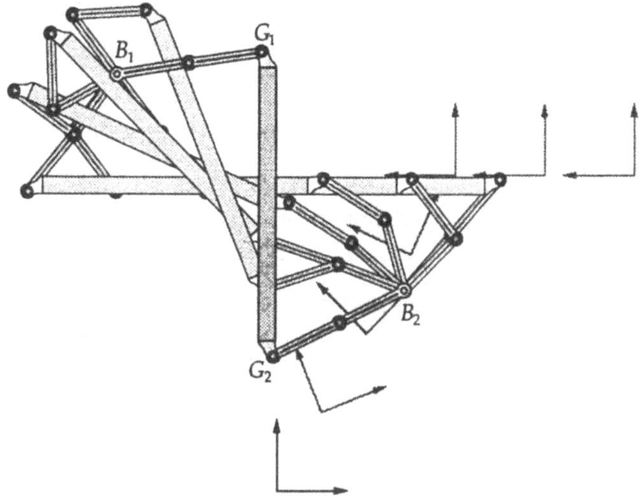

Figure 7: Two cooperating robots of minimal size for 7 specified positions.

VI. Conclusion

In this paper a numerically efficient method for planning possible grasp points and base locations for single robots and cooperating robot systems was introduced. For a simple robot workspace and neglecting collisions, we show how to compute the minimal robot size to perform an n position planar movement, and where the base and grasp points on the workpiece may be located. It is shown that the boundaries of both regions are convex, which turns out to be beneficial for various planing tasks. Though the results presented here are for planar robots, similar algorithms have been developed for spatial robots with spherical workspaces. The result is the ability to plan the grasp and base points for a given motion interactively on graphical systems. Furthermore, this provides a framework for a more detailed analysis of the coordinated motion and collision avoidance of general robot systems.

References

[1] Alt, H., Zur Synthese der ebenen Mechanismen. *ZAMM.* 1(5): 373-398, 1921.

[2] Bottema, O., and Roth, B., *Theoretical Kinematics.* North-Holland Pub. Co., Amsterdam, New York, 1979.

[3] Brent, R. P., *Algorithms for Minimization without Derivatives.* Englewood Cliffs, N.J., Prentice-Hall, 1973.

[4] Kerr, J., and Roth, B., Analysis of Multifingered Hands. *Int. J. of Robotics Research.* 4(4): 3-17, 1986.

[5] McCarthy, J. M., and Bodduluri, R. M. C., Planning Workpiece Grasp Points For Cooperating Robot Movements. *Proc. of the CISM Symposium on Robotics and Manipulators.* Udine, Italy, 1992.

[6] McCarthy, J. M., *An Introduction to Theoretical Kinematics.* MIT Press, Cambridge, MA, 1990.

[7] Megiddo, N., Linear-Time Algorithms for Linear Programming in $\Re^3$ and Related Problems. *SIAM J. Comput..* 12:759-776, 1983.

[8] Seidel, R., Linear programming and convex hulls made easy. *Proc. 6th Annual ACM Symposium on Computational Geometry.* 211-215, 1990.

[9] Welzl, E., Smallest enclosing disks (balls and ellipsoids). *Lecture Notes in Computer Science.* 555:359-370, 1991.

Vision-Based Robot Path Planning

Aleš Ude Rüdiger Dillmann

Institute for Real-Time Computer Systems and Robotics, Department of Computer Science, University of Karlsruhe, Kaiserstr. 12, 76128 Karlsruhe, Germany

Abstract – We present a vision-based approach to robot path planning. The programmer demonstrates a desired motion by moving an object to be manipulated with his own hand. His performance is measured with a stereo vision system. The demonstrated motion is reconstructed with the help of a non-parametric regression technique which is resistant to a considerable amount of measurement noise. The generated path is given as a linear combination of natural vector splines. The proposed methodology requires neither a geometrical model of a robot workcell, which is used by classical motion planners, nor a robot manipulator, which is needed for teaching by guiding, for the specification of the robot path.

I. Introduction

The most widespread approach to robot path specification in the industry is teaching by guiding which involves guiding the robot through a sequence of poses which are recorded by the robot's internal sensors. This simple method, however, suffers from a number of difficulties. It forces the user to use a teach-pendant, which is tiring and cumbersome and often leads to mistakes in the movements, especially if fine positioning is needed. Furthermore, the use of real manipulators can be dangerous for the programmer.

Many recent approaches to robot path specification fall in the category of geometrical motion planning. The basic motion planning problem is defined as follows:

- given an initial pose and a goal pose, generate a path specifying a continuous sequence of poses avoiding contact with the obstacles in the robot workspace, starting at the initial pose and terminating at the goal pose.

There exist many extensions to the basic motion planning problem [1]. Classical motion planning algorithms suffer from computational complexity and dependency on the model of the robot workspace. Their conversion into commercial systems has turned out to be an extremely difficult task. There is an adjustment problem due to inconsistencies between the model and the real environment and there are many problems in which the robot's environment is only partially known. The generation of geometrical models is usually carried out with the help of CAD systems and is a tedious task. When the model of the robot workcell is not available, classical motion planning algorithms cannot be used. The robot is forced to use local obstacle avoidance strategies based on on-line sensing to move from its initial to its final configuration. However, on-line sensing is computationally expensive and often cannot be performed in real-time. Furthermore, the lack of global knowledge results in non-optimal motions.

Obstacle avoidance is not always the only criterion which the robot path must fulfill. Especially in adaptive manufacturing processes such as spray painting, the robot end-effector is required to follow a well defined path in the robot workspace. In this case,

A. J. Lenarčič and B. B. Ravani (eds.), Advances in Robot Kinematics and Computationed Geometry, 505–512.

the primary goal of the planner is to specify such path. It is difficult to tackle this kind of problems with the help of classical motion planners because every task must fulfill its own criterion. Different motion planners must be used for different tasks as a consequence. Another approach is to exploit the human programmer's implicit knowledge about the path to be programmed. The programmer usually has some idea about the course of the desired robot motion but lacks of any means to communicate this path to the robot. A CAD oriented trajectory design-editor was proposed to support the explicit programming of robot paths [2], but such an editor again depends on the geometrical model of the workcell. In this paper we employ a *teaching by showing* paradigm [3] as an efficient approach to explicit programming of robot paths.

Methods for the specification of robot poses and paths which utilize teaching by showing were proposed in [4, 5, 6, 7]. In [5] vision was employed to teach the poses on the desired path at some important passing points whereas in [4, 6] vision was used to specify overall paths. In these systems, the path specification was accomplished by moving a specially designed teaching tool with attached LEDs along the desired path. Various heuristics were developed in order to reduce the noise in it. However, no attempt was made to assure the optimality of the reconstructed path according to a suitable optimization criterion and to explicitly consider the noise of the sensors. By designing a tool which can be easily recognized by the image processing system, the object tracking can be made faster and more robust. But no teaching tool can be appropriate to show all possible paths and it is sometimes easier to show the desired path by moving the actual object which should be manipulated. The inability to track more general objects is a serious disadvantage of both methods.

In [7, 8] we described a theoretically well supported approach to programming of robot trajectories based on teaching by showing. The user demonstrates the motion by moving the actual object to be manipulated along the desired path. The proposed methodology employs a non-parametric regression technique based on *natural vector splines* to reconstruct the shown path. The use of non-parametric techniques is essential if one wants to reconstruct paths of all possible forms. In this paper we extend our previous results.

II. Measurement Acquisition

Our vision algorithms are based on a CAD model of the object to be manipulated. We would like to stress here that, unlike in geometrical motion planning systems, which require the overall geometry of the robot workspace to be known, we have to model only the geometry of this object. Such models are often already available in the industry. In a preprocessing phase, which can be viewed as a learning phase, model features are gathered in groups according to relations which remain invariant under Euclidean motion and perspective projection. Groups used in the system are planar polygons and sets of more than two straight line segments which intersect in one point.

A stereo vision system [9] is employed to measure discrete object poses on the shown path. At each measurement instant, a stereo image pair is taken and a list of straight line segments is extracted. Nearly parallel, adjacent segments are concatenated and added to the lists in order to compensate for the errors in the segmentation of edges into straight line segments. This symbolic data is used as an input to the matching procedure which starts by grouping the extracted line segments into groups

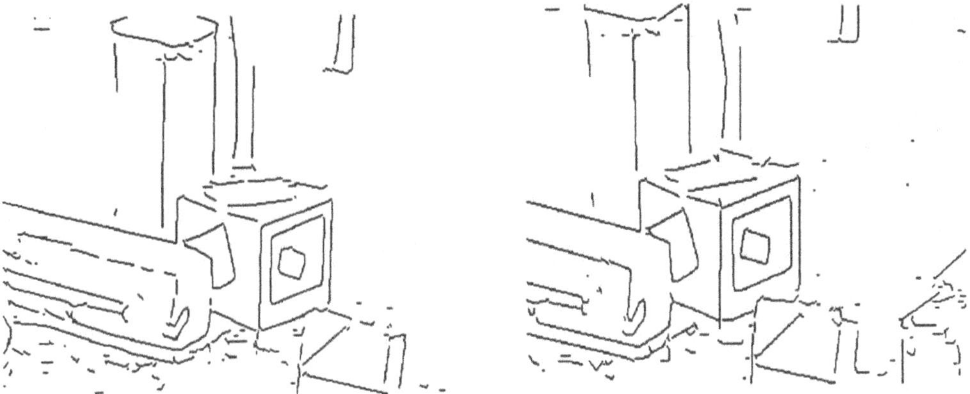

Figure 1: Extracted straight line segments

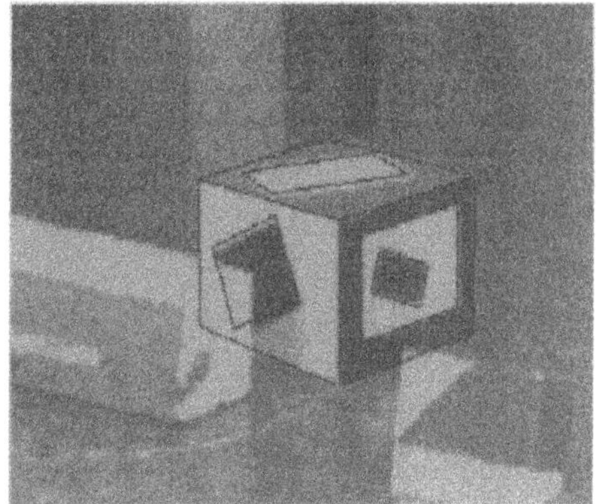

Figure 2: Visible model edges projected onto the original stereo image pair from the calculated pose

which could correspond to the model groups. The hypothesize-and-verify paradigm is utilized in order to determine the best set of correspondences. Hypotheses about possible correspondences between the 2-D image and the 3-D model groups are established by use of Euclidean, projective and affine invariants. Projective and affine invariants can be used to establish the correspondences between the image and the model groups directly whereas Euclidean invariants require the establishment of stereo correspondences between the image groups and the calculation of their 3-D poses first. For each of the hypothesized matches, the object pose is computed with the help of the stereo information. Model edges which are visible at the hypothesized pose are then projected onto both image planes in order to find more correspondences between the model and the image segments. This enables the refinement of the measured pose. The procedure is aborted when either a quality measure associated with every hypothesized pose crosses a prespecified threshold or after a certain number of "best hypotheses" have been tested. At the next time instant, the object pose is predicted by use of a *random acceleration kinematic model* for the object motion and *Kalman filtering*. This prediction leads the search for correspondences at the next computing step in order to speed up the matching process. Typical inputs and outputs of the localizer are shown in Figs. 1 and 2.

III. Robot-Oriented Path Approximation

After the tracking procedure is successfully accomplished, we obtain a sequence of pose measurements $\boldsymbol{p}_j \in \mathbf{R}^D$, $j = 1, \ldots, M$, at the time instants ξ_j, together with their variance-covariance matrices $\sigma^2 \boldsymbol{\Gamma}_j$. It is essential to model the uncertainties as full D-dimensional distributions because measurements obtained by a stereo camera depend on both the distance between the object and the cameras and the viewing direction [10], which can both change while the object is moving. The conversion of uncertainties in the image features into uncertainties in the measured poses is described in [7]. A robot path specified in Cartesian coordinates can be divided in two parts: the translational and the rotational. Sometimes, the orientation is not important and can be omitted ($D = 3$). If the object to be manipulated has an axis of symmetry, e.g. a cylindrical pin, the orientation has two degrees of freedom and can be parameterized by spherical coordinates ($D = 5$). In the most general case, the orientation has three degrees of freedom and can be parameterized, for instance, by Euler angles ($D = 6$). If one has an interface to the robot controller at the joint level, then it is normally preferred to specify the path directly at this level. When the number of degrees of freedom coincides with D, this can be easily done by converting the measurements $\boldsymbol{p}_i$ into the joint coordinates $\boldsymbol{q}_i = \boldsymbol{k}^{-1}(\boldsymbol{p}_i)$, where $\boldsymbol{k}$ and $\boldsymbol{k}^{-1}$ respectively represent the robot direct and inverse kinematics. The variance-covariance matrix of the transformed measurement $\boldsymbol{q}_i$ can be estimated by $\sigma^2 \boldsymbol{\Gamma}_{qi}$, where

$$\boldsymbol{\Gamma}_{qi} = \boldsymbol{J}_i \boldsymbol{\Gamma}_{pi} \boldsymbol{J}_i^T, \quad \boldsymbol{J}_i = \frac{\partial \boldsymbol{k}^{-1}}{\partial \boldsymbol{p}}(\boldsymbol{p}_i) = \left(\frac{\partial \boldsymbol{k}}{\partial \boldsymbol{q}}(\boldsymbol{q}_i) \right)^{-1}. \tag{1}$$

Let us assume now that $\boldsymbol{\vartheta}$ is the true path of the moving object. We have arrived to the following *regression model*

$$\boldsymbol{p}_j = \boldsymbol{\vartheta}(\xi_j) + \boldsymbol{\varepsilon}_j, \tag{2}$$

where $\boldsymbol{\varepsilon}_j$ are assumed to be zero mean, mutually independent random vectors with variance-covariance matrices $\sigma^2 \boldsymbol{\Gamma}_j$. The parametric form of the path $\boldsymbol{\vartheta}$ is in general unknown. The aim of the reconstruction is to find a path which the robot's tip will follow. For this reason we must seek for the desired path in the class of smooth curves. If the measurements were simply interpolated, this requirement would not be fulfilled. On the other hand, the reconstructed path must stay close to the measurements. Hence we must keep the balance between the two conflicting goals: *goodness-of-fit* and *smoothness*. The goodness-of-fit for any path $\boldsymbol{\mu}$ can be defined as $f(\boldsymbol{\mu})/\sigma^2$, where

$$f(\boldsymbol{\mu}) = \frac{1}{M} \sum_{j=1}^{M} (\boldsymbol{p}_j - \boldsymbol{\mu}(\xi_j))^T \boldsymbol{\Gamma}_j^{-1} (\boldsymbol{p}_j - \boldsymbol{\mu}(\xi_j)), \tag{3}$$

whereas the measurement of smoothness associated with each component of $\boldsymbol{\mu}$ is

$$g_k(\boldsymbol{\mu}) = \int_a^b (\mu_k^{(m)}(t))^2\, dt, \; k = 1, \ldots, D, \tag{4}$$

where m is the required degree of smoothness, D is the dimension of the space in which the path should be specified, $a = \xi_1$ and $b = \xi_M$.

We would like to minimize the criteria $f(\boldsymbol{\mu})/\sigma^2$ and $g_k(\boldsymbol{\mu})$ simultaneously, but these are conflicting goals in general. A standard non-parametric solution is to minimize the following combined criterion [11]

$$\boldsymbol{\mu}_\lambda = \arg\min_{\boldsymbol{\mu}} f(\boldsymbol{\mu}) + \sum_{k=1}^{D} \lambda_k g_k(\boldsymbol{\mu}). \tag{5}$$

This criterion function must be minimized over a space of vector functions which are "smooth enough" so that criteria (4) can be evaluated. See [7] for some mathematical details. It is obvious that the scale parameter σ can be left out because it can be included in the initially unknown smoothing parameter $\boldsymbol{\lambda} = [\lambda_1, \ldots, \lambda_D]^T$, $\lambda_k > 0$. This parameter controls the tradeoff between goodness-of-fit and smoothness. It can be determined automatically from the data by use of a cross-validation or a generalized cross-validation criterion [7], but we omit the details here.

If one has no other knowledge about the path to be reconstructed and no other requests by the application, then the criterion (5) should be minimized over a class of smooth vector functions. In robotics, constraints on the robot motion are often set. For instance, when programming a pick and place operation, some important passing points such as approach-points, grip-points and withdraw-points must be reached by the end-effector with a greater precision than other points on the path. Such points can be obtained either from a world model or by a more precise measurement of the object pose at these positions. Thus the possibility to set equality constraints on the robot pose must be included in our method

$$\boldsymbol{\mu}(\tau_i) = \boldsymbol{c}_i. \tag{6}$$

If the path is specified directly in joint coordinates, one can limit the joints to lie in the robot workspace

$$\boldsymbol{c}^l \leq \boldsymbol{\mu}(t) \leq \boldsymbol{c}^u. \tag{7}$$

The expression $\boldsymbol{a} \leq \boldsymbol{b}$ is meant componentwise. These constraints are given in a continuous form which is unfortunately difficult to handle numerically. Therefore we decided to discretize them

$$\boldsymbol{c}_i^l \leq \boldsymbol{\mu}(\tau_i) \leq \boldsymbol{c}_i^u, \; i = 1, \ldots, L. \tag{8}$$

Note that equality constraints can be included in this expression by taking $\boldsymbol{c}_i^l = \boldsymbol{c}_i^u$ and that some of the τ_j can be equal to some of the ξ_j. Discrete constraints represent a good approximation to the continuous constraints if the time interval between a pair of consecutive constraints is not too large. If slightly conservative bounds are used in the calculations, the reconstructed path fulfills the continuous constraints as well.

The interface between the user and the industrial robot is usually given in the form of a robot programming language. The problem which arises here is that present robot programming languages do not provide any means to transfer functionally defined motions, i.e. motions specified as general functions of time. As a consequence, the temporal course of the shown path is lost when the path is conveyed to the robot. It is not sensible to consider the robot's kinematic and dynamic limitations during the determination of geometry of motion in this case. Furthermore, since the robot dynamics is highly nonlinear, it is an extremely difficult task to account for it during the generation of the geometrical path. For these reasons, most of the present geometrical motion planning algorithms concentrate on the reconstruction of the robot path without considering the full nonlinear dynamics of the manipulator. The path is normally generated as a purely geometrical entity. Once the geometry of the path is specified, the motion along it takes place with one degree of freedom only. This degree can be used to fulfill the robot's dynamic constraints and to track the path in the minimum possible time [12]. Note that the constraints of the type (8), which are imposed by robot joints on the relative motion of links they connect, are purely geometrical and cannot be considered in this way. They must be accounted for already during the specification of geometry of motion.

In order to find the path which is smooth, well adapted to the measurements and which fulfills the geometrical constraints, one has to minimize (5) subject to (8) over a class of smooth vector functions. Let $\boldsymbol{\mu}_\lambda$ describe the minimum of (5) subject to (8), let $\{t_j\}_{j=1}^{N}$ be the sequence of knot points combining $\{\xi_j\}$ and $\{\tau_j\}$ ordered in a strictly ascending order and let $a = t_1$ and $b = t_N$. It can be shown (see [7] for the unconstrained case) that the optimal path $\boldsymbol{\mu}_\lambda$ belongs to the space of natural vector splines of order $2m$ defined on the knot sequence $\{t_j\}$

$$\boldsymbol{\mu}_\lambda(s) = \sum_{i=1}^{N} \boldsymbol{\alpha}_j N_j(s), \tag{9}$$

where

$$\begin{aligned} N_j(s) &= [t_1, \ldots, t_{m+j}](\cdot - s)_+^{2m-1}, \; j = 1, \ldots, m, \\ N_j(s) &= (t_{j+m} - t_{j-m})[t_{j-m}, \ldots, t_{j+m}](\cdot - s)_+^{2m-1}, \; j = m+1, \ldots, N-m, \\ N_j(s) &= (-1)^{N+m-j}[t_{j-m}, \ldots, t_N](\cdot - s)_+^{2m-1}, \; j = N-m+1, \ldots, N. \end{aligned} \tag{10}$$

$[t_i, \ldots, t_{i+2m}]f(\cdot)$ denotes the divided difference of order $2m$. A natural vector spline of order $2m$ is a piecewise polynomial of degree $2m-1$ on each subinterval $[t_j, t_{j+1})$. It has $2m-2$ continuous derivatives and is a polynomial of degree $m-1$ outside of $[a, b]$. The functions of the form (10) are nothing else but B-splines which are modified to fulfill the natural boundary conditions. The coefficients $\boldsymbol{\alpha}_j \in \mathbf{R}^D$ can be determined by solving a quadratic program which is obtained by inserting $\boldsymbol{\mu}_\lambda$ of the form (9) into Eqs. (5,8). As the modified B-splines (10) have small support, the matrices in the resulting quadratic program are sparse, which considerably reduces the computational complexity of the proposed algorithm.

Figs. 3 and 4 show a path reconstructed in Cartesian coordinates as a function of time. The true path consisted of straight line segments and parabolic segments. In

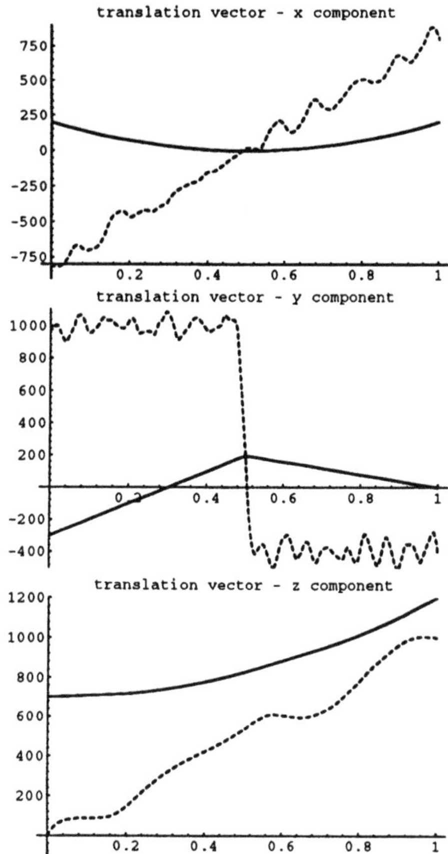

Figure 3: Translation vector - solid line, translational velocity - dashed line

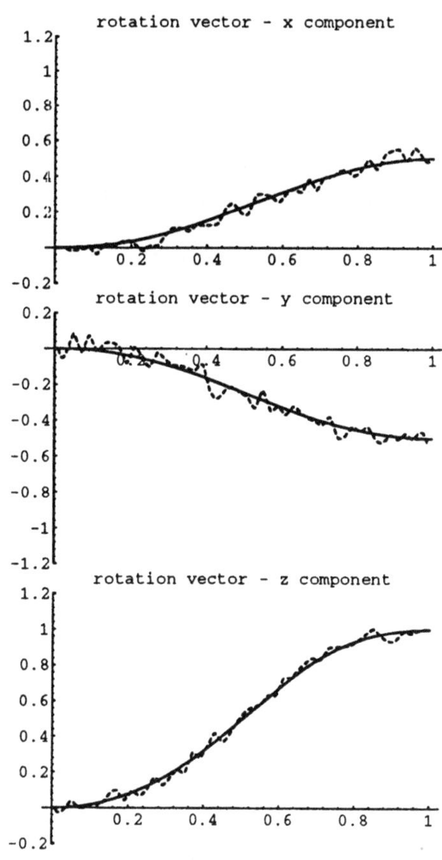

Figure 4: Rotation vector: smoothing spline - solid line, interpolating spline - dashed line.

Fig. 3 the components of a translation vector and their velocities are depicted. As one could have expected is the estimation of velocities much more sensitive. Fig. 4 shows how important it is to smooth the measurements instead of simply interpolating them. Both figures show that our algorithm for the automatic determination of the smoothing parameter λ finds the smoothing parameter which is "about right".

While the shown path is normally collision-free for the end-effector's tip, this cannot be guaranteed for the manipulator's links, especially when the path is designed by an inexperienced user. If the manipulator is redundant with respect to the given task, one can split the planner in two parts; the specification of the robot's tip path in the three dimensional workspace followed by the local obstacle avoidance algorithm [13] based on on-line sensing. Only the obstacle avoidance part was considered in [13]. The combination of both methods can be viewed as a general method for the specification of collision-free paths for redundant robots. If the manipulator is hyper-redundant, the manipulator links can adapt to the shown path, i.e. the shown path plays the role of a "backbone curve", which results in a collision-free path even without on-line sensing.

IV. Conclusion

A new procedure, using stereo vision and natural vector splines, has been developed for generating robot paths. The proposed method enables explicit consideration of uncertainties in the measured poses and can handle a considerable amount of measurement noise. Furthermore, our system allows the user to demonstrate the desired trajectory by moving a fairly general object. Of course, the option to construct a teaching tool which can be tracked more reliably still exists. A two-stage method can be used to account for the robot's physical limitations.

Acknowledgment - This work has been performed at the Institute for Real-Time Computer Systems and Robotics, Prof.Dr.-Ing. U. Rembold, Prof.Dr.-Ing. R. Dillmann, University of Karlsruhe, Germany, in cooperation with the Department of Automation, Biocybernetics and Robot Systems, Jožef Stefan Institute, Ljubljana, Slovenia.

V. References

[1] J.-C. Latombe, *Robot Motion Planning*, Kluwer, Boston (1991).

[2] R. Dillmann and S. Schneider, "A CAD oriented trajectory design-editor", *Robotersysteme*, 4(3):161–171 (1988, in German).

[3] Y. Kuniyoshi, M. Inaba, and H. Inoue, "Teaching by Showing: Generating Robot Programs by Visual Observation of Human Performance", *Proc. 20th Int. Symp. Industrial Robots*, 119–126, Tokyo, Japan (1989).

[4] H. El-Zorkany, R. Liscano, B. Tondu, and G. Sawatzky, "Sensor-Based Location and Trajectory Specification and Correction in Robot Programming", *Proc. 16th Int. Symp. Industrial Robots*, 643–655, Brussels, Belgium, (1986).

[5] M. Ishii, S. Sakane, M. Kakikura, and Y. Mikami, "A 3-D Sensor System for Teaching Robot Paths and Environments", *Int. J. Robotics Res.*, 6(2):45–59 (1987).

[6] S. K. Tso and K. P. Liu, "Visual programming for capturing of human manipulation skill", *Proc. IEEE/RSJ Int. Conf. Intelligent Robots and Systems*, 42–48, Yokohama, Japan (1993).

[7] A. Ude, "Trajectory generation from noisy positions of object features for teaching robot paths", *Robotics and Autonomous Systems*, 11(2):113–127 (1993).

[8] A. Ude and R. Dillmann, "Trajectory Reconstruction from Stereo Image Sequences for Teaching Robot Paths", In *Proc. 24th Int. Symp. Industrial Robots*, 407–414, Tokyo, Japan (1993).

[9] F. Wallner, P. Weckesser, and R. Dillmann, "Calibration of the active stereo vision system KASTOR with standardized perspective matrices", *Optical 3-D Measurement Techniques II* (A. Gruen and H. Kahmen, Eds.), 98–105, Wichmann, Karlsruhe (1993).

[10] L. Matthies and S. A. Shafer, "Error Modeling in Stereo Navigation", *IEEE J. Robotics Automat.*, 3(3):239–248 (1987).

[11] G. Wahba, *Spline Models for Observational Data*, SIAM, Philadelphia (1990).

[12] F. Pfeiffer and R. Johanni, "A Concept for Manipulator Trajectory Planning", *IEEE J. Robotics Automat.*, 3(2):115–123 (1987).

[13] A. A. Maciejewski and C. A. Klein, "Obstacle Avoidance for Kinematically Redundant Manipulators in Dynamically Varying Environments", *Int. J. Robotics Res.*, 4(3):109–117 (1985).

Author Index

GPSR Compliance
The European Union's (EU) General Product Safety Regulation (GPSR) is a set of rules that requires consumer products to be safe and our obligations to ensure this.

If you have any concerns about our products, you can contact us on

ProductSafety@springernature.com

In case Publisher is established outside the EU, the EU authorized representative is:

Springer Nature Customer Service Center GmbH
Europaplatz 3
69115 Heidelberg, Germany

www.ingramcontent.com/pod-product-compliance
Ingram Content Group UK Ltd.
Pitfield, Milton Keynes, MK11 3LW, UK
UKHW021901190726

13853UKWH00003B/1363

* 9 7 8 9 4 0 1 5 8 3 4 9 7 *